i. n. herstein
University of Chicago

TOPICS IN ALGEBRA

2nd edition

John Wiley & Sons, Inc.
New York • Santa Barbara • London • Sydney • Toronto

To Marianne

ISBN 0 471 01090 1
Library of Congress Catalog Card Number: 74-82577
Printed in the United States of America.

10 9 8 7 6 5

Preface to the Second Edition

I approached revising *Topics in Algebra* with a certain amount of trepidation. On the whole, I was satisfied with the first edition and did not want to tamper with it. However, there were certain changes I felt should be made, changes which would not affect the general style or content, but which would make the book a little more complete. I hope that I have achieved this objective in the present version.

For the most part, the major changes take place in the chapter on group theory. When the first edition was written it was fairly uncommon for a student learning abstract algebra to have had any previous exposure to linear algebra. Nowadays quite the opposite is true; many students, perhaps even a majority, have learned something about 2×2 matrices at this stage. Thus I felt free here to draw on 2×2 matrices for examples and problems. These parts, which depend on some knowledge of linear algebra, are indicated with a #.

In the chapter on groups I have largely expanded one section, that on Sylow's theorem, and added two others, one on direct products and one on the structure of finite abelian groups.

In the previous treatment of Sylow's theorem, only the existence of a Sylow subgroup was shown. This was done following the proof of Wielandt. The conjugacy of the Sylow subgroups and their number were developed in a series of exercises, but not in the text proper. Now all the parts of Sylow's theorem are done in the text material.

In addition to the proof previously given for the existence, two other proofs of existence are carried out. One could accuse me of overkill at this point, probably rightfully so. The fact of the matter is that Sylow's theorem is important, that each proof illustrates a different aspect of group theory and, above all, that I love Sylow's theorem. The proof of the conjugacy and number of Sylow subgroups exploits double cosets. A by-product of this development is that a means is given for finding Sylow subgroups in a large set of symmetric groups.

For some mysterious reason known only to myself, I had omitted direct products in the first edition. Why is beyond me. The material is easy, straightforward, and important. This lacuna is now filled in the section treating direct products. With this in hand, I go on in the next section to prove the decomposition of a finite abelian group as a direct product of cyclic groups and also prove the uniqueness of the invariants associated with this decomposition. In point of fact, this decomposition was already in the first edition, at the end of the chapter on vector spaces, as a consequence of the structure of finitely generated modules over Euclidean rings. However, the case of a finite group is of great importance by itself; the section on finite abelian groups underlines this importance. Its presence in the chapter on groups, an early chapter, makes it more likely that it will be taught.

One other entire section has been added at the end of the chapter on field theory. I felt that the student should see an explicit polynomial over an explicit field whose Galois group was the symmetric group of degree 5, hence one whose roots could not be expressed by radicals. In order to do so, a theorem is first proved which gives a criterion that an irreducible polynomial of degree p, p a prime, over the rational field have S_p as its Galois group. As an application of this criterion, an irreducible polynomial of degree 5 is given, over the rational field, whose Galois group is the symmetric group of degree 5.

There are several other additions. More than 150 new problems are to be found here. They are of varying degrees of difficulty. Many are routine and computational, many are very difficult. Furthermore, some interpolatory remarks are made about problems that have given readers a great deal of difficulty. Some paragraphs have been inserted, others rewritten, at places where the writing had previously been obscure or too terse.

Above I have described what I have added. What gave me greater difficulty about the revision was, perhaps, that which I have not added. I debated for a long time with myself whether or not to add a chapter on category theory and some elementary functors, whether or not to enlarge the material on modules substantially. After a great deal of thought and soul-searching, I decided not to do so. The book, as stands, has a certain concreteness about it with which this new material would not blend. It could be made to blend, but this would require a complete reworking of the material of the book and a complete change in its philosophy—something I did not want to do. A mere addition of this new material, as an adjunct with no applications and no discernible goals, would have violated my guiding principle that all matters discussed should lead to some clearly defined objectives, to some highlight, to some exciting theorems. Thus I decided to omit the additional topics.

Many people wrote me about the first edition pointing out typographical mistakes or making suggestions on how to improve the book. I should like to take this opportunity to thank them for their help and kindness.

Preface to the First Edition

The idea to write this book, and more important the desire to do so, is a direct outgrowth of a course I gave in the academic year 1959–1960 at Cornell University. The class taking this course consisted, in large part, of the most gifted sophomores in mathematics at Cornell. It was my desire to experiment by presenting to them material a little beyond that which is usually taught in algebra at the junior-senior level.

I have aimed this book to be, both in content and degree of sophistication, about halfway between two great classics, *A Survey of Modern Algebra*, by Birkhoff and MacLane, and *Modern Algebra*, by Van der Waerden.

The last few years have seen marked changes in the instruction given in mathematics at the American universities. This change is most notable at the upper undergraduate and beginning graduate levels. Topics that a few years ago were considered proper subject matter for semiadvanced graduate courses in algebra have filtered down to, and are being taught in, the very first course in abstract algebra. Convinced that this filtration will continue and will become intensified in the next few years, I have put into this book, which is designed to be used as the student's first introduction to algebra, material which hitherto has been considered a little advanced for that stage of the game.

There is always a great danger when treating abstract ideas to introduce them too suddenly and without a sufficient base of examples to render them credible or natural. In order to try to mitigate this, I have tried to motivate the concepts beforehand and to illustrate them in concrete situations. One of the most telling proofs of the worth of an abstract concept is what it, and the results about it, tells us in familiar situations. In almost every chapter an attempt is made to bring out the significance of the general results by applying them to particular problems. For instance, in the chapter on rings, the two-square theorem of Fermat is exhibited as a direct consequence of the theory developed for Euclidean rings.

The subject matter chosen for discussion has been picked not only because it has become standard to present it at this level or because it is important in the whole general development but also with an eye to this "concreteness." For this reason I chose to omit the Jordan-Hölder theorem, which certainly could have easily been included in the results derived about groups. However, to appreciate this result for its own sake requires a great deal of hindsight and to see it used effectively would require too great a digression. True, one could develop the whole theory of dimension of a vector space as one of its corollaries, but, for the first time around, this seems like a much too fancy and unnatural approach to something so basic and down-to-earth. Likewise, there is no mention of tensor products or related constructions. There is so much time and opportunity to become abstract; why rush it at the beginning?

A word about the problems. There are a great number of them. It would be an extraordinary student indeed who could solve them all. Some are present merely to complete proofs in the text material, others to illustrate and to give practice in the results obtained. Many are introduced not so much to be solved as to be tackled. The value of a problem is not so much in coming up with the answer as in the ideas and attempted ideas it forces on the would-be solver. Others are included in anticipation of material to be developed later, the hope and rationale for this being both to lay the groundwork for the subsequent theory and also to make more natural ideas, definitions, and arguments as they are introduced. Several problems appear more than once. Problems that for some reason or other seem difficult to me are often starred (sometimes with two stars). However, even here there will be no agreement among mathematicians; many will feel that some unstarred problems should be starred and vice versa.

Naturally, I am indebted to many people for suggestions, comments and criticisms. To mention just a few of these: Charles Curtis, Marshall Hall, Nathan Jacobson, Arthur Mattuck, and Maxwell Rosenlicht. I owe a great deal to Daniel Gorenstein and Irving Kaplansky for the numerous conversations we have had about the book, its material and its approach. Above all, I thank George Seligman for the many incisive suggestions and remarks that he has made about the presentation both as to its style and to its content. I am also grateful to Francis McNary of the staff of Ginn and Company for his help and cooperation. Finally, I should like to express my thanks to the John Simon Guggenheim Memorial Foundation; this book was in part written with their support while the author was in Rome as a Guggenheim Fellow.

Contents

1

Preliminary Notions

One of the amazing features of twentieth century mathematics has been its recognition of the power of the abstract approach. This has given rise to a large body of new results and problems and has, in fact, led us to open up whole new areas of mathematics whose very existence had not even been suspected.

In the wake of these developments has come not only a new mathematics but a fresh outlook, and along with this, simple new proofs of difficult classical results. The isolation of a problem into its basic essentials has often revealed for us the proper setting, in the whole scheme of things, of results considered to have been special and apart and has shown us interrelations between areas previously thought to have been unconnected.

The algebra which has evolved as an outgrowth of all this is not only a subject with an independent life and vigor—it is one of the important current research areas in mathematics—but it also serves as the unifying thread which interlaces almost all of mathematics—geometry, number theory, analysis, topology, and even applied mathematics.

This book is intended as an introduction to that part of mathematics that today goes by the name of abstract algebra. The term "abstract" is a highly subjective one; what is abstract to one person is very often concrete and down-to-earth to another, and vice versa. In relation to the current research activity in algebra, it could be described as "not too abstract"; from the point of view of someone schooled in the calculus and who is seeing the present material for the first time, it may very well be described as "quite abstract."

Be that as it may, we shall concern ourselves with the introduction and development of some of the important algebraic systems—groups, rings, vector spaces, fields. An algebraic system can be described as a set of objects together with some operations for combining them.

Prior to studying sets restricted in any way whatever—for instance, with operations—it will be necessary to consider sets in general and some notions about them. At the other end of the spectrum, we shall need some information about the particular set, the set of integers. It is the purpose of this chapter to discuss these and to derive some results about them which we can call upon, as the occasions arise, later in the book.

1.1 Set Theory

We shall not attempt a formal definition of a set nor shall we try to lay the groundwork for an axiomatic theory of sets. Instead we shall take the operational and intuitive approach that a set is some given collection of objects. In most of our applications we shall be dealing with rather specific things, and the nebulous notion of a set, in these, will emerge as something quite recognizable. For those whose tastes run more to the formal and abstract side, we can consider a set as a primitive notion which one does not define.

A few remarks about notation and terminology. Given a set S we shall use the notation throughout $a \in S$ *to read* "*a is an element of* S." In the same vein, $a \notin S$ will read "a is *not* an element of S." The set A will be said to be a *subset* of the set S if every element in A is an element of S, that is, if $a \in A$ implies $a \in S$. We shall write this as $A \subset S$ (or, sometimes, as $S \supset A$), which may be read "A is contained in S" (or, S contains A). This notation is not meant to preclude the possibility that $A = S$. By the way, what is meant by the equality of two sets? For us this will always mean that they contain the same elements, that is, every element which is in one is in the other, and vice versa. In terms of the symbol for the containing relation, the two sets A and B are equal, written $A = B$, if both $A \subset B$ and $B \subset A$. The standard device for proving the equality of two sets, something we shall be required to do often, is to demonstrate that the two opposite containing relations hold for them. A subset A of S will be called a *proper* subset of S if $A \subset S$ but $A \neq S$ (A is not equal to S).

The *null set* is the set having no elements; it is a subset of every set. We shall often describe that a set S is the null set by saying it is *empty*.

One final, purely notational remark: Given a set S we shall constantly use the notation $A = \{a \in S \mid P(a)\}$ to read "A is the set of all elements in S for which the property P holds." For instance, if S is the set of integers and if A is the subset of positive integers, then we can describe A as $A = \{a \in S \mid a > 0\}$. Another example of this: If S is the set consisting of the objects (1), (2), . . . , (10), then the subset A consisting of (1), (4), (7), (10) could be described by $A = \{(i) \in S \mid i = 3n + 1, n = 0, 1, 2, 3\}$.

Given two sets we can combine them to form new sets. There is nothing sacred or particular about this number two; we can carry out the same procedure for any number of sets, finite or infinite, and in fact we shall. We do so for two first because it illustrates the general construction but is not obscured by the additional notational difficulties.

DEFINITION The *union* of the two sets A and B, written as $A \cup B$, is the set $\{x \mid x \in A \text{ or } x \in B\}$.

A word about the use of "or." In ordinary English when we say that something is one or the other we imply that it is not both. The mathematical "or" is quite different, at least when we are speaking about set theory. *For when we say that x is in A or x is in B we mean x is in at least one of A or B, and may be in both.*

Let us consider a few examples of the union of two sets. For any set A, $A \cup A = A$; in fact, whenever B is a subset of A, $A \cup B = A$. If A is the set $\{x_1, x_2, x_3\}$ (i.e., the set whose elements are x_1, x_2, x_3) and if B is the set $\{y_1, y_2, x_1\}$, then $A \cup B = \{x_1, x_2, x_3, y_1, y_2\}$. If A is the set of all blonde-haired people and if B is the set of all people who smoke, then $A \cup B$ consists of all the people who either have blonde hair or smoke or both. Pictorially we can illustrate the union of the two sets A and B by

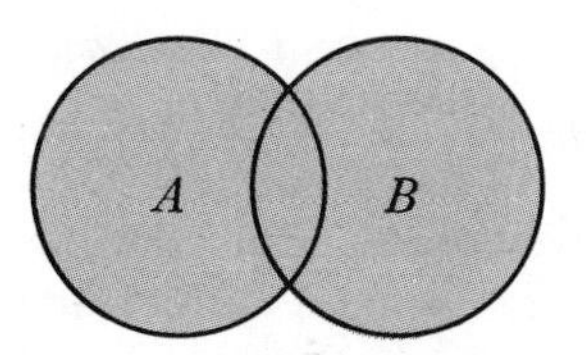

Here, A is the circle on the left, B that on the right, and $A \cup B$ is the shaded part.

DEFINITION The *intersection* of the two sets A and B, written as $A \cap B$, is the set $\{x \mid x \in A \text{ and } x \in B\}$.

The intersection of A and B is thus the set of all elements which are both in A and in B. In analogy with the examples used to illustrate the union of two sets, let us see what the intersections are in those very examples. For any set A, $A \cap A = A$; in fact, if B is any subset of A, then $A \cap B = B$. If A is the set $\{x_1, x_2, x_3\}$ and B the set $\{y_1, y_2, x_1\}$, then $A \cap B = \{x_1\}$ (we are supposing no y is an x). If A is the set of all blonde-haired people and if B is the set of all people that smoke, then $A \cap B$ is the set of all blonde-haired people who smoke. Pictorially we can illustrate the intersection of the two sets A and B by

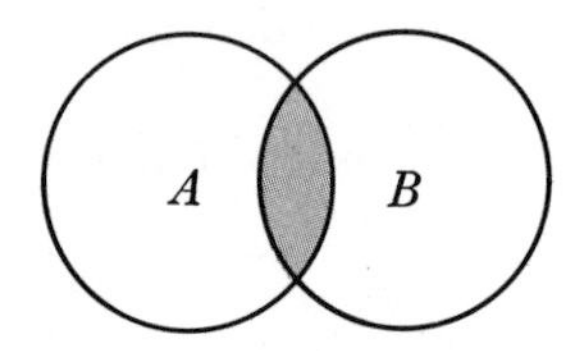

Here A is the circle on the left, B that on the right, while their intersection is the shaded part.

Two sets are said to be *disjoint* if their intersection is empty, that is, is the null set. For instance, if A is the set of positive integers and B the set of negative integers, then A and B are disjoint. Note however that if C is the set of nonnegative integers and if D is the set of nonpositive integers, then they are not disjoint, for their intersection consists of the integer 0, and so is not empty.

Before we generalize union and intersection from two sets to an arbitrary number of them, we should like to prove a little proposition interrelating union and intersection. This is the first of a whole host of such results that can be proved; some of these can be found in the problems at the end of this section.

PROPOSITION *For any three sets, A, B, C we have*

$$A \cap (B \cup C) = (A \cap B) \cup (A \cap C).$$

Proof. The proof will consist of showing, to begin with, the relation $(A \cap B) \cup (A \cap C) \subset A \cap (B \cup C)$ and then the converse relation $A \cap (B \cup C) \subset (A \cap B) \cup (A \cap C)$.

We first dispose of $(A \cap B) \cup (A \cap C) \subset A \cap (B \cup C)$. Because $B \subset B \cup C$, it is immediate that $A \cap B \subset A \cap (B \cup C)$. In a similar manner, $A \cap C \subset A \cap (B \cup C)$. Therefore

$$(A \cap B) \cup (A \cap C) \subset (A \cap (B \cup C)) \cup (A \cap (B \cup C)) = A \cap (B \cup C).$$

Now for the other direction. Given an element $x \in A \cap (B \cup C)$, first of all it must be an element of A. Secondly, as an element in $B \cup C$ it is either in B or in C. Suppose the former; then as an element both of A and of B, x must be in $A \cap B$. The second possibility, namely, $x \in C$, leads us to $x \in A \cap C$. Thus in either eventuality $x \in (A \cap B) \cup (A \cap C)$, whence $A \cap (B \cup C) \subset (A \cap B) \cup (A \cap C)$.

The two opposite containing relations combine to give us the equality asserted in the proposition.

We continue the discussion of sets to extend the notion of union and of intersection to arbitrary collections of sets.

Given a set T we say that T serves as an *index set* for the family $\mathscr{F} = \{A_\alpha\}$ of sets if for every $\alpha \in T$ there exists a set of A_α in the family $\mathscr{F}$. The index set T can be any set, finite or infinite. Very often we use the set of non-negative integers as an index set, but, we repeat, T can be any (nonempty) set.

By the *union* of the sets A_α, where α is in T, we mean the set $\{x \mid x \in A_\alpha$ for at least one α in $T\}$. We shall denote it by $\bigcup_{\alpha \in T} A_\alpha$. By the *intersection* of the sets A_α, where α is in T, we mean the set $\{x \mid x \in A_\alpha$ for every $\alpha \in T\}$; we shall denote it by $\bigcap_{\alpha \in T} A_\alpha$. The sets A_α are *mutually disjoint* if for $\alpha \neq \beta$, $A_\alpha \cap A_\beta$ is the null set.

For instance, if S is the set of real numbers, and if T is the set of rational numbers, let, for $\alpha \in T$, $A_\alpha = \{x \in S \mid x \geq \alpha\}$. It is an easy exercise to see that $\bigcup_{\alpha \in T} A_\alpha = S$ whereas $\bigcap_{\alpha \in T} A_\alpha$ is the null set. The sets A_α are not mutually disjoint.

DEFINITION Given the two sets A, B then the *difference set*, $A - B$, is the set $\{x \in A \mid x \notin B\}$.

Returning to our little pictures, if A is the circle on the left, B that on the right, then $A - B$ is the shaded area.

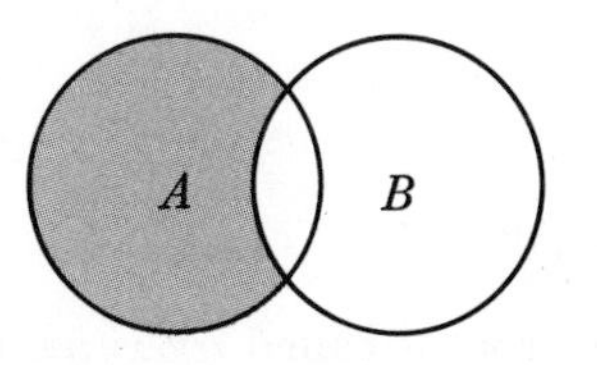

Note that for any set B, the set A satisfies $A = (A \cap B) \cup (A - B)$. (Prove!) Note further that $B \cap (A - B)$ is the null set. A particular case of interest of the difference of two sets is when one of these is a subset of the other. In that case, when B is a subset of A, we call $A - B$ the *complement of B in A*.

We still want one more construct of two given sets A and B, their *Cartesian product* $A \times B$. This set $A \times B$ is defined as the set of all ordered pairs (a, b) where $a \in A$ and $b \in B$ and where we declare the pair (a_1, b_1) to be equal to (a_2, b_2) if and only if $a_1 = a_2$ and $b_1 = b_2$.

A few remarks about the Cartesian product. Given the two sets A and B we could construct the sets $A \times B$ and $B \times A$ from them. As sets these are distinct, yet we feel that they must be closely related. Given three sets A, B, C we can construct many Cartesian products from them: for instance, the set $A \times D$, where $D = B \times C$; the set $E \times C$, where $E = A \times B$; and also the set of all ordered triples (a, b, c) where $a \in A$, $b \in B$, and $c \in C$. These give us three distinct sets, yet here, also, we feel that these sets must be closely related. Of course, we can continue this process with more and more sets. To see the exact relation between them we shall have to wait until the next section, where we discuss one-to-one correspondences.

Given any index set T we could define the Cartesian product of the sets A_α as α varies over T; since we shall not need so general a product, we do not bother to define it.

Finally, we can consider the Cartesian product of a set A with itself, $A \times A$. Note that if the set A is a finite set having n elements, then the set $A \times A$ is also a finite set, but has n^2 elements. The set of elements (a, a) in $A \times A$ is called the *diagonal* of $A \times A$.

A subset R of $A \times A$ is said to define an *equivalence relation* on A if

1. $(a, a) \in R$ for all $a \in A$.
2. $(a, b) \in R$ implies $(b, a) \in R$.
3. $(a, b) \in R$ and $(b, c) \in R$ imply that $(a, c) \in R$.

Instead of speaking about subsets of $A \times A$ we can speak about a binary relation (one between two elements of A) on A itself, defining b to be related to a if $(a, b) \in R$. The properties 1, 2, 3 of the subset R immediately translate into the properties 1, 2, 3 of the definition below.

DEFINITION The binary relation $\sim$ on A is said to be an *equivalence relation* on A if for all a, b, c in A

1. $a \sim a$.
2. $a \sim b$ implies $b \sim a$.
3. $a \sim b$ and $b \sim c$ imply $a \sim c$.

The first of these properties is called *reflexivity*, the second, *symmetry*, and the third, *transitivity*.

The concept of an equivalence relation is an extremely important one and plays a central role in all of mathematics. We illustrate it with a few examples.

Example 1.1.1 Let S be any set and define $a \sim b$, for $a, b \in S$, if and only if $a = b$. This clearly defines an equivalence relation on S. In fact, an equivalence relation is a generalization of equality, measuring equality up to some property.

Example 1.1.2 Let S be the set of all integers. Given $a, b \in S$, define $a \sim b$ if $a - b$ is an even integer. We verify that this defines an equivalence relation of S.

1. Since $0 = a - a$ is even, $a \sim a$.
2. If $a \sim b$, that is, if $a - b$ is even, then $b - a = -(a - b)$ is also even, whence $b \sim a$.
3. If $a \sim b$ and $b \sim c$, then both $a - b$ and $b - c$ are even, whence $a - c = (a - b) + (b - c)$ is also even, proving that $a \sim c$.

Example 1.1.3 Let S be the set of all integers and let $n > 1$ be a fixed integer. Define for $a, b \in S$, $a \sim b$ if $a - b$ is a multiple of n. We leave it as an exercise to prove that this defines an equivalence relation on S.

Example 1.1.4 Let S be the set of all triangles in the plane. Two triangles are defined to be equivalent if they are similar (i.e., have corresponding angles equal). This defines an equivalence relation on S.

Example 1.1.5 Let S be the set of points in the plane. Two points a and b are defined to be equivalent if they are equidistant from the origin. A simple check verifies that this defines an equivalence relation on S.

There are many more equivalence relations; we shall encounter a few as we proceed in the book.

DEFINITION If A is a set and if $\sim$ is an equivalence relation on A, then the *equivalence class* of $a \in A$ is the set $\{x \in A \mid a \sim x\}$. We write it as $\text{cl}(a)$.

In the examples just discussed, what are the equivalence classes? In Example 1.1.1, the equivalence class of a consists merely of a itself. In Example 1.1.2 the equivalence class of a consists of all the integers of the form $a + 2m$, where $m = 0, \pm 1, \pm 2, \ldots$; in this example there are only two distinct equivalence classes, namely, $\text{cl}(0)$ and $\text{cl}(1)$. In Example 1.1.3, the equivalence class of a consists of all integers of the form $a + kn$ where $k = 0, \pm 1, \pm 2, \ldots$; here there are n distinct equivalence classes, namely $\text{cl}(0), \text{cl}(1), \ldots, \text{cl}(n - 1)$. In Example 1.1.5, the equivalence class of a consists of all the points in the plane which lie on the circle which has its center at the origin and passes through a.

Although we have made quite a few definitions, introduced some concepts, and have even established a simple little proposition, one could say in all fairness that up to this point we have not proved any result of real substance. We are now about to prove the first genuine result in the book. The proof of this theorem is not very difficult—actually it is quite easy—but nonetheless the result it embodies will be of great use to us.

THEOREM 1.1.1 *The distinct equivalence classes of an equivalence relation on A provide us with a decomposition of A as a union of mutually disjoint subsets. Conversely, given a decomposition of A as a union of mutually disjoint, nonempty subsets, we can define an equivalence relation on A for which these subsets are the distinct equivalence classes.*

Proof. Let the equivalence relation on A be denoted by $\sim$.

We first note that since for any $a \in A$, $a \sim a$, a must be in $\text{cl}(a)$, whence the union of the $\text{cl}(a)$'s is all of A. We now assert that given two equivalence classes they are either equal or disjoint. For, suppose that $\text{cl}(a)$ and $\text{cl}(b)$ are not disjoint; then there is an element $x \in \text{cl}(a) \cap \text{cl}(b)$. Since $x \in \text{cl}(a)$, $a \sim x$; since $x \in \text{cl}(b)$, $b \sim x$, whence by the symmetry of the relation, $x \sim b$. However, $a \sim x$ and $x \sim b$ by the transitivity of the relation forces $a \sim b$. Suppose, now that $y \in \text{cl}(b)$; thus $b \sim y$. However, from $a \sim b$ and $b \sim y$, we deduce that $a \sim y$, that is, that $y \in \text{cl}(a)$. Therefore, every element in $\text{cl}(b)$ is in $\text{cl}(a)$, which proves that $\text{cl}(b) \subset \text{cl}(a)$. The argument is clearly symmetric, whence we conclude that $\text{cl}(a) \subset \text{cl}(b)$. The two opposite containing relations imply that $\text{cl}(a) = \text{cl}(b)$.

We have thus shown that the distinct $\text{cl}(a)$'s are mutually disjoint and that their union is A. This proves the first half of the theorem. Now for the other half!

Suppose that $A = \bigcup A_\alpha$ where the A_α are mutually disjoint, nonempty sets (α is in some index set T). How shall we use them to define an equivalence relation? The way is clear; given an element a in A it is in *exactly one* A_α. We define for $a, b \in A$, $a \sim b$ if a and b are in the same A_α. We leave it as an exercise to prove that this is an equivalence relation on A and that the distinct equivalence classes are the A_α's.

Problems

1. (a) If A is a subset of B and B is a subset of C, prove that A is a subset of C.
   (b) If $B \subset A$, prove that $A \cup B = A$, and conversely.
   (c) If $B \subset A$, prove that for any set C both $B \cup C \subset A \cup C$ and $B \cap C \subset A \cap C$.
2. (a) Prove that $A \cap B = B \cap A$ and $A \cup B = B \cup A$.
   (b) Prove that $(A \cap B) \cap C = A \cap (B \cap C)$.
3. Prove that $A \cup (B \cap C) = (A \cup B) \cap (A \cup C)$.
4. For a subset C of S let C' denote the complement of C in S. For any two subsets A, B of S prove the *De Morgan rules*:
   (a) $(A \cap B)' = A' \cup B'$.
   (b) $(A \cup B)' = A' \cap B'$.
5. For a finite set C let $o(C)$ indicate the number of elements in C. If A and B are finite sets prove $o(A \cup B) = o(A) + o(B) - o(A \cap B)$.

6. If A is a finite set having n elements, prove that A has exactly 2^n distinct subsets.
7. A survey shows that 63% of the American people like cheese whereas 76% like apples. What can you say about the percentage of the American people that like both cheese and apples? (The given statistics are not meant to be accurate.)
8. Given two sets A and B their *symmetric difference* is defined to be $(A - B) \cup (B - A)$. Prove that the symmetric difference of A and B equals $(A \cup B) - (A \cap B)$.
9. Let S be a set and let S^* be the set whose elements are the various subsets of S. In S^* we define an addition and multiplication as follows: If $A, B \in S^*$ (remember, this means that they are subsets of S):
   (1) $A + B = (A - B) \cup (B - A)$.
   (2) $A \cdot B = A \cap B$.
   Prove the following laws that govern these operations:
   (a) $(A + B) + C = A + (B + C)$.
   (b) $A \cdot (B + C) = A \cdot B + A \cdot C$.
   (c) $A \cdot A = A$.
   (d) $A + A =$ null set.
   (e) If $A + B = A + C$ then $B = C$.
   (The system just described is an example of a *Boolean algebra*.)
10. For the given set and relation below determine which define equivalence relations.
    (a) S is the set of all people in the world today, $a \sim b$ if a and b have an ancestor in common.
    (b) S is the set of all people in the world today, $a \sim b$ if a lives within 100 miles of b.
    (c) S is the set of all people in the world today, $a \sim b$ if a and b have the same father.
    (d) S is the set of real numbers, $a \sim b$ if $a = \pm b$.
    (e) S is the set of integers, $a \sim b$ if both $a > b$ and $b > a$.
    (f) S is the set of all straight lines in the plane, $a \sim b$ if a is parallel to b.
11. (a) Property 2 of an equivalence relation states that if $a \sim b$ then $b \sim a$; property 3 states that if $a \sim b$ and $b \sim c$ then $a \sim c$. What is wrong with the following proof that properties 2 and 3 imply property 1? Let $a \sim b$; then $b \sim a$, whence, by property 3 (using $a = c$), $a \sim a$.
    (b) Can you suggest an alternative of property 1 which will insure us that properties 2 and 3 do imply property 1?
12. In Example 1.1.3 of an equivalence relation given in the text, prove that the relation defined is an equivalence relation and that there are exactly n distinct equivalence classes, namely, $\text{cl}(0), \text{cl}(1), \ldots, \text{cl}(n - 1)$.
13. Complete the proof of the second half of Theorem 1.1.1.

1.2 Mappings

We are about to introduce the concept of a mapping of one set into another. Without exaggeration this is probably the single most important and universal notion that runs through all of mathematics. It is hardly a new thing to any of us, for we have been considering mappings from the very earliest days of our mathematical training. When we were asked to plot the relation $y = x^2$ we were simply being asked to study the particular mapping which takes every real number onto its square.

Loosely speaking, a mapping from one set, S, into another, T, is a "rule" (whatever that may mean) that associates with each element in S a *unique* element t in T. We shall define a mapping somewhat more formally and precisely but the purpose of the definition is to allow us to think and speak in the above terms. We should think of them as rules or devices or mechanisms that transport us from one set to another.

Let us motivate a little the definition that we will make. The point of view we take is to consider the mapping to be defined by its "graph." We illustrate this with the familiar example $y = x^2$ defined on the real numbers S and taking its values also in S. For this set S, $S \times S$, the set of all pairs (a, b) can be viewed as the plane, the pair (a, b) corresponding to the point whose coordinates are a and b, respectively. In this plane we single out all those points whose coordinates are of the form (x, x^2) and call this set of points the graph of $y = x^2$. We even represent this set pictorially as

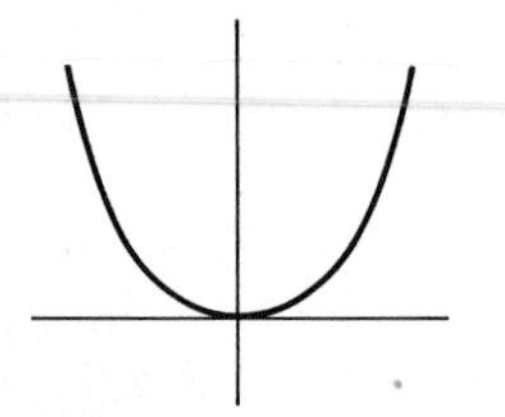

To find the "value" of the function or mapping at the point $x = a$, we look at the point in the graph whose first coordinate is a and read off the second coordinate as the value of the function at $x = a$.

This is, no more or less, the approach we take in the general setting to define a mapping from one set into another.

DEFINITION If S and T are nonempty sets, then a *mapping* from S to T is a subset, M, of $S \times T$ such that for *every* $s \in S$ there is a *unique* $t \in T$ such that the ordered pair (s, t) is in M.

This definition serves to make the concept of a mapping precise for us but we shall almost never use it in this form. Instead we do prefer to think of a mapping as a rule which associates with any element s in S some element t in T, *the rule being, associate (or map)* $s \in S$ *with* $t \in T$ *if and only if* $(s, t) \in M$. We shall say that t is the *image* of s under the mapping.

Now for some notation for these things. Let σ be a mapping from S to T; we often denote this by writing $\sigma: S \rightarrow T$ or $S \xrightarrow{\sigma} T$. If t is the image of s under σ we shall sometimes write this as $\sigma: s \rightarrow t$; more often, we shall represent this fact by $t = s\sigma$. Note that we write the mapping σ on the *right*. There is no overall consistency in this usage; many people would write it as $t = \sigma(s)$. Algebraists often write mappings on the right; other mathematicians write them on the left. In fact, we shall not be absolutely consistent in this ourselves; when we shall want to emphasize the functional nature of σ we may very well write $t = \sigma(s)$.

Examples of Mappings

In all the examples the sets are assumed to be nonempty.

Example 1.2.1 Let S be any set; define $\iota: S \rightarrow S$ by $s = s\iota$ for any $s \in S$. This mapping ι is called the *identity mapping* of S.

Example 1.2.2 Let S and T be any sets and let t_0 be an element of T. Define $\tau: S \rightarrow T$ by $\tau: s \rightarrow t_0$ for every $s \in S$.

Example 1.2.3 Let S be the set of positive rational numbers and let $T = J \times J$ where J is the set of integers. Given a rational number s we can write it as $s = m/n$, where m and n have no common factor. Define $\tau: S \rightarrow T$ by $s\tau = (m, n)$.

Example 1.2.4 Let J be the set of integers and $S = \{(m, n) \in J \times J \mid n \neq 0\}$; let T be the set of rational numbers; define $\tau: S \rightarrow T$ by $(m, n)\tau = m/n$ for every (m, n) in S.

Example 1.2.5 Let J be the set of integers and $S = J \times J$. Define $\tau: S \rightarrow J$ by $(m, n)\tau = m + n$.

Note that in Example 1.2.5 the addition in J itself can be represented in terms of a mapping of $J \times J$ into J. Given an arbitrary set S we call a mapping of $S \times S$ into S a *binary operation* on S. Given such a mapping $\tau: S \times S \rightarrow S$ we could use it to define a "product" $*$ in S by declaring $a * b = c$ if $(a, b)\tau = c$.

Example 1.2.6 Let S and T be any sets; define $\tau: S \times T \rightarrow S$ by $(a, b)\tau = a$ for any $(a, b) \in S \times T$. This τ is called the *projection* of $S \times T$ on S. We could similarly define the projection of $S \times T$ on T.

Example 1.2.7 Let S be the set consisting of the elements x_1, x_2, x_3. Define $\tau:S \to S$ by $x_1\tau = x_2$, $x_2\tau = x_3$, $x_3\tau = x_1$.

Example 1.2.8 Let S be the set of integers and let T be the set consisting of the elements E and 0. Define $\tau:S \to T$ by declaring $n\tau = E$ if n is even and $n\tau = 0$ if n is odd.

If S is any set, let $\{x_1, \ldots, x_n\}$ be its subset consisting of the elements $x_1, x_2, \ldots, x_n$ of S. In particular, $\{x\}$ is the subset of S whose only element is x. Given S we can use it to construct a new set S^*, the set whose elements are the subsets of S. We call S^* the *set of subsets* of S. Thus for instance, if $S = \{x_1, x_2\}$ then S^* has exactly four elements, namely, $a_1 =$ null set, $a_2 =$ the subset, S, of S, $a_3 = \{x_1\}$, $a_4 = \{x_2\}$. The relation of S to S^*, in general, is a very interesting one; some of its properties are examined in the problems.

Example 1.2.9 Let S be a set, $T = S^*$; define $\tau:S \to T$ by $s\tau =$ complement of $\{s\}$ in $S = S - \{s\}$.

Example 1.2.10 Let S be a set with an equivalence relation, and let T be the set of equivalence classes in S (note that T is a subset of S^*). Define $\tau:S \to T$ by $s\tau = \text{cl}(s)$.

We leave the examples to continue the general discussion. Given a mapping $\tau:S \to T$ we define for $t \in T$, the *inverse image* of t with respect to τ to be the set $\{s \in S \mid t = s\tau\}$. In Example 1.2.8, the inverse image of E is the subset of S consisting of the even integers. It may happen that for some t in T that its inverse image with respect to τ is empty; that is, t is not the image under τ of any element in S. In Example 1.2.3, the element (4, 2) is not the image of any element in S under the τ used; in Example 1.2.9, S, as an element in S^*, is not the image under the τ used of any element in S.

DEFINITION The mapping τ of S into T is said to be *onto* T if given $t \in T$ there exists an element $s \in S$ such that $t = s\tau$.

If we call the subset $S\tau = \{x \in T \mid x = s\tau \text{ for some } s \in S\}$ the *image* of S under τ, then τ is onto if the image of S under τ is all of T. Note that in Examples 1.2.1, 1.2.4–1.2.8, and 1.2.10 the mappings used are all onto.

Another special type of mapping arises often and is important: the one-to-one mapping.

DEFINITION The mapping τ of S into T is said to be a *one-to-one mapping* if whenever $s_1 \neq s_2$, then $s_1\tau \neq s_2\tau$.

In terms of inverse images, the mapping τ is one-to-one if for any $t \in T$ the inverse image of t is either empty or is a set consisting of one element. In the examples discussed, the mappings in Examples 1.2.1, 1.2.3, 1.2.7, and 1.2.9 are all one-to-one.

When should we say that two mappings from S to T are equal? A natural definition for this is that they should have the same effect on every element of S; that is, the image of any element in S under each of these mappings should be the same. In a little more formal manner:

DEFINITION The two mappings σ and τ of S into T are said to be *equal* if $s\sigma = s\tau$ for every $s \in S$.

Consider the following situation: We have a mapping σ from S to T and another mapping τ from T to U. Can we compound these mappings to produce a mapping from S to U? The most natural and obvious way of doing this is to send a given element s, in S, in two stages into U, first by applying σ to s and then applying τ to the resulting element $s\sigma$ in T. This is the basis of the

DEFINITION If $\sigma:S \to T$ and $\tau:T \to U$ then the *composition* of σ and τ (also called their *product*) is the mapping $\sigma \circ \tau:S \to U$ defined by means of $s(\sigma \circ \tau) = (s\sigma)\tau$ for every $s \in S$.

Note that the order of events reads from left to right; $\sigma \circ \tau$ reads: first perform σ and then follow it up with τ. Here, too, the left-right business is not a uniform one. Mathematicians who write their mappings on the left would read $\sigma \circ \tau$ to mean first perform τ and then σ. Accordingly, in reading a given book in mathematics one must make absolutely sure as to what convention is being followed in writing the product of two mappings. We reiterate, *for us $\sigma \circ \tau$ will always mean: first apply σ and then τ.*

We illustrate the composition of σ and τ with a few examples.

Example 1.2.11 Let $S = \{x_1, x_2, x_3\}$ and let $T = S$. Let $\sigma:S \to S$ be defined by

$$x_1\sigma = x_2,$$
$$x_2\sigma = x_3,$$
$$x_3\sigma = x_1;$$

and $\tau:S \to S$ by

$$x_1\tau = x_1,$$
$$x_2\tau = x_3,$$
$$x_3\tau = x_2.$$

Thus

$$x_1(\sigma \circ \tau) = (x_1\sigma)\tau = x_2\tau = x_3,$$
$$x_2(\sigma \circ \tau) = (x_2\sigma)\tau = x_3\tau = x_2,$$
$$x_3(\sigma \circ \tau) = (x_3\sigma)\tau = x_1\tau = x_1.$$

At the same time we can compute $\tau \circ \sigma$, because in this case it also makes sense. Now

$$x_1(\tau \circ \sigma) = (x_1\tau)\sigma = (x_1\sigma) = x_2,$$
$$x_2(\tau \circ \sigma) = (x_2\tau)\sigma = x_3\sigma = x_1,$$
$$x_3(\tau \circ \sigma) = (x_3\tau)\sigma = x_2\sigma = x_3.$$

Note that $x_2 = x_1(\tau \circ \sigma)$, whereas $x_3 = x_1(\sigma \circ \tau)$ whence $\sigma \circ \tau \neq \tau \circ \sigma$.

Example 1.2.12 Let S be the set of integers, T the set $S \times S$, and suppose $\sigma : S \to T$ is defined by $m\sigma = (m - 1, 1)$. Let $U = S$ and suppose that $\tau : T \to U(= S)$ is defined by $(m, n)\tau = m + n$. Thus $\sigma \circ \tau : S \to S$ whereas $\tau \circ \sigma : T \to T$; even to speak about the equality of $\sigma \circ \tau$ and $\tau \circ \sigma$ would make no sense since they do not act on the same space. We now compute $\sigma \circ \tau$ as a mapping of S into itself and then $\tau \circ \sigma$ as one on T into itself.

Given $m \in S$, $m\sigma = (m - 1, 1)$ whence $m(\sigma \circ \tau) = (m\sigma)\tau = (m - 1, 1)\tau = (m - 1) + 1 = m$. Thus $\sigma \circ \tau$ is the identity mapping of S into itself. What about $\tau \circ \sigma$? Given $(m, n) \in T$, $(m, n)\tau = m + n$, whereby $(m, n)(\tau \circ \sigma) = ((m, n)\tau)\sigma = (m + n)\sigma = (m + n - 1, 1)$. Note that $\tau \circ \sigma$ is *not* the identity map of T into itself; it is not even an onto mapping of T.

Example 1.2.13 Let S be the set of real numbers, T the set of integers, and $U = \{E, 0\}$. Define $\sigma : S \to T$ by $s\sigma$ = largest integer less than or equal to s, and $\tau : T \to U$ defined by $n\tau = E$ if n is even, $n\tau = 0$ if n is odd. Note that in this case $\tau \circ \sigma$ cannot be defined. We compute $\sigma \circ \tau$ for two real numbers $s = \frac{8}{3}$ and $s = \pi$. Now since $\frac{8}{3} = 2 + \frac{2}{3}$, $(\frac{8}{3})\sigma = 2$, whence $(\frac{8}{3})(\sigma \circ \tau) = (\frac{8}{3}\sigma)\tau = (2)\tau = E$; $(\pi)\sigma = 3$, whence $\pi(\sigma \circ \tau) = (\pi\sigma)\tau = (3)\tau = 0$.

For mappings of sets, provided the requisite products make sense, a general *associative law* holds. This is the content of

LEMMA 1.2.1 (Associative Law) *If $\sigma : S \to T$, $\tau : T \to U$, and $\mu : U \to V$, then $(\sigma \circ \tau) \circ \mu = \sigma \circ (\tau \circ \mu)$.*

Proof. Note first that $\sigma \circ \tau$ makes sense and takes S into U, thus $(\sigma \circ \tau) \circ \mu$ also makes sense and takes S into V. Similarly $\sigma \circ (\tau \circ \mu)$ is meaningful and takes S into V. Thus we can speak about the equality, or lack of equality, of $(\sigma \circ \tau) \circ \mu$ and $\sigma \circ (\tau \circ \mu)$.

To prove the asserted equality we merely must show that for any $s \in S$, $s((\sigma \circ \tau) \circ \mu) = s(\sigma \circ (\tau \circ \mu))$. Now by the very definition of the composition of maps, $s((\sigma \circ \tau) \circ \mu) = (s(\sigma \circ \tau))\mu = ((s\sigma)\tau)\mu$ whereas $s(\sigma \circ (\tau \circ \mu)) = (s\sigma)(\tau \circ \mu) = ((s\sigma)\tau)\mu$. Thus, the elements $s((\sigma \circ \tau) \circ \mu)$ and $s(\sigma \circ (\tau \circ \mu))$ are indeed equal. This proves the lemma.

We should like to show that if two mappings σ and τ are properly conditioned the very same conditions carry over to $\sigma \circ \tau$.

LEMMA 1.2.2 *Let $\sigma : S \to T$ and $\tau : T \to U$; then*

1. *$\sigma \circ \tau$ is onto if each of σ and τ is onto.*
2. *$\sigma \circ \tau$ is one-to-one if each of σ and τ is one-to-one.*

Proof. We prove only part 2, leaving the proof of part 1 as an exercise.
Suppose that $s_1, s_2 \in S$ and that $s_1 \neq s_2$. By the one-to-one nature of σ, $s_1\sigma \neq s_2\sigma$. Since τ is one-to-one and $s_1\sigma$ and $s_2\sigma$ are distinct elements of T, $(s_1\sigma)\tau \neq (s_2\sigma)\tau$ whence $s_1(\sigma \circ \tau) = (s_1\sigma)\tau \neq (s_2\sigma)\tau = s_2(\sigma \circ \tau)$, proving that $\sigma \circ \tau$ is indeed one-to-one, and establishing the lemma.

Suppose that σ is a one-to-one mapping of S *onto* T; we call σ a *one-to-one correspondence* between S and T. Given any $t \in T$, by the "onto-ness" of σ there exists an element $s \in S$ such that $t = s\sigma$; by the "one-to-oneness" of σ this s is unique. We define the mapping $\sigma^{-1} : T \to S$ by $s = t\sigma^{-1}$ if and only if $t = s\sigma$. The mapping σ^{-1} is called the *inverse* of σ. Let us compute $\sigma \circ \sigma^{-1}$ which maps S into itself. Given $s \in S$, let $t = s\sigma$, whence by definition $s = t\sigma^{-1}$; thus $s(\sigma \circ \sigma^{-1}) = (s\sigma)\sigma^{-1} = t\sigma^{-1} = s$. We have shown that $\sigma \circ \sigma^{-1}$ is the identity mapping of S onto itself. A similar computation reveals that $\sigma^{-1} \circ \sigma$ is the identity mapping of T onto itself.

Conversely, if $\sigma : S \to T$ is such that there exists a $\mu : T \to S$ with the property that $\sigma \circ \mu$ and $\mu \circ \sigma$ are the identity mappings on S and T, respectively, then we claim that σ is a one-to-one correspondence between S and T. First observe that σ is onto for, given $t \in T$, $t = t(\mu \circ \sigma) = (t\mu)\sigma$ (since $\mu \circ \sigma$ is the identity on T) and so t is the image under σ of the element $t\mu$ in S. Next observe that σ is one-to-one, for if $s_1\sigma = s_2\sigma$, using that $\sigma \circ \mu$ is the identity on S, we have $s_1 = s_1(\sigma \circ \mu) = (s_1\sigma)\mu = (s_2\sigma)\,\mu = s_2(\sigma \circ \mu) = s_2$. We have now proved

LEMMA 1.2.3 *The mapping $\sigma : S \to T$ is a one-to-one correspondence between S and T if and only if there exists a mapping $\mu : T \to S$ such that $\sigma \circ \mu$ and $\mu \circ \sigma$ are the identity mappings on S and T, respectively.*

DEFINITION If S is a nonempty set then $A(S)$ is the *set of all one-to-one mappings* of S onto itself.

Aside from its own intrinsic interest $A(S)$ plays a central and universal type of role in considering the mathematical system known as a group

(Chapter 2). For this reason we state the next theorem concerning its nature. All the constituent parts of the theorem have already been proved in the various lemmas, so we state the theorem without proof.

THEOREM 1.2.1 *If σ, τ, μ are elements of $A(S)$, then*

1. *$\sigma \circ \tau$ is in $A(S)$.*
2. *$(\sigma \circ \tau) \circ \mu = \sigma \circ (\tau \circ \mu)$.*
3. *There exists an element ι (the identity map) in $A(S)$ such that $\sigma \circ \iota = \iota \circ \sigma = \sigma$.*
4. *There exists an element $\sigma^{-1} \in A(S)$ such that $\sigma \circ \sigma^{-1} = \sigma^{-1} \circ \sigma = \iota$.*

We close the section with a remark about $A(S)$. Suppose that S has more than two elements; let x_1, x_2, x_3 be three distinct elements in S; define the mapping $\sigma:S \to S$ by $x_1\sigma = x_2$, $x_2\sigma = x_3$, $x_3\sigma = x_1$, $s\sigma = s$ for any $s \in S$ different from x_1, x_2, x_3. Define the mapping $\tau:S \to S$ by $x_2\tau = x_3$, $x_3\tau = x_2$, and $s\tau = s$ for any $s \in S$ different from x_2, x_3. Clearly both σ and τ are in $A(S)$. A simple computation shows that $x_1(\sigma \circ \tau) = x_3$ but that $x_1(\tau \circ \sigma) = x_2 \neq x_3$. Thus $\sigma \circ \tau \neq \tau \circ \sigma$. This is

LEMMA 1.2.4 *If S has more that two elements we can find two elements σ, τ in $A(S)$ such that $\sigma \circ \tau \neq \tau \circ \sigma$.*

Problems

1. In the following, where $\sigma:S \to T$, determine whether the σ is onto and/or one-to-one and determine the inverse image of any $t \in T$ under σ.
   (a) S = set of real numbers, T = set of nonnegative real numbers, $s\sigma = s^2$.
   (b) S = set of nonnegative real numbers, T = set of nonnegative real numbers, $s\sigma = s^2$.
   (c) S = set of integers, T = set of integers, $s\sigma = s^2$.
   (d) S = set of integers, T = set of integers, $s\sigma = 2s$.
2. If S and T are nonempty sets, prove that there exists a one-to-one correspondence between $S \times T$ and $T \times S$.
3. If S, T, U are nonempty sets, prove that there exists a one-to-one correspondence between
   (a) $(S \times T) \times U$ and $S \times (T \times U)$.
   (b) Either set in part (a) and the set of ordered triples (s, t, u) where $s \in S$, $t \in T$, $u \in U$.
4. (a) If there is a one-to-one correspondence between S and T, prove that there exists a one-to-one correspondence between T and S.

(b) If there is a one-to-one correspondence between S and T and between T and U, prove that there is a one-to-one correspondence between S and U.

5. If ι is the identity mapping on S, prove that for any $\sigma \in A(S)$, $\sigma \circ \iota = \iota \circ \sigma = \sigma$.

*6. If S is any set, prove that it is *impossible* to find a mapping of S *onto* S^*.

7. If the set S has n elements, prove that $A(S)$ has $n!$ (n factorial) elements.

8. If the set S has a finite number of elements, prove the following:
   (a) If σ maps S onto S, then σ is one-to-one.
   (b) If σ is a one-to-one mapping of S onto itself, then σ is onto.
   (c) Prove, by example, that both part (a) and part (b) are false if S does not have a finite number of elements.

9. Prove that the converse to both parts of Lemma 1.2.2 are false; namely,
   (a) If $\sigma \circ \tau$ is onto, it need not be that both σ and τ are onto.
   (b) If $\sigma \circ \tau$ is one-to-one, it need not be that both σ and τ are one-to-one.

10. Prove that there is a one-to-one correspondence between the set of integers and the set of rational numbers.

11. If $\sigma: S \to T$ and if A is a subset of S, the *restriction of* σ *to* A, σ_A, is defined by $a\sigma_A = a\sigma$ for any $a \in A$. Prove
    (a) σ_A defines a mapping of A into T.
    (b) σ_A is one-to-one if σ is.
    (c) σ_A may very well be one-to-one even if σ is not.

12. If $\sigma: S \to S$ and A is a subset of S such that $A\sigma \subset A$, prove that $(\sigma \circ \sigma)_A = \sigma_A \circ \sigma_A$.

13. A set S is said to be *infinite* if there is a one-to-one correspondence between S and a proper subset of S. Prove
    (a) The set of integers is infinite.
    (b) The set of real numbers is infinite.
    (c) If a set S has a subset A which is infinite, then S must be infinite.
    (*Note:* By the result of Problem 8, a set finite in the usual sense is not infinite.)

*14. If S is infinite and can be brought into one-to-one correspondence with the set of integers, prove that there is one-to-one correspondence between S and $S \times S$.

*15. Given two sets S and T we declare $S < T$ (S is smaller than T) if there is a mapping of T *onto* S but *no* mapping of S *onto* T. Prove that if $S < T$ and $T < U$ then $S < U$.

16. If S and T are finite sets having m and n elements, respectively, prove that if $m < n$ then $S < T$.

1.3 The Integers

We close this chapter with a brief discussion of the set of integers. We shall make no attempt to construct them axiomatically, assuming instead that we already have the set of integers and that we know many of the elementary facts about them. In this number we include the principle of mathematical induction (which will be used freely throughout the book) and the fact that a nonempty set of positive integers always contains a smallest element. As to notation, the familiar symbols: $a > b$, $a \leq b$, $|a|$, etc., will occur with their usual meaning. To avoid repeating that something is an integer, we make the assumption *that all symbols, in this section, written as lowercase Latin letters will be integers.*

Given a and b, with $b \neq 0$, we can divide a by b to get a nonnegative remainder r which is smaller in size than b; that is, we can find m and r such that $a = mb + r$ where $0 \leq r < |b|$. This fact is known as the *Euclidean algorithm* and we assume familiarity with it.

We say that $b \neq 0$ *divides* a if $a = mb$ for some m. We denote that b divides a by $b \mid a$, and that b does not divide a by $b \nmid a$. Note that if $a \mid 1$ then $a = \pm 1$, that when both $a \mid b$ and $b \mid a$, then $a = \pm b$, and that any b divides 0. If $b \mid a$, we call b a *divisor* of a. Note that if b is a divisor of g and of h, then it is a divisor of $mg + nh$ for arbitrary integers m and n. We leave the verification of these remarks as exercises.

DEFINITION The positive integer c is said to be the *greatest common divisor* of a and b if

1. c is a divisor of a and of b.
2. Any divisor of a and b is a divisor of c.

We shall use the notation (a, b) for the greatest common divisor of a and b. Since we insist that the greatest common divisor be positive, $(a, b) = (a, -b) = (-a, b) = (-a, -b)$. For instance, $(60, 24) = (60, -24) = 12$. Another comment: The mere fact that we have defined what is to be meant by the greatest common divisor does not guarantee that it exists. This will have to be proved. However, we can say that if it exists then it is unique, for, if we had c_1 and c_2 satisfying both conditions of the definition above, then $c_1 \mid c_2$ and $c_2 \mid c_1$, whence we would have $c_1 = \pm c_2$; the insistence on positivity would then force $c_1 = c_2$. Our first business at hand then is to dispose of the existence of (a, b). In doing so, in the next lemma, we actually prove a little more, namely that (a, b) must have a particular form.

LEMMA 1.3.1 *If a and b are integers, not both 0, then (a, b) exists; moreover, we can find integers m_0 and n_0 such that $(a, b) = m_0 a + n_0 b$.*

Proof. Let $\mathcal{M}$ be the set of all integers of the form $ma + nb$, where m and n range freely over the set of integers. Since one of a or b is not 0, there are nonzero integers in $\mathcal{M}$. Because $x = ma + nb$ is in $\mathcal{M}$, $-x = (-m)a + (-n)b$ is also in $\mathcal{M}$; therefore, $\mathcal{M}$ always has in it some positive integers. But then there is a smallest positive integer, c, in $\mathcal{M}$; being in $\mathcal{M}$, c has the form $c = m_0a + n_0b$. We claim that $c = (a, b)$.

Note first that if $d \mid a$ and $d \mid b$, the $d \mid (m_0a + n_0b)$, whence $d \mid c$. We now must show that $c \mid a$ and $c \mid b$. Given any element $x = ma + nb$ in $\mathcal{M}$, then by the Euclidean algorithm, $x = tc + r$ where $0 \leq r < c$. Writing this out explicitly, $ma + nb = t(m_0a + n_0b) + r$, whence $r = (m - tm_0)a + (n - tn_0)b$ and so must be in $\mathcal{M}$. Since $0 \leq r$ and $r < c$, by the choice of c, $r = 0$. Thus $x = tc$; we have proved that $c \mid x$ for any $x \in \mathcal{M}$. But $a = 1a + 0b \in \mathcal{M}$ and $b = 0a + 1b \in \mathcal{M}$, whence $c \mid a$ and $c \mid b$.

We have shown that c satisfies the requisite properties to be (a, b) and so we have proved the lemma.

DEFINITION The integers a and b are *relatively prime* if $(a, b) = 1$.

As an immediate consequence of Lemma 1.3.1, we have the

COROLLARY *If a and b are relatively prime, we can find integers m and n such that $ma + nb = 1$.*

We introduce another familiar notion, that of prime number. By this we shall mean an integer which has no nontrivial factorization. For technical reasons, we exclude 1 from the set of prime numbers. The sequence 2, 3, 5, 7, 11, ... are all prime numbers; equally, $-2, -3, -5, \ldots$ are prime numbers. Since, in factoring, the negative introduces no essential differences, for us prime numbers will always be positive.

DEFINITION The integer $p > 1$ is a *prime number* if its only divisors are $\pm 1, \pm p$.

Another way of putting this is to say that an integer p (larger than 1) is a prime number if and only if given any other integer n then either $(p, n) = 1$ or $p \mid n$. As we shall soon see, the prime numbers are the building blocks of the integers. But first we need the important observation,

LEMMA 1.3.2 *If a is relatively prime to b but $a \mid bc$, then $a \mid c$.*

Proof. Since a and b are relatively prime, by the corollary to Lemma 1.3.1, we can find integers m and n such that $ma + nb = 1$. Thus $mac + nbc = c$. Now $a \mid mac$ and, by assumption, $a \mid nbc$; consequently,

$a \mid (mac + nbc)$. Since $mac + nbc = c$, we conclude that $a \mid c$, which is precisely the assertion of the lemma.

Following immediately from the lemma and the definition of prime number is the important

COROLLARY *If a prime number divides the product of certain integers it must divide at least one of these integers.*

We leave the proof of the corollary to the reader.

We have asserted that the prime numbers serve as the building blocks for the set of integers. The precise statement of this is the *unique factorization theorem*:

THEOREM 1.3.1 *Any positive integer $a > 1$ can be factored in a unique way as $a = p_1^{\alpha_1} p_2^{\alpha_2} \cdots p_t^{\alpha_t}$, where $p_1 > p_2 > \cdots > p_t$ are prime numbers and where each $\alpha_i > 0$.*

Proof. The theorem as stated actually consists of two distinct subtheorems; the first asserts the possibility of factoring the given integer as a product of prime powers; the second assures us that this decomposition is unique. We shall prove the theorem itself by proving each of these subtheorems separately.

An immediate question presents itself: How shall we go about proving the theorem? A natural method of attack is to use mathematical induction. A short word about this; we shall use the following version of mathematical induction: If the proposition $P(m_0)$ is true and if the truth of $P(r)$ for all r such that $m_0 \leq r < k$ implies the truth of $P(k)$, then $P(n)$ is true for all $n \geq m_0$. This variant of induction can be shown to be a consequence of the basic property of the integers which asserts that any nonempty set of positive integers has a minimal element (see Problem 10).

We first prove that every integer $a > 1$ can be factored as a product of prime powers; our approach is via mathematical induction.

Certainly $m_0 = 2$, being a prime number, has a representation as a product of prime powers.

Suppose that any integer r, $2 \leq r < k$ can be factored as a product of prime powers. If k itself is a prime number, then it is a product of prime powers. If k is not a prime number, then $k = uv$, where $1 < u < k$ and $1 < v < k$. By the induction hypothesis, since both u and v are less than k, each of these can be factored as a product of prime powers. Thus $k = uv$ is also such a product. We have shown that the truth of the proposition for all integers r, $2 \leq r < k$, implies its truth for k. Consequently, by the basic induction principle, the proposition is true for all integers $n \geq m_0 = 2$; that is, every integer $n \geq 2$ is a product of prime powers.

Now for the uniqueness. Here, too, we shall use mathematical induction, and in the form used above. Suppose that

$$a = p_1^{\alpha_1}p_2^{\alpha_2}\cdots p_r^{\alpha_r} = q_1^{\beta_1}q_2^{\beta_2}\cdots q_s^{\beta_s},$$

where $p_1 > p_2 > \cdots p_r$, $q_1 > q_2 > \cdots > q_s$ are prime numbers, and where each $\alpha_i > 0$ and each $\beta_i > 0$. Our object is to prove

1. $r = s$.
2. $p_1 = q_1, p_2 = q_2, \ldots, p_r = q_r$.
3. $\alpha_1 = \beta_1, \alpha_2 = \beta_2, \ldots, \alpha_r = \beta_r$.

For $a = 2$ this is clearly true. Proceeding by induction we suppose it to be true for all integers u, $2 \leq u < a$. Now, since

$$a = p_1^{\alpha_1}\cdots p_r^{\alpha_r} = q_1^{\beta_1}\cdots q_s^{\beta_s}$$

and since $\alpha_1 > 0$, $p_1 \mid a$, hence $p_1 \mid q_1^{\beta_1}\cdots q_s^{\beta_s}$. However, since p_1 is a prime number, by the corollary to Lemma 1.3.2, it follows easily that $p_1 = q_i$ for some i. Thus $q_1 \geq q_i = p_1$. Similarly, since $q_1 \mid a$ we get $q_1 = p_j$ for some j, whence $p_1 \geq p_j = q_1$. In short, we have shown that $p_1 = q_1$. Therefore $a = p_1^{\alpha_1}p_2^{\alpha_2}\cdots p_r^{\alpha_r} = p_1^{\beta_1}q_2^{\beta_2}\cdots q_s^{\beta_s}$. We claim that this forces $\alpha_1 = \beta_1$. (Prove!) But then

$$b = \frac{a}{p^{\alpha_1}} = p_2^{\alpha_2}\cdots p_r^{\alpha_r} = q_2^{\beta_2}\cdots q_s^{\beta_s}.$$

If $b = 1$, then $\alpha_2 = \cdots = \alpha_r = 0$ and $\beta_2 = \cdots = \beta_s = 0$; that is, $r = s = 1$, and we are done. If $b > 1$, then since $b < a$ we can apply our induction hypothesis to b to get

1. The number of distinct prime power factors (in b) on both sides is equal, that is, $r - 1 = s - 1$, hence $r = s$.
2. $\alpha_2 = \beta_2, \ldots, \alpha_r = \beta_r$.
3. $p_2 = q_2, \ldots, p_r = q_r$.

Together with the information we already have obtained, namely, $p_1 = q_1$ and $\alpha_1 = \beta_1$, this is precisely what we were trying to prove. Thus we see that the assumption of the uniqueness of factorization for the integers less than a implied the uniqueness of factorization for a. In consequence, the induction is completed and the assertion of unique factorization is established.

We change direction a little to study the important notion of congruence modulo a given integer. As we shall see later, the relation that we now introduce is a special case of a much more general one that can be defined in a much broader context.

DEFINITION Let $n > 0$ be a fixed integer. We define $a \equiv b \bmod n$ if $n \mid (a - b)$.

The relation is referred to as *congruence modulo n*, n is called the *modulus* of the relation, and we read $a \equiv b \bmod n$ as "a is congruent to b modulo n." Note, for example, that $73 \equiv 4 \bmod 23$, $21 \equiv -9 \bmod 10$, etc.

This congruence relation enjoys the following basic properties:

LEMMA 1.3.3

1. *The relation congruence modulo n defines an equivalence relation on the set of integers.*
2. *This equivalence relation has n distinct equivalence classes.*
3. *If* $a \equiv b \bmod n$ *and* $c \equiv d \bmod n$, *then* $a + c \equiv b + d \bmod n$ *and* $ac \equiv bd \bmod n$.
4. *If* $ab \equiv ac \bmod n$ *and a is relatively prime to n, then* $b \equiv c \bmod n$.

Proof. We first verify that the relation congruence modulo n is an equivalence relation. Since $n \mid 0$, we indeed have that $n \mid (a - a)$ whence $a \equiv a \bmod n$ for every a. Further, if $a \equiv b \bmod n$ then $n \mid (a - b)$, and so $n \mid (b - a) = -(a - b)$; thus $b \equiv a \bmod n$. Finally, if $a \equiv b \bmod n$ and $b \equiv c \bmod n$, then $n \mid (a - b)$ and $n \mid (b - c)$ whence $n \mid \{(a - b) + (b - c)\}$, that is, $n \mid (a - c)$. This, of course, implies that $a \equiv c \bmod n$.

Let the equivalence class, under this relation, of a be denoted by $[a]$; we call it the *congruence class* (mod n) of a. Given any integer a, by the Euclidean algorithm, $a = kn + r$ where $0 \leq r < n$. But then, $a \in [r]$ and so $[a] = [r]$. Thus there are at most n distinct congruence classes; namely, $[0], [1], \ldots, [n - 1]$. However, these are distinct, for if $[i] = [j]$ with, say, $0 \leq i < j < n$, then $n \mid (j - i)$ where $j - i$ is a positive integer less than n, which is obviously impossible. Consequently, there are exactly the n distinct congruence classes $[0], [1], \ldots, [n - 1]$. We have now proved assertions 1 and 2 of the lemma.

We now prove part 3. Suppose that $a \equiv b \bmod n$ and $c \equiv d \bmod n$; therefore, $n \mid (a - b)$ and $n \mid (c - d)$ whence $n \mid \{(a - d) + (c - d)\}$, and so $n \mid \{(a + c) - (b + d)\}$. But then $a + c \equiv b + d \bmod n$. In addition, $n \mid \{(a - b)c + (c - d)b\} = ac - bd$, whence $ac \equiv bd \bmod n$.

Finally, notice that if $ab \equiv ac \bmod n$ and if a is relatively prime to n, then the fact that $n \mid a(b - c)$, by Lemma 1.3.2, implies that $n \mid (b - c)$ and so $b \equiv c \bmod n$.

If a is not relatively prime to n, the result of part 4 may be false; for instance, $2.3 \equiv 4.3 \bmod 6$, yet $2 \not\equiv 4 \bmod 6$.

Lemma 1.3.3 opens certain interesting possibilities for us. Let J_n be the set of the congruence classes mod n; that is, $J_n = \{[0], [1], \ldots, [n-1]\}$. Given two elements, $[i]$ and $[j]$ in J_n, let us define

$$[i] + [j] = [i + j]; \tag{a}$$

$$[i][j] = [ij]. \tag{b}$$

We assert that the lemma assures us that this "addition" and "multiplication" are *well defined*; that is, if $[i] = [i']$ and $[j] = [j']$, then $[i] + [j] = [i + j] = [i' + j'] = [i'] + [j']$ and that $[i][j] = [i'][j']$. (Verify!) These operations in J_n have the following interesting properties (whose proofs we leave as exercises): for any $[i]$, $[j]$, $[k]$ in J_n,

1. $[i] + [j] = [j] + [i]$ } commutative laws.
2. $[i][j] = [j][i]$ }
3. $([i] + [j]) + [k] = [i] + ([j] + [k])$ } associative laws.
4. $([i][j])[k] = [i]([j][k])$ }
5. $[i]([j] + [k]) = [i][j] + [i][k]$ distributive law.
6. $[0] + [i] = [i]$.
7. $[1][i] = [i]$.

One more remark: if $n = p$ is a prime number and if $[a] \neq [0]$ is in J_p, then there is an element $[b]$ in J_p such that $[a][b] = [1]$.

The set J_n plays an important role in algebra and number theory. It is called the set of *integers* mod n; before we proceed much further we will have become well acquainted with it.

Problems

1. If $a \mid b$ and $b \mid a$, show that $a = \pm b$.
2. If b is a divisor of g and of h, show it is a divisor of $mg + nh$.
3. If a and b are integers, the *least common multiple* of a and b, written as $[a, b]$, is defined as that positive integer d such that
   (a) $a \mid d$ and $b \mid d$.
   (b) Whenever $a \mid x$ and $b \mid x$ then $d \mid x$.
   Prove that $[a, b]$ exists and that $[a, b] = ab/(a, b)$, if $a > 0$, $b > 0$.
4. If $a \mid x$ and $b \mid x$ and $(a, b) = 1$ prove that $(ab) \mid x$.
5. If $a = p_1^{\alpha_1} \cdots p_k^{\alpha_k}$ and $b = p_1^{\beta_1} \cdots p_k^{\beta_k}$ where the p_i are distinct prime numbers and where each $\alpha_i \geq 0$, $\beta_i \geq 0$, prove
   (a) $(a, b) = p_1^{\delta_1} \cdots p_k^{\delta_k}$ where $\delta_i =$ minimum of α_i and β_i for each i.
   (b) $[a, b] = p_1^{\gamma_1} \cdots p_k^{\gamma_k}$ where $\gamma_i =$ maximum of α_i and β_i for each i.

6. Given a, b, on applying the Euclidean algorithm successively we have

$$\begin{aligned} a &= q_0 b + r_1, & 0 &\le r_1 < |b|, \\ b &= q_1 r_1 + r_2, & 0 &\le r_2 < r_1, \\ r_1 &= q_2 r_2 + r_3, & 0 &\le r_3 < r_2, \\ &\vdots \\ r_k &= q_{k+1} r_{k+1} + r_{k+2}, & 0 &\le r_{k+2} < r_{k+1}. \end{aligned}$$

Since the integers r_k are decreasing and are all nonnegative, there is a first integer n such that $r_{n+1} = 0$. Prove that $r_n = (a, b)$. (We consider, here, $r_0 = |b|$.)

7. Use the method in Problem 6 to calculate
(a) (1128, 33). (b) (6540, 1206).

8. To check that n is a prime number, prove that it is sufficient to show that it is not divisible by any prime number p, such that $p \le \sqrt{n}$.

9. Show that $n > 1$ is a prime number if and only if for any a either $(a, n) = 1$ or $n \mid a$.

10. Assuming that any nonempty set of positive integers has a minimal element, prove
(a) If the proposition P is such that
 (1) $P(m_0)$ is true,
 (2) the truth of $P(m - 1)$ implies the truth of $P(m)$,
 then $P(n)$ is true for all $n \ge m_0$.
(b) If the proposition P is such that
 (1) $P(m_0)$ is true,
 (2) $P(m)$ is true whenever $P(a)$ is true for all a such that $m_0 \le a < m$,
 then $P(n)$ is true for all $n \ge m_0$.

11. Prove that the addition and multiplication used in J_n are well defined.

12. Prove the properties 1–7 for the addition and multiplication in J_n.

13. If $(a, n) = 1$, prove that one can find $[b] \in J_n$ such that $[a][b] = [1]$ in J_n.

*14. If p is a prime number, prove that for any integer a, $a^p \equiv a \bmod p$.

15. If $(m, n) = 1$, given a and b, prove that there exists an x such that $x \equiv a \bmod m$ and $x \equiv b \bmod n$.

16. Prove the corollary to Lemma 1.3.2.

17. Prove that n is a prime number if and only if in J_n, $[a][b] = [0]$ implies that $[a] = [b] = [0]$.

Supplementary Reading

For sets and cardinal numbers:

BIRKHOFF, G., and MACLANE, S., *A Brief Survey of Modern Algebra*, 2nd ed. New York: The Macmillan Company, 1965.

2

Group Theory

In this chapter we shall embark on the study of the algebraic object known as a group which serves as one of the fundamental building blocks for the subject today called abstract algebra. In later chapters we shall have a look at some of the others such as rings, fields, vector spaces, and linear algebras. Aside from the fact that it has become traditional to consider groups at the outset, there are natural, cogent reasons for this choice. To begin with, groups, being one-operational systems, lend themselves to the simplest formal description. Yet despite this simplicity of description the fundamental algebraic concepts such as homomorphism, quotient construction, and the like, which play such an important role in all algebraic structures—in fact, in all of mathematics—already enter here in a pure and revealing form.

At this point, before we become weighted down with details, let us take a quick look ahead. In abstract algebra we have certain basic systems which, in the history and development of mathematics, have achieved positions of paramount importance. These are usually sets on whose elements we can operate algebraically—by this we mean that we can combine two elements of the set, perhaps in several ways, to obtain a third element of the set—and, in addition, we assume that these algebraic operations are subject to certain rules, which are explicitly spelled out in what we call the axioms or postulates defining the system. In this abstract setting we then attempt to prove theorems about these very general structures, always hoping that when these results are applied to a particular, concrete realization of the abstract system there will flow out facts and insights into the example at hand which would have been obscured from us by the mass of inessential information available to us in the particular, special case.

We should like to stress that these algebraic systems and the axioms which define them must have a certain naturality about them. They must come from the experience of looking at many examples; they should be rich in meaningful results. One does not just sit down, list a few axioms, and then proceed to study the system so described. This, admittedly, is done by some, but most mathematicians would dismiss these attempts as poor mathematics. The systems chosen for study are chosen because particular cases of these structures have appeared time and time again, because someone finally noted that these special cases were indeed special instances of a general phenomenon, because one notices analogies between two highly disparate mathematical objects and so is led to a search for the root of these analogies. To cite an example, case after case after case of the special object, which we know today as groups, was studied toward the end of the eighteenth, and at the beginning of the nineteenth, century, yet it was not until relatively late in the nineteenth century that the notion of an abstract group was introduced. The only algebraic structures, so far encountered, that have stood the test of time and have survived to become of importance, have been those based on a broad and tall pillar of special cases. Amongst mathematicians neither the beauty nor the significance of the first example which we have chosen to discuss—groups—is disputed.

2.1 Definition of a Group

At this juncture it is advisable to recall a situation discussed in the first chapter. For an arbitrary nonempty set S we defined $A(S)$ to be the set of all *one-to-one* mappings of the set S *onto* itself. For any two elements σ, $\tau \in A(S)$ we introduced a product, denoted by $\sigma \circ \tau$, and on further investigation it turned out that the following facts were true for the elements of $A(S)$ subject to this product:

1. Whenever $\sigma, \tau \in A(S)$, then it follows that $\sigma \circ \tau$ is also in $A(S)$. This is described by saying that $A(S)$ is *closed* under the product (or, sometimes, as closed under multiplication).
2. For any three elements $\sigma, \tau, \mu \in A(S)$, $\sigma \circ (\tau \circ \mu) = (\sigma \circ \tau) \circ \mu$. This relation is called the *associative law.*
3. There is a very special element $\iota \in A(S)$ which satisfies $\iota \circ \sigma = \sigma \circ \iota = \sigma$ for all $\sigma \in A(S)$. Such an element is called an *identity element* for $A(S)$.
4. For every $\sigma \in A(S)$ there is an element, written as σ^{-1}, also in $A(S)$, such that $\sigma \circ \sigma^{-1} = \sigma^{-1} \circ \sigma = \iota$. This is usually described by saying that every element in $A(S)$ has an *inverse* in $A(S)$.

One other fact about $A(S)$ stands out, namely, that whenever S has three or more elements we can find two elements $\alpha, \beta \in A(S)$ such that $\alpha \circ \beta \neq \beta \circ \alpha$. This possibility, which runs counter to our usual experience and intuition in mathematics so far, introduces a richness into $A(S)$ which would have not been present except for it.

With this example as a model, and with a great deal of hindsight, we abstract and make the

DEFINITION A nonempty set of elements G is said to form a *group* if in G there is defined a binary operation, called the product and denoted by $\cdot$, such that

1. $a, b \in G$ implies that $a \cdot b \in G$ (closed).
2. $a, b, c \in G$ implies that $a \cdot (b \cdot c) = (a \cdot b) \cdot c$ (associative law).
3. There exists an element $e \in G$ such that $a \cdot e = e \cdot a = a$ for all $a \in G$ (the existence of an identity element in G).
4. For every $a \in G$ there exists an element $a^{-1} \in G$ such that $a \cdot a^{-1} = a^{-1} \cdot a = e$ (the existence of inverses in G).

Considering the source of this definition it is not surprising that for every nonempty set S the set $A(S)$ is a group. Thus we already have presented to us an infinite source of interesting, concrete groups. We shall see later (in a theorem due to Cayley) that these $A(S)$'s constitute, in some sense, a universal family of groups. If S has three or more elements, recall that we can find elements $\sigma, \tau \in A(S)$ such that $\sigma \circ \tau \neq \tau \circ \sigma$. This prompts us to single out a highly special, but very important, class of groups as in the next definition.

DEFINITION A group G is said to be *abelian* (or *commutative*) if for every $a, b \in G$, $a \cdot b = b \cdot a$.

A group which is not abelian is called, naturally enough, *non-abelian*; having seen a family of examples of such groups we know that non-abelian groups do indeed exist.

Another natural characteristic of a group G is the number of elements it contains. We call this the *order* of G and denote it by $o(G)$. This number is, of course, most interesting when it is finite. In that case we say that G is a *finite group*.

To see that finite groups which are not trivial do exist just note that if the set S contains n elements, then the group $A(S)$ has $n!$ elements. (Prove!) This highly important example will be denoted by S_n whenever it appears in this book, and will be called the *symmetric group* of degree n. In the next section we shall more or less dissect S_3, which is a non-abelian group of order 6.

2.2 Some Examples of Groups

Example 2.2.1 Let G consist of the integers $0, \pm 1, \pm 2, \ldots$ where we mean by $a \cdot b$ for $a, b \in G$ the usual sum of integers, that is, $a \cdot b = a + b$. Then the reader can quickly verify that G is an infinite abelian group in which 0 plays the role of e and $-a$ that of a^{-1}.

Example 2.2.2 Let G consist of the real numbers $1, -1$ under the multiplication of real numbers. G is then an abelian group of order 2.

Example 2.2.3 Let $G = S_3$, the group of all 1–1 mappings of the set $\{x_1, x_2, x_3\}$ onto itself, under the product which we defined in Chapter 1. G is a group of order 6. We digress a little before returning to S_3.

For a neater notation, not just in S_3, but in any group G, let us define for any $a \in G$, $a^0 = e$, $a^1 = a$, $a^2 = a \cdot a$, $a^3 = a \cdot a^2, \ldots, a^k = a \cdot a^{k-1}$, and $a^{-2} = (a^{-1})^2$, $a^{-3} = (a^{-1})^3$, etc. The reader may verify that the usual rules of exponents prevail; namely, for any two integers (positive, negative, or zero) m, n,

$$a^m \cdot a^n = a^{m+n}, \tag{1}$$

$$(a^m)^n = a^{mn}. \tag{2}$$

(It is worthwhile noting that, in this notation, if G is the group of Example 2.2.1, a^n means the integer na).

With this notation at our disposal let us examine S_3 more closely. Consider the mapping ϕ defined on the set x_1, x_2, x_3 by

$$\phi: \quad \begin{array}{l} x_1 \to x_2 \\ x_2 \to x_1 \\ x_3 \to x_3, \end{array}$$

and the mapping

$$\psi: \quad \begin{array}{l} x_1 \to x_2 \\ x_2 \to x_3 \\ x_3 \to x_1. \end{array}$$

Checking, we readily see that $\phi^2 = e$, $\psi^3 = e$, and that

$$\phi \cdot \psi: \quad \begin{array}{l} x_1 \to x_3 \\ x_2 \to x_2 \\ x_3 \to x_1, \end{array}$$

whereas

$$\psi \cdot \phi: \quad \begin{array}{l} x_1 \to x_1 \\ x_2 \to x_3 \\ x_3 \to x_2. \end{array}$$

It is clear that $\phi \cdot \psi \neq \psi \cdot \phi$ for they do not take x_1 into the same image. Since $\psi^3 = e$, it follows that $\psi^{-1} = \psi^2$. Let us now compute the action of $\psi^{-1} \cdot \phi$ on x_1, x_2, x_3. Since $\psi^{-1} = \psi^2$ and

$$\psi^2: \quad \begin{array}{l} x_1 \to x_3 \\ x_2 \to x_1 \\ x_3 \to x_2, \end{array}$$

we have that

$$\psi^{-1} \cdot \phi: \quad \begin{array}{l} x_1 \to x_3 \\ x_2 \to x_2 \\ x_3 \to x_1. \end{array}$$

In other words, $\phi \cdot \psi = \psi^{-1} \cdot \phi$. Consider the elements $e, \phi, \psi, \psi^2, \phi \cdot \psi, \psi \cdot \phi$; these are all distinct and are in G (since G is closed), which only has six elements. Thus this list enumerates all the elements of G. One might ask, for instance, What is the entry in the list for $\psi \cdot (\phi \cdot \psi)$? Using $\phi \cdot \psi = \psi^{-1} \cdot \phi$, we see that $\psi \cdot (\phi \cdot \psi) = \psi \cdot (\psi^{-1} \cdot \phi) = (\psi \cdot \psi^{-1}) \cdot \phi = e \cdot \phi = \phi$. Of more interest is the form of $(\phi \cdot \psi) \cdot (\psi \cdot \phi) = \phi \cdot (\psi \cdot (\psi \cdot \phi)) = \phi \cdot (\psi^2 \cdot \phi) = \phi \cdot (\psi^{-1} \cdot \phi) = \phi \cdot (\phi \cdot \psi) = \phi^2 \cdot \psi = e \cdot \psi = \psi$. (The reader should not be frightened by the long, wearisome chain of equalities here. It is the last time we shall be so boringly conscientious.) Using the same techniques as we have used, the reader can compute to his heart's content others of the 25 products which do not involve e. Some of these will appear in the exercises.

Example 2.2.4 Let n be any integer. We construct a group of order n as follows: G will consist of all symbols $a^i, i = 0, 1, 2, \ldots, n-1$ where we insist that $a^0 = a^n = e$, $a^i \cdot a^j = a^{i+j}$ if $i + j \leq n$ and $a^i \cdot a^j = a^{i+j-n}$ if $i + j > n$. The reader may verify that this is a group. It is called a *cyclic group* of order n.

A geometric realization of the group in Example 2.2.4 may be achieved as follows: Let S be the circle, in the plane, of radius 1, and let ρ_n be a rotation through an angle of $2\pi/n$. Then $\rho_n \in A(S)$ and ρ_n in $A(S)$ generates a group of order n, namely, $\{e, \rho_n, \rho_n^{\,2}, \ldots, \rho_n^{\,n-1}\}$.

Example 2.2.5 Let S be the set of integers and, as usual, let $A(S)$ be the set of all one-to-one mappings of S onto itself. Let G be the set of all elements in $A(S)$ which move only a *finite* number of elements of S; that is, $\sigma \in G$ if and only if the number of x in S such that $x\sigma \neq x$ is finite. If $\sigma, \tau \in G$, let $\sigma \cdot \tau$ be the product of σ and τ as elements of $A(S)$. We claim that G is a group relative to this operation. We verify this now.

To begin with, if $\sigma, \tau \in G$, then σ and τ each moves only a finite number of elements of S. In consequence, $\sigma \cdot \tau$ can possibly move only those elements in S which are moved by at least one of σ or τ. Hence $\sigma \cdot \tau$ moves only a finite number of elements in S; this puts $\sigma \cdot \tau$ in G. The identity element, ι, of $A(S)$ moves no element of S; thus ι certainly must be in G. Since the associative law holds universally in $A(S)$, it holds for elements of G. Finally, if $\sigma \in G$ and $x\sigma^{-1} \neq x$ for some $x \in S$, then $(x\sigma^{-1})\sigma \neq x\sigma$, which is to say, $x(\sigma^{-1} \cdot \sigma) \neq x\sigma$. This works out to say merely that $x \neq x\sigma$. In other words, σ^{-1} moves only those elements of S which are moved by σ. Because σ only moves a finite number of elements of S, this is also true for σ^{-1}. Therefore σ^{-1} must be in G.

We have verified that G satisfies the requisite four axioms which define a group, relative to the operation we specified. Thus G is a group. The reader should verify that G is an infinite, non-abelian group.

#Example 2.2.6 Let G be the set of all 2×2 matrices $\begin{pmatrix} a & b \\ c & d \end{pmatrix}$ where a, b, c, d are real numbers, such that $ad - bc \neq 0$. For the operation in G we use the multiplication of matrices; that is,

$$\begin{pmatrix} a & b \\ c & d \end{pmatrix} \cdot \begin{pmatrix} w & x \\ y & z \end{pmatrix} = \begin{pmatrix} aw + by & ax + bz \\ cw + dy & cx + dz \end{pmatrix}.$$

The entries of this 2×2 matrix are clearly real. To see that this matrix is in G we merely must show that

$$(aw + by)(cx + dz) - (ax + bz)(cw + dy) \neq 0$$

(this is the required relation on the entries of a matrix which puts it in G). A short computation reveals that

$$(aw + by)(cx + dz) - (ax + bz)(cw + dy) = (ad - bc)(wz - xy) \neq 0$$

since both

$$\begin{pmatrix} a & b \\ c & d \end{pmatrix} \text{ and } \begin{pmatrix} w & x \\ y & z \end{pmatrix}$$

are in G. The associative law of multiplication holds in matrices; therefore it holds in G. The element

$$I = \begin{pmatrix} 1 & 0 \\ 0 & 1 \end{pmatrix}$$

is in G, since $1 \cdot 1 - 0 \cdot 0 = 1 \neq 0$; moreover, as the reader knows, or can verify, I acts as an identity element relative to the operation of G. Finally, if $\begin{pmatrix} a & b \\ c & d \end{pmatrix} \in G$ then, since $ad - bc \neq 0$, the matrix

$$\begin{pmatrix} \dfrac{d}{ad - bc} & \dfrac{-b}{ad - bc} \\ \dfrac{-c}{ad - bc} & \dfrac{a}{ad - bc} \end{pmatrix}$$

makes sense. Moreover,

$$\left(\frac{d}{ad - bc}\right)\left(\frac{a}{ad - bc}\right) - \left(\frac{-b}{ad - bc}\right)\left(\frac{-c}{ad - bc}\right) = \frac{ad - bc}{(ad - bc)^2} = \frac{1}{ad - bc} \neq 0,$$

hence the matrix

$$\begin{pmatrix} \dfrac{d}{ad - bc} & \dfrac{-b}{ad - bc} \\ \dfrac{-c}{ad - bc} & \dfrac{a}{ad - bc} \end{pmatrix}$$

is in G. An easy computation shows that

$$\begin{pmatrix} a & b \\ c & d \end{pmatrix}\begin{pmatrix} \dfrac{d}{ad - bc} & \dfrac{-b}{ad - bc} \\ \dfrac{-c}{ad - bc} & \dfrac{a}{ad - bc} \end{pmatrix} = \begin{pmatrix} 1 & 0 \\ 0 & 1 \end{pmatrix} = \begin{pmatrix} \dfrac{d}{ad - bc} & \dfrac{-b}{ad - bc} \\ \dfrac{-c}{ad - bc} & \dfrac{a}{ad - bc} \end{pmatrix}\begin{pmatrix} a & b \\ c & d \end{pmatrix};$$

thus this element of G acts as the inverse of $\begin{pmatrix} a & b \\ c & d \end{pmatrix}$. In short, G is a group. It is easy to see that G is an infinite, non-abelian group.

#Example 2.2.7 Let G be the set of all 2×2 matrices $\begin{pmatrix} a & b \\ c & d \end{pmatrix}$, where a, b, c, d are real numbers such that $ad - bc = 1$. Define the operation $\cdot$ in G, as we did in Example 2.2.6, via the multiplication of matrices. We leave it to the reader to verify that G is a group. It is, in fact, an infinite, non-abelian group.

One should make a comment about the relationship of the group in Example 2.2.7 to that in Example 2.2.6. Clearly, the group of Example 2.2.7 is a subset of that in Example 2.2.6. However, more is true. Relative to the same operation, as an entity in its own right, it forms a group. One could describe the situation by declaring it to be a *subgroup* of the group of Example 2.2.6. We shall see much more about the concept of subgroup in a few pages.

#Example 2.2.8 Let G be the set of all 2×2 matrices $\begin{pmatrix} a & b \\ -b & a \end{pmatrix}$, where a and b are real numbers, not both 0. (We can state this more succinctly by saying that $a^2 + b^2 \neq 0$.) Using the same operation as in the preceding two examples, we can easily show that G becomes a group. In fact, G is an infinite, abelian group.

Does the multiplication in G remind you of anything? Write $\begin{pmatrix} a & b \\ -b & a \end{pmatrix}$ as $aI + bJ$ where $J = \begin{pmatrix} 0 & 1 \\ -1 & 0 \end{pmatrix}$ and compute the product in these terms. Perhaps that will ring a bell with you.

#Example 2.2.9 Let G be the set of all 2×2 matrices $\begin{pmatrix} a & b \\ c & d \end{pmatrix}$ where a, b, c, d are integers modulo p, p a prime number, such that $ad - bc \neq 0$. Define the multiplication in G as we did in Example 2.2.6, understanding the multiplication and addition of the entries to be those modulo p. We leave it to the reader to verify that G is a non-abelian *finite* group.

In fact, how many elements does G have? Perhaps it might be instructive for the reader to try the early cases $p = 2$ and $p = 3$. Here one can write down all the elements of G explicitly. (A word of warning! For $p = 3$, G already has 48 elements.) To get the case of a general prime, p will require an idea rather than a direct hacking-out of the answer. Try it!

2.3 Some Preliminary Lemmas

We have now been exposed to the theory of groups for several pages and as yet not a single, solitary fact has been proved about groups. It is high time to remedy this situation. Although the first few results we demonstrate are, admittedly, not very exciting (in fact, they are rather dull) they will be extremely useful. Learning the alphabet was probably not the most interesting part of our childhood education, yet, once this hurdle was cleared, fascinating vistas were opened before us.

We begin with

LEMMA 2.3.1 *If G is a group, then*

a. The identity element of G is unique.
b. Every $a \in G$ has a unique inverse in G.
c. For every $a \in G$, $(a^{-1})^{-1} = a$.
d. For all $a, b \in G$, $(a \cdot b)^{-1} = b^{-1} \cdot a^{-1}$.

Proof. Before we proceed with the proof itself it might be advisable to see what it is that we are going to prove. In part (a) we want to show that if two elements e and f in G enjoy the property that for every $a \in G$, $a = a \cdot e = e \cdot a = a \cdot f = f \cdot a$, then $e = f$. In part (b) our aim is to show that if $x \cdot a = a \cdot x = e$ and $y \cdot a = a \cdot y = e$, where all of a, x, y are in G, then $x = y$.

First let us consider part (a). Since $e \cdot a = a$ for every $a \in G$, then, in particular, $e \cdot f = f$. But, on the other hand, since $b \cdot f = b$ for every $b \in G$, we must have that $e \cdot f = e$. Piecing these two bits of information together we obtain $f = e \cdot f = e$, and so $e = f$.

Rather than proving part (b), we shall prove something stronger which immediately will imply part (b) as a consequence. Suppose that for a in G, $a \cdot x = e$ and $a \cdot y = e$; then, obviously, $a \cdot x = a \cdot y$. Let us make this our starting point, that is, assume that $a \cdot x = a \cdot y$ for a, x, y in G. There is an element $b \in G$ such that $b \cdot a = e$ (as far as we know yet there may be several such b's). Thus $b \cdot (a \cdot x) = b \cdot (a \cdot y)$; using the associative law this leads to

$$x = e \cdot x = (b \cdot a) \cdot x = b \cdot (a \cdot x) = b \cdot (a \cdot y) = (b \cdot a) \cdot y = e \cdot y = y.$$

We have, in fact, proved that $a \cdot x = a \cdot y$ in a group G forces $x = y$. Similarly we can prove that $x \cdot a = y \cdot a$ implies that $x = y$. This says that we can cancel, from the same side, in equations in groups. A note of caution, however, for we cannot conclude that $a \cdot x = y \cdot a$ implies $x = y$ for we have no way of knowing whether $a \cdot x = x \cdot a$. This is illustrated in S_3 with $a = \phi$, $x = \psi$, $y = \psi^{-1}$.

Part (c) follows from this by noting that $a^{-1} \cdot (a^{-1})^{-1} = e = a^{-1} \cdot a$; canceling off the a^{-1} on the left leaves us with $(a^{-1})^{-1} = a$. This is the analog in general groups of the familiar result $-(-5) = 5$, say, in the group of real numbers under addition.

Part (d) is the most trivial of these, for

$$(a \cdot b) \cdot (b^{-1} \cdot a^{-1}) = a \cdot ((b \cdot b^{-1}) \cdot a^{-1}) = a \cdot (e \cdot a^{-1}) = a \cdot a^{-1} = e,$$

and so by the very definition of the inverse, $(a \cdot b)^{-1} = b^{-1} \cdot a^{-1}$.

Certain results obtained in the proof just given are important enough to single out and we do so now in

LEMMA 2.3.2 *Given a, b in the group G, then the equations $a \cdot x = b$ and $y \cdot a = b$ have unique solutions for x and y in G. In particular, the two cancellation laws,*

$$a \cdot u = a \cdot w \text{ implies } u = w$$

and

$$u \cdot a = w \cdot a \text{ implies } u = w$$

hold in G.

The few details needed for the proof of this lemma are left to the reader.

Problems

1. In the following determine whether the systems described are groups. If they are not, point out which of the group axioms fail to hold.
   (a) G = set of all integers, $a \cdot b \equiv a - b$.
   (b) G = set of all positive integers, $a \cdot b = ab$, the usual product of integers.
   (c) $G = a_0, a_1, \ldots, a_6$ where

$$a_i \cdot a_j = a_{i+j} \quad \text{if} \quad i + j < 7,$$

$$a_i \cdot a_j = a_{i+j-7} \quad \text{if} \quad i + j \geq 7$$

   (for instance, $a_5 \cdot a_4 = a_{5+4-7} = a_2$ since $5 + 4 = 9 > 7$).
   (d) G = set of all rational numbers with odd denominators, $a \cdot b \equiv a + b$, the usual addition of rational numbers.
2. Prove that if G is an abelian group, then for all $a, b \in G$ and all integers n, $(a \cdot b)^n = a^n \cdot b^n$.
3. If G is a group such that $(a \cdot b)^2 = a^2 \cdot b^2$ for all $a, b \in G$, show that G must be abelian.

*4. If G is a group in which $(a \cdot b)^i = a^i \cdot b^i$ for three consecutive integers i for all $a, b \in G$, show that G is abelian.

5. Show that the conclusion of Problem 4 does not follow if we assume the relation $(a \cdot b)^i = a^i \cdot b^i$ for just two consecutive integers.
6. In S_3 give an example of two elements x, y such that $(x \cdot y)^2 \neq x^2 \cdot y^2$.
7. In S_3 show that there are four elements satisfying $x^2 = e$ and three elements satisfying $y^3 = e$.
8. If G is a finite group, show that there exists a positive integer N such that $a^N = e$ for all $a \in G$.
9. (a) If the group G has three elements, show it must be abelian.
   (b) Do part (a) if G has four elements.
   (c) Do part (a) if G has five elements.
10. Show that if every element of the group G is its own inverse, then G is abelian.
11. If G is a group of even order, prove it has an element $a \neq e$ satisfying $a^2 = e$.
12. Let G be a nonempty set closed under an associative product, which in addition satisfies:
    (a) There exists an $e \in G$ such that $a \cdot e = a$ for all $a \in G$.
    (b) Give $a \in G$, there exists an element $y(a) \in G$ such that $a \cdot y(a) = e$.
    Prove that G must be a group under this product.

13. Prove, by an example, that the conclusion of Problem 12 is false if we assume instead:
    (a′) There exists an $e \in G$ such that $a \cdot e = a$ for all $a \in G$.
    (b′) Given $a \in G$, there exists $y(a) \in G$ such that $y(a) \cdot a = e$.
14. Suppose a *finite* set G is closed under an associative product and that both cancellation laws hold in G. Prove that G must be a group.
15. (a) Using the result of Problem 14, prove that the nonzero integers modulo p, p a prime number, form a group under multiplication mod p.
    (b) Do part (a) for the nonzero integers relatively prime to n under multiplication mod n.
16. In Problem 14 show by an example that if one just assumed one of the cancellation laws, then the conclusion need not follow.
17. Prove that in Problem 14 infinite examples exist, satisfying the conditions, which are not groups.
18. For any $n > 2$ construct a non-abelian group of order $2n$. (*Hint:* imitate the relations in S_3.)
19. If S is a set closed under an associative operation, prove that no matter how you bracket $a_1 a_2 \cdots a_n$, retaining the order of the elements, you get the same element in S (e.g., $(a_1 \cdot a_2) \cdot (a_3 \cdot a_4) = a_1 \cdot (a_2 \cdot (a_3 \cdot a_4))$; use induction on n).

#20. Let G be the set of all real 2×2 matrices $\begin{pmatrix} a & b \\ c & d \end{pmatrix}$, where $ad - bc \neq 0$ is a rational number. Prove that G forms a group under matrix multiplication.

#21. Let G be the set of all real 2×2 matrices $\begin{pmatrix} a & b \\ 0 & d \end{pmatrix}$ where $ad \neq 0$. Prove that G forms a group under matrix multiplication. Is G abelian?

#22. Let G be the set of all real 2×2 matrices $\begin{pmatrix} a & 0 \\ 0 & a^{-1} \end{pmatrix}$ where $a \neq 0$. Prove that G is an abelian group under matrix multiplication.

#23. Construct in the G of Problem 21 a subgroup of order 4.

#24. Let G be the set of all 2×2 matrices $\begin{pmatrix} a & b \\ c & d \end{pmatrix}$ where a, b, c, d are integers modulo 2, such that $ad - bc \neq 0$. Using matrix multiplication as the operation in G, prove that G is a group of order 6.

#25. (a) Let G be the group of all 2×2 matrices $\begin{pmatrix} a & b \\ c & d \end{pmatrix}$ where $ad - bc \neq 0$ and a, b, c, d are integers modulo 3, relative to matrix multiplication. Show that $o(G) = 48$.

(b) If we modify the example of G in part (a) by insisting that $ad - bc = 1$, then what is $o(G)$?

#*26. (a) Let G be the group of all 2×2 matrices $\begin{pmatrix} a & b \\ c & d \end{pmatrix}$ where a, b, c, d are integers modulo p, p a prime number, such that $ad - bc \neq 0$. G forms a group relative to matrix multiplication. What is $o(G)$?

(b) Let H be the subgroup of the G of part (a) defined by

$$H = \left\{ \begin{pmatrix} a & b \\ c & d \end{pmatrix} \in G \mid ad - bc = 1 \right\}.$$

What is $o(H)$?

2.4 Subgroups

Before turning to the study of groups we should like to change our notation slightly. It is cumbersome to keep using the $\cdot$ for the group operation; henceforth we shall drop it and instead of writing $a \cdot b$ for $a, b \in G$ we shall simply denote this product as ab.

In general we shall not be interested in arbitrary subsets of a group G for they do not reflect the fact that G has an algebraic structure imposed on it. Whatever subsets we do consider will be those endowed with algebraic properties derived from those of G. The most natural such subsets are introduced in the

DEFINITION A nonempty subset H of a group G is said to be a *subgroup* of G if, under the product in G, H itself forms a group.

The following remark is clear: if H is a subgroup of G and K is a subgroup of H, then K is a subgroup of G.

It would be useful to have some criterion for deciding whether a given subset of a group is a subgroup. This is the purpose of the next two lemmas.

LEMMA 2.4.1 *A nonempty subset H of the group G is a subgroup of G if and only if*

1. *$a, b \in H$ implies that $ab \in H$.*
2. *$a \in H$ implies that $a^{-1} \in H$.*

Proof. If H is a subgroup of G, then it is obvious that (1) and (2) must hold.

Suppose conversely that H is a subset of G for which (1) and (2) hold. In order to establish that H is a subgroup, all that is needed is to verify that $e \in H$ and that the associative law holds for elements of H. Since the associative law does hold for G, it holds all the more so for H, which is a subset of G. If $a \in H$, by part 2, $a^{-1} \in H$ and so by part 1, $e = aa^{-1} \in H$. This completes the proof.

In the special case of a finite group the situation becomes even nicer for there we can dispense with part 2.

LEMMA 2.4.2 *If H is a nonempty finite subset of a group G and H is closed under multiplication, then H is a subgroup of G.*

Proof. In light of Lemma 2.4.1 we need but show that whenever $a \in H$, then $a^{-1} \in H$. Suppose that $a \in H$; thus $a^2 = aa \in H$, $a^3 = a^2a \in H$, $\ldots$, $a^m \in H, \ldots$ since H is closed. Thus the infinite collection of elements $a, a^2, \ldots, a^m, \ldots$ must all fit into H, which is a finite subset of G. Thus there must be repetitions in this collection of elements; that is, for some integers r, s with $r > s > 0$, $a^r = a^s$. By the cancellation in G, $a^{r-s} = e$ (whence e is in H); since $r - s - 1 \geq 0$, $a^{r-s-1} \in H$ and $a^{-1} = a^{r-s-1}$ since $aa^{r-s-1} = a^{r-s} = e$. Thus $a^{-1} \in H$, completing the proof of the lemma.

The lemma tells us that to check whether a subset of a finite group is a subgroup we just see whether or not it is closed under multiplication.

We should, perhaps, now see some groups and some of their subgroups. G is always a subgroup of itself; likewise the set consisting of e is a subgroup of G. Neither is particularly interesting in the role of a subgroup, so we describe them as trivial subgroups. The subgroups between these two extremes we call nontrivial subgroups and it is in these we shall exhibit the most interest.

Example 2.4.1 Let G be the group of integers under addition, H the subset consisting of all the multiples of 5. The student should check that H is a subgroup.

In this example there is nothing extraordinary about 5; we could similarly define the subgroup H_n as the subset of G consisting of all the multiples of n. H_n is then a subgroup for every n. What can one say about $H_n \cap H_m$? It might be wise to try it for $H_6 \cap H_9$.

Example 2.4.2 Let S be any set, $A(S)$ the set of one-to-one mappings of S onto itself, made into a group under the composition of mappings. If $x_0 \in S$, let $H(x_0) = \{\phi \in A(S) \mid x_0\phi = x_0\}$. $H(x_0)$ is a subgroup of $A(S)$. If for $x_1 \neq x_0 \in S$ we similarly define $H(x_1)$, what is $H(x_0) \cap H(x_1)$?

Example 2.4.3 Let G be any group, $a \in G$. Let $(a) = \{a^i \mid i = 0, \pm 1, \pm 2, \ldots\}$. (a) is a subgroup of G (verify!); it is called the *cyclic subgroup generated by a*. This provides us with a ready means of producing subgroups of G. If for some choice of a, $G = (a)$, then G is said to be a *cyclic group*. Such groups are very special but they play a very important role in the theory of groups, especially in that part which deals with abelian groups. Of course, cyclic groups are abelian, but the converse is false.

Example 2.4.4 Let G be a group, W a subset of G. Let (W) be the set of all elements of G representable as a product of elements of W raised to positive, zero, or negative integer exponents. (W) is the *subgroup of G generated by W* and is the smallest subgroup of G containing W. In fact, (W) is the intersection of all the subgroups of G which contain W (this intersection is not vacuous since G is a subgroup of G which contains W).

Example 2.4.5 Let G be the group of nonzero real numbers under multiplication, and let H be the subset of positive rational numbers. Then H is a subgroup of G.

Example 2.4.6 Let G be the group of all real numbers under addition, and let H be the set of all integers. Then H is a subgroup of G.

#Example 2.4.7 Let G be the group of all real 2×2 matrices $\begin{pmatrix} a & b \\ c & d \end{pmatrix}$ with $ad - bc \neq 0$ under matrix multiplication. Let

$$H = \left\{ \begin{pmatrix} a & b \\ 0 & d \end{pmatrix} \in G \mid ad \neq 0 \right\}.$$

Then, as is easily verified, H is a subgroup of G.

#Example 2.4.8 Let H be the group of Example 2.4.7, and let $K = \left\{ \begin{pmatrix} 1 & b \\ 0 & 1 \end{pmatrix} \right\}$. Then K is a subgroup of H.

Example 2.4.9 Let G be the group of all nonzero complex numbers $a + bi$ (a, b real, not both 0) under multiplication, and let

$$H = \{a + bi \in G \mid a^2 + b^2 = 1\}.$$

Verify that H is a subgroup of G.

DEFINITION Let G be a group, H a subgroup of G; for $a, b \in G$ we say *a is congruent to b* mod H, written as $a \equiv b \bmod H$ if $ab^{-1} \in H$.

LEMMA 2.4.3 *The relation $a \equiv b \bmod H$ is an equivalence relation.*

Proof. If we look back in Chapter 1, we see that to prove Lemma 2.4.3 we must verify the following three conditions: For all $a, b, c \in G$,

1. $a \equiv a \bmod H$.
2. $a \equiv b \bmod H$ implies $b \equiv a \bmod H$.
3. $a \equiv b \bmod H$, $b \equiv c \bmod H$ implies $a \equiv c \bmod H$.

Let's go through each of these in turn.

1. To show that $a \equiv a \bmod H$ we must prove, using the very definition of congruence mod H, that $aa^{-1} \in H$. Since H is a subgroup of G, $e \in H$, and since $aa^{-1} = e$, $aa^{-1} \in H$, which is what we were required to demonstrate.

2. Suppose that $a \equiv b \bmod H$, that is, suppose $ab^{-1} \in H$; we want to get from this $b \equiv a \bmod H$, or, equivalently, $ba^{-1} \in H$. Since $ab^{-1} \in H$, which is a subgroup of G, $(ab^{-1})^{-1} \in H$; but, by Lemma 2.3.1, $(ab^{-1})^{-1} = (b^{-1})^{-1}a^{-1} = ba^{-1}$, and so $ba^{-1} \in H$ and $b \equiv a \bmod H$.

3. Finally we require that $a \equiv b \bmod H$ and $b \equiv c \bmod H$ forces $a \equiv c \bmod H$. The first congruence translates into $ab^{-1} \in H$, the second into $bc^{-1} \in H$; using that H is a subgroup of G, $(ab^{-1})(bc^{-1}) \in H$. However, $ac^{-1} = aec^{-1} = a(b^{-1}b)c^{-1} = (ab^{-1})(bc^{-1})$; hence $ac^{-1} \in H$, from which it follows that $a \equiv c \bmod H$.

This establishes that congruence mod H is a bona fide equivalence relation as defined in Chapter 1, and all results about equivalence relations have become available to us to be used in examining this particular relation.

A word about the notation we used. If G were the group of integers under addition, and $H = H_n$ were the subgroup consisting of all multiples of n, then in G, the relation $a \equiv b \bmod H$, that is, $ab^{-1} \in H$, under the additive notation, reads "$a - b$ is a multiple of n." This is the usual number theoretic congruence mod n. In other words, the relation we defined using an arbitrary group and subgroup is the natural generalization of a familiar relation in a familiar group.

DEFINITION If H is a subgroup of G, $a \in G$, then $Ha = \{ha \mid h \in H\}$. Ha is called a *right coset* of H in G.

LEMMA 2.4.4 *For all* $a \in G$,

$$Ha = \{x \in G \mid a \equiv x \bmod H\}.$$

Proof. Let $[a] = \{x \in G \mid a \equiv x \bmod H\}$. We first show that $Ha \subset [a]$. For, if $h \in H$, then $a(ha)^{-1} = a(a^{-1}h^{-1}) = h^{-1} \in H$ since H is a subgroup of G. By the definition of congruence mod H this implies that $ha \in [a]$ for every $h \in H$, and so $Ha \subset [a]$.

Suppose, now, that $x \in [a]$. Thus $ax^{-1} \in H$, so $(ax^{-1})^{-1} = xa^{-1}$ is also in H. That is, $xa^{-1} = h$ for some $h \in H$. Multiplying both sides by a from the right we come up with $x = ha$, and so $x \in Ha$. Thus $[a] \subset Ha$. Having proved the two inclusions $[a] \subset Ha$ and $Ha \subset [a]$, we can conclude that $[a] = Ha$, which is the assertion of the lemma.

In the terminology of Chapter 1, $[a]$, and thus Ha, is the equivalence class of a in G. By Theorem 1.1.1 these equivalence classes yield a decomposition of G into disjoint subsets. *Thus any two right cosets of H in G either are identical or have no element in common.*

We now claim that between any two right cosets Ha and Hb of H in G there exists a one-to-one correspondence, namely, with any element $ha \in Ha$, where $h \in H$, associate the element $hb \in Hb$. Clearly this mapping is onto Hb. We aver that it is a one-to-one correspondence, for if $h_1 b = h_2 b$, with $h_1, h_2 \in H$, then by the cancellation law in G, $h_1 = h_2$ and so $h_1 a = h_2 a$. This proves

LEMMA 2.4.5 *There is a one-to-one correspondence between any two right cosets of H in G.*

Lemma 2.4.5 is of most interest when H is a finite group, for then it merely states that any two right cosets of H have the same number of elements. How many elements does a right coset of H have? Well, note that $H = He$ is itself a right coset of H, so any right coset of H in G has $o(H)$ elements. Suppose now that G is a finite group, and let k be the number of distinct right cosets of H in G. By Lemmas 2.4.4 and 2.4.5 any two distinct right cosets of H in G have no element in common, and each has $o(H)$ elements. Since any $a \in G$ is in the unique right coset Ha, the right cosets fill out G. Thus if k represents the number of distinct right cosets of H in G we must have that $ko(H) = o(G)$. We have proved the famous theorem due to Lagrange, namely,

THEOREM 2.4.1 *If G is a finite group and H is a subgroup of G, then $o(H)$ is a divisor of $o(G)$.*

DEFINITION If H is a subgroup of G, the *index of H in G* is the number of distinct right cosets of H in G.

We shall denote it by $i_G(H)$. In case G is a finite group, $i_G(H) = o(G)/o(H)$, as became clear in the proof of Lagrange's theorem. It is quite possible for an infinite group G to have a subgroup $H \neq G$ which is of finite index in G.

It might be difficult, at this point, for the student to see the extreme importance of this result. As the subject is penetrated more deeply one will become more and more aware of its basic character. Because the theorem is of such stature it merits a little closer scrutiny, a little more analysis, and so we give, below, a slightly different way of looking at its proof. In truth, the procedure outlined below is no different from the one already given. The introduction of the congruence mod H smooths out the listing of elements used below, and obviates the need for checking that the new elements introduced at each stage did not appear before.

So suppose again that G is a finite group and that H is a subgroup of G. Let $h_1, h_2, \ldots, h_r$ be a complete list of the elements of H, $r = o(H)$. If $H = G$, there is nothing to prove. Suppose, then, that $H \neq G$; thus there is an $a \in G$, $a \notin H$. List all the elements so far in two rows as

$$\begin{gathered} h_1, h_2, \ldots, h_r, \\ h_1a, h_2a, \ldots, h_ra. \end{gathered}$$

We claim that all the entries in the second line are different from each other and are different from the entries in the first line. If any two in the second line were equal, then $h_ia = h_ja$ with $i \neq j$, but by the cancellation law this would lead to $h_i = h_j$, a contradiction. If an entry in the second line were equal to one in the first line, then $h_ia = h_j$, resulting in $a = h_i^{-1}h_j \in H$ since H is a subgroup of G; this violates $a \notin H$.

Thus we have, so far, listed $2o(H)$ elements; if these elements account for all the elements of G, we are done. If not, there is a $b \in G$ which did not occur in these two lines. Consider the new list

$$\begin{gathered} h_1, h_2, \ldots, h_r, \\ h_1a, h_2a, \ldots, h_ra, \\ h_1b, h_2b, \ldots, h_rb. \end{gathered}$$

As before (we are now waving our hands) we could show that no two entries in the third line are equal to each other, and that no entry in the third line occurs in the first or second line. Thus we have listed $3o(H)$ elements. Continuing in this way, every new element introduced, in fact, produces $o(H)$ new elements. Since G is a finite group, we must eventually exhaust all the elements of G. But if we ended up using k lines to list all the elements of the group, we would have written down $ko(H)$ distinct elements, and so $ko(H) = o(G)$.

It is essential to point out that the converse to Lagrange's theorem is false—a group G need not have a subgroup of order m if m is a divisor of $o(G)$. For instance, a group of order 12 exists which has no subgroup of order 6. The reader might try to find an example of this phenomenon; the place to look is in S_4, the symmetric group of degree 4 which has a subgroup of order 12, which will fulfill our requirement.

Lagrange's theorem has some very important corollaries. Before we present these we make one definition.

DEFINITION If G is a group and $a \in G$, the *order* (or *period*) of a is the least positive integer m such that $a^m = e$.

If no such integer exists we say that a is of infinite order. We use the notation $o(a)$ for the order of a. Recall our other notation: for two integers u, v, $u \mid v$ reads "u is a divisor of v."

COROLLARY 1 *If G is a finite group and $a \in G$, then $o(a) \mid o(G)$.*

Proof. With Lagrange's theorem already in hand, it seems most natural to prove the corollary by exhibiting a subgroup of G whose order is $o(a)$. The element a itself furnishes us with this subgroup by considering the cyclic subgroup, (a), of G generated by a; (a) consists of $e, a, a^2, \ldots$. How many elements are there in (a)? We assert that this number is the order of a. Clearly, since $a^{o(a)} = e$, this subgroup has at most $o(a)$ elements. If it should actually have fewer than this number of elements, then $a^i = a^j$ for some integers $0 \leq i < j < o(a)$. Then $a^{j-i} = e$, yet $0 < j - i < o(a)$ which would contradict the very meaning of $o(a)$. Thus the cyclic subgroup generated by a has $o(a)$ elements, whence, by Lagrange's theorem, $o(a) \mid o(G)$.

COROLLARY 2 *If G is a finite group and $a \in G$, then $a^{o(G)} = e$.*

Proof. By Corollary 1, $o(a) \mid o(G)$; thus $o(G) = mo(a)$. Therefore, $a^{o(G)} = a^{mo(a)} = (a^{o(a)})^m = e^m = e$.

A particular case of Corollary 2 is of great interest in number theory. The Euler ϕ-function, $\phi(n)$, is defined for all integers n by the following: $\phi(1) = 1$; for $n > 1$, $\phi(n) =$ number of positive integers less than n and relatively prime to n. Thus, for instance, $\phi(8) = 4$ since only 1, 3, 5, 7 are the numbers less than 8 which are relatively prime to 8. In Problem 15(b) at the end of Section 2.3 the reader was asked to prove that the numbers less than n and relatively prime to n formed a group under multiplication mod n. This group has order $\phi(n)$. If we apply Corollary 2 to this group we obtain

COROLLARY 3 (EULER) *If n is a positive integer and a is relatively prime to n, then $a^{\phi(n)} \equiv 1 \bmod n$.*

In order to apply Corollary 2 one should replace a by its remainder on division by n. If n should be a prime number p, then $\phi(p) = p - 1$. If a is an integer relatively prime to p, then by Corollary 3, $a^{p-1} \equiv 1 \bmod p$, whence $a^p \equiv a \bmod p$. If, on the other hand, a is not relatively prime to p, since p is a prime number, we must have that $p \mid a$, so that $a \equiv 0 \bmod p$; hence $0 \equiv a^p \equiv a \bmod p$ here also. Thus

COROLLARY 4 (FERMAT) *If p is a prime number and a is any integer, then $a^p \equiv a \bmod p$.*

COROLLARY 5 *If G is a finite group whose order is a prime number p, then G is a cyclic group.*

Proof. First we claim that G has no nontrivial subgroups H; for $o(H)$ must divide $o(G) = p$ leaving only two possibilities, namely, $o(H) = 1$ or $o(H) = p$. The first of these implies $H = (e)$, whereas the second implies that $H = G$. Suppose now that $a \neq e \in G$, and let $H = (a)$. H is a subgroup of G, $H \neq (e)$ since $a \neq e \in H$. Thus $H = G$. This says that G is cyclic and that every element in G is a power of a.

This section is of great importance in all that comes later, not only for its results but also because the spirit of the proofs occurring here are genuinely group-theoretic. The student can expect to encounter other arguments having a similar flavor. It would be wise to assimilate the material and approach thoroughly, now, rather than a few theorems later when it will be too late.

2.5 A Counting Principle

As we have defined earlier, if H is a subgroup of G and $a \in G$, then Ha consists of all elements in G of the form ha where $h \in H$. Let us generalize this notion. If H, K are two subgroups of G, let

$$HK = \{x \in G \mid x = hk, h \in H, k \in K\}.$$

Let's pause and look at an example; in S_3 let $H = \{e, \phi\}$, $K = \{e, \phi\psi\}$. Since $\phi^2 = (\phi\psi)^2 = e$, both H and K are subgroups. What can we say about HK? Just using the definition of HK we can see that HK consists of the elements e, ϕ, $\phi\psi$, $\phi^2\psi = \psi$. Since HK consists of four elements and 4 is not a divisor of 6, the order of S_3 by Lagrange's theorem HK could not be a subgroup of S_3. (Of course, we could verify this directly but it does not hurt to keep recalling Lagrange's theorem.) We might try to find out why HK is not a subgroup. Note that $KH = \{e, \phi, \phi\psi, \phi\psi\phi = \psi^{-1}\} \neq HK$. This is precisely why HK fails to be a subgroup, as we see in the next lemma.

LEMMA 2.5.1 *HK is a subgroup of G if and only if $HK = KH$.*

Proof. Suppose, first, that $HK = KH$; that is, if $h \in H$ and $k \in K$, then $hk = k_1h_1$ for some $k_1 \in K$, $h_1 \in H$ (it need not be that $k_1 = k$ or

$h_1 = h$!). To prove that HK is a subgroup we must verify that it is closed and every element in HK has its inverse in HK. Let's show the closure first; so suppose $x = hk \in HK$ and $y = h'k' \in HK$. Then $xy = hkh'k'$, but since $kh' \in KH = HK$, $kh' = h_2k_2$ with $h_2 \in H$, $k_2 \in K$. Hence $xy = h(h_2k_2)k' = (hh_2)(k_2k') \in HK$, and HK is closed. Also $x^{-1} = (hk)^{-1} = k^{-1}h^{-1} \in KH = HK$, so $x^{-1} \in HK$. Thus HK is a subgroup of G.

On the other hand, if HK is a subgroup of G, then for any $h \in H$, $k \in K$, $h^{-1}k^{-1} \in HK$ and so $kh = (h^{-1}k^{-1})^{-1} \in HK$. Thus $KH \subset HK$. Now if x is any element of HK, $x^{-1} = hk \in HK$ and so $x = (x^{-1})^{-1} = (hk)^{-1} = k^{-1}h^{-1} \in KH$, so $HK \subset KH$. Thus $HK = KH$.

An interesting special case is the situation when G is an abelian group for in that case trivially $HK = KH$. Thus as a consequence we have the

COROLLARY *If H, K are subgroups of the abelian group G, then HK is a subgroup of G.*

If H, K are subgroups of a group G, we have seen that the subset HK need not be a subgroup of G. Yet it is a perfect meaningful question to ask: How many distinct elements are there in the subset HK? If we denote this number by $o(HK)$, we prove

THEOREM 2.5.1 *If H and K are finite subgroups of G of orders $o(H)$ and $o(K)$, respectively, then*

$$o(HK) = \frac{o(H)o(K)}{o(H \cap K)}.$$

Proof. Although there is no need to pay special attention to the particular case in which $H \cap K = (e)$, looking at this case, which is devoid of some of the complexity of the general situation, is quite revealing. Here we should seek to show that $o(HK) = o(H)o(K)$. One should ask oneself: How could this fail to happen? The answer clearly must be that if we list all the elements hk, $h \in H$, $k \in K$ there should be some collapsing; that is, some element in the list must appear at least twice. Equivalently, for some $h \neq h_1 \in H$, $hk = h_1k_1$. But then $h_1^{-1}h = k_1k^{-1}$; now since $h_1 \in H$, h_1^{-1} must also be in H, thus $h_1^{-1}h \in H$. Similarly, $k_1k^{-1} \in K$. Since $h_1^{-1}h = k_1k^{-1}$, $h_1^{-1}h \in H \cap K = (e)$, so $h_1^{-1}h = e$, whence $h = h_1$, a contradiction. We have proved that no collapsing can occur, and so, here, $o(HK)$ is indeed $o(H)o(K)$.

With this experience behind us we are ready to attack the general case. As above we must ask: How often does a given element hk appear as a product in the list of HK? We assert it must appear $o(H \cap K)$ times! To see this we first remark that if $h_1 \in H \cap K$, then

$$hk = (hh_1)(h_1^{-1}k), \tag{1}$$

where $hh_1 \in H$, since $h \in H$, $h_1 \in H \cap K \subset H$ and $h_1^{-1}k \in K$ since $h_1^{-1} \in H \cap K \subset K$ and $k \in K$. Thus hk is duplicated in the product at least $o(H \cap K)$ times. However, if $hk = h'k'$, then $h^{-1}h' = k(k')^{-1} = u$, and $u \in H \cap K$, and so $h' = hu$, $k' = u^{-1}k$; thus all duplications were accounted for in (1). Consequently hk appears in the list of HK exactly $o(H \cap K)$ times. Thus the number of distinct elements in HK is the total number in the listing of HK, that is, $o(H)o(K)$ divided by the number of times a given element appears, namely, $o(H \cap K)$. This proves the theorem.

Suppose H, K are subgroups of the finite group G and $o(H) > \sqrt{o(G)}$, $o(K) > \sqrt{o(G)}$. Since $HK \subset G$, $o(HK) \leq o(G)$. However,

$$o(G) \geq o(HK) = \frac{o(H)o(K)}{o(H \cap K)} > \frac{\sqrt{o(G)}\sqrt{o(G)}}{o(H \cap K)} = \frac{o(G)}{o(H \cap K)},$$

thus $o(H \cap K) > 1$. Therefore, $H \cap K \neq (e)$. We have proved the

COROLLARY *If H and K are subgroups of G and $o(H) > \sqrt{o(G)}$, $o(K) > \sqrt{o(G)}$, then $H \cap K \neq (e)$.*

We apply this corollary to a very special group. Suppose G is a finite group of order pq where p and q are prime numbers with $p > q$. We claim that G can have at most one subgroup of order p. For suppose H, K are subgroups of order p. By the corollary, $H \cap K \neq (e)$, and being a subgroup of H, which having prime order has no nontrivial subgroups, we must conclude that $H \cap K = H$, and so $H \subset H \cap K \subset K$. Similarly $K \subset H$, whence $H = K$, proving that there is at most one subgroup of order p. Later on we shall see that there is at least one subgroup of order p, which, combined with the above, will tell us there is exactly one subgroup of order p in G. From this we shall be able to determine completely the structure of G.

Problems

1. If H and K are subgroups of G, show that $H \cap K$ is a subgroup of G. (Can you see that the same proof shows that the intersection of any number of subgroups of G, finite or infinite, is again a subgroup of G?)
2. Let G be a group such that the intersection of all its subgroups which are different from (e) is a subgroup different from (e). Prove that every element in G has finite order.
3. If G has no nontrivial subgroups, show that G must be finite of prime order.

4. (a) If H is a subgroup of G, and $a \in G$ let $aHa^{-1} = \{aha^{-1} \mid h \in H\}$. Show that aHa^{-1} is a subgroup of G.
   (b) If H is finite, what is $o(aHa^{-1})$?
5. For a subgroup H of G define the left coset aH of H in G as the set of all elements of the form ah, $h \in H$. Show that there is a one-to-one correspondence between the set of left cosets of H in G and the set of right cosets of H in G.
6. Write out all the right cosets of H in G where
   (a) $G = (a)$ is a cyclic group of order 10 and $H = (a^2)$ is the subgroup of G generated by a^2.
   (b) G as in part (a), $H = (a^5)$ is the subgroup of G generated by a^5.
   (c) $G = A(S)$, $S = \{x_1, x_2, x_3\}$, and $H = \{\sigma \in G \mid x_1\sigma = x_1\}$.
7. Write out all the left cosets of H in G for H and G as in parts (a), (b), (c) of Problem 6.
8. Is every right coset of H in G a left coset of H in G in the groups of Problem 6?
9. Suppose that H is a subgroup of G such that whenever $Ha \neq Hb$ then $aH \neq bH$. Prove that $gHg^{-1} \subset H$ for all $g \in G$.
10. Let G be the group of integers under addition, H_n the subgroup consisting of all multiples of a fixed integer n in G. Determine the index of H_n in G and write out all the right cosets of H_n in G.
11. In Problem 10, what is $H_n \cap H_m$?
12. If G is a group and H, K are two subgroups of finite index in G, prove that $H \cap K$ is of finite index in G. Can you find an upper bound for the index of $H \cap K$ in G?
13. If $a \in G$, define $N(a) = \{x \in G \mid xa = ax\}$. Show that $N(a)$ is a subgroup of G. $N(a)$ is usually called the *normalizer* or *centralizer* of a in G.
14. If H is a subgroup of G, then by the centralizer $C(H)$ of H we mean the set $\{x \in G \mid xh = hx \text{ all } h \in H\}$. Prove that $C(H)$ is a subgroup of G.
15. The *center* Z of a group G is defined by $Z = \{z \in G \mid zx = xz \text{ all } x \in G\}$. Prove that Z is a subgroup of G. Can you recognize Z as $C(T)$ for some subgroup T of G?
16. If H is a subgroup of G, let $N(H) = \{a \in G \mid aHa^{-1} = H\}$ [see Problem 4(a)]. Prove that
   (a) $N(H)$ is a subgroup of G. (b) $N(H) \supset H$.
17. Give an example of a group G and a subgroup H such that $N(H) \neq C(H)$. Is there any containing relation between $N(H)$ and $C(H)$?

18. If H is a subgroup of G let

$$N = \bigcap_{x \in G} xHx^{-1}.$$

Prove that N is a subgroup of G such that $aNa^{-1} = N$ for all $a \in G$.

*19. If H is a subgroup of finite index in G, prove that there is only a finite number of distinct subgroups in G of the form aHa^{-1}.

*20. If H is of finite index in G prove that there is a subgroup N of G, contained in H, and of finite index in G such that $aNa^{-1} = N$ for all $a \in G$. Can you give an upper bound for the index of this N in G?

21. Let the mapping τ_{ab} for a, b real numbers, map the reals into the reals by the rule $\tau_{ab}: x \to ax + b$. Let $G = \{\tau_{ab} \mid a \neq 0\}$. Prove that G is a group under the composition of mappings. Find the formula for $\tau_{ab}\tau_{cd}$.

22. In Problem 21, let $H = \{\tau_{ab} \in G \mid a \text{ is rational}\}$. Show that H is a subgroup of G. List all the right cosets of H in G, and all the left cosets of H in G. From this show that every left coset of H in G is a right coset of H in G.

23. In the group G of Problem 21, let $N = \{\tau_{1b} \in G\}$. Prove
    (a) N is a subgroup of G.
    (b) If $a \in G$, $n \in N$, then $ana^{-1} \in N$.

*24. Let G be a finite group whose order is *not* divisible by 3. Suppose that $(ab)^3 = a^3b^3$ for all $a, b \in G$. Prove that G must be abelian.

*25. Let G be an abelian group and suppose that G has elements of orders m and n, respectively. Prove that G has an element whose order is the least common multiple of m and n.

**26. If an abelian group has subgroups of orders m and n, respectively, then show it has a subgroup whose order is the least common multiple of m and n. (Don't be discouraged if you don't get this problem with what you know about group theory up to this stage. I don't know anybody, including myself, who has done it subject to the restriction of using material developed so far in the text. But it is fun to try. I've had more correspondence about this problem than about any other point in the whole book.)

27. Prove that any subgroup of a cyclic group is itself a cyclic group.

28. How many generators does a cyclic group of order n have? ($b \in G$ is a generator if $(b) = G$.)

Let U_n denote the integers relatively prime to n under multiplication mod n. In Problem 15(b), Section 2.3, it is indicated that U_n is a group.

In the next few problems we look at the nature of U_n as a group for some specific values of n.

29. Show that U_8 is not a cyclic group.
30. Show that U_9 is a cyclic group. What are all its generators?
31. Show that U_{17} is a cyclic group. What are all its generators?
32. Show that U_{18} is a cyclic group.
33. Show that U_{20} is not a cyclic group.
34. Show that both U_{25} and U_{27} are cyclic groups.
35. Hazard a guess at what all the n such that U_n is cyclic are. (You can verify your guess by looking in any reasonable book on number theory.)
36. If $a \in G$ and $a^m = e$, prove that $o(a) \mid m$.
37. If in the group G, $a^5 = e$, $aba^{-1} = b^2$ for some $a, b \in G$, find $o(b)$.

*38. Let G be a finite abelian group in which the number of solutions in G of the equation $x^n = e$ is at most n for every positive integer n. Prove that G must be a cyclic group.

39. Let G be a group and A, B subgroups of G. If $x, y \in G$ define $x \sim y$ if $y = axb$ for some $a \in A$, $b \in B$. Prove
    (a) The relation so defined is an equivalence relation.
    (b) The equivalence class of x is $AxB = \{axb \mid a \in A, b \in B\}$. ($AxB$ is called a *double coset* of A and B in G.)
40. If G is a finite group, show that the number of elements in the double coset AxB is

$$\frac{o(A)o(B)}{o(A \cap xBx^{-1})}.$$

41. If G is a finite group and A is a subgroup of G such that all double cosets AxA have the same number of elements, show that $gAg^{-1} = A$ for all $g \in G$.

2.6 Normal Subgroups and Quotient Groups

Let G be the group S_3 and let H be the subgroup $\{e, \phi\}$. Since the index of H in G is 3, there are three right cosets of H in G and three left cosets of H in G. We list them:

Right Cosets	Left Cosets
$H = \{e, \phi\}$	$H = \{e, \phi\}$
$H\psi = \{\psi, \phi\psi\}$	$\psi H = \{\psi, \psi\phi = \phi\psi^2\}$
$H\psi^2 = \{\psi^2, \phi\psi^2\}$	$\psi^2 H = \{\psi^2, \psi^2\phi = \phi\psi\}$

A quick inspection yields the interesting fact that the right coset $H\psi$ is not a left coset. Thus, at least for this subgroup, the notions of left and right coset need not coincide.

In $G = S_3$ let us consider the subgroup $N = \{e, \psi, \psi^2\}$. Since the index of N in G is 2 there are two left cosets and two right cosets of N in G. We list these:

Right Cosets	Left Cosets
$N = \{e, \psi, \psi^2\}$	$N = \{e, \psi, \psi^2\}$
$N\phi = \{\phi, \psi\phi, \psi^2\phi\}$	$\phi N = \{\phi, \phi\psi, \phi\psi^2\}$
	$= \{\phi, \psi^2\phi, \psi\phi\}$

A quick inspection here reveals that every left coset of N in G is a right coset in G and conversely. Thus we see that for some subgroups the notion of left coset coincides with that of right coset, whereas for some subgroups these concepts differ.

It is a tribute to the genius of Galois that he recognized that those subgroups for which the left and right cosets coincide are distinguished ones. Very often in mathematics the crucial problem is to recognize and to discover what are the relevant concepts; once this is accomplished the job may be more than half done.

We shall define this special class of subgroups in a slightly different way, which we shall then show to be equivalent to the remarks in the above paragraph.

DEFINITION A subgroup N of G is said to be a *normal subgroup* of G if for every $g \in G$ and $n \in N$, $gng^{-1} \in N$.

Equivalently, if by gNg^{-1} we mean the set of all gng^{-1}, $n \in N$, then N is a normal subgroup of G if and only if $gNg^{-1} \subset N$ for every $g \in G$.

LEMMA 2.6.1 *N is a normal subgroup of G if and only if $gNg^{-1} = N$ for every $g \in G$.*

Proof. If $gNg^{-1} = N$ for every $g \in G$, certainly $gNg^{-1} \subset N$, so N is normal in G.

Suppose that N is normal in G. Thus if $g \in G$, $gNg^{-1} \subset N$ and $g^{-1}Ng = g^{-1}N(g^{-1})^{-1} \subset N$. Now, since $g^{-1}Ng \subset N$, $N = g(g^{-1}Ng)g^{-1} \subset gNg^{-1} \subset N$, whence $N = gNg^{-1}$.

In order to avoid a point of confusion here let us stress that Lemma 2.6.1 *does not* say that for every $n \in N$ and every $g \in G$, $gng^{-1} = n$. No! This can be false. Take, for instance, the group G to be S_3 and N to be the subgroup $\{e, \psi, \psi^2\}$. If we compute $\phi N \phi^{-1}$ we obtain $\{e, \phi\psi\phi^{-1}, \phi\psi^2\phi^{-1}\} = \{e, \psi^2, \psi\}$, yet $\phi\psi\phi^{-1} \neq \psi$. All we require is that the *set* of elements gNg^{-1} be the same as the *set* of elements N.

We now can return to the question of the equality of left cosets and right cosets.

LEMMA 2.6.2 *The subgroup N of G is a normal subgroup of G if and only if every left coset of N in G is a right coset of N in G.*

Proof. If N is a normal subgroup of G, then for every $g \in G$, $gNg^{-1} = N$, whence $(gNg^{-1})g = Ng$; equivalently $gN = Ng$, and so the left coset gN is the right coset Ng.

Suppose, conversely, that every left coset of N in G is a right coset of N in G. Thus, for $g \in G$, gN, being a left coset, must be a right coset. What right coset can it be?

Since $g = ge \in gN$, whatever right coset gN turns out to be, it must contain the element g; however, g is in the right coset Ng, and two distinct right cosets have no element in common. (Remember the proof of Lagrange's theorem?) So this right coset is unique. Thus $gN = Ng$ follows. In other words, $gNg^{-1} = Ngg^{-1} = N$, and so N is a normal subgroup of G.

We have already defined what is meant by HK whenever H, K are subgroups of G. We can easily extend this definition to arbitrary subsets, and we do so by defining, for two subsets, A and B, of G, $AB = \{x \in G \mid x = ab, a \in A, b \in B\}$. As a special case, what can we say when $A = B = H$, a subgroup of G? $HH = \{h_1h_2 \mid h_1, h_2 \in H\} \subset H$ since H is closed under multiplication. But $HH \supset He = H$ since $e \in H$. Thus $HH = H$.

Suppose that N is a normal subgroup of G, and that $a, b \in G$. Consider $(Na)(Nb)$; since N is normal in G, $aN = Na$, and so

$$NaNb = N(aN)b = N(Na)b = NNab = Nab.$$

What a world of possibilities this little formula opens! But before we get carried away, for emphasis and future reference we record this as

LEMMA 2.6.3 *A subgroup N of G is a normal subgroup of G if and only if the product of two right cosets of N in G is again a right coset of N in G.*

Proof. If N is normal in G we have just proved the result. The proof of the other half is one of the problems at the end of this section.

Suppose that N is a normal subgroup of G. The formula $NaNb = Nab$, for $a, b \in G$ is highly suggestive; the product of right cosets is a right coset. Can we use this product to make the collection of right cosets into a group? Indeed we can! This type of construction, often occurring in mathematics and usually called forming a *quotient structure*, is of the utmost importance.

Let G/N denote the collection of right cosets of N in G (that is, the elements of G/N are certain subsets of G) and we use the product of subsets of G to yield for us a product in G/N.

For this product we claim

1. $X, Y \in G/N$ implies $XY \in G/N$; for $X = Na$, $Y = Nb$ for some $a, b \in G$, and $XY = NaNb = Nab \in G/N$.
2. $X, Y, Z \in G/N$, then $X = Na$, $Y = Nb$, $Z = Nc$ with $a, b, c \in G$, and so $(XY)Z = (NaNb)Nc = N(ab)Nc = N(ab)c = Na(bc)$ (since G is associative) $= Na(Nbc) = Na(NbNc) = X(YZ)$. Thus the product in G/N satisfies the associative law.
3. Consider the element $N = Ne \in G/N$. If $X \in G/N$, $X = Na$, $a \in G$, so $XN = NaNe = Nae = Na = X$, and similarly $NX = X$. Consequently, Ne is an identity element for G/N.
4. Suppose $X = Na \in G/N$ (where $a \in G$); thus $Na^{-1} \in G/N$, and $NaNa^{-1} = Naa^{-1} = Ne$. Similarly $Na^{-1}Na = Ne$. Hence Na^{-1} is the inverse of Na in G/N.

But a system which satisfies 1, 2, 3, 4 is exactly what we called a group. That is,

THEOREM 2.6.1 *If G is a group, N a normal subgroup of G, then G/N is also a group. It is called the quotient group or factor group of G by N.*

If, in addition, G is a finite group, what is the order of G/N? Since G/N has as its elements the right cosets of N in G, and since there are precisely $i_G(N) = o(G)/o(N)$ such cosets, we can say

LEMMA 2.6.4 *If G is a finite group and N is a normal subgroup of G, then $o(G/N) = o(G)/o(N)$.*

We close this section with an example.

Let G be the group of integers under addition and let N be the set of all multiplies of 3. Since the operation in G is addition we shall write the cosets of N in G as $N + a$ rather than as Na. Consider the three cosets N, $N + 1$, $N + 2$. We claim that these are all the cosets of N in G. For, given $a \in G$, $a = 3b + c$ where $b \in G$ and $c = 0, 1,$ or 2 (c is the remainder of a on division by 3). Thus $N + a = N + 3b + c = (N + 3b) + c = N + c$ since $3b \in N$. Thus every coset is, as we stated, one of N, $N + 1$, or $N + 2$, and $G/N = \{N, N + 1, N + 2\}$. How do we add elements in G/N? Our formula $NaNb = Nab$ translates into: $(N + 1) + (N + 2) = N + 3 = N$ since $3 \in N$; $(N + 2) + (N + 2) = N + 4 = N + 1$ and so on. Without being specific one feels that G/N is closely related to the integers mod 3 under addition. Clearly what we did for 3 we could emulate for any integer n, in which case the factor group should suggest a relation to the integers mod n under addition. This type of relation will be clarified in the next section.

Problems

1. If H is a subgroup of G such that the product of two right cosets of H in G is again a right coset of H in G, prove that H is normal in G.
2. If G is a group and H is a subgroup of index 2 in G, prove that H is a normal subgroup of G.
3. If N is a normal subgroup of G and H is any subgroup of G, prove that NH is a subgroup of G.
4. Show that the intersection of two normal subgroups of G is a normal subgroup of G.
5. If H is a subgroup of G and N is a normal subgroup of G, show that $H \cap N$ is a normal subgroup of H.
6. Show that every subgroup of an abelian group is normal.

*7. Is the converse of Problem 6 true? If yes, prove it, if no, give an example of a non-abelian group all of whose subgroups are normal.

8. Give an example of a group G, subgroup H, and an element $a \in G$ such that $aHa^{-1} \subset H$ but $aHa^{-1} \neq H$.
9. Suppose H is the only subgroup of order $o(H)$ in the finite group G. Prove that H is a normal subgroup of G.
10. If H is a subgroup of G, let $N(H) = \{g \in G \mid gHg^{-1} = H\}$. Prove
    (a) $N(H)$ is a subgroup of G.
    (b) H is normal in $N(H)$.
    (c) If H is a normal subgroup of the subgroup K in G, then $K \subset N(H)$ (that is, $N(H)$ is the largest subgroup of G in which H is normal).
    (d) H is normal in G if and only if $N(H) = G$.
11. If N and M are normal subgroups of G, prove that NM is also a normal subgroup of G.

*12. Suppose that N and M are two normal subgroups of G and that $N \cap M = (e)$. Show that for any $n \in N$, $m \in M$, $nm = mn$.

13. If a cyclic subgroup T of G is normal in G, then show that every subgroup of T is normal in G.

*14. Prove, by an example, that we can find three groups $E \subset F \subset G$, where E is normal in F, F is normal in G, but E is *not* normal in G.

15. If N is normal in G and $a \in G$ is of order $o(a)$, prove that the order, m, of Na in G/N is a divisor of $o(a)$.

16. If N is a normal subgroup in the finite group such that $i_G(N)$ and $o(N)$ are relatively prime, show that any element $x \in G$ satisfying $x^{o(N)} = e$ must be in N.
17. Let G be defined as all formal symbols $x^i y^j$, $i = 0$, $i, j = 0, 1, 2, \ldots, n - 1$ where we assume

$$x^i y^j = x^{i'} y^{j'} \text{ if and only if } i = i', j = j'$$

$$x^2 = y^n = e, \quad n > 2$$

$$xy = y^{-1}x.$$

(a) Find the form of the product $(x^i y^j)(x^k y^l)$ as $x^\alpha y^\beta$.
(b) Using this, prove that G is a non-abelian group of order $2n$.
(c) If n is odd, prove that the center of G is (e), while if n is even the center of G is larger than (e).

This group is known as a *dihedral* group. A geometric realization of this is obtained as follows: let y be a rotation of the Euclidean plane about the origin through an angle of $2\pi/n$, and x the reflection about the vertical axis. G is the group of motions of the plane generated by y and x.

18. Let G be a group in which, for some integer $n > 1$, $(ab)^n = a^n b^n$ for all $a, b \in G$. Show that
(a) $G^{(n)} = \{x^n \mid x \in G\}$ is a normal subgroup of G.
(b) $G^{(n-1)} = \{x^{n-1} \mid x \in G\}$ is a normal subgroup of G.
19. Let G be as in Problem 18. Show
(a) $a^{n-1}b^n = b^n a^{n-1}$ for all $a, b \in G$.
(b) $(aba^{-1}b^{-1})^{n(n-1)} = e$ for all $a, b \in G$.
20. Let G be a group such that $(ab)^p = a^p b^p$ for all $a, b \in G$, where p is a prime number. Let $S = \{x \in G \mid x^{p^m} = e$ for some m depending on $x\}$. Prove
(a) S is a normal subgroup of G.
(b) If $\bar{G} = G/S$ and if $\bar{x} \in \bar{G}$ is such that $\bar{x}^p = \bar{e}$ then $\bar{x} = \bar{e}$.

#21. Let G be the set of all real 2×2 matrices $\begin{pmatrix} a & b \\ 0 & d \end{pmatrix}$ where $ad \neq 0$, under matrix multiplication. Let $N = \left\{\begin{pmatrix} 1 & b \\ 0 & 1 \end{pmatrix}\right\}$. Prove that
(a) N is a normal subgroup of G.
(b) G/N is abelian.

2.7 Homomorphisms

The ideas and results in this section are closely interwoven with those of the preceding one. If there is one central idea which is common to all aspects of modern algebra it is the notion of homomorphism. By this one means a mapping from one algebraic system to a like algebraic system which preserves structure. We make this precise, for groups, in the next definition.

DEFINITION A mapping ϕ from a group G into a group $\bar{G}$ is said to be a *homomorphism* if for all $a, b \in G$, $\phi(ab) = \phi(a)\phi(b)$.

Notice that on the left side of this relation, namely, in the term $\phi(ab)$, the product ab is computed in G using the product of elements of G, whereas on the right side of this relation, namely, in the term $\phi(a)\phi(b)$, the product is that of elements in $\bar{G}$.

Example 2.7.0 $\phi(x) = e$ all $x \in G$. This is trivially a homomorphism. Likewise $\phi(x) = x$ for every $x \in G$ is a homomorphism.

Example 2.7.1 Let G be the group of all real numbers under addition (i.e., ab for $a, b \in G$ is really the real number $a + b$) and let $\bar{G}$ be the group of nonzero real numbers with the product being ordinary multiplication of real numbers. Define $\phi: G \to \bar{G}$ by $\phi(a) = 2^a$. In order to verify that this mapping is a homomorphism we must check to see whether $\phi(ab) = \phi(a)\phi(b)$, remembering that by the product on the left side we mean the operation in G (namely, addition), that is, we must check if $2^{a+b} = 2^a 2^b$, which indeed is true. Since 2^a is always positive, the image of ϕ is not all of $\bar{G}$, so ϕ is a homomorphism of G into $\bar{G}$, but not onto $\bar{G}$.

Example 2.7.2 Let $G = S_3 = \{e, \phi, \psi, \psi^2, \phi\psi, \phi\psi^2\}$ and $\bar{G} = \{e, \phi\}$. Define the mapping $f: G \to \bar{G}$ by $f(\phi^i\psi^j) = \phi^i$. Thus $f(e) = e$, $f(\phi) = \phi$, $f(\psi) = e$, $f(\psi^2) = e$, $f(\phi\psi) = \phi$, $f(\phi\psi^2) = \phi$. The reader should verify that f so defined is a homomorphism.

Example 2.7.3 Let G be the group of integers under addition and let $\bar{G} = G$. For the integer $x \in G$ define ϕ by $\phi(x) = 2x$. That ϕ is a homomorphism then follows from $\phi(x + y) = 2(x + y) = 2x + 2y = \phi(x) + \phi(y)$.

Example 2.7.4 Let G be the group of nonzero real numbers under multiplication, $\bar{G} = \{1, -1\}$, where $1.1 = 1$, $(-1)(-1) = 1$, $1(-1) = (-1)1 = -1$. Define $\phi: G \to \bar{G}$ by $\phi(x) = 1$ if x is positive, $\phi(x) = -1$ if x is negative. The fact that ϕ is a homomorphism is equivalent to the statements: positive times positive is positive, positive times negative is negative, negative times negative is positive.

Example 2.7.5 Let G be the group of integers under addition, let $\bar{G}_n$ be the group of integers under addition modulo n. Define ϕ by $\phi(x) =$ remainder of x on division by n. One can easily verify this is a homomorphism.

Example 2.7.6 Let G be the group of positive real numbers under multiplication and let $\bar{G}$ be the group of all real numbers under addition. Define $\phi: G \to G$ by $\phi(x) = \log_{10} x$. Thus

$$\phi(xy) = \log_{10}(xy) = \log_{10}(x) + \log_{10}(y) = \phi(x)\phi(y)$$

since the operation, on the right side, in $\bar{G}$ is in fact addition. Thus ϕ is a homomorphism of G into $\bar{G}$. In fact, not only is ϕ a homomorphism but, in addition, it is one-to-one and onto.

#Example 2.7.7 Let G be the group of all real 2×2 matrices $\begin{pmatrix} a & b \\ c & d \end{pmatrix}$ such that $ad - bc \neq 0$, under matrix multiplication. Let $\bar{G}$ be the group of all nonzero real numbers under multiplication. Define $\phi: G \to \bar{G}$ by

$$\phi\begin{pmatrix} a & b \\ c & d \end{pmatrix} = ad - bc.$$

We leave it to the reader to check that ϕ is a homomorphism of G onto $\bar{G}$.

The result of the following lemma yields, for us, an infinite class of examples of homomorphisms. When we prove Theorem 2.7.1 it will turn out that in some sense this provides us with the most general example of a homomorphism.

LEMMA 2.7.1 *Suppose G is a group, N a normal subgroup of G; define the mapping ϕ from G to G/N by $\phi(x) = Nx$ for all $x \in G$. Then ϕ is a homomorphism of G onto G/N.*

Proof. In actuality, there is nothing to prove, for we already have proved this fact several times. But for the sake of emphasis we repeat it.

That ϕ is onto is trivial, for every element $X \in G/N$ is of the form $X = Ny$, $y \in G$, so $X = \phi(y)$. To verify the multiplicative property required in order that ϕ be a homomorphism, one just notes that if $x, y \in G$,

$$\phi(xy) = Nxy = NxNy = \phi(x)\phi(y).$$

In Lemma 2.7.1 and in the examples preceding it, a fact which comes through is that a homomorphism need not be one-to-one; but there is a certain uniformity in this process of deviating from one-to-oneness. This will become apparent in a few lines.

DEFINITION If ϕ is a homomorphism of G into $\bar{G}$, the *kernel of* ϕ, K_ϕ, is defined by $K_\phi = \{x \in G \mid \phi(x) = \bar{e}, \bar{e} = \text{identity element of } \bar{G}\}$.

Before investigating any properties of K_ϕ it is advisable to establish that, as a set, K_ϕ is not empty. This is furnished us by the first part of

LEMMA 2.7.2 *If ϕ is a homomorphism of G into $\bar{G}$, then*

1. $\phi(e) = \bar{e}$, *the unit element of* $\bar{G}$.
2. $\phi(x^{-1}) = \phi(x)^{-1}$ *for all* $x \in G$.

Proof. To prove (1) we merely calculate $\phi(x)\bar{e} = \phi(x) = \phi(xe) = \phi(x)\phi(e)$, so by the cancellation property in $\bar{G}$ we have that $\phi(e) = \bar{e}$.

To establish (2) one notes that $\bar{e} = \phi(e) = \phi(xx^{-1}) = \phi(x)\phi(x^{-1})$, so by the very definition of $\phi(x)^{-1}$ in $\bar{G}$ we obtain the result that $\phi(x^{-1}) = \phi(x)^{-1}$.

The argument used in the proof of Lemma 2.7.2 should remind any reader who has been exposed to a development of logarithms of the argument used in proving the familiar results that $\log 1 = 0$ and $\log (1/x) = -\log x$; this is no coincidence, for the mapping $\phi: x \to \log x$ is a homomorphism of the group of positive real numbers under multiplication into the group of real numbers under addition, as we have seen in Example 2.7.6.

Lemma 2.7.2 shows that e is in the kernel of any homomorphism, so any such kernel is not empty. But we can say even more.

LEMMA 2.7.3 *If ϕ is a homomorphism of G into $\bar{G}$ with kernel K, then K is a normal subgroup of G.*

Proof. First we must check whether K is a subgroup of G. To see this one must show that K is closed under multiplication and has inverses in it for every element belonging to K.

If $x, y \in K$, then $\phi(x) = \bar{e}$, $\phi(y) = \bar{e}$, where $\bar{e}$ is the identity element of $\bar{G}$, and so $\phi(xy) = \phi(x)\phi(y) = \bar{e}\bar{e} = \bar{e}$, whence $xy \in K$. Also, if $x \in K$, $\phi(x) = \bar{e}$, so, by Lemma 2.7.2, $\phi(x^{-1}) = \phi(x)^{-1} = \bar{e}^{-1} = \bar{e}$; thus $x^{-1} \in K$. K is, accordingly, a subgroup of G.

To prove the normality of K one must establish that for any $g \in G$, $k \in K$, $gkg^{-1} \in K$; in other words, one must prove that $\phi(gkg^{-1}) = \bar{e}$ whenever $\phi(k) = \bar{e}$. But $\phi(gkg^{-1}) = \phi(g)\phi(k)\phi(g^{-1}) = \phi(g)\bar{e}\phi(g)^{-1} = \phi(g)\phi(g)^{-1} = \bar{e}$. This completes the proof of Lemma 2.7.3.

Let ϕ now be a homomorphism of the group G *onto* the group $\bar{G}$, and suppose that K is the kernel of ϕ. If $\bar{g} \in \bar{G}$, we say an element $x \in G$ is an *inverse image* of $\bar{g}$ under ϕ if $\phi(x) = \bar{g}$. What are all the inverse images of $\bar{g}$? For $\bar{g} = \bar{e}$ we have the answer, namely (by its very definition) K. What about elements $\bar{g} \neq \bar{e}$? Well, suppose $x \in G$ is one inverse image of $\bar{g}$; can we write down others? Clearly yes, for if $k \in K$, and if $y = kx$, then $\phi(y) = \phi(kx) = \phi(k)\phi(x) = \bar{e}\bar{g} = \bar{g}$. Thus all the elements Kx are in the inverse image of $\bar{g}$ whenever x is. Can there be others? Let us suppose that $\phi(z) = \bar{g} = \phi(x)$. Ignoring the middle term we are left with $\phi(z) = \phi(x)$, and so $\phi(z)\phi(x)^{-1} = \bar{e}$. But $\phi(x)^{-1} = \phi(x^{-1})$, whence

$\bar{e} = \phi(z)\phi(x)^{-1} = \phi(z)\phi(x^{-1}) = \phi(zx^{-1})$, in consequence of which $zx^{-1} \in K$; thus $z \in Kx$. In other words, we have shown that Kx accounts for exactly all the inverse images of $\bar{g}$ whenever x is a single such inverse image. We record this as

LEMMA 2.7.4 *If ϕ is a homomorphism of G onto $\bar{G}$ with kernel K, then the set of all inverse images of $\bar{g} \in \bar{G}$ under ϕ in G is given by Kx, where x is any particular inverse image of $\bar{g}$ in G.*

A special case immediately presents itself, namely, the situation when $K = (e)$. But here, by Lemma 2.7.4, any $\bar{g} \in \bar{G}$ has exactly one inverse image. That is, ϕ is a one-to-one mapping. The converse is trivially true, namely, if ϕ is a one-to-one homomorphism of G into (not even onto) G, its kernel must consist exactly of e.

DEFINITION A homomorphism ϕ from G into $\bar{G}$ is said to be an *isomorphism* if ϕ is one-to-one.

DEFINITION Two groups G, G^* are said to be *isomorphic* if there is an isomorphism of G *onto* G^*. In this case we write $G \approx G^*$.

We leave to the reader to verify the following three facts:

1. $G \approx G$.
2. $G \approx G^*$ implies $G^* \approx G$.
3. $G \approx G^*$, $G^* \approx G^{**}$ implies $G \approx G^{**}$.

When two groups are isomorphic, then, in some sense, they are equal. They differ in that their elements are labeled differently. The isomorphism gives us the key to the labeling, and with it, knowing a given computation in one group, we can carry out the analogous computation in the other. The isomorphism is like a dictionary which enables one to translate a sentence in one language into a sentence, of the same meaning, in another language. (Unfortunately no such perfect dictionary exists, for in languages words do not have single meanings, and nuances do not come through in a literal translation.) But merely to say that a given sentence in one language can be expressed in another is of little consequence; one needs the dictionary to carry out the translation. Similarly it might be of little consequence to know that two groups are isomorphic; the object of interest might very well be the isomorphism itself. So, whenever we prove two groups to be isomorphic, we shall endeavor to exhibit the precise mapping which yields this isomorphism.

Returning to Lemma 2.7.4 for a moment, we see in it a means of characterizing in terms of the kernel when a homomorphism is actually an isomorphism.

COROLLARY *A homomorphism ϕ of G into $\bar{G}$ with kernel K_ϕ is an isomorphism of G into $\bar{G}$ if and only if $K_\phi = (e)$.*

This corollary provides us with a standard technique for proving two groups to be isomorphic. First we find a homomorphism of one onto the other, and then prove the kernel of this homomorphism consists only of the identity element. This method will be illustrated for us in the proof of the very important

THEOREM 2.7.1 *Let ϕ be a homomorphism of G onto $\bar{G}$ with kernel K. Then $G/K \approx \bar{G}$.*

Proof. Consider the diagram

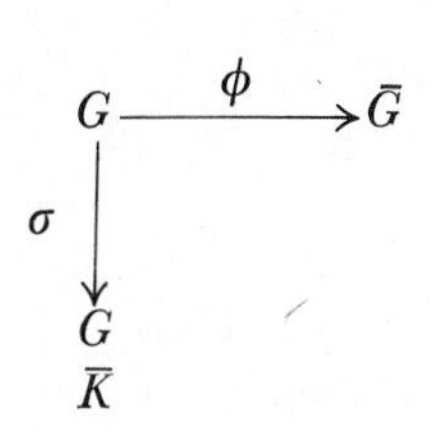

where $\sigma(g) = Kg$.

We should like to complete this to

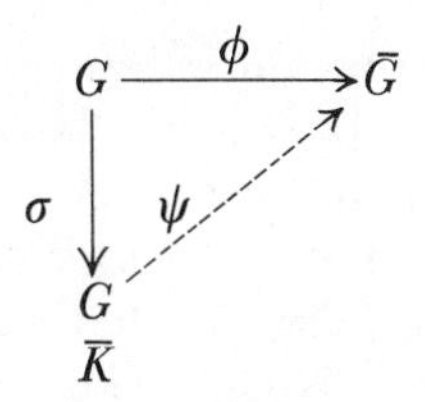

It seems clear that, in order to construct the mapping ψ from G/K to $\bar{G}$, we should use G as an intermediary, and also that this construction should be relatively uncomplicated. What is more natural than to complete the diagram using

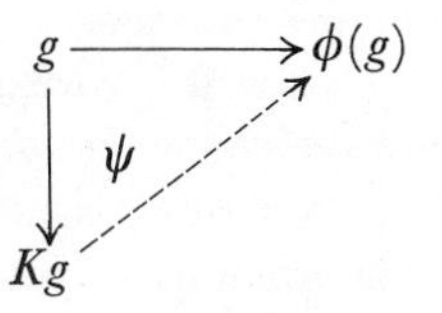

With this preamble we formally define the mapping ψ from G/K to $\bar{G}$ by: if $X \in G/K$, $X = Kg$, then $\psi(X) = \phi(g)$. A problem immediately arises: is this mapping well defined? If $X \in G/K$, it can be written as Kg in several ways (for instance, $Kg = Kkg$, $k \in K$); but if $X = Kg = Kg'$, $g, g' \in G$, then on one hand $\psi(X) = \phi(g)$, and on the other, $\psi(X) = \phi(g')$. For the mapping ψ to make sense it had better be true that $\phi(g) = \phi(g')$. So, suppose $Kg = Kg'$; then $g = kg'$, where $k \in K$, hence $\phi(g) = \phi(kg') = \phi(k)\phi(g') = \bar{e}\phi(g') = \phi(g')$ since $k \in K$, the kernel of ϕ.

We next determine that ψ is onto. For, if $\bar{x} \in \bar{G}$, $\bar{x} = \phi(g)$, $g \in G$ (since ϕ is onto) so $\bar{x} = \phi(g) = \psi(Kg)$.

If $X, Y \in G/K$, $X = Kg$, $Y = Kf$, $g, f \in G$, then $XY = KgKf = Kgf$, so that $\psi(XY) = \psi(Kgf) = \phi(gf) = \phi(g)\phi(f)$ since ϕ is a homomorphism of G onto $\bar{G}$. But $\psi(X) = \psi(Kg) = \phi(g)$, $\psi(Y) = \psi(Kf) = \phi(f)$, so we see that $\psi(XY) = \psi(X)\psi(Y)$, and ψ is a homomorphism of G/K onto $\bar{G}$.

To prove that ψ is an isomorphism of G/K onto $\bar{G}$ all that remains is to demonstrate that the kernel of ψ is the unit element of G/K. Since the unit element of G/K is $K = Ke$, we must show that if $\psi(Kg) = \bar{e}$, then $Kg = Ke = K$. This is now easy, for $\bar{e} = \psi(Kg) = \phi(g)$, so that $\phi(g) = \bar{e}$, whence g is in the kernel of ϕ, namely K. But then $Kg = K$ since K is a subgroup of G. All the pieces have been put together. We have exhibited a one-to-one homomorphism of G/K onto $\bar{G}$. Thus $G/K \approx \bar{G}$, and Theorem 2.7.1 is established.

Theorem 2.7.1 is important, for it tells us precisely what groups can be expected to arise as homomorphic images of a given group. These must be expressible in the form G/K, where K is normal in G. But, by Lemma 2.7.1, for any normal subgroup N of G, G/N is a homomorphic image of G. Thus there is a one-to-one correspondence between homomorphic images of G and normal subgroups of G. If one were to seek all homomorphic images of G one could do it by never leaving G as follows: find all normal subgroups N of G and construct all groups G/N. The set of groups so constructed yields all homomorphic images of G (up to isomorphisms).

A group is said to be *simple* if it has no nontrivial homomorphic images, that is, if it has no nontrivial normal subgroups. A famous, long-standing conjecture was that a non-abelian simple group of finite order has an even number of elements. This important result has been proved by the two American mathematicians, Walter Feit and John Thompson.

We have stated that the concept of a homomorphism is a very important one. To strengthen this statement we shall now show how the methods and results of this section can be used to prove nontrivial facts about groups. When we construct the group G/N, where N is normal in G, if we should happen to know the structure of G/N we would know that of G "up to N." True, we blot out a certain amount of information about G, but often enough is left so that from facts about G/N we can ascertain certain ones about G. When we photograph a certain scene we transfer a three-dimensional object to a two-dimensional representation of it. Yet, looking at the picture we can derive a great deal of information about the scene photographed.

In the two applications of the ideas developed so far, which are given below, the proofs given are not the best possible. In fact, a little later in this chapter these results will be proved in a more general situation in an easier manner. We use the presentation here because it does illustrate effectively many group-theoretic concepts.

APPLICATION 1 (Cauchy's Theorem for Abelian Groups) *Suppose G is a finite abelian group and $p \mid o(G)$, where p is a prime number. Then there is an element $a \neq e \in G$ such that $a^p = e$.*

Proof. We proceed by induction over $o(G)$. In other words, we assume that the theorem is true for all abelian groups having fewer elements than G. From this we wish to prove that the result holds for G. To start the induction we note that the theorem is vacuously true for groups having a single element.

If G has no subgroups $H \neq (e), G$, by the result of a problem earlier in the chapter, G must be cyclic of prime order. This prime must be p, and G certainly has $p - 1$ elements $a \neq e$ satisfying $a^p = a^{o(G)} = e$.

So suppose G has a subgroup $N \neq (e), G$. If $p \mid o(N)$, by our induction hypothesis, since $o(N) < o(G)$ and N is abelian, there is an element $b \in N$, $b \neq e$, satisfying $b^p = e$; since $b \in N \subset G$ we would have exhibited an element of the type required. So we may assume that $p \nmid o(N)$. Since G is abelian, N is a normal subgroup of G, so G/N is a group. Moreover, $o(G/N) = o(G)/o(N)$, and since $p \nmid o(N)$,

$$p \,\Big|\, \frac{o(G)}{o(N)} < o(G).$$

Also, since G is abelian, G/N is abelian. Thus by our induction hypothesis there is an element $X \in G/N$ satisfying $X^p = e_1$, the unit element of G/N, $X \neq e_1$. By the very form of the elements of G/N, $X = Nb$, $b \in G$, so that $X^p = (Nb)^p = Nb^p$. Since $e_1 = Ne$, $X^p = e_1$, $X \neq e_1$ translates into $Nb^p = N$, $Nb \neq N$. Thus $b^p \in N$, $b \notin N$. Using one of the corollaries to Lagrange's theorem, $(b^p)^{o(N)} = e$. That is, $b^{o(N)p} = e$. Let $c = b^{o(N)}$. Certainly $c^p = e$. In order to show that c is an element that satisfies the conclusion of the theorem we must finally show that $c \neq e$. However, if $c = e$, $b^{o(N)} = e$, and so $(Nb)^{o(N)} = N$. Combining this with $(Nb)^p = N$, $p \nmid o(N)$, p a prime number, we find that $Nb = N$, and so $b \in N$, a contradiction. Thus $c \neq e$, $c^p = e$, and we have completed the induction. This proves the result.

APPLICATION 2 (SYLOW'S THEOREM FOR ABELIAN GROUPS) *If G is an abelian group of order $o(G)$, and if p is a prime number, such that $p^\alpha \mid o(G)$, $p^{\alpha+1} \nmid o(G)$, then G has a subgroup of order p^α.*

Proof. If $\alpha = 0$, the subgroup (e) satisfies the conclusion of the result. So suppose $\alpha \neq 0$. Then $p \mid o(G)$. By Application 1, there is an element $a \neq e \in G$ satisfying $a^p = e$. Let $S = \{x \in G \mid x^{p^n} = e \text{ some integer } n\}$. Since $a \in S$, $a \neq e$, it follows that $S \neq (e)$. We now assert that S is a subgroup of G. Since G is finite we must only verify that S is closed. If $x, y \in S$, $x^{p^n} = e$, $y^{p^m} = e$, so that $(xy)^{p^{n+m}} = x^{p^{n+m}}y^{p^{n+m}} = e$ (we have used that G is abelian), proving that $xy \in S$.

We next claim that $o(S) = p^\beta$ with β an integer $0 < \beta \leq \alpha$. For, if some prime $q \mid o(S)$, $q \neq p$, by the result of Application 1 there is an element $c \in S$, $c \neq e$, satisfying $c^q = e$. However, $c^{p^n} = e$ for some n since $c \in S$. Since p^n, q are relatively prime, we can find integers λ, μ such that $\lambda q + \mu p^n = 1$, so that $c = c^1 = c^{\lambda q + \mu p^n} = (c^q)^\lambda (c^{p^n})^\mu = e$, contradicting $c \neq e$. By Lagrange's theorem $o(S) \mid o(G)$, so that $\beta \leq \alpha$. Suppose that $\beta < \alpha$; consider the abelian group G/S. Since $\beta < \alpha$ and $o(G/S) = o(G)/o(S)$, $p \mid o(G/S)$, there is an element Sx, $(x \in G)$ in G/S satisfying $Sx \neq S$, $(Sx)^{p^n} = S$ for some integer $n > 0$. But $S = (Sx)^{p^n} = Sx^{p^n}$, and so $x^{p^n} \in S$; consequently $e = (x^{p^n})^{o(S)} = (x^{p^n})^{p^\beta} = x^{p^{n+\beta}}$. Therefore, x satisfies the exact requirements needed to put it in S; in other words, $x \in S$. Consequently $Sx = S$ contradicting $Sx \neq S$. Thus $\beta < \alpha$ is impossible and we are left with the only alternative, namely, that $\beta = \alpha$. S is the required subgroup of order p^α.

We strengthen the application slightly. Suppose T is another subgroup of G of order p^α, $T \neq S$. Since G is abelian $ST = TS$, so that ST is a subgroup of G. By Theorem 2.5.1

$$o(ST) = \frac{o(S)o(T)}{o(S \cap T)} = \frac{p^\alpha p^\alpha}{o(S \cap T)}$$

and since $S \neq T$, $o(S \cap T) < p^\alpha$, leaving us with $o(ST) = p^\gamma$, $\gamma > \alpha$. Since ST is a subgroup of G, $o(ST) \mid o(G)$; thus $p^\gamma \mid o(G)$ violating the fact that α is the largest power of p which divides $o(G)$. Thus no such subgroup T exists, and S is the unique subgroup of order p^α. We have proved the

COROLLARY *If G is abelian of order $o(G)$ and $p^\alpha \mid o(G)$, $p^{\alpha+1} \nmid o(G)$, there is a unique subgroup of G of order p^α.*

If we look at $G = S_3$, which is non-abelian, $o(G) = 2.3$, we see that G has 3 distinct subgroups of order 2, namely, $\{e, \phi\}$, $\{e, \phi\psi\}$, $\{e, \phi\psi^2\}$, so that the corollary asserting the uniqueness does not carry over to non-abelian groups. But Sylow's theorem holds for all finite groups.

We leave the application and return to the general development. Suppcse ϕ is a homomorphism of G onto $\bar{G}$ with kernel K, and suppose that $\bar{H}$ is a subgroup of $\bar{G}$. Let $H = \{x \in G \mid \phi(x) \in \bar{H}\}$. We assert that H is a subgroup of G and that $H \supset K$. That $H \supset K$ is trivial, for if $x \in K$, $\phi(x) = \bar{e}$ is in $\bar{H}$, so that $K \subset H$ follows. Suppose now that $x, y \in H$; hence $\phi(x) \in \bar{H}$, $\phi(y) \in \bar{H}$ from which we deduce that $\phi(xy) = \phi(x)\phi(y) \in \bar{H}$. Therefore, $xy \in H$ and H is closed under the product in G. Furthermore, if $x \in H$, $\phi(x) \in \bar{H}$ and so $\phi(x^{-1}) = \phi(x)^{-1} \in \bar{H}$ from which it follows that $x^{-1} \in H$. All in all, our assertion has been established. What can we say in addition in case $\bar{H}$ is normal in $\bar{G}$? Let $g \in G$, $h \in H$; then $\phi(h) \in \bar{H}$, whence $\phi(ghg^{-1}) = \phi(g)\phi(h)\phi(g)^{-1} \in \bar{H}$, since $\bar{H}$ is normal in G. Otherwise stated, $ghg^{-1} \in H$, from which it follows that H is normal in G. One other point should be noted, namely, that the homomorphism ϕ from G onto $\bar{G}$, when just considered on elements of H, induces a homomorphism of H onto $\bar{H}$, with kernel exactly K, since $K \subset H$; by Theorem 2.7.1 we have that $\bar{H} \approx H/K$.

Suppose, conversely, that L is a subgroup of G and $K \subset L$. Let $\bar{L} = \{\bar{x} \in \bar{G} \mid \bar{x} = \phi(l), l \in L\}$. The reader should verify that $\bar{L}$ is a subgroup of $\bar{G}$. Can we explicitly describe the subgroup $T = \{y \in G \mid \phi(y) \in \bar{L}\}$? Clearly $L \subset T$. Is there any element $t \in T$ which is not in L? So, suppose $t \in T$; thus $\phi(t) \in \bar{L}$, so by the very definition of $\bar{L}$, $\phi(t) = \phi(l)$ for some $l \in L$. Thus $\phi(tl^{-1}) = \phi(t)\phi(l)^{-1} = \bar{e}$, whence $tl^{-1} \in K \subset L$, thus t is in $Ll = L$. Equivalently we have proved that $T \subset L$, which, combined with $L \subset T$, yields that $L = T$.

Thus we have set up a one-to-one correspondence between the set of all subgroups of $\bar{G}$ and the set of all subgroups of G which contain K. Moreover, in this correspondence, a normal subgroup of G corresponds to a normal subgroup of $\bar{G}$.

We summarize these few paragraphs in

LEMMA 2.7.5 *Let ϕ be a homomorphism of G onto $\bar{G}$ with kernel K. For $\bar{H}$ a subgroup of $\bar{G}$ let H be defined by $H = \{x \in G \mid \phi(x) \in \bar{H}\}$. Then H is a subgroup of G and $H \supset K$; if $\bar{H}$ is normal in $\bar{G}$, then H is normal in G. Moreover, this association sets up a one-to-one mapping from the set of all subgroups of $\bar{G}$ onto the set of all subgroups of G which contain K.*

We wish to prove one more general theorem about the relation of two groups which are homomorphic.

THEOREM 2.7.2. *Let ϕ be a homomorphism of G onto $\bar{G}$ with kernel K, and let $\bar{N}$ be a normal subgroup of $\bar{G}$, $N = \{x \in G \mid \phi(x) \in \bar{N}\}$. Then $G/N \approx \bar{G}/\bar{N}$. Equivalently, $G/N \approx (G/K)/(N/K)$.*

Proof. As we already know, there is a homomorphism θ of $\bar{G}$ onto $\bar{G}/\bar{N}$ defined by $\theta(\bar{g}) = \bar{N}\bar{g}$. We define the mapping $\psi:G \to \bar{G}/\bar{N}$ by $\psi(g) = \bar{N}\phi(g)$ for all $g \in G$. To begin with, ψ is onto, for if $\bar{g} \in \bar{G}$, $\bar{g} = \phi(g)$ for some $g \in G$, since ϕ is onto, so the typical element $\bar{N}\bar{g}$ in $\bar{G}/\bar{N}$ can be represented as $\bar{N}\phi(g) = \psi(g)$.

If $a, b \in G$, $\psi(ab) = \bar{N}\phi(ab)$ by the definition of the mapping ψ. However, since ϕ is a homomorphism, $\phi(ab) = \phi(a)\phi(b)$. Thus $\psi(ab) = \bar{N}\phi(a)\phi(b) = \bar{N}\phi(a)\bar{N}\phi(b) = \psi(a)\psi(b)$. So far we have shown that ψ is a homomorphism of G onto $\bar{G}/\bar{N}$. What is the kernel, T, of ψ? Firstly, if $n \in N$, $\phi(n) \in \bar{N}$, so that $\psi(n) = \bar{N}\phi(n) = \bar{N}$, the identity element of $\bar{G}/\bar{N}$, proving that $N \subset T$. On the other hand, if $t \in T$, $\psi(t) =$ identity element of $\bar{G}/\bar{N} = \bar{N}$; but $\psi(t) = \bar{N}\phi(t)$. Comparing these two evaluations of $\psi(t)$, we arrive at $\bar{N} = \bar{N}\phi(t)$, which forces $\phi(t) \in \bar{N}$; but this places t in N by definition of N. That is, $T \subset N$. The kernel of ψ has been proved to be equal to N. But then ψ is a homomorphism of G onto $\bar{G}/\bar{N}$ with kernel N. By Theorem 2.7.1 $G/N \approx \bar{G}/\bar{N}$, which is the first part of the theorem. The last statement in the theorem is immediate from the observation (following as a consequence of Theorem 2.7.1) that $\bar{G} \approx G/K$, $\bar{N} \approx N/K$, $\bar{G}/\bar{N} \approx (G/K)/(N/K)$.

Problems

1. In the following, verify if the mappings defined are homomorphisms, and in those cases in which they are homomorphisms, determine the kernel.
   (a) G is the group of nonzero real numbers under multiplication, $\bar{G} = G$, $\phi(x) = x^2$ all $x \in G$.
   (b) G, $\bar{G}$ as in (a), $\phi(x) = 2^x$.
   (c) G is the group of real numbers under addition, $\bar{G} = G$, $\phi(x) = x + 1$ all $x \in G$.
   (d) G, $\bar{G}$ as in (c), $\phi(x) = 13x$ for $x \in G$.
   (e) G is any abelian group, $\bar{G} = G$, $\phi(x) = x^5$ all $x \in G$.
2. Let G be any group, g a fixed element in G. Define $\phi:G \to G$ by $\phi(x) = gxg^{-1}$. Prove that ϕ is an isomorphism of G onto G.
3. Let G be a finite abelian group of order $o(G)$ and suppose the integer n is relatively prime to $o(G)$. Prove that every $g \in G$ can be written as $g = x^n$ with $x \in G$. (*Hint:* Consider the mapping $\phi:G \to G$ defined by $\phi(y) = y^n$, and prove this mapping is an isomorphism of G onto G.)
4. (a) Given any group G and a subset U, let $\hat{U}$ be the smallest subgroup of G which contains U. Prove there is such a subgroup $\hat{U}$ in G. ($\hat{U}$ is called the *subgroup generated by* U.)

(b) If $gug^{-1} \in U$ for all $g \in G$, $u \in U$, prove that $\hat{U}$ is a normal subgroup of G.

5. Let $U = \{xyx^{-1}y^{-1} \mid x, y \in G\}$. In this case $\hat{U}$ is usually written as G' and is called the *commutator subgroup of* G.
   (a) Prove that G' is normal in G.
   (b) Prove that G/G' is abelian.
   (c) If G/N is abelian, prove that $N \supset G'$.
   (d) Prove that if H is a subgroup of G and $H \supset G'$, then H is normal in G.

6. If N, M are normal subgroups of G, prove that $NM/M \approx N/N \cap M$.

7. Let V be the set of real numbers, and for a, b real, $a \neq 0$ let $\tau_{ab}: V \to V$ defined by $\tau_{ab}(x) = ax + b$. Let $G = \{\tau_{ab} \mid a, b \text{ real}, a \neq 0\}$ and let $N = \{\tau_{1b} \in G\}$. Prove that N is a normal subgroup of G and that $G/N \approx$ group of nonzero real numbers under multiplication.

8. Let G be the dihedral group defined as the set of all formal symbols $x^i y^j$, $i = 0, 1$, $j = 0, 1, \ldots, n - 1$, where $x^2 = e$, $y^n = e$, $xy = y^{-1}x$. Prove
   (a) The subgroup $N = \{e, y, y^2, \ldots, y^{n-1}\}$ is normal in G.
   (b) That $G/N \approx W$, where $W = \{1, -1\}$ is the group under the multiplication of the real numbers.

9. Prove that the center of a group is always a normal subgroup.

10. Prove that a group of order 9 is abelian.

11. If G is a non-abelian group of order 6, prove that $G \approx S_3$.

12. If G is abelian and if N is any subgroup of G, prove that G/N is abelian.

13. Let G be the dihedral group defined in Problem 8. Find the center of G.

14. Let G be as in Problem 13. Find G', the commutator subgroup of G.

15. Let G be the group of nonzero complex numbers under multiplication and let N be the set of complex numbers of absolute value 1 (that is, $a + bi \in N$ if $a^2 + b^2 = 1$). Show that G/N is isomorphic to the group of all positive real numbers under multiplication.

#16. Let G be the group of all nonzero complex numbers under multiplication and let $\bar{G}$ be the group of all real 2×2 matrices of the form $\begin{pmatrix} a & b \\ -b & a \end{pmatrix}$, where not both a and b are 0, under matrix multiplication. Show that G and $\bar{G}$ are isomorphic by exhibiting an isomorphism of G onto $\bar{G}$.

*17. Let G be the group of real numbers under addition and let N be the subgroup of G consisting of all the integers. Prove that G/N is isomorphic to the group of all complex numbers of absolute value 1 under multiplication.

#18. Let G be the group of all real 2×2 matrices $\begin{pmatrix} a & b \\ c & d \end{pmatrix}$, with $ad - bc \neq 0$, under matrix multiplication, and let

$$N = \left\{ \begin{pmatrix} a & b \\ c & d \end{pmatrix} \in G \mid ad - bc = 1 \right\}.$$

Prove that $N \supset G'$, the commutator subgroup of G.

*#19. In Problem 18 show, in fact, that $N = G'$.

#20. Let G be the group of all real 2×2 matrices of the form $\begin{pmatrix} a & b \\ 0 & d \end{pmatrix}$, where $ad \neq 0$, under matrix multiplication. Show that G' is precisely the set of all matrices of the form $\begin{pmatrix} 1 & x \\ 0 & 1 \end{pmatrix}$.

21. Let S_1 and S_2 be two sets. Suppose that there exists a one-to-one mapping ψ of S_1 into S_2. Show that there exists an isomorphism of $A(S_1)$ into $A(S_2)$, where $A(S)$ means the set of all one-to-one mappings of S onto itself.

2.8 Automorphisms

In the preceding section the concept of an isomorphism of one group into another was defined and examined. The special case in which the isomorphism maps a given group into itself should obviously be of some importance. We use the word "into" advisedly, for groups G do exist which have isomorphisms mapping G into, and not onto, itself. The easiest such example is the following: Let G be the group of integers under addition and define $\phi: G \to G$ by $\phi: x \to 2x$ for every $x \in G$. Since $\phi: x + y \to 2(x + y) = 2x + 2y$, ϕ is a homomorphism. Also if the image of x and y under ϕ are equal, then $2x = 2y$ whence $x = y$. ϕ is thus an isomorphism. Yet ϕ is not onto, for the image of any integer under ϕ is an even integer, so, for instance, 1 does not appear an image under ϕ of any element of G. Of greatest interest to us will be the isomorphisms of a group *onto* itself.

DEFINITION By an *automorphism* of a group G we shall mean an isomorphism of G onto itself.

As we mentioned in Chapter 1, whenever we talk about mappings of a set into itself we shall write the mappings on the right side, thus if $T: S \to S$, $x \in S$, then xT is the image of x under T.

Let I be the mapping of G which sends every element onto itself, that is, $xI = x$ for all $x \in G$. Trivially I is an automorphism of G. Let $\mathscr{A}(G)$ denote the set of all automorphisms of G; being a subset of $A(G)$, the set of one-to-one mappings of G onto itself, for elements of $\mathscr{A}(G)$ we can use the product of $A(G)$, namely, composition of mappings. This product then satisfies the associative law in $A(G)$, and so, *a fortiori*, in $\mathscr{A}(G)$. Also I, the unit element of $A(G)$, is in $\mathscr{A}(G)$, so $\mathscr{A}(G)$ is not empty.

An obvious fact that we should try to establish is that $\mathscr{A}(G)$ is a subgroup of $A(G)$, and so, in its own rights, $\mathscr{A}(G)$ should be a group. If T_1, T_2 are in $\mathscr{A}(G)$ we already know that $T_1T_2 \in A(G)$. We want it to be in the smaller set $\mathscr{A}(G)$. We proceed to verify this. For all $x, y \in G$,

$$(xy)T_1 = (xT_1)(yT_1),$$
$$(xy)T_2 = (xT_2)(yT_2),$$

therefore

$$\begin{aligned}(xy)T_1T_2 &= ((xy)T_1)T_2 = ((xT_1)(yT_1))T_2 \\ &= ((xT_1)T_2)((yT_1)T_2) = (xT_1T_2)(yT_1T_2).\end{aligned}$$

That is, $T_1T_2 \in \mathscr{A}(G)$. There is only one other fact that needs verifying in order that $\mathscr{A}(G)$ be a subgroup of $A(G)$, namely, that if $T \in \mathscr{A}(G)$, then $T^{-1} \in \mathscr{A}(G)$. If $x, y \in G$, then

$$((xT^{-1})(yT^{-1}))T = ((xT^{-1})T)((yT^{-1})T) = (xI)(yI) = xy,$$

thus

$$(xT^{-1})(yT^{-1}) = (xy)T^{-1},$$

placing T^{-1} in $\mathscr{A}(G)$. Summarizing these remarks, we have proved

LEMMA 2.8.1 *If G is a group, then $\mathscr{A}(G)$, the set of automorphisms of G, is also a group.*

Of course, as yet, we have no way of knowing that $\mathscr{A}(G)$, in general, has elements other than I. If G is a group having only two elements, the reader should convince himself that $\mathscr{A}(G)$ consists only of I. For groups G with more than two elements, $\mathscr{A}(G)$ always has more than one element.

What we should like is a richer sample of automorphisms than the ones we have (namely, I). If the group G is abelian and there is some element $x_0 \in G$ satisfying $x_0 \neq x_0^{-1}$, we can write down an explicit automorphism, the mapping T defined by $xT = x^{-1}$ for all $x \in G$. For any group G, T is onto; for any abelian G, $(xy)T = (xy)^{-1} = y^{-1}x^{-1} = x^{-1}y^{-1} = (xT)(yT)$. Also $x_0T = x_0^{-1} \neq x_0$, so $T \neq I$.

However, the class of abelian groups is a little limited, and we should like to have some automorphisms of non-abelian groups. Strangely enough the task of finding automorphisms for such groups is easier than for abelian groups.

Let G be a group; for $g \in G$ define $T_g:G \to G$ by $xT_g = g^{-1}xg$ for all $x \in G$. We claim that T_g is an automorphism of G. First, T_g is onto, for given $y \in G$, let $x = gyg^{-1}$. Then $xT_g = g^{-1}(x)g = g^{-1}(gyg^{-1})g = y$, so T_g is onto. Now consider, for $x, y \in G$, $(xy)T_g = g^{-1}(xy)g = g^{-1}(xgg^{-1}y)g = (g^{-1}xg)(g^{-1}yg) = (xT_g)(yT_g)$. Consequently T_g is a homomorphism of G onto itself. We further assert that T_g is one-to-one, for if $xT_g = yT_g$, then $g^{-1}xg = g^{-1}yg$, so by the cancellation laws in G, $x = y$. T_g is called the *inner automorphism* corresponding to g. If G is non-abelian, there is a pair $a, b \in G$ such that $ab \neq ba$; but then $bT_a = a^{-1}ba \neq b$, so that $T_a \neq I$. Thus for a non-abelian group G there always exist nontrivial automorphisms.

Let $\mathscr{I}(G) = \{T_g \in \mathscr{A}(G) \mid g \in G\}$. The computation of T_{gh}, for $g, h \in G$, might be of some interest. So, suppose $x \in G$; by definition,

$$xT_{gh} = (gh)^{-1}x(gh) = h^{-1}g^{-1}xgh = (g^{-1}xg)T_h = (xT_g)T_h = xT_gT_h.$$

Looking at the start and finish of this chain of equalities we find that $T_{gh} = T_gT_h$. This little remark is both interesting and suggestive. It is of interest because it immediately yields that $\mathscr{I}(G)$ is a subgroup of $\mathscr{A}(G)$. (Verify!) $\mathscr{I}(G)$ is usually called *the group of inner automorphisms of G*. It is suggestive, for if we consider the mapping $\psi:G \to \mathscr{A}(G)$ defined by $\psi(g) = T_g$ for every $g \in G$, then $\psi(gh) = T_{gh} = T_gT_h = \psi(g)\psi(h)$. That is, ψ is a homomorphism of G into $\mathscr{A}(G)$ whose image is $\mathscr{I}(G)$. What is the kernel of ψ? Suppose we call it K, and suppose $g_0 \in K$. Then $\psi(g_0) = I$, or, equivalently, $T_{g_0} = I$. But this says that for any $x \in G$, $xT_{g_0} = x$; however, $xT_{g_0} = g_0^{-1}xg_0$, and so $x = g_0^{-1}xg_0$ for all $x \in G$. Thus $g_0x = g_0g_0^{-1}xg_0 = xg_0$; g_0 must commute with all elements of G. But the center of G, Z, was defined to be precisely all elements in G which commute with every element of G. (See Problem 15, Section 2.5.) Thus $K \subset Z$. However, if $z \in Z$, then $xT_z = z^{-1}xz = z^{-1}(zx)$ (since $zx = xz$) $= x$, whence $T_z = I$ and so $z \in K$. Therefore, $Z \subset K$. Having proved both $K \subset Z$ and $Z \subset K$ we have that $Z = K$. Summarizing, ψ is a homomorphism of G into $\mathscr{A}(G)$ with image $\mathscr{I}(G)$ and kernel Z. By Theorem 2.7.1 $\mathscr{I}(G) \approx G/Z$. In order to emphasize this general result we record it as

LEMMA 2.8.2 *$\mathscr{I}(G) \approx G/Z$, where $\mathscr{I}(G)$ is the group of inner automorphisms of G, and Z is the center of G.*

Suppose that ϕ is an automorphisms of a group G, and suppose that $a \in G$ has order n (that is, $a^n = e$ *but for no lower positive power*). Then $\phi(a)^n = \phi(a^n) = \phi(e) = e$, hence $\phi(a)^n = e$. If $\phi(a)^m = e$ for some $0 < m < n$, then $\phi(a^m) = \phi(a)^m = e$, which implies, since ϕ is one-to-one, that $a^m = e$, a contradiction. Thus

LEMMA 2.8.3 *Let G be a group and ϕ an automorphism of G. If $a \in G$ is of order $o(a) > 0$, then $o(\phi(a)) = o(a)$.*

Automorphisms of groups can be used as a means of constructing new groups from the original group. Before explaining this abstractly, we consider a particular example.

Let G be a cyclic group of order 7, that is, G consists of all a^i, where we assume $a^7 = e$. The mapping $\phi: a^i \to a^{2i}$, as can be checked trivially, is an automorphism of G of order 3, that is, $\phi^3 = I$. Let x be a symbol which we formally subject to the following conditions: $x^3 = e$, $x^{-1}a^ix = \phi(a^i) = a^{2i}$, and consider all formal symbols x^ia^j, where $i = 0, 1, 2$ and $j = 0, 1, 2, \ldots, 6$. We declare that $x^ia^j = x^ka^l$ if and only if $i \equiv k \bmod 3$ and $j \equiv l \bmod 7$. We multiply these symbols using the rules $x^3 = a^7 = e$, $x^{-1}ax = a^2$. For instance, $(xa)(xa^2) = x(ax)a^2 = x(xa^2)a^2 = x^2a^4$. The reader can verify that one obtains, in this way, a non-abelian group of order 21.

Generally, if G is a group, T an automorphism of order r of G which is not an inner automorphism, pick a symbol x and consider all elements x^ig, $i = 0, \pm 1, \pm 2, \ldots$, $g \in G$ subject to $x^ig = x^{i'}g'$ if and only if $i \equiv i' \bmod r$, $g = g'$ and $x^{-1}g^ix = gT^i$ for all i. This way we obtain a larger group $\{G, T\}$; G is normal in $\{G, T\}$ and $\{G, T\}/G \approx$ group generated by $T =$ cyclic group of order r.

We close the section by determining $\mathscr{A}(G)$ for all cyclic groups.

Example 2.8.1 Let G be a finite cyclic group of order r, $G = (a)$, $a^r = e$. Suppose T is an automorphism of G. If aT is known, since $a^iT = (aT)^i$, a^iT is determined, so gT is determined for all $g \in G = (a)$. Thus we need consider only possible images of a under T. Since $aT \in G$, and since every element in G is a power of a, $aT = a^t$ for some integer $0 < t < r$. However, since T is an automorphism, aT must have the same order as a (Lemma 2.8.3), and this condition, we claim, forces t to be relatively prime to r. For if $d \mid t$, $d \mid r$, then $(aT)^{r/d} = a^{t(r/d)} = a^{r(t/d)} = (a^r)^{t/d} = e$; thus aT has order a divisor of r/d, which, combined with the fact that aT has order r, leads us to $d = 1$. Conversely, for any $0 < s < r$ and relatively prime to r, the mapping $S: a^i \to a^{si}$ is an automorphism of G. Thus $\mathscr{A}(G)$ is in one-to-one correspondence with the group U_r of integers less than r and relatively prime to r under multiplication modulo r. We claim not only is there such a one-to-one correspondence, but there is one which furthermore is an isomorphism. Let us label the elements of $\mathscr{A}(G)$ as T_i where $T_i: a \to a^i$, $0 < i < r$ and relatively prime to r; $T_iT_j: a \to a^i \to (a^i)^j = a^{ij}$, thus $T_iT_j = T_{ij}$. The mapping $i \to T_i$ exhibits the isomorphism of U_r onto $\mathscr{A}(G)$. Here then, $\mathscr{A}(G) \approx U_r$.

Example 2.8.2 G is an infinite cyclic group. That is, G consists of all a^i, $i = 0, \pm 1, \pm 2, \ldots$, where we assume that $a^i = e$ if and only if $i = 0$. Suppose that T is an automorphism of G. As in Example 2.8.1, $aT = a^t$.

The question now becomes, What values of t are possible? Since T is an automorphism of G, it maps G onto itself, so that $a = gT$ for some $g \in G$. Thus $a = a^i T = (aT)^i$ for some integer i. Since $aT = a^t$, we must have that $a = a^{ti}$, so that $a^{ti-1} = e$. Hence $ti - 1 = 0$; that is, $ti = 1$. Clearly, since t and i are integers, this must force $t = \pm 1$, and each of these gives rise to an automorphism, $t = 1$ yielding the identity automorphism I, $t = -1$ giving rise to the automorphism $T: g \to g^{-1}$ for every g in the cyclic group G. Thus here, $\mathscr{A}(G) \approx$ cyclic group of order 2.

Problems

1. Are the following mappings automorphisms of their respective groups?
   (a) G group of integers under addition, $T: x \to -x$.
   (b) G group of positive reals under multiplication, $T: x \to x^2$.
   (c) G cyclic group of order 12, $T: x \to x^3$.
   (d) G is the group S_3, $T: x \to x^{-1}$.
2. Let G be a group, H a subgroup of G, T an automorphism of G. Let $(H)T = \{hT \mid h \in H\}$. Prove $(H)T$ is a subgroup of G.
3. Let G be a group, T an automorphism of G, N a normal subgroup of G. Prove that $(N)T$ is a normal subgroup of G.
4. For $G = S_3$ prove that $G \approx \mathscr{I}(G)$.
5. For any group G prove that $\mathscr{I}(G)$ is a normal subgroup of $\mathscr{A}(G)$ (the group $\mathscr{A}(G)/\mathscr{I}(G)$ is called the *group of outer automorphisms* of G).
6. Let G be a group of order 4, $G = \{e, a, b, ab\}$, $a^2 = b^2 = e$, $ab = ba$. Determine $\mathscr{A}(G)$.
7. (a) A subgroup C of G is said to be a *characteristic subgroup* of G if $(C)T \subset C$ for all automorphisms T of G. Prove a characteristic subgroup of G must be a normal subgroup of G.
   (b) Prove that the converse of (a) is false.
8. For any group G, prove that the commutator subgroup G' is a characteristic subgroup of G. (See Problem 5, Section 2.7).
9. If G is a group, N a normal subgroup of G, M a characteristic subgroup of N, prove that M is a normal subgroup of G.
10. Let G be a finite group, T an automorphism of G with the property that $xT = x$ for $x \in G$ if and only if $x = e$. Prove that every $g \in G$ can be represented as $g = x^{-1}(xT)$ for some $x \in G$.
11. Let G be a finite group, T an automorphism of G with the property that $xT = x$ if and only if $x = e$. Suppose further that $T^2 = I$. Prove that G must be abelian.

*12. Let G be a finite group and suppose the automorphism T sends more than three-quarters of the elements of G onto their inverses. Prove that $xT = x^{-1}$ for all $x \in G$ and that G is abelian.

13. In Problem 12, can you find an example of a finite group which is non-abelian and which has an automorphism which maps exactly three-quarters of the elements of G onto their inverses?

*14. Prove that every finite group having more than two elements has a nontrivial automorphism.

*15. Let G be a group of order $2n$. Suppose that half of the elements of G are of order 2, and the other half form a subgroup H of order n. Prove that H is of odd order and is an abelian subgroup of G.

*16. Let $\phi(n)$ be the Euler ϕ-function. If $a > 1$ is an integer, prove that $n \mid \phi(a^n - 1)$.

17. Let G be a group and Z the center of G. If T is any automorphism of G, prove that $(Z)T \subset Z$.

18. Let G be a group and T an automorphism of G. If, for $a \in G$, $N(a) = \{x \in G \mid xa = ax\}$, prove that $N(aT) = (N(a))T$.

19. Let G be a group and T an automorphism of G. If N is a normal subgroup of G such that $(N)T \subset N$, show how you could use T to define an automorphism of G/N.

20. Use the discussion following Lemma 2.8.3 to construct
    (a) a non-abelian group of order 55.
    (b) a non-abelian group of order 203.

21. Let G be the group of order 9 generated by elements a, b, where $a^3 = b^3 = e$. Find all the automorphisms of G.

2.9 Cayley's Theorem

When groups first arose in mathematics they usually came from some specific source and in some very concrete form. Very often it was in the form of a set of transformations of some particular mathematical object. In fact, most finite groups appeared as groups of permutations, that is, as subgroups of S_n. ($S_n = A(S)$ when S is a finite set with n elements.) The English mathematician Cayley first noted that every group could be realized as a subgroup of $A(S)$ for some S. Our concern, in this section, will be with a presentation of Cayley's theorem and some related results.

THEOREM 2.9.1 (CAYLEY) *Every group is isomorphic to a subgroup of $A(S)$ for some appropriate S.*

Proof. Let G be a group. For the set S we will use the elements of G; that is, put $S = G$. If $g \in G$, define $\tau_g : S (= G) \to S (= G)$ by $x\tau_g = xg$

for every $x \in G$. If $y \in G$, then $y = (yg^{-1})g = (yg^{-1})\tau_g$, so that τ_g maps S onto itself. Moreover, τ_g is one-to-one, for if $x, y \in S$ and $x\tau_g = y\tau_g$, then $xg = yg$, which, by the cancellation property of groups, implies that $x = y$. We have proved that for every $g \in G$, $\tau_g \in A(S)$.

If $g, h \in G$, consider τ_{gh}. For any $x \in S = G$, $x\tau_{gh} = x(gh) = (xg)h = (x\tau_g)\tau_h = x\tau_g\tau_h$. Note that we used the associative law in a very essential way here. From $x\tau_{gh} = x\tau_g\tau_h$ we deduce that $\tau_{gh} = \tau_g\tau_h$. Therefore, if $\psi: G \to A(S)$ is defined by $\psi(g) = \tau_g$, the relation $\tau_{gh} = \tau_g\tau_h$ tells us that ψ is a homomorphism. What is the kernel K of ψ? If $g_0 \in K$, then $\psi(g_0) = \tau_{g_0}$ is the identity map on S, so that for $x \in G$, and, in particular, for $e \in G$, $e\tau_{g_0} = e$. But $e\tau_{g_0} = eg_0 = g_0$. Thus comparing these two expressions for $e\tau_{g_0}$ we conclude that $g_0 = e$, whence $K = (e)$. Thus by the corollary to Lemma 2.7.4 ψ is an isomorphism of G into $A(S)$, proving the theorem.

The theorem enables us to exhibit any abstract group as a more concrete object, namely, as a group of mappings. However, it has its shortcomings; for if G is a finite group of order $o(G)$, then, using $S = G$, as in our proof, $A(S)$ has $o(G)!$ elements. Our group G of order $o(G)$ is somewhat lost in the group $A(S)$ which, with its $o(G)!$ elements, is huge in comparison to G. We ask: Can we find a more economical S, one for which $A(S)$ is smaller? This we now attempt to accomplish.

Let G be a group, H a subgroup of G. Let S be the set whose elements are the right cosets of H in G. That is, $S = \{Hg \mid g \in G\}$. S need not be a group itself, in fact, it would be a group only if H were a normal subgroup of G. However, we can make our group G act on S in the following natural way: for $g \in G$ let $t_g: S \to S$ be defined by $(Hx)t_g = Hxg$. Emulating the proof of Theorem 2.9.1 we can easily prove

1. $t_g \in A(S)$ for every $g \in G$.
2. $t_{gh} = t_g t_h$.

Thus the mapping $\theta: G \to A(S)$ defined by $\theta(g) = t_g$ is a homomorphism of G into $A(S)$. Can one always say that θ is an isomorphism? Suppose that K is the kernel of θ. If $g_0 \in K$, then $\theta(g_0) = t_{g_0}$ is the identity map on S, so that for every $X \in S$, $Xt_{g_0} = X$. Since every element of S is a right coset of H in G, we must have that $Hat_{g_0} = Ha$ for every $a \in G$, and using the definition of t_{g_0}, namely, $Hat_{g_0} = Hag_0$, we arrive at the identity $Hag_0 = Ha$ for every $a \in G$. On the other hand, if $b \in G$ is such that $Hxb = Hx$ for every $x \in G$, retracing our argument we could show that $b \in K$. Thus $K = \{b \in G \mid Hxb = Hx \text{ all } x \in G\}$. We claim that from this characterization of K, K must be the largest normal subgroup of G which is contained in H. We first explain the use of the word largest; by this we mean that if N is a normal subgroup of G which is contained in H, then N must be contained in K. We wish to show this is the case. That K is a normal subgroup of G follows from the fact that it is the kernel of a homomorphism of G. Now we assert that $K \subset H$, for if $b \in K$, $Hab = Ha$ for every $a \in G$, so, in particular, $Hb = Heb = He = H$, whence $b \in H$. Finally, if N is a normal subgroup of G which is contained in H, if $n \in N$, $a \in G$, then $ana^{-1} \in N \subset H$, so that $Hana^{-1} = H$; thus $Han = Ha$ for all $a \in G$. Therefore, $n \in K$ by our characterization of K.

We have proved

THEOREM 2.9.2 *If G is a group, H a subgroup of G, and S is the set of all right cosets of H in G, then there is a homomorphism θ of G into $A(S)$ and the kernel of θ is the largest normal subgroup of G which is contained in H.*

The case $H = (e)$ just yields Cayley's theorem (Theorem 2.9.1). If H should happen to have no normal subgroup of G other than (e) in it, then θ must be an isomorphism of G into $A(S)$. In this case we would have cut down the size of the S used in proving Theorem 2.9.1. This is interesting mostly for finite groups. For we shall use this observation both as a means of proving certain finite groups have nontrivial normal subgroups, and also as a means of representing certain finite groups as permutation groups on small sets.

We examine these remarks a little more closely. Suppose that G has a subgroup H whose index $i(H)$ (that is, the number of right cosets of H in G) satisfies $i(H)! < o(G)$. Let S be the set of all right cosets of H in G. The mapping, θ, of Theorem 2.9.2 cannot be an isomorphism, for if it were, $\theta(G)$ would have $o(G)$ elements and yet would be a subgroup of $A(S)$ which has $i(H)! < o(G)$ elements. Therefore the kernel of θ must be larger than (e); this kernel being the largest normal subgroup of G which is contained in H, we can conclude that H contains a nontrivial normal subgroup of G.

However, the argument used above has implications even when $i(H)!$ is not less than $o(G)$. If $o(G)$ does not divide $i(H)!$ then by invoking Lagrange's theorem we know that $A(S)$ can have no subgroup of order $o(G)$, hence no subgroup isomorphic to G. However, $A(S)$ does contain $\theta(G)$, whence $\theta(G)$ cannot be isomorphic to G; that is, θ cannot be an isomorphism. But then, as above, H must contain a nontrivial normal subgroup of G.

We summarize this as

LEMMA 2.9.1 *If G is a finite group, and $H \neq G$ is a subgroup of G such that $o(G) \nmid i(H)!$ then H must contain a nontrivial normal subgroup of G. In particular, G cannot be simple.*

APPLICATIONS

1. Let G be a group of order 36. Suppose that G has a subgroup H of order 9 (we shall see later that this is always the case). Then $i(H) = 4$,

$4! = 24 < 36 = o(G)$ so that in H there must be a normal subgroup $N \neq (e)$, of G, of order a divisor of 9, that is, of order 3 or 9.

2. Let G be a group of order 99 and suppose that H is a subgroup of G of order 11 (we shall also see, later, that this must be true). Then $i(H) = 9$, and since $99 \nmid 9!$ there is a nontrivial normal subgroup $N \neq (e)$ of G in H. Since H is of order 11, which is a prime, its only subgroup other than (e) is itself, implying that $N = H$. That is, H itself is a normal subgroup of G.

3. Let G be a non-abelian group of order 6. By Problem 11, Section 2.3, there is an $a \neq e \in G$ satisfying $a^2 = e$. Thus the subgroup $H = \{e, a\}$ is of order 2, and $i(H) = 3$. Suppose, for the moment, that we know that H is not normal in G. Since H has only itself and (e) as subgroups, H has no nontrivial normal subgroups of G in it. Thus G is isomorphic to a subgroup T of order 6 in $A(S)$, where S is the set of right cosets of H in G. Since $o(A(S)) = i(H)! = 3! = 6$, $T = S$. In other words, $G \approx A(S) = S_3$. We would have proved that any non-abelian group of order 6 is isomorphic to S_3. All that remains is to show that H is not normal in G. Since it might be of some interest we go through a detailed proof of this. If $H = \{e, a\}$ were normal in G, then for every $g \in G$, since $gag^{-1} \in H$ and $gag^{-1} \neq e$, we would have that $gag^{-1} = a$, or, equivalently, that $ga = ag$ for every $g \in G$. Let $b \in G$, $b \notin H$, and consider $N(b) = \{x \in G \mid xb = bx\}$. By an earlier problem, $N(b)$ is a subgroup of G, and $N(b) \supset H$; $N(b) \neq H$ since $b \in N(b)$, $b \notin H$. Since H is a subgroup of $N(b)$, $o(H) \mid o(N(b)) \mid 6$. The only even number n, $2 < n \leq 6$ which divides 6 is 6. So $o(N(b)) = 6$; whence b commutes with all elements of G. Thus every element of G commutes with every other element of G, making G into an abelian group, contrary to assumption. Thus H could not have been normal in G. This proof is somewhat long-winded, but it illustrates some of the ideas already developed.

Problems

1. Let G be a group; consider the mappings of G into itself, λ_g, defined for $g \in G$ by $x\lambda_g = gx$ for all $x \in G$. Prove that λ_g is one-to-one and onto, and that $\lambda_{gh} = \lambda_h \lambda_g$.
2. Let λ_g be defined as in Problem 1, τ_g as in the proof of Theorem 2.9.1. Prove that for any $g, h \in G$, the mappings λ_g, τ_h satisfy $\lambda_g \tau_h = \tau_h \lambda_g$. (*Hint:* For $x \in G$ consider $x(\lambda_g \tau_h)$ and $x(\tau_h \lambda_g)$.)
3. If θ is a one-to-one mapping of G onto itself such that $\lambda_g \theta = \theta \lambda_g$ for all $g \in G$, prove that $\theta = \tau_h$ for some $h \in G$.
4. (a) If H is a subgroup of G show that for every $g \in G$, gHg^{-1} is a subgroup of G.

(b) Prove that W = intersection of all gHg^{-1} is a normal subgroup of G.

5. Using Lemma 2.9.1 prove that a group of order p^2, where p is a prime number, must have a normal subgroup of order p.
6. Show that in a group G of order p^2 any normal subgroup of order p must lie in the center of G.
7. Using the result of Problem 6, prove that any group of order p^2 is abelian.
8. If p is a prime number, prove that any group G of order $2p$ must have a subgroup of order p, and that this subgroup is normal in G.
9. If $o(G)$ is pq where p and q are distinct prime numbers and if G has a normal subgroup of order p and a normal subgroup of order q, prove that G is cyclic.

*10. Let $o(G)$ be pq, $p > q$ are primes, prove
   (a) G has a subgroup of order p and a subgroup of order q.
   (b) If $q \nmid p - 1$, then G is cyclic.
   (c) Given two primes p, q, $q \mid p - 1$, there exists a non-abelian group of order pq.
   (d) Any two non-abelian groups of order pq are isomorphic.

2.10 Permutation Groups

We have seen that every group can be represented isomorphically as a subgroup of $A(S)$ for some set S, and, in particular, a finite group G can be represented as a subgroup of S_n, for some n, where S_n is the symmetric group of degree n. This clearly shows that the groups S_n themselves merit closer examination.

Suppose that S is a finite set having n elements $x_1, x_2, \ldots, x_n$. If $\phi \in A(S) = S_n$, then ϕ is a one-to-one mapping of S onto itself, and we could write ϕ out by showing what it does to every element, e.g., $\phi: x_1 \to x_2$, $x_2 \to x_4$, $x_4 \to x_3$, $x_3 \to x_1$. But this is very cumbersome. One short cut might be to write ϕ out as

$$\begin{pmatrix} x_1 & x_2 & x_3 & \cdots & x_n \\ x_{i_1} & x_{i_2} & x_{i_3} & \cdots & x_{i_n} \end{pmatrix},$$

where x_{i_k} is the image of x_i under ϕ. Returning to our example just above, ϕ might be represented by

$$\begin{pmatrix} x_1 & x_2 & x_3 & x_4 \\ x_2 & x_4 & x_1 & x_3 \end{pmatrix}.$$

While this notation is a little handier there still is waste in it, for there seems to be no purpose served by the symbol x. We could equally well represent the permutation as

$$\begin{pmatrix} 1 & 2 & \cdots & n \\ i_1 & i_2 & \cdots & i_n \end{pmatrix}.$$

Our specific example would read

$$\begin{pmatrix} 1 & 2 & 3 & 4 \\ 2 & 4 & 1 & 3 \end{pmatrix}.$$

Given two permutations θ, ψ in S_n, using this symbolic representation of θ and ψ, what would the representation of $\theta\psi$ be? To compute it we could start and see what $\theta\psi$ does to x_1 (henceforth written as 1). θ takes 1 into i_1, while ψ takes i_1 into k, say, then $\theta\psi$ takes 1 into k. Then repeat this procedure for $2, 3, \ldots, n$. For instance, if θ is the permutation represented by

$$\begin{pmatrix} 1 & 2 & 3 & 4 \\ 3 & 1 & 2 & 4 \end{pmatrix}$$

and ψ by

$$\begin{pmatrix} 1 & 2 & 3 & 4 \\ 1 & 3 & 2 & 4 \end{pmatrix},$$

then $i_1 = 3$ and ψ takes 3 into 2, so $k = 2$ and $\theta\psi$ takes 1 into 2. Similarly $\theta\psi: 2 \to 1,\ 3 \to 3,\ 4 \to 4$. That is, the representation for $\theta\psi$ is

$$\begin{pmatrix} 1 & 2 & 3 & 4 \\ 2 & 1 & 3 & 4 \end{pmatrix}.$$

If we write

$$\theta = \begin{pmatrix} 1 & 2 & 3 & 4 \\ 3 & 1 & 2 & 4 \end{pmatrix}$$

and

$$\psi = \begin{pmatrix} 1 & 2 & 3 & 4 \\ 1 & 3 & 2 & 4 \end{pmatrix},$$

then

$$\theta\psi = \begin{pmatrix} 1 & 2 & 3 & 4 \\ 3 & 1 & 2 & 4 \end{pmatrix}\begin{pmatrix} 1 & 2 & 3 & 4 \\ 1 & 3 & 2 & 4 \end{pmatrix} = \begin{pmatrix} 1 & 2 & 3 & 4 \\ 2 & 1 & 3 & 4 \end{pmatrix}.$$

This is the way we shall multiply the symbols of the form

$$\begin{pmatrix} 1 & 2 & \cdots & n \\ i_1 & i_2 & \cdots & i_n \end{pmatrix}, \quad \begin{pmatrix} 1 & 2 & \cdots & n \\ k_1 & k_2 & \cdots & k_n \end{pmatrix}.$$

Let S be a set and $\theta \in A(S)$. Given two elements $a, b \in S$ we define $a \equiv {}_{\theta} b$ if and only if $b = a\theta^i$ for some integer i (i can be positive, negative, or 0). We claim this defines an equivalence relation on S. For

1. $a \equiv {}_{\theta} a$ since $a = a\theta^0 = ae$.
2. If $a \equiv {}_{\theta} b$, then $b = a\theta^i$, so that $a = b\theta^{-i}$, whence $b \equiv {}_{\theta} a$.
3. If $a \equiv {}_{\theta} b$, $b \equiv {}_{\theta} c$, then $b = a\theta^i$, $c = b\theta^j = (a\theta^i)\theta^j = a\theta^{i+j}$, which implies that $a \equiv {}_{\theta} c$.

This equivalence relation by Theorem 1.1.1 induces a decomposition of S into disjoint subsets, namely, the equivalence classes. We call the equivalence class of an element $s \in S$ the *orbit* of s under θ; thus the orbit of s under θ consists of all the elements $s\theta^i$, $i = 0, \pm 1, \pm 2, \ldots$.

In particular, if S is a finite set and $s \in S$, there is a smallest positive integer $l = l(s)$ depending on s such that $s\theta^l = s$. The orbit of s under θ then consists of the elements $s, s\theta, s\theta^2, \ldots, s\theta^{l-1}$. By a *cycle* of θ we mean the *ordered* set $(s, s\theta, s\theta^2, \ldots, s\theta^{l-1})$. If we know all the cycles of θ we clearly know θ since we would know the image of any element under θ. Before proceeding we illustrate these ideas with an example. Let

$$\theta = \begin{pmatrix} 1 & 2 & 3 & 4 & 5 & 6 \\ 2 & 1 & 3 & 5 & 6 & 4 \end{pmatrix},$$

where S consists of the elements $1, 2, \ldots, 6$ (remember 1 stands for x_1, 2 for x_2, etc.). Starting with 1, then the orbit of 1 consists of $1 = 1\theta^0$, $1\theta^1 = 2$, $1\theta^2 = 2\theta = 1$, so the orbit of 1 is the set of elements 1 and 2. This tells us the orbit of 2 is the same set. The orbit of 3 consists just of 3; that of 4 consists of the elements 4, $4\theta = 5$, $4\theta^2 = 5\theta = 6$, $4\theta^3 = 6\theta = 4$. The cycles of θ are (1, 2), (3), (4, 5, 6).

We digress for a moment, leaving our particular θ. Suppose that by the cycle $(i_1, i_2, \ldots, i_r)$ we mean the permutation ψ which sends i_1 into i_2, i_2 into $i_3 \cdots i_{r-1}$ into i_r and i_r into i_1, and leaves all other elements of S fixed. Thus, for instance, if S consists of the elements $1, 2, \ldots, 9$, then the symbol (1, 3, 4, 2, 6) means the permutation

$$\begin{pmatrix} 1 & 2 & 3 & 4 & 5 & 6 & 7 & 8 & 9 \\ 3 & 6 & 4 & 2 & 5 & 1 & 7 & 8 & 9 \end{pmatrix}.$$

We multiply cycles by multiplying the permutations they represent. Thus again, if S has 9 elements,

$$(1 \ \ 2 \ \ 3)(5 \ \ 6 \ \ 4 \ \ 1 \ \ 8)$$

$$= \begin{pmatrix} 1 & 2 & 3 & 4 & 5 & 6 & 7 & 8 & 9 \\ 2 & 3 & 1 & 4 & 5 & 6 & 7 & 8 & 9 \end{pmatrix} \begin{pmatrix} 1 & 2 & 3 & 4 & 5 & 6 & 7 & 8 & 9 \\ 8 & 2 & 3 & 1 & 6 & 4 & 7 & 5 & 9 \end{pmatrix}$$

$$= \begin{pmatrix} 1 & 2 & 3 & 4 & 5 & 6 & 7 & 8 & 9 \\ 2 & 3 & 8 & 1 & 6 & 4 & 7 & 5 & 9 \end{pmatrix}.$$

Let us return to the ideas of the paragraph preceding the last one, and ask: Given the permutation

$$\theta = \begin{pmatrix} 1 & 2 & 3 & 4 & 5 & 6 & 7 & 8 & 9 \\ 2 & 3 & 8 & 1 & 6 & 4 & 7 & 5 & 9 \end{pmatrix},$$

what are the cycles of θ? We first find the orbit of 1; namely, $1, 1\theta = 2$, $1\theta^2 = 2\theta = 3$, $1\theta^3 = 3\theta = 8$, $1\theta^4 = 8\theta = 5$, $1\theta^5 = 5\theta = 6$, $1\theta^6 = 6\theta = 4$, $1\theta^7 = 4\theta = 1$. That is, the orbit of 1 is the set $\{1, 2, 3, 8, 5, 6, 4\}$. The orbits of 7 and 9 can be found to be $\{7\}$, $\{9\}$, respectively. The cycles of θ thus are (7), (9), $(1, 1\theta, 1\theta^2, \ldots, 1\theta^6) = (1, 2, 3, 8, 5, 6, 4)$. The reader should now verify that if he takes the product (as defined in the last paragraph) of $(1, 2, 3, 8, 5, 6, 4)$, (7), (9) he will obtain θ. That is, at least in this case, θ is the product of its cycles.

But this is no accident for it is now trivial to prove

LEMMA 2.10.1 *Every permutation is the product of its cycles.*

Proof. Let θ be the permutation. Then its cycles are of the form $(s, s\theta, \ldots, s\theta^{l-1})$. By the multiplication of cycles, as defined above, and since the cycles of θ are disjoint, the image of $s' \in S$ under θ, which is $s'\theta$, is the same as the image of s' under the product, ψ, of all the distinct cycles of θ. So θ, ψ have the same effect on every element of S, hence $\theta = \psi$, which is what we sought to prove.

If the remarks above are still not transparent at this point, the reader should take a given permutation, find its cycles, take their product, and verify the lemma. In doing so the lemma itself will become obvious.

Lemma 2.10.1 is usually stated in the form *every permutation can be uniquely expressed as a product of disjoint cycles.*

Consider the m-cycle $(1, 2, \ldots, m)$. A simple computation shows that $(1, 2, \ldots, m) = (1, 2)(1, 3) \cdots (1, m)$. More generally the m-cycle $(a_1, a_2, \ldots, a_m) = (a_1, a_2)(a_1, a_3) \cdots (a_1, a_m)$. This decomposition is not unique; by this we mean that an m-cycle can be written as a product of 2-cycles in more than one way. For instance, $(1, 2, 3) = (1, 2)(1, 3) = (3, 1)(3, 2)$. Now, since every permutation is a product of disjoint cycles and every cycle is a product of 2-cycles, we have proved

LEMMA 2.10.2 *Every permutation is a product of 2-cycles.*

We shall refer to 2-cycles as *transpositions.*

DEFINITION A permutation $\theta \in S_n$ is said to be an *even permutation* if it can be represented as a product of an even number of transpositions.

The definition given just insists that θ have one representation as a product of an even number of transpositions. Perhaps it has other representations as a product of an odd number of transpositions. We first want to show that this cannot happen. Frankly, we are not happy with the proof we give of this fact for it introduces a polynomial which seems extraneous to the matter at hand.

Consider the polynomial in n-variables

$$p(x_1, \ldots, x_n) = \prod_{i<j} (x_i - x_j).$$

If $\theta \in S_n$ let θ act on the polynomial $p(x_1, \ldots, x_n)$ by

$$\theta : p(x_1, \ldots, x_n) = \prod_{i<j} (x_i - x_j) \rightarrow \prod_{i<j} (x_{\theta(i)} - x_{\theta(j)}).$$

It is clear that $\theta : p(x_1, \ldots, x_n) \rightarrow \pm p(x_1, \ldots, x_n)$. For instance, in S_5, $\theta = (134)(25)$ takes

$$p(x_1, \ldots, x_5) = (x_1 - x_2)(x_1 - x_3)(x_1 - x_4)(x_1 - x_5)(x_2 - x_3) \times (x_2 - x_4)(x_2 - x_5)(x_3 - x_4)(x_3 - x_5)(x_4 - x_5)$$

into

$$(x_3 - x_5)(x_3 - x_4)(x_3 - x_1)(x_3 - x_2)(x_5 - x_4)(x_5 - x_1) \times (x_5 - x_2)(x_4 - x_1)(x_4 - x_2)(x_1 - x_2),$$

which can easily be verified to be $-p(x_1, \ldots, x_5)$.

If, in particular, θ is a transposition, $\theta : p(x_1, \ldots, x_n) \rightarrow -p(x_1, \ldots, x_n)$. (Verify!) Thus if a permutation Π can be represented as a product of an even number of transpositions in one representation, Π must leave $p(x_1, \ldots, x_n)$ fixed, so that any representation of Π as a product of transposition must be such that it leaves $p(x_1, \ldots, x_n)$ fixed; that is, in any representation it is a product of an even number of transpositions. This establishes that the definition given for an even permutation is a significant one. We call a permutation *odd* if it is not an even permutation.

The following facts are now clear:

1. The product of two even permutations is an even permutation.
2. The product of an even permutation and an odd one is odd (likewise for the product of an odd and even permutation).
3. The product of two odd permutations is an even permutation.

The rule for combining even and odd permutations is like that of combining even and odd numbers under addition. This is not a coincidence since this latter rule is used in establishing 1, 2, and 3.

Let A_n be the subset of S_n consisting of all even permutations. Since the product of two even permutations is even, A_n must be a subgroup of S_n. We claim it is normal in S_n. Perhaps the best way of seeing this is as follows:

let W be the group of real numbers 1 and -1 under multiplication. Define $\psi: S_n \to W$ by $\psi(s) = 1$ if s is an even permutation, $\psi(s) = -1$ if s is an odd permutation. By the rules 1, 2, 3 above ψ is a homomorphism onto W. The kernel of ψ is precisely A_n; being the kernel of a homomorphism A_n is a normal subgroup of S_n. By Theorem 2.7.1 $S_n/A_n \approx W$, so, since

$$2 = o(W) = o\left(\frac{S_n}{A_n}\right) = \frac{o(S_n)}{o(A_n)},$$

we see that $o(A_n) = \frac{1}{2}n!$. A_n is called the *alternating group* of degree n. We summarize our remarks in

LEMMA 2.10.3 *S_n has as a normal subgroup of index 2 the alternating group, A_n, consisting of all even permutations.*

At the end of the next section we shall return to S_n again.

Problems

1. Find the orbits and cycles of the following permutations:
   (a) $\begin{pmatrix} 1 & 2 & 3 & 4 & 5 & 6 & 7 & 8 & 9 \\ 2 & 3 & 4 & 5 & 1 & 6 & 7 & 9 & 8 \end{pmatrix}$.
   (b) $\begin{pmatrix} 1 & 2 & 3 & 4 & 5 & 6 \\ 6 & 5 & 4 & 3 & 1 & 2 \end{pmatrix}$.
2. Write the permutations in Problem 1 as the product of disjoint cycles.
3. Express as the product of disjoint cycles:
   (a) $(1, 2, 3)(4, 5)(1, 6, 7, 8, 9)(1, 5)$.
   (b) $(1, 2)(1, 2, 3)(1, 2)$.
4. Prove that $(1, 2, \ldots, n)^{-1} = (n, n-1, n-2, \ldots, 2, 1)$.
5. Find the cycle structure of all the powers of $(1, 2, \ldots, 8)$.
6. (a) What is the order of an n-cycle?
   (b) What is the order of the product of the disjoint cycles of lengths $m_1, m_2, \ldots, m_k$?
   (c) How do you find the order of a given permutation?
7. Compute $a^{-1}ba$, where
   (1) $a = (1, 3, 5)(1, 2)$, $b = (1, 5, 7, 9)$.
   (2) $a = (5, 7, 9)$, $b = (1, 2, 3)$.
8. (a) Given the permutation $x = (1, 2)(3, 4)$, $y = (5, 6)(1, 3)$, find a permutation a such that $a^{-1}xa = y$.
   (b) Prove that there is no a such that $a^{-1}(1, 2, 3)a = (1, 3)(5, 7, 8)$.
   (c) Prove that there is no permutation a such that $a^{-1}(1, 2)a = (3, 4)(1, 5)$.
9. Determine for what m an m-cycle is an even permutation.

10. Determine which of the following are even permutations:
    (a) $(1, 2, 3)(1, 2)$.
    (b) $(1, 2, 3, 4, 5)(1, 2, 3)(4, 5)$.
    (c) $(1, 2)(1, 3)(1, 4)(2, 5)$.
11. Prove that the smallest subgroup of S_n containing $(1, 2)$ and $(1, 2, \ldots, n)$ is S_n. (In other words, these generate S_n.)
*12. Prove that for $n \geq 3$ the subgroup generated by the 3-cycles is A_n.
*13. Prove that if a normal subgroup of A_n contains even a single 3-cycle it must be all of A_n.
*14. Prove that A_5 has no normal subgroups $N \neq (e), A_5$.
15. Assuming the result of Problem 14, prove that any subgroup of A_5 has order at most 12.
16. Find all the normal subgroups in S_4.
*17. If $n \geq 5$ prove that A_n is the only nontrivial normal subgroup in S_n.

Cayley's theorem (Theorem 2.9.1) asserts that every group is isomorphic to a subgroup of $A(S)$ for some S. In particular, it says that every finite group can be realized as a group of permutations. Let us call the realization of the group as a group of permutations as given in the proof of Theorem 2.9.1 the *permutation representation* of G.

18. Find the permutation representation of a cyclic group of order n.
19. Let G be the group $\{e, a, b, ab\}$ of order 4, where $a^2 = b^2 = e$, $ab = ba$. Find the permutation representation of G.
20. Let G be the group S_3. Find the permutation representation of S_3. (*Note:* This gives an isomorphism of S_3 into S_6.)
21. Let G be the group $\{e, \theta, a, b, c, \theta a, \theta b, \theta c\}$, where $a^2 = b^2 = c^2 = \theta$, $\theta^2 = e$, $ab = \theta ba = c$, $bc = \theta cb = a$, $ca = \theta ac = b$.
    (a) Show that θ is in the center Z of G, and that $Z = \{e, \theta\}$.
    (b) Find the commutator subgroup of G.
    (c) Show that every subgroup of G is normal.
    (d) Find the permutation representation of G.
    (*Note:* G is often called the group of *quaternion units*; it, and algebraic systems constructed from it, will reappear in the book.)
22. Let G be the dihedral group of order $2n$ (see Problem 17, Section 2.6). Find the permutation representation of G.

Let us call the realization of a group G as a set of permutations given in Problem 1, Section 2.9 the *second permutation representation* of G.

23. Show that if G is an abelian group, then the permutation representation of G coincides with the second permutation representation of G (i.e., in the notation of the previous section, $\lambda_g = \tau_g$ for all $g \in G$.)

24. Find the second permutation representation of S_3. Verify directly from the permutations obtained here and in Problem 20 that $\lambda_a \tau_b = \tau_b \lambda_a$ for all $a, b \in S_3$.

25. Find the second permutation representation of the group G defined in Problem 21.

26. Find the second permutation representation of the dihedral group of order $2n$.

If H is a subgroup of G, let us call the mapping $\{t_g \mid g \in G\}$ defined in the discussion preceding Theorem 2.9.2 the *coset representation* of G by H. This also realizes G as a group of permutations, but not necessarily isomorphically, merely homomorphically (see Theorem 2.9.2).

27. Let $G = (a)$ be a cyclic group of order 8 and let $H = (a^4)$ be its subgroup of order 2. Find the coset representation of G by H.

28. Let G be the dihedral group of order $2n$ generated by elements a, b such that $a^2 = b^n = e$, $ab = b^{-1}a$. Let $H = \{e, a\}$. Find the coset representation of G by H.

29. Let G be the group of Problem 21 and let $H = \{e, \theta\}$. Find the coset representation of G by H.

30. Let G be S_n, the symmetric group of order n, acting as permutations on the set $\{1, 2, \ldots, n\}$. Let $H = \{\sigma \in G \mid n\sigma = n\}$.
    (a) Prove that H is isomorphic to S_{n-1}.
    (b) Find a set of elements $a_1, \ldots, a_n \in G$ such that $Ha_1, \ldots, Ha_n$ give all the right cosets of H in G.
    (c) Find the coset representation of G by H.

2.11 Another Counting Principle

Mathematics is rich in technique and arguments. In this great variety one of the most basic tools is counting. Yet, strangely enough, it is one of the most difficult. Of course, by counting we do not mean the creation of tables of logarithms or addition tables; rather, we mean the process of precisely accounting for all possibilities in highly complex situations. This can sometimes be done by a brute force case-by-case exhaustion, but such a routine is invariably dull and violates a mathematician's sense of aesthetics. One prefers the light, deft, delicate touch to the hammer blow. But the most serious objection to case-by-case division is that it works far too rarely. Thus in various phases of mathematics we find neat counting devices which tell us exactly how many elements, in some fairly broad context, satisfy certain conditions. A great favorite with mathematicians is the process of counting up a given situation in two different ways; the comparison of the two counts is then used as a means of drawing conclusions. Generally speaking, one introduces an equivalence relation on a finite set, measures the size of the equivalence classes under this relation, and then equates the number of elements in the set to the sum of the orders of these equivalence classes. This kind of an approach will be illustrated in this section. We shall introduce a relation, prove it is an equivalence relation, and then find a neat algebraic description for the size of each equivalence class. From this simple description there will flow a stream of beautiful and powerful results about finite groups.

DEFINITION If $a, b \in G$, then b is said to be a *conjugate* of a in G if there exists an element $c \in G$ such that $b = c^{-1}ac$.

We shall write, for this, $a \sim b$ and shall refer to this relation as *conjugacy*.

LEMMA 2.11.1 *Conjugacy is an equivalence relation on G.*

Proof. As usual, in order to establish this, we must prove that

1. $a \sim a$;
2. $a \sim b$ implies that $b \sim a$;
3. $a \sim b$, $b \sim c$ implies that $a \sim c$

for all a, b, c in G.

We prove each of these in turn.

1. Since $a = e^{-1}ae$, $a \sim a$, with $c = e$ serving as the c in the definition of conjugacy.
2. If $a \sim b$, then $b = x^{-1}ax$ for some $x \in G$, hence, $a = (x^{-1})^{-1}b(x^{-1})$, and since $y = x^{-1} \in G$ and $a = y^{-1}by$, $b \sim a$ follows.
3. Suppose that $a \sim b$ and $b \sim c$ where $a, b, c \in G$. Then $b = x^{-1}ax$, $c = y^{-1}by$ for some $x, y \in G$. Substituting for b in the expression for c we obtain $c = y^{-1}(x^{-1}ax)y = (xy)^{-1}a(xy)$; since $xy \in G$, $a \sim c$ is a consequence.

For $a \in G$ let $C(a) = \{x \in G \mid a \sim x\}$. $C(a)$, the equivalence class of a in G under our relation, is usually called the *conjugate class* of a in G; it consists of the set of all distinct elements of the form $y^{-1}ay$ as y ranges over G.

Our attention now narrows to the case in which G is a finite group. Suppose that $C(a)$ has c_a elements. We seek an alternative description of c_a. Before doing so, note that $o(G) = \sum c_a$ where the sum runs over a set of $a \in G$ using one a from each conjugate class. This remark is, of course, merely a restatement of the fact that our equivalence relation—conjugacy—

induces a decomposition of G into disjoint equivalence classes—the conjugate classes. Of paramount interest now is an evaluation of c_a.

In order to carry this out we recall a concept introduced in Problem 13, Section 2.5. Since this concept is important—far too important to leave to the off-chance that the student solved the particular problem—we go over what may very well be familiar ground to many of the readers.

DEFINITION If $a \in G$, then $N(a)$, the *normalizer of a in G*, is the set $N(a) = \{x \in G \mid xa = ax\}$.

$N(a)$ consists of precisely those elements in G which commute with a.

LEMMA 2.11.2 *$N(a)$ is a subgroup of G.*

Proof. In this result the order of G, whether it be finite or infinite, is of no relevance, and so we put no restrictions on the order of G.

Suppose that $x, y \in N(a)$. Thus $xa = ax$ and $ya = ay$. Therefore, $(xy)a = x(ya) = x(ay) = (xa)y = (ax)y = a(xy)$, in consequence of which $xy \in N(a)$. From $ax = xa$ it follows that $x^{-1}a = x^{-1}(ax)x^{-1} = x^{-1}(xa)x^{-1} = ax^{-1}$, so that x^{-1} is also in $N(a)$. But then $N(a)$ has been demonstrated to be a subgroup of G.

We are now in a position to enunciate our counting principle.

THEOREM 2.11.1 *If G is a finite group, then $c_a = o(G)/o(N(a))$; in other words, the number of elements conjugate to a in G is the index of the normalizer of a in G.*

Proof. To begin with, the conjugate class of a in G, $C(a)$, consists exactly of all the elements $x^{-1}ax$ as x ranges over G. c_a measures the number of distinct $x^{-1}ax$'s. Our method of proof will be to show that two elements in the same right coset of $N(a)$ in G yield the same conjugate of a whereas two elements in different right cosets of $N(a)$ in G give rise to different conjugates of a. In this way we shall have a one-to-one correspondence between conjugates of a and right cosets of $N(a)$.

Suppose that $x, y \in G$ are in the same right coset of $N(a)$ in G. Thus $y = nx$, where $n \in N(a)$, and so $na = an$. Therefore, since $y^{-1} = (nx)^{-1} = x^{-1}n^{-1}$, $y^{-1}ay = x^{-1}n^{-1}anx = x^{-1}n^{-1}nax = x^{-1}ax$, whence x and y result in the same conjugate of a.

If, on the other hand, x and y are in different right cosets of $N(a)$ in G we claim that $x^{-1}ax \neq y^{-1}ay$. Were this not the case, from $x^{-1}ax = y^{-1}ay$ we would deduce that $yx^{-1}a = ayx^{-1}$; this in turn would imply that $yx^{-1} \in N(a)$. However, this declares x and y to be in the same right coset of $N(a)$ in G, contradicting the fact that they are in different cosets. The proof is now complete.

COROLLARY

$$o(G) = \sum \frac{o(G)}{o(N(a))}$$

where this sum runs over one element a in each conjugate class.

Proof. Since $o(G) = \sum c_a$, using the theorem the corollary becomes immediate.

The equation in this corollary is usually referred to as the *class equation* of G.

Before going on to the applications of these results let us examine these concepts in some specific group. There is no point in looking at abelian groups because there two elements are conjugate if and only if they are equal (that is, $c_a = 1$ for every a). So we turn to our familiar friend, the group S_3. Its elements are e, $(1, 2)$, $(1, 3)$, $(2, 3)$, $(1, 2, 3)$, $(1, 3, 2)$. We enumerate the conjugate classes:

$$C(e) = \{e\}$$

$$\begin{aligned} C(1,2) &= \{(1,2), (1,3)^{-1}(1,2)(1,3), (2,3)^{-1}(1,2)(2,3), \\ &\qquad (1,2,3)^{-1}(1,2)(1,2,3), (1,3,2)^{-1}(1,2)(1,3,2)\} \\ &= \{(1,2), (1,3), (2,3)\} \quad \text{(Verify!)} \end{aligned}$$

$$C(1,2,3) = \{(1,2,3), (1,3,2)\} \quad \text{(after another verification)}.$$

The student should verify that $N((1, 2)) = \{e, (1, 2)\}$ and $N((1, 2, 3)) = \{e, (1, 2, 3), (1, 3, 2)\}$, so that $c_{(1,2)} = \frac{6}{2} = 3$, $c_{(1,2,3)} = \frac{6}{3} = 2$.

Applications of Theorem 2.11.1

Theorem 2.11.1 lends itself to immediate and powerful application. We need no artificial constructs to illustrate its use, for the results below which reveal the strength of the theorem are themselves theorems of stature and importance.

Let us recall that the center $Z(G)$ of a group G is the set of all $a \in G$ such that $ax = xa$ for all $x \in G$. Note the

SUBLEMMA *$a \in Z$ if and only if $N(a) = G$. If G is finite, $a \in Z$ if and only if $o(N(a)) = o(G)$.*

Proof. If $a \in Z$, $xa = ax$ for all $x \in G$, whence $N(a) = G$. If, conversely, $N(a) = G$, $xa = ax$ for all $x \in G$, so that $a \in Z$. If G is finite, $o(N(a)) = o(G)$ is equivalent to $N(a) = G$.

APPLICATION 1

THEOREM 2.11.2 *If $o(G) = p^n$ where p is a prime number, then $Z(G) \neq (e)$.*

Proof. If $a \in G$, since $N(a)$ is a subgroup of G, $o(N(a))$, being a divisor of $o(G) = p^n$, must be of the form $o(N(a)) = p^{n_a}$; $a \in Z(G)$ if and only if $n_a = n$. Write out the class equation for this G, letting $z = o(Z(G))$. We get $p^n = o(G) = \sum (p^n/p^{n_a})$; however, since there are exactly z elements such that $n_a = n$, we find that

$$p^n = z + \sum_{n_a < n} \frac{p^n}{p^{n_a}}.$$

Now look at this! p is a divisor of the left-hand side; since $n_a < n$ for each term in the $\sum$ of the right side,

$$p \,\Big|\, \frac{p^n}{p^{n_a}} = p^{n-n_a}$$

so that p is a divisor of each term of this sum, hence a divisor of this sum. Therefore,

$$p \,\Bigg|\, \left(p^n - \sum_{n_a < n} \frac{p^n}{p^{n_a}}\right) = z.$$

Since $e \in Z(G)$, $z \neq 0$; thus z is a positive integer divisible by the prime p. Therefore, $z > 1$! But then there must be an element, besides e, in $Z(G)$! This is the contention of the theorem.

Rephrasing, the theorem states that a group of prime-power order must always have a nontrivial center.

We can now simply prove, as a corollary for this, a result given in an earlier problem.

COROLLARY *If $o(G) = p^2$ where p is a prime number, then G is abelian.*

Proof. Our aim is to show that $Z(G) = G$. At any rate, we already know that $Z(G) \neq (e)$ is a subgroup of G so that $o(Z(G)) = p$ or p^2. If $o(Z(G)) = p^2$, then $Z(G) = G$ and we are done. Suppose that $o(Z(G)) = p$; let $a \in G$, $a \notin Z(G)$. Thus $N(a)$ is a subgroup of G, $Z(G) \subset N(a)$, $a \in N(a)$, so that $o(N(a)) > p$, yet by Lagrange's theorem $o(N(a)) \mid o(G) = p^2$. The only way out is for $o(N(a)) = p^2$, implying that $a \in Z(G)$, a contradiction. Thus $o(Z(G)) = p$ is not an actual possibility.

APPLICATION 2 We now use Theorem 2.11.1 to prove an important theorem due to Cauchy. The reader may remember that this theorem was already proved for abelian groups as an application of the results developed in the section on homomorphisms. In fact, we shall make use of this special case in the proof below. But, to be frank, we shall prove, in the very next section, a much stronger result, due to Sylow, which has Cauchy's theorem as an immediate corollary, in a manner which completely avoids Theorem 2.11.1. To continue our candor, were Cauchy's theorem itself our ultimate and only goal, we could prove it, using the barest essentials of group theory, in a few lines. [The reader should look up the charming, one-paragraph proof of Cauchy's theorem found by McKay and published in the *American Mathematical Monthly*, Vol. 66 (1959), page 119.] Yet, despite all these counter-arguments we present Cauchy's theorem here as a striking illustration of Theorem 2.11.1.

THEOREM 2.11.3 (Cauchy) *If p is a prime number and $p \mid o(G)$, then G has an element of order p.*

Proof. We seek an element $a \neq e \in G$ satisfying $a^p = e$. To prove its existence we proceed by induction on $o(G)$; that is, we assume the theorem to be true for all groups T such that $o(T) < o(G)$. We need not worry about starting the induction for the result is vacuously true for groups of order 1.

If for any subgroup W of G, $W \neq G$, were it to happen that $p \mid o(W)$, then by our induction hypothesis there would exist an element of order p in W, and thus there would be such an element in G. Thus we may assume that p is not a divisor of the order of any proper subgroup of G. In particular, if $a \notin Z(G)$, since $N(a) \neq G$, $p \nmid o(N(a))$. Let us write down the class equation:

$$o(G) = o(Z(G)) + \sum_{N(a) \neq G} \frac{o(G)}{o(N(a))}.$$

Since $p \mid o(G)$, $p \nmid o(N(a))$ we have that

$$p \,\Big|\, \frac{o(G)}{o(N(a))},$$

and so

$$p \,\Big|\, \sum_{N(a) \neq G} \frac{o(G)}{o(N(a))};$$

since we also have that $p \mid o(G)$, we conclude that

$$p \,\Big|\, \left(o(G) - \sum_{N(a) \neq G} \frac{o(G)}{o(N(a))} \right) = o(Z(G)).$$

$Z(G)$ is thus a subgroup of G whose order is divisible by p. But, after all, we have assumed that p is not a divisor of the order of any proper subgroup of G, so that $Z(G)$ cannot be a proper subgroup of G. We are forced to accept the only possibility left us, namely, that $Z(G) = G$. But then G is abelian; now we invoke the result already established for abelian groups to complete the induction. This proves the theorem.

We conclude this section with a consideration of the conjugacy relation in a specific class of groups, namely, the symmetric groups S_n.

Given the integer n we say the sequence of positive integers $n_1, n_2, \ldots, n_r$, $n_1 \leq n_2 \leq \cdots \leq n_r$ constitute a *partition* of n if $n = n_1 + n_2 + \cdots + n_r$. Let $p(n)$ denote the number of partitions of n. Let us determine $p(n)$ for small values of n:

$$p(1) = 1 \text{ since } 1 = 1 \text{ is the only partition of } 1,$$

$$p(2) = 2 \text{ since } 2 = 2 \text{ and } 2 = 1 + 1,$$

$$p(3) = 3 \text{ since } 3 = 3,\ 3 = 1 + 2,\ 3 = 1 + 1 + 1,$$

$$\begin{aligned} p(4) = 5 \text{ since } 4 &= 4,\ 4 = 1 + 3,\ 4 = 1 + 1 + 2, \\ 4 &= 1 + 1 + 1 + 1,\ 4 = 2 + 2. \end{aligned}$$

Some others are $p(5) = 7$, $p(6) = 11$, $p(61) = 1{,}121{,}505$. There is a large mathematical literature on $p(n)$.

Every time we break a given permutation in S_n into a product of disjoint cycles we obtain a partition of n; for if the cycles appearing have lengths $n_1, n_2, \ldots, n_r$, respectively, $n_1 \leq n_2 \leq \cdots \leq n_r$, then $n = n_1 + n_2 + \cdots + n_r$. We shall say a permutation $\sigma \in S_n$ has the cycle decomposition $\{n_1, n_2, \ldots, n_r\}$ if it can be written as the product of disjoint cycles of lengths $n_1, n_2, \ldots, n_r$, $n_1 \leq n_2 \leq \cdots \leq n_r$. Thus in S_9

$$\sigma = \begin{pmatrix} 1 & 2 & 3 & 4 & 5 & 6 & 7 & 8 & 9 \\ 1 & 3 & 2 & 5 & 6 & 4 & 7 & 9 & 8 \end{pmatrix} = (1)(2.\ 3)(4, 5, 6)(7)(8, 9)$$

has cycle decomposition $\{1, 1, 2, 2, 3\}$; note that $1 + 1 + 2 + 2 + 3 = 9$. We now aim to prove that two permutations in S_n are conjugate if and only if they have the same cycle decomposition. Once this is proved, then S_n will have exactly $p(n)$ conjugate classes.

To reach our goal we exhibit a very simple rule for computing the conjugate of a given permutation. Suppose that $\sigma \in S_n$ and that σ sends $i \to j$. How do we find $\theta^{-1}\sigma\theta$ where $\theta \in S_n$? Suppose that θ sends $i \to s$ and $j \to t$; then $\theta^{-1}\sigma\theta$ sends $s \to t$. *In other words, to compute* $\theta^{-1}\sigma\theta$ *replace every symbol in* σ *by its image under* θ. For example, to determine $\theta^{-1}\sigma\theta$ where $\theta = (1, 2, 3)(4, 7)$ and $\sigma = (5, 6, 7)(3, 4, 2)$, then, since $\theta{:}5 \to 5$, $6 \to 6$, $7 \to 4$, $3 \to 1$, $4 \to 7$, $2 \to 3$, $\theta^{-1}\sigma\theta$ is obtained from σ by replacing in σ, 5 by 5, 6 by 6, 7 by 4, 3 by 1, 4 by 7, and 2 by 3, so that $\theta^{-1}\sigma\theta = (5, 6, 4)(1, 7, 3)$.

With this algorithm for computing conjugates it becomes clear that two permutations having the same cycle decomposition are conjugate. For if

$\sigma = (a_1, a_2, \ldots, a_{n_1})(b_1, b_2, \ldots, b_{n_2}) \cdots (x_1, x_2, \ldots, x_{n_r})$ and $\tau = (\alpha_1, \alpha_2, \ldots, \alpha_{n_1})(\beta_1, \beta_2, \ldots, \beta_{n_2}) \cdots (\chi_1, \chi_2, \ldots, \chi_{n_r})$, then $\tau = \theta^{-1}\sigma\theta$, where one could use as θ the permutation

$$\begin{pmatrix} a_1 & a_2 & \cdots & a_{n_1} & b_1 & \cdots & b_{n_2} & \cdots & x_1 & \cdots & x_{n_r} \\ \alpha_1 & \alpha_2 & \cdots & \alpha_{n_1} & \beta_1 & \cdots & \beta_{n_2} & \cdots & \chi_1 & \cdots & \chi_{n_r} \end{pmatrix}.$$

Thus, for instance, $(1, 2)(3, 4, 5)(6, 7, 8)$ and $(7, 5)(1, 3, 6)(2, 4, 8)$ can be exhibited as conjugates by using the conjugating permutation

$$\begin{pmatrix} 1 & 2 & 3 & 4 & 5 & 6 & 7 & 8 \\ 7 & 5 & 1 & 3 & 6 & 2 & 4 & 8 \end{pmatrix}.$$

That two conjugates have the same cycle decomposition is now trivial for, by our rule, to compute a conjugate, replace every element in a given cycle by its image under the conjugating permutation.

We restate the result proved in the previous discussion as

LEMMA 2.11.3 *The number of conjugate classes in S_n is $p(n)$, the number of partitions of n.*

Since we have such an explicit description of the conjugate classes in S_n we can find all the elements commuting with a given permutation. We illustrate this with a very special and simple case.

Given the permutation $(1, 2)$ in S_n, what elements commute with it? Certainly any permutation leaving both 1 and 2 fixed does. There are $(n - 2)!$ such. Also $(1, 2)$ commutes with itself. This way we get $2(n - 2)!$ elements in the group generated by $(1, 2)$ and the $(n - 2)!$ permutations leaving 1 and 2 fixed. Are there others? There are $n(n - 1)/2$ transpositions and these are precisely all the conjugates of $(1, 2)$. Thus the conjugate class of $(1, 2)$ has in it $n(n - 1)/2$ elements. If the order of the normalizer of $(1, 2)$ is r, then, by our counting principle,

$$\frac{n(n-1)}{2} = \frac{o(S_n)}{r} = \frac{n!}{r}.$$

Thus $r = 2(n - 2)!$. That is, the order of the normalizer of $(1, 2)$ is $2(n - 2)!$. But we exhibited $2(n - 2)!$ elements which commute with $(1, 2)$; thus the general element σ commuting with $(1, 2)$ is $\sigma = (1, 2)^i\tau$, where $i = 0$ or 1, τ is a permutation leaving both 1 and 2 fixed.

As another application consider the permutation $(1, 2, 3, \ldots, n) \in S_n$. We claim this element commutes only with its powers. Certainly it does commute with all its powers, and this gives rise to n elements. Now, any n-cycle is conjugate to $(1, 2, \ldots, n)$ and there are $(n - 1)!$ distinct n-cycles in S_n. Thus if u denotes the order of the normalizer of $(1, 2, \ldots, n)$

in S_n, since $o(S_n)/u$ = number of conjugates of $(1, 2, \ldots, n)$ in $S_n = (n-1)!$,

$$u = \frac{n!}{(n-1)!} = n.$$

So the order of the normalizer of $(1, 2, \ldots, n)$ in S_n is n. The powers of $(1, 2, \ldots, n)$ having given us n such elements, there is no room left for others and we have proved our contention.

Problems

1. List all the conjugate classes in S_3, find the c_a's, and verify the class equation.
2. List all the conjugate classes in S_4, find the c_a's and verify the class equation.
3. List all the conjugate classes in the group of quaternion units (see Problem 21, Section 2.10), find the c_a's and verify the class equation.
4. List all the conjugate classes in the dihedral group of order $2n$, find the c_a's and verify the class equation. Notice how the answer depends on the parity of n.
5. (a) In S_n prove that there are $\frac{1}{r}\frac{n!}{(n-r)!}$ distinct r cycles.
   (b) Using this, find the number of conjugates that the r-cycle $(1, 2, \ldots, r)$ has in S_n.
   (c) Prove that any element σ in S_n which commutes with $(1, 2, \ldots, r)$ is of the form $\sigma = (1, 2, \ldots, r)^i\tau$, where $i = 0, 1, 2, \ldots, r$, τ is a permutation leaving all of $1, 2, \ldots, r$ fixed.
6. (a) Find the number of conjugates of $(1, 2)(3, 4)$ in S_n, $n \geq 4$.
   (b) Find the form of all elements commuting with $(1, 2)(3, 4)$ in S_n.
7. If p is a prime number, show that in S_p there are $(p-1)! + 1$ elements x satisfying $x^p = e$.
8. If in a finite group G an element a has exactly two conjugates, prove that G has a normal subgroup $N \neq (e), G$.
9. (a) Find two elements in A_5, the alternating group of degree 5, which are conjugate in S_5 but not in A_5.
   (b) Find all the conjugate classes in A_5 and the number of elements in each conjugate class.
10. (a) If N is a normal subgroup of G and $a \in N$, show that every conjugate of a in G is also in N.
    (b) Prove that $o(N) = \sum c_a$ for some choices of a in N.

(c) Using this and the result for Problem 9(b), prove that in A_5 there is no normal subgroup N other than (e) and A_5.

11. Using Theorem 2.11.2 as a tool, prove that if $o(G) = p^n$, p a prime number, then G has a subgroup of order p^α for all $0 \leq \alpha \leq n$.
12. If $o(G) = p^n$, p a prime number, prove that there exist subgroups N_i, $i = 0, 1, \ldots, r$ (for some r) such that $G = N_0 \supset N_1 \supset N_2 \supset \cdots \supset N_r = (e)$ where N_i is a normal subgroup of N_{i-1} and where N_{i-1}/N_i is abelian.
13. If $o(G) = p^n$, p a prime number, and $H \neq G$ is a subgroup of G, show that there exists an $x \in G$, $x \notin H$ such that $x^{-1}Hx = H$.
14. Prove that any subgroup of order p^{n-1} in a group G of order p^n, p a prime number, is normal in G.

*15. If $o(G) = p^n$, p a prime number, and if $N \neq (e)$ is a normal subgroup of G, prove that $N \cap Z \neq (e)$, where Z is the center of G.

16. If G is a group, Z its center, and if G/Z is cyclic, prove that G must be abelian.
17. Prove that any group of order 15 is cyclic.
18. Prove that a group of order 28 has a normal subgroup of order 7.
19. Prove that if a group G of order 28 has a normal subgroup of order 4, then G is abelian.

2.12 Sylow's Theorem

Lagrange's theorem tells us that the order of a subgroup of a finite group is a divisor of the order of that group. The converse, however, is false. There are very few theorems which assert the existence of subgroups of prescribed order in arbitrary finite groups. The most basic, and widely used, is a classic theorem due to the Norwegian mathematician Sylow.

We present here three proofs of this result of Sylow. The first is a very elegant and elementary argument due to Wielandt. It appeared in the journal *Archiv der Matematik*, Vol. 10 (1959), pages 401–402. The basic elements in Wielandt's proof are number-theoretic and combinatorial. It has the advantage, aside from its elegance and simplicity, of producing the subgroup we are seeking. The second proof is based on an exploitation of induction in an interplay with the class equation. It is one of the standard classical proofs, and is a nice illustration of combining many of the ideals developed so far in the text to derive this very important cornerstone due to Sylow. The third proof is of a completely different philosophy. The basic idea there is to show that if a larger group than the one we are considering satisfies the conclusion of Sylow's theorem, then our group also must.

This forces us to prove Sylow's theorem for a special family of groups—the symmetric groups. By invoking Cayley's theorem (Theorem 2.9.1) we are then able to deduce Sylow's theorem for all finite groups. Apart from this strange approach—to prove something for a given group, first prove it for a much larger one—this third proof has its own advantages. Exploiting the ideas used, we easily derive the so-called second and third parts of Sylow's theorem.

One might wonder: why give three proofs of the same result when, clearly, one suffices? The answer is simple. Sylow's theorem is *that* important that it merits this multifront approach. Add to this the completely diverse nature of the three proofs and the nice application each gives of different things that we have learned, the justification for the whole affair becomes persuasive (at least to the author). Be that as it may, we state Sylow's theorem and get on with Wielandt's proof.

THEOREM 2.12.1 (Sylow) *If p is a prime number and $p^\alpha \mid o(G)$, then G has a subgroup of order p^α.*

Before entering the first proof of the theorem we digress slightly to a brief number-theoretic and combinatorial discussion.

The number of ways of picking a subset of k elements from a set of n elements can easily be shown to be

$$\binom{n}{k} = \frac{n!}{k!(n-k)!}.$$

If $n = p^\alpha m$ where p is a prime number, and if $p^r \mid m$ but $p^{r+1} \nmid m$, consider

$$\binom{p^\alpha m}{p^\alpha} = \frac{(p^\alpha m)!}{(p^\alpha)!(p^\alpha m - p^\alpha)!}$$

$$= \frac{p^\alpha m(p^\alpha m - 1)\cdots(p^\alpha m - i)\cdots(p^\alpha m - p^\alpha + 1)}{p^\alpha(p^\alpha - 1)\cdots(p^\alpha - i)\cdots(p^\alpha - p^\alpha + 1)}.$$

The question is, What power of p divides $\binom{p^\alpha m}{p^\alpha}$? Looking at this number, written out as we have written it out, one can see that except for the term m in the numerator, the power of p dividing $(p^\alpha m - i)$ is the same as that dividing $p^\alpha - i$, so all powers of p cancel out except the power which divides m. Thus

$$p^r \mid \binom{p^\alpha m}{p^\alpha} \quad \text{but} \quad p^{r+1} \nmid \binom{p^\alpha m}{p^\alpha}.$$

First Proof of the Theorem. Let $\mathcal{M}$ be the set of all subsets of G which have p^α elements. Thus $\mathcal{M}$ has $\binom{p^\alpha m}{p^\alpha}$ elements. Given $M_1, M_2 \in \mathcal{M}$ (M is a subset of G having p^α elements, and likewise so is M_2) define $M_1 \sim M_2$ if there exists an element $g \in G$ such that $M_1 = M_2 g$. It is immediate to verify that this defines an equivalence relation on $\mathcal{M}$. We claim that there is at least one equivalence class of elements in $\mathcal{M}$ such that the number of elements in this class is not a multiple of p^{r+1}, for if p^{r+1} is a divisor of the size of each equivalence class, then p^{r+1} would be a divisor of the number of elements in $\mathcal{M}$. Since $\mathcal{M}$ has $\binom{p^\alpha m}{p^\alpha}$ elements and $p^{r+1} \nmid \binom{p^\alpha m}{p^\alpha}$, this cannot be the case. Let $\{M_1, \ldots, M_n\}$ be such an equivalence class in $\mathcal{M}$ where $p^{r+1} \nmid n$. By our very definition of equivalence in $\mathcal{M}$, if $g \in G$, for each $i = 1, \ldots, n$, $M_i g = M_j$ for some j, $1 \leq j \leq n$. We let $H = \{g \in G \mid M_1 g = M_1\}$. Clearly H is a subgroup of G, for if $a, b \in H$, then $M_1 a = M_1$, $M_1 b = M_1$ whence $M_1 ab = (M_1 a)b = M_1 b = M_1$. We shall be vitally concerned with $o(H)$. We claim that $no(H) = o(G)$; we leave the proof to the reader, but suggest the argument used in the counting principle in Section 2.11. Now $no(H) = o(G) = p^\alpha m$; since $p^{r+1} \nmid n$ and $p^{\alpha+r} \mid p^\alpha m = no(H)$, it must follow that $p^\alpha \mid o(H)$, and so $o(H) \geq p^\alpha$. However, if $m_1 \in M_1$, then for all $h \in H$, $m_1 h \in M_1$. Thus M_1 has at least $o(H)$ distinct elements. However, M_1 was a subset of G containing p^α elements. Thus $p^\alpha \geq o(H)$. Combined with $o(H) \geq p^\alpha$ we have that $o(H) = p^\alpha$. *But then we have exhibited a subgroup of G having exactly p^α elements, namely H.* This proves the theorem; it actually has done more—it has constructed the required subgroup before our very eyes!

What is usually known as Sylow's theorem is a special case of Theorem 2.12.1, namely that

COROLLARY *If $p^m \mid o(G)$, $p^{m+1} \nmid o(G)$, then G has a subgroup of order p^m.*

A subgroup of G of order p^m, where $p^m \mid o(G)$ but $p^{m+1} \nmid o(G)$, is called a *p-Sylow subgroup of G.* The corollary above asserts that a finite group has p-Sylow subgroups for every prime p dividing its order. Of course the conjugate of a p-Sylow subgroup is a p-Sylow subgroup. In a short while we shall see how any two p-Sylow subgroups of G—for the same prime p—are related. We shall also get some information on how many p-Sylow subgroups there are in G for a given prime p. Before passing to this, we want to give two other proofs of Sylow's theorem.

We begin with a remark. As we observed just prior to the corollary, the corollary is a special case of the theorem. However, we claim that the theorem is easily derivable from the corollary. That is, if we know that G possesses a subgroup of order p^m, where $p^m \mid o(G)$ but $p^{m+1} \nmid o(G)$, then we know that G has a subgroup of order p^α for any α such that $p^\alpha \mid o(G)$. This follows from the result of Problem 11, Section 2.11. This result states that any group of order p^m, p a prime, has subgroups of order p^α for any $0 \leq \alpha \leq m$. Thus to prove Theorem 2.12.1—as we shall proceed to do, again, in two more ways—it is enough for us to prove the existence of p-Sylow subgroups of G, for every prime p dividing the order of G.

Second Proof of Sylow's Theorem. We prove, by induction on the order of the group G, that for every prime p dividing the order of G, G has a p-Sylow subgroup.

If the order of the group is 2, the only relevant prime is 2 and the group certainly has a subgroup of order 2, namely itself.

So we suppose the result to be correct for all groups of order less than $o(G)$. From this we want to show that the result is valid for G. Suppose, then, that $p^m \mid o(G)$, $p^{m+1} \nmid o(G)$, where p is a prime, $m \geq 1$. If $p^m \mid o(H)$ for any subgroup H of G, where $H \neq G$, then by the induction hypothesis, H would have a subgroup T of order p^m. However, since T is a subgroup of H, and H is a subgroup of G, T too is a subgroup of G. But then T would be the sought-after subgroup of order p^m.

We therefore may assume that $p^m \nmid o(H)$ for *any* subgroup H of G, where $H \neq G$. We restrict our attention to a limited set of such subgroups. Recall that if $a \in G$ then $N(a) = \{x \in G \mid xa = ax\}$ is a subgroup of G; moreover, if $a \notin Z$, the center of G, then $N(a) \neq G$. Recall, too, that the class equation of G states that

$$o(G) = \sum \frac{o(G)}{o(N(a))},$$

where this sum runs over one element a from each conjugate class. We separate this sum into two pieces: those a which lie in Z, and those which don't. This gives

$$o(G) = z + \sum_{a \notin Z} \frac{o(G)}{o(N(a))},$$

where $z = o(Z)$. Now invoke the reduction we have made, namely, that $p^m \nmid o(H)$ for any subgroup $H \neq G$ of G, to those subgroups $N(a)$ for $a \notin Z$. Since in this case, $p^m \mid o(G)$ and $p^m \nmid o(N(a))$, we must have that

$$p \,\Big|\, \frac{o(G)}{o(N(a))}.$$

Restating this result,

$$p \,\Big|\, \frac{o(G)}{o(N(a))}$$

for every $a \in G$ where $a \notin Z$. Look at the class equation with this information in hand. Since $p^m \mid o(G)$, we have that $p \mid o(G)$; also

$$p \;\Big|\; \sum_{a \notin Z} \frac{o(G)}{o(N(a))}.$$

Thus the class equation gives us that $p \mid z$. Since $p \mid z = o(Z)$, by Cauchy's theorem (Theorem 2.11.3), Z has an element $b \neq e$ of order p. Let $B = (b)$, the subgroup of G generated by b. B is of order p; moreover, *since $b \in Z$, B must be normal* in G. Hence we can form the quotient group $\bar{G} = G/B$. We look at $\bar{G}$. First of all, its order is $o(G)/o(B) = o(G)/p$, hence is certainly less than $o(G)$. Secondly, we have $p^{m-1} \mid o(\bar{G})$, but $p^m \nmid o(\bar{G})$. Thus, by the induction hypothesis, $\bar{G}$ has a subgroup $\bar{P}$ of order p^{m-1}. Let $P = \{x \in G \mid xB \in \bar{P}\}$; by Lemma 2.7.5, P is a subgroup of G. Moreover, $\bar{P} \approx P/B$ (Prove!); thus

$$p^{m-1} = o(\bar{P}) = \frac{o(P)}{o(B)} = \frac{o(P)}{p}.$$

This results in $o(P) = p^m$. Therefore P is the required p-Sylow subgroup of G. This completes the induction and so proves the theorem.

With this we have finished the second proof of Sylow's theorem. Note that this second proof can easily be adapted to prove that if $p^\alpha \mid o(G)$, then G has a subgroup of order p^α directly, without first passing to the existence of a p-Sylow subgroup. (This is Problem 1 of the problems at the end of this section.)

We now proceed to the third proof of Sylow's theorem.

Third Proof of Sylow's Theorem. Before going into the details of the proof proper, we outline its basic strategy. We will first show that the symmetric groups S_{p^r}, p a prime, all have p-Sylow subgroups. The next step will be to show that if G is contained in M and M has a p-Sylow subgroup, then G has a p-Sylow subgroup. Finally we will show, via Cayley's theorem, that we can use S_{p^k}, for large enough k, as our M. With this we will have all the pieces, and the theorem will drop out.

In carrying out this program in detail, we will have to know how large a p-Sylow subgroup of S_{p^r} should be. This will necessitate knowing what power of p divides $(p^r)!$. This will be easy. To produce the p-Sylow subgroup of S_{p^r} will be harder. To carry out another vital step in this rough sketch, it will be necessary to introduce a new equivalence relation in groups, and the corresponding equivalence classes known as *double cosets*. This will have several payoffs, not only in pushing through the proof of Sylow's theorem, but also in getting us the second and third parts of the full Sylow theorem.

So we get down to our first task, that of finding what power of a prime p exactly divides $(p^k)!$. Actually, it is quite easy to do this for $n!$ for any integer n (see Problem 2). But, for our purposes, it will be clearer and will suffice to do it only for $(p^k)!$.

Let $n(k)$ be defined by $p^{n(k)} \mid (p^k)!$ but $p^{n(k)+1} \nmid (p^k)!$.

LEMMA 2.12.1 $n(k) = 1 + p + \cdots + p^{k-1}$.

Proof. If $k = 1$ then, since $p! = 1 \cdot 2 \cdots (p-1) \cdot p$, it is clear that $p \mid p!$ but $p^2 \nmid p!$. Hence $n(1) = 1$, as it should be.

What terms in the expansion of $(p^k)!$ can contribute to powers of p dividing $(p^k)!$? Clearly, only the multiples of p; that is, $p, 2p, \ldots, p^{k-1}p$. In other words $n(k)$ must be the power of p which divides $p(2p)(3p)\cdots(p^{k-1}p) = p^{p^{k-1}}(p^{k-1})!$. But then $n(k) = p^{k-1} + n(k-1)$. Similarly, $n(k-1) = n(k-2) + p^{k-2}$, and so on. Write these out as

$$\begin{aligned} n(k) - n(k-1) &= p^{k-1}, \\ n(k-1) - n(k-2) &= p^{k-2}, \\ &\vdots \\ n(2) - n(1) &= p, \\ n(1) &= 1. \end{aligned}$$

Adding these up, with the cross-cancellation that we get, we obtain $n(k) = 1 + p + p^2 + \cdots + p^{k-1}$. This is what was claimed in the lemma, so we are done.

We are now ready to show that S_{p^k} has a p-Sylow subgroup; that is, we shall show (in fact, produce) a subgroup of order $p^{n(k)}$ in S_{p^k}.

LEMMA 2.12.2 *S_{p^k} has a p-Sylow subgroup.*

Proof. We go by induction on k. If $k = 1$, then the element $(1\ 2\ \ldots\ p)$, in S_p is of order p, so generated a subgroup of order p. Since $n(1) = 1$, the result certainly checks out for $k = 1$.

Suppose that the result is correct for $k - 1$; we want to show that it then must follow for k. Divide the integers $1, 2, \ldots, p^k$ into p clumps, each with p^{k-1} elements as follows:

$$\{1, 2, \ldots, p^{k-1}\}, \{p^{k-1}+1, p^{k-1}+2, \ldots, 2p^{k-1}\}, \ldots, \{(p-1)p^{k-1}+1, \ldots, p^k\}.$$

The permutation σ defined by $\sigma = (1, p^{k-1}+1, 2p^{k-1}+1, \ldots, (p-1)p^{k-1}+1) \cdots (j, p^{k-1}+j, 2p^{k-1}+j, \ldots, (p-1)p^{k-1}+1+j) \cdots (p^{k-1}, 2p^{k-1}, \ldots, (p-1)p^{k-1}, p^k)$ has the following properties:

1. $\sigma^p = e$.

2. If τ is a permutation that leaves all i fixed for $i > p^{k-1}$ (hence, affects only $1, 2, \ldots, p^{k-1}$), then $\sigma^{-1}\tau\sigma$ moves only elements in $\{p^{k-1} + 1, p^{k-1} + 2, \ldots, 2p^{k-1}\}$, and more generally, $\sigma^{-j}\tau\sigma^{j}$ moves only elements in $\{jp^{k-1} + 1, jp^{k-1} + 2, \ldots, (j + 1)p^{k-1}\}$.

Consider $A = \{\tau \in S_{p^k} \mid \tau(i) = i \text{ if } i > p^{k-1}\}$. A is a subgroup of S_{p^k} and elements in A can carry out any permutation on $1, 2, \ldots, p^{k-1}$. From this it follows easily that $A \approx S_{p^{k-1}}$. By induction, A has a subgroup P_1 of order $p^{n(k-1)}$.

Let $T = P_1(\sigma^{-1}P_1\sigma)(\sigma^{-2}P_1\sigma^2)\cdots(\sigma^{-(p-1)}P_1\sigma^{p-1}) = P_1P_2\cdots P_{n-1}$, where $P_i = \sigma^{-i}P_1\sigma^i$. Each P_i is isomorphic to P_1 so has order $p^{n(k-1)}$. Also elements in distinct P_i's influence nonoverlapping sets of integers, *hence commute*. Thus T is a subgroup of S_{p^k}. What is its order? Since $P_i \cap P_j = (e)$ if $0 \leq i \neq j \leq p - 1$, we see that $o(T) = o(P_1)^p = p^{pn(k-1)}$. We are not quite there yet. T is not the p-Sylow subgroup we seek!

Since $\sigma^p = e$ and $\sigma^{-i}P_1\sigma^i = P_i$ we have $\sigma^{-1}T\sigma = T$. Let $P = \{\sigma^j t \mid t \in T, 0 \leq j \leq p - 1\}$. Since $\sigma \notin T$ and $\sigma^{-1}T\sigma = T$ we have two things: firstly, T is a subgroup of S_{p^k} and, furthermore, $o(P) = p \cdot o(T) = p \cdot p^{n(k-1)p} = p^{n(k-1)p+1}$. Now we are finally there! P is the sought-after p-Sylow subgroup of S_{p^k}.

Why? Well, what is its order? It is $p^{n(k-1)p+1}$. But $n(k - 1) = 1 + p + \cdots + p^{k-2}$, hence $pn(k - 1) + 1 = 1 + p + \cdots + p^{k-1} = n(k)$. Since now $o(P) = p^{n(k)}$, P is indeed a p-Sylow subgroup of S_{p^k}.

Note something about the proof. Not only does it prove the lemma, it actually allows us to construct the p-Sylow subgroup inductively. We follow the procedure of the proof to construct a 2-Sylow subgroup in S_4.

Divide 1, 2, 3, 4 into $\{1, 2\}$ and $\{3, 4\}$. Let $P_1 = ((1\ 2))$ and $\sigma = (1\ 3)(2\ 4)$. Then $P_2 = \sigma^{-1}P_1\sigma = (3\ 4)$. Our 2-Sylow subgroup is then the group generated by $(1\ 3)(2\ 4)$ and

$$T = P_1P_2 = \{(1\ 2), (3\ 4), (1\ 2)(3\ 4), e\}.$$

In order to carry out the program of the third proof that we outlined, we now introduce a new equivalence relation in groups (see Problem 39, Section 2.5).

DEFINITION Let G be a group, A, B subgroups of G. If $x, y \in G$ define $x \sim y$ if $y = axb$ for some $a \in A$, $b \in B$.

We leave to the reader the verification—it is easy—of

LEMMA 2.12.3 *The relation defined above is an equivalence relation on G. The equivalence class of $x \in G$ is the set $AxB = \{axb \mid a \in A, b \in B\}$.*

We call the set AxB a *double coset* of A, B in G.

If A, B are finite subgroups of G, how many elements are there in the double coset AxB? To begin with, the mapping $T:AxB \to AxBx^{-1}$ given by $(axb)T = axbx^{-1}$ is one-to-one and onto (verify). Thus $o(AxB) = o(AxBx^{-1})$. Since xBx^{-1} is a subgroup of G, of order $o(B)$, by Theorem 2.5.1,

$$o(AxB) = o(AxBx^{-1}) = \frac{o(A)o(xBx^{-1})}{o(A \cap xBx^{-1})} = \frac{o(A)o(B)}{o(A \cap xBx^{-1})}.$$

We summarize this in

LEMMA 2.12.4 *If A, B are finite subgroups of G then*

$$o(AxB) = \frac{o(A)o(B)}{o(A \cap xBx^{-1})}.$$

We now come to the gut step in this third proof of Sylow's theorem.

LEMMA 2.12.5 *Let G be a finite group and suppose that G is a subgroup of the finite group M. Suppose further that M has a p-Sylow subgroup Q. Then G has a p-Sylow subgroup P. In fact, $P = G \cap xQx^{-1}$ for some $x \in M$.*

Proof. Before starting the details of the proof, we translate the hypotheses somewhat. Suppose that $p^m \mid o(M)$, $p^{m+1} \nmid o(M)$, Q is a subgroup of M of order p^m. Let $o(G) = p^n t$ where $p \nmid t$. We want to produce a subgroup P in G of order p^n.

Consider the double coset decomposition of M given by G and Q; $M = \bigcup GxQ$. By Lemma 2.12.4,

$$o(GxQ) = \frac{o(G)o(Q)}{o(G \cap xQx^{-1})} = \frac{p^n t p^m}{o(G \cap xQx^{-1})}.$$

Since $G \cap xQx^{-1}$ is a subgroup of xQx^{-1}, its order is p^{m_x}. We claim that $m_x = n$ for some $x \in M$. If not, then

$$o(GxQ) = \frac{p^n t p^m}{p^{m_x}} = tp^{m+n-m_x},$$

so is divisible by p^{m+1}. Now, since $M = \bigcup GxQ$, and this is disjoint union, $o(M) = \sum o(GxQ)$, the sum running over one element from each double coset. But $p^{m+1} \mid o(GxQ)$; hence $p^{m+1} \mid o(M)$. This contradicts $p^{m+1} \nmid o(M)$. Thus $m_x = n$ for some $x \in M$. But then $o(G \cap xQx^{-1}) = p^n$. Since $G \cap xQx^{-1} = P$ is a subgroup of G and has order p^n, the lemma is proved.

We now can easily prove Sylow's theorem. By Cayley's theorem (Theorem 2.9.1) we can isomorphically embed our finite group G in S_n, the symmetric group of degree n. Pick k so that $n < p^k$; then we can isomorphically embed S_n in S_{p^k} (by acting on $1, 2, \ldots, n$ only in the set

$1, 2, \ldots, n, \ldots, p^k$), hence G is isomorphically embedded in S_{p^k}. By Lemma 2.12.2, S_{p^k} has a p-Sylow subgroup. Hence, by Lemma 2.12.5, G must have a p-Sylow subgroup. This finishes the third proof of Sylow's theorem.

This third proof has given us quite a bit more. From it we have the machinery to get the other parts of Sylow's theorem.

THEOREM 2.12.2 (SECOND PART OF SYLOW'S THEOREM) *If G is a finite group, p a prime and $p^n \mid o(G)$ but $p^{n+1} \nmid o(G)$, then any two subgroups of G of order p^n are conjugate.*

Proof. Let A, B be subgroups of G, each of order p^n. We want to show that $A = gBg^{-1}$ for some $g \in G$.

Decompose G into double cosets of A and B; $G = \bigcup AxB$. Now, by Lemma 2.12.4,

$$o(AxB) = \frac{o(A)o(B)}{o(A \cap xBx^{-1})}.$$

If $A \neq xBx^{-1}$ for every $x \in G$ then $o(A \cap xBx^{-1}) = p^m$ where $m < n$. Thus

$$o(AxB) = \frac{o(A)o(B)}{p^m} = \frac{p^{2n}}{p^m} = p^{2n-m}$$

and $2n - m \geq n + 1$. Since $p^{n+1} \mid o(AxB)$ for every x and since $o(G) = \sum o(AxB)$, we would get the contradiction $p^{n+1} \mid o(G)$. Thus $A = gBg^{-1}$ for some $g \in G$. This is the assertion of the theorem.

Knowing that for a given prime p all p-Sylow subgroups of G are conjugate allows us to count up precisely how many such p-Sylow subgroups there are in G. The argument is exactly as that given in proving Theorem 2.11.1. In some earlier problems (see, in particular, Problem 16, Section 2.5) we discussed the normalizer $N(H)$, of a subgroup, defined by $N(H) = \{x \in G \mid xHx^{-1} = H\}$. Then, as in the proof of Theorem 2.11.1, we have that *the number of distinct conjugates, xHx^{-1}, of H in G is the index of $N(H)$ in G.* Since all p-Sylow subgroups are conjugate we have

LEMMA 2.12.6 *The number of p-Sylow subgroups in G equals $o(G)/o(N(P))$, where P is any p-Sylow subgroup of G. In particular, this number is a divisor of $o(G)$.*

However, much more can be said about the number of p-Sylow subgroups there are, for a given prime p, in G. We go into this now. The technique will involve double cosets again.

THEOREM 2.12.3 (Third Part of Sylow's Theorem) *The number of p-Sylow subgroups in G, for a given prime, is of the form $1 + kp$.*

Proof. Let P be a p-Sylow subgroup of G. We decompose G into double cosets of P and P. Thus $G = \bigcup PxP$. We now ask: How many elements are there in PxP? By Lemma 2.12.4 we know the answer:

$$o(PxP) = \frac{o(P)^2}{o(P \cap xPx^{-1})}.$$

Thus, if $P \cap xPx^{-1} \neq P$ then $p^{n+1} \mid o(PxP)$, where $p^n = o(P)$. Paraphrasing this: if $x \notin N(P)$ then $p^{n+1} \mid o(PxP)$. Also, if $x \in N(P)$, then $PxP = P(Px) = P^2x = Px$, so $o(PxP) = p^n$ in this case.

Now

$$o(G) = \sum_{x \in N(P)} o(PxP) + \sum_{x \notin N(P)} o(PxP),$$

where each sum runs over one element from each double coset. However, if $x \in N(P)$, since $PxP = Px$, the first sum is merely $\sum_{x \in N(P)} o(Px)$ over the *distinct cosets* of P in $N(P)$. Thus this first sum is just $o(N(P))$. What about the second sum? We saw that each of its constituent terms is divisible by p^{n+1}, hence

$$p^{n+1} \,\Big|\, \sum_{x \notin N(P)} o(PxP).$$

We can thus write this second sum as

$$\sum_{x \notin N(P)} o(PxP) = p^{n+1}u.$$

Therefore $o(G) = o(N(P)) + p^{n+1}u$, so

$$\frac{o(G)}{o(N(P))} = 1 + \frac{p^{n+1}u}{o(N(P))}.$$

Now $o(N(P)) \mid o(G)$ since $N(P)$ is a subgroup of G, hence $p^{n+1}u/o(N(P))$ is an integer. Also, since $p^{n+1} \nmid o(G)$, p^{n+1} can't divide $o(N(P))$. But then $p^{n+1}u/o(N(P))$ *must be divisible by p*, so we can write $p^{n+1}u/o(N(P))$ as kp, where k is an integer. Feeding this information back into our equation above, we have

$$\frac{o(G)}{o(N(P))} = 1 + kp.$$

Recalling that $o(G)/o(N(P))$ is the number of p-Sylow subgroups in G, we have the theorem.

In Problems 20–24 in the Supplementary Problems at the end of this chapter, there is outlined another approach to proving the second and third parts of Sylow's theorem.

We close this section by demonstrating how the various parts of Sylow's theorem can be used to gain a great deal of information about finite groups.

Let G be a group of order $11^2 \cdot 13^2$. We want to determine how many 11-Sylow subgroups and how many 13-Sylow subgroups there are in G. The number of 11-Sylow subgroups, by Theorem 2.12.13, is of the form $1 + 11k$. By Lemma 2.12.5, this must divide $11^2 \cdot 13^2$; being prime to 11, it must divide 13^2. Can 13^2 have a factor of the form $1 + 11k$? Clearly no, other than 1 itself. Thus $1 + 11k = 1$, and so there must be only one 11-Sylow subgroup in G. Since all 11-Sylow subgroups are conjugate (Theorem 2.12.2) we conclude that the 11-Sylow subgroup is *normal* in G.

What about the 13-Sylow subgroups? Their number is of the form $1 + 13k$ and must divide $11^2 \cdot 13^2$, hence must divide 11^2. Here, too, we conclude that there can be only one 13-Sylow subgroup in G, and it must be normal.

We now know that G has a normal subgroup A of order 11^2 and a normal subgroup B of order 13^2. By the corollary to Theorem 2.11.2, any group of order p^2 is abelian; hence A and B are both abelian. Since $A \cap B = (e)$, we easily get $AB = G$. Finally, if $a \in A$, $b \in B$, then $aba^{-1}b^{-1} = a(ba^{-1}b^{-1}) \in A$ since A is normal, and $aba^{-1}b^{-1} = (aba^{-1})b^{-1} \in B$ since B is normal. Thus $aba^{-1}b^{-1} \in A \cap B = (e)$. This gives us $aba^{-1}b^{-1} = e$, and so $ab = ba$ for $a \in A$, $b \in B$. This, together with $AB = G$, A, B abelian, allows us to conclude that G *is abelian*. Hence any group of order $11^2 \cdot 13^2$ must be abelian.

We give one other illustration of the use of the various parts of Sylow's theorem. Let G be a group of order 72; $o(G) = 2^3 3^2$. How many 3-Sylow subgroups can there be in G? If this number is t, then, according to Theorem 2.12.3, $t = 1 + 3k$. According to Lemma 2.12.5, $t \mid 72$, and since t is prime to 3, we must have $t \mid 8$. The only factors of 8 of the form $1 + 3k$ are 1 and 4; hence $t = 1$ or $t = 4$ are the only possibilities. In other words G has either one 3-Sylow subgroup or 4 such.

If G has only one 3-Sylow subgroup, since all 3-Sylow subgroups are conjugate, this 3-Sylow subgroup must be normal in G. In this case G would certainly contain a nontrivial normal subgroup. On the other hand if the number of 3-Sylow subgroups of G is 4, by Lemma 2.12.5 the index of N in G is 4, where N is the normalizer of a 3-Sylow subgroup. But $72 \nmid 4! = (i(N))!$. By Lemma 2.9.1 N must contain a nontrivial normal subgroup of G (of order at least 3). Thus here again we can conclude that G contains a nontrivial normal subgroup. The upshot of the discussion is that any group of order 72 must have a nontrivial normal subgroup, hence cannot be simple.

Problems

1. Adapt the second proof given of Sylow's theorem to prove directly that if p is a prime and $p^\alpha \mid o(G)$, then G has a subgroup of order p^α.

2. If $x > 0$ is a real number, define $[x]$ to be m, where m is that integer such that $m \leq x < m + 1$. If p is a prime, show that the power of p which exactly divides $n!$ is given by

$$\left[\frac{n}{p}\right] + \left[\frac{n}{p^2}\right] + \cdots + \left[\frac{n}{p^r}\right] + \cdots.$$

3. Use the method for constructing the p-Sylow subgroup of S_{p^k} to find generators for
(a) a 2-Sylow subgroup in S_8. (b) a 3-Sylow subgroup in S_9.

4. Adopt the method used in Problem 3 to find generators for
(a) a 2-Sylow subgroup of S_6. (b) a 3-Sylow subgroup of S_6.

5. If p is a prime number, give explicit generators for a p-Sylow subgroup of S_{p^2}.

6. Discuss the number and nature of the 3-Sylow subgroups and 5-Sylow subgroups of a group of order $3^2 \cdot 5^2$.

7. Let G be a group of order 30.
(a) Show that a 3-Sylow subgroup or a 5-Sylow subgroup of G must be normal in G.
(b) From part (a) show that every 3-Sylow subgroup and every 5-Sylow subgroup of G must be normal in G.
(c) Show that G has a normal subgroup of order 15.
(d) From part (c) classify all groups of order 30.
(e) How many different nonisomorphic groups of order 30 are there?

8. If G is a group of order 231, prove that the 11-Sylow subgroup is in the center of G.

9. If G is a group of order 385 show that its 11-Sylow subgroup is normal and its 7-Sylow subgroup is in the center of G.

10. If G is of order 108 show that G has a normal subgroup of order 3^k, where $k \geq 2$.

11. If $o(G) = pq$, p and q distinct primes, $p < q$, show
(a) if $p \nmid (q - 1)$, then G is cyclic.
*(b) if $p \mid (q - 1)$, then there exists a unique non-abelian group of order pq.

*12. Let G be a group of order pqr, $p < q < r$ primes. Prove
(a) the r-Sylow subgroup is normal in G.
(b) G has a normal subgroup of order qr.
(c) if $q \nmid (r - 1)$, the q-Sylow subgroup of G is normal in G.

13. If G is of order p^2q, p, q primes, prove that G has a nontrivial normal subgroup.

*14. If G is of order p^2q, p, q primes, prove that either a p-Sylow subgroup or a q-Sylow subgroup of G must be normal in G.

15. Let G be a finite group in which $(ab)^p = a^pb^p$ for every $a, b \in G$, where p is a prime dividing $o(G)$. Prove
   (a) The p-Sylow subgroup of G is normal in G.
   *(b) If P is the p-Sylow subgroup of G, then there exists a normal subgroup N of G with $P \cap N = (e)$ and $PN = G$.
   (c) G has a nontrivial center.

**16. If G is a finite group and its p-Sylow subgroup P lies in the center of G, prove that there exists a normal subgroup N of G with $P \cap N = (e)$ and $PN = G$.

*17. If H is a subgroup of G, recall that $N(H) = \{x \in G \mid xHx^{-1} = H\}$. If P is a p-Sylow subgroup of G, prove that $N(N(P)) = N(P)$.

*18. Let P be a p-Sylow subgroup of G and suppose a, b are in the center of P. Suppose further that $a = xbx^{-1}$ for some $x \in G$. Prove that there exists a $y \in N(P)$ such that $a = yby^{-1}$.

**19. Let G be a finite group and suppose that ϕ is an automorphism of G such that ϕ^3 is the identity automorphism. Suppose further that $\phi(x) = x$ implies that $x = e$. Prove that for every prime p which divides $o(G)$, the p-Sylow subgroup is normal in G.

#20. Let G be the group of $n \times n$ matrices over the integers modulo p, p a prime, which are invertible. Find a p-Sylow subgroup of G.

21. Find the possible number of 11-Sylow subgroups, 7-Sylow subgroups, and 5-Sylow subgroups in a group of order $5^2 \cdot 7 \cdot 11$.

22. If G is S_3 and $A = ((1\ 2))$ in G, find all the double cosets AxA of A in G.

23. If G is S_4 and $A = ((1\ 2\ 3\ 4))$, $B = ((1\ 2))$, find all the double cosets AxB of A, B in G.

24. If G is the dihedral group of order 18 generated by $a^2 = b^9 = e$, $ab = b^{-1}a$, find the double cosets for H, K in G, where $H = (a)$ and $K = (b^3)$.

2.13 Direct Products

On several occasions in this chapter we have had a need for constructing a new group from some groups we already had on hand. For instance, towards the end of Section 2.8, we built up a new group using a given group and one of its automorphisms. A special case of this type of construction has been seen earlier in the recurring example of the dihedral group.

However, no attempt had been made for some systematic device for constructing new groups from old. We shall do so now. The method represents the most simple-minded, straightforward way of combining groups to get other groups.

We first do it for two groups—not that two is sacrosanct. However, with this experience behind us, we shall be able to handle the case of any finite number easily and with dispatch. Not that any finite number is sacrosanct either; we could equally well carry out the discussion in the wider setting of any number of groups. However, we shall have no need for so general a situation here, so we settle for the case of any finite number of groups as our ultimate goal.

Let A and B be any two groups and consider the Cartesian product (which we discussed in Chapter 1) $G = A \times B$ of A and B. G consists of all ordered pairs (a, b), where $a \in A$ and $b \in B$. Can we use the operations in A and B to endow G with a product in such a way that G is a group? Why not try the obvious? Multiply componentwise. That is, let us define, for (a_1, b_1) and (a_2, b_2) in G, their product via $(a_1, b_1)(a_2, b_2) = (a_1a_2, b_1b_2)$. Here, the product a_1a_2 in the first component is the product of the elements a_1 and a_2 as calculated in the group A. The product b_1b_2 in the second component is that of b_1 and b_2 as elements in the group B.

With this definition we at least have a product defined in G. Is G a group relative to this product? The answer is yes, and is easy to verify. We do so now.

First we do the associative law. Let (a_1, b_1), (a_2, b_2), and (a_3, b_3) be three elements of G. Then $((a_1, b_1)(a_2, b_2))(a_3, b_3) = (a_1a_2, b_1b_2)(a_3, b_3) = ((a_1a_2)a_3, (b_1b_2)b_3)$, while $(a_1, b_1)((a_2, b_2)(a_3, b_3)) = (a_1, b_1)(a_2a_3, b_2b_3) = (a_1(a_2a_3), b_1(b_2b_3))$. The associativity of the product in A and in B then show us that our product in G is indeed associative.

Now to the unit element. What would be more natural than to try (e, f), where e is the unit element of A and f that of B, as the proposed unit element for G? We have $(a, b)(e, f) = (ae, bf) = (a, b)$ and $(e, f)(a, b) = (ea, fb) = (a, b)$. Thus (e, f) acts as a unit element in G.

Finally, we need the inverse in G for any element of G. Here, too, why not try the obvious? Let $(a, b) \in G$; try (a^{-1}, b^{-1}) as its inverse. Now $(a, b)(a^{-1}, b^{-1}) = (aa^{-1}, bb^{-1}) = (e, f)$ and $(a^{-1}, b^{-1})(a, b) = (a^{-1}a, b^{-1}b) = (e, f)$, so that (a^{-1}, b^{-1}) does serve as the inverse for (a, b).

With this we have verified that $G = A \times B$ is a group. We call it the *external direct product* of A and B.

Since $G = A \times B$ has been built up from A and B in such a trivial manner, we would expect that the structure of A and B would reflect heavily in that of G. This is indeed the case. Knowing A and B completely gives us complete information, structurally, about $A \times B$.

The construction of $G = A \times B$ has been from the outside, external. Now we want to turn the affair around and try to carry it out internally in G.

Consider $\bar{A} = \{(a, f) \in G \mid a \in A\} \subset G = A \times B$, where f is the unit element of B. What would one expect of $\bar{A}$? Answer: $\bar{A}$ is a subgroup of G and is isomorphic to A. To effect this isomorphism, define $\phi: A \to \bar{A}$ by $\phi(a) = (a, f)$ for $a \in A$. It is trivial that ϕ is an isomorphism of A onto $\bar{A}$. It is equally trivial that $\bar{A}$ is a subgroup of G. Furthermore, $\bar{A}$ is normal in G. For if $(a, f) \in \bar{A}$ and $(a_1, b_1) \in G$, then $(a_1, b_1)(a, f)(a_1, b_1)^{-1} = (a_1, b_1)(a, f)(a_1^{-1}, b_1^{-1}) = (a_1 a a_1^{-1}, b_1 f b^{1-1}) = (a_1 a a_1^{-1}, f) \in \bar{A}$. So we have an isomorphic copy, $\bar{A}$, of A in G which is a normal subgroup of G.

What we did for A we can also do for B. If $\bar{B} = \{(e, b) \in G \mid b \in B\}$, then $\bar{B}$ is isomorphic to B and is a normal subgroup of G.

We claim a little more, namely $G = \bar{A}\bar{B}$ and every $g \in G$ has a unique decomposition in the form $g = \bar{a}\bar{b}$ with $\bar{a} \in \bar{A}$ and $\bar{b} \in \bar{B}$. For, $g = (a, b) = (a, f)(e, b)$ and, since $(a, f) \in \bar{A}$ and $(e, b) \in \bar{B}$, we do have $g = \bar{a}\bar{b}$ with $\bar{a} = (a, f)$ and $\bar{b} = (e, b)$. Why is this unique? If $(a, b) = \bar{x}\bar{y}$, where $\bar{x} \in \bar{A}$ and $\bar{y} \in \bar{B}$, then $\bar{x} = (x, f)$, $x \in A$ and $\bar{y} = (e, y)$, $y \in B$; thus $(a, b) = \bar{x}\bar{y} = (x, f)(e, y) = (x, y)$. This gives $x = a$ and $y = b$, and so $\bar{x} = \bar{a}$ and $\bar{y} = \bar{b}$.

Thus we have realized G as an internal product $\bar{A}\bar{B}$ of two normal subgroups, $\bar{A}$ isomorphic to A, $\bar{B}$ to B in such a way that every element $g \in G$ has a unique representation in the form $g = \bar{a}\bar{b}$, with $\bar{a} \in \bar{A}$ and $\bar{b} \in \bar{B}$.

We leave the discussion of the product of two groups and go to the case of n groups, $n > 1$ any integer.

Let $G_1, G_2, \ldots, G_n$ be any n groups. Let $G = G_1 \times G_2 \times \cdots \times G_n = \{(g_1, g_2, \ldots, g_n) \mid g_i \in G_i\}$ be the set of all ordered n-tuples, that is, the Cartesian product of $G_1, G_2, \ldots, G_n$. We define a product in G via $(g_1, g_2, \ldots, g_n)(g_1', g_2', \ldots, g_n') = (g_1g_1', g_2g_2', \ldots, g_ng_n')$, that is, via componentwise multiplication. The product in the ith component is carried in the group G_i. Then G is a group in which $(e_1, e_2, \ldots, e_n)$ is the unit element, where each e_i is the unit element of G_i, and where $(g_1, g_2, \ldots, g_n)^{-1} = (g_1^{-1}, g_2^{-1}, \ldots, g_n^{-1})$. We call this group G the *external direct product* of $G_1, G_2, \ldots, G_n$.

In $G = G_1 \times G_2 \times \cdots \times G_n$ let $\bar{G}_i = \{(e_1, e_2, \ldots, e_{i-1}, g_i, e_{i+1}, \ldots, e_n) \mid g_i \in G_i\}$. Then $\bar{G}_i$ is a normal subgroup of G and is isomorphic to G_i. Moreover, $G = \bar{G}_1\bar{G}_2 \cdots \bar{G}_n$ and every $g \in G$ has a unique decomposition $g = \bar{g}_1\bar{g}_2 \cdots \bar{g}_n$, where $\bar{g}_1 \in \bar{G}_1, \ldots, \bar{g}_n \in \bar{G}_n$. We leave the verification of these facts to the reader.

Here, too, as in the case $A \times B$, we have realized the group G *internally* as the product of normal subgroups $\bar{G}_1, \ldots, \bar{G}_n$ in such a way that every element is uniquely representable as a product of elements $\bar{g}_1 \cdots \bar{g}_n$, where each $\bar{g}_i \in \bar{G}_i$. With this motivation we make the

DEFINITION Let G be a group and $N_1, N_2, \ldots, N_n$ *normal* subgroups of G such that

1. $G = N_1 N_2 \cdots N_n$.
2. Given $g \in G$ then $g = m_1 m_2 \cdots m_n$, $m_i \in N_i$ in a unique way.

We then say that G is the *internal direct product* of $N_1, N_2, \ldots, N_n$.

Before proceeding let's look at an example of a group G which is the internal direct product of some of its subgroups. Let G be a finite *abelian* group of order $p_1^{\alpha_1} p_2^{\alpha_2} \cdots p_k^{\alpha_k}$ where $p_1, p_2, \ldots, p_k$ are distinct primes and each $\alpha_i > 0$. If $P_1, \ldots, P_k$ are the p_1-Sylow subgroup, $\ldots$, p_k-Sylow subgroup respectively of G, then G is the internal direct product of $P_1, P_2, \ldots, P_k$ (see Problem 5).

We continue with the general discussion. Suppose that G is the internal direct product of the normal subgroups $N_1, \ldots, N_n$. The $N_1, \ldots, N_n$ are groups in their own right—forget that they are normal subgroups of G for the moment. Thus we can form the group $T = N_1 \times N_2 \times \cdots \times N_n$, the external direct product of $N_1, \ldots, N_n$. One feels that G and T should be related. Our aim, in fact, is to show that G is isomorphic to T. If we could establish this then we could abolish the prefix external and internal in the phrases external direct product, internal direct product—after all these would be the same group up to isomorphism—and just talk about the direct product.

We start with

LEMMA 2.13.1 *Suppose that G is the internal direct product of $N_1, \ldots, N_n$. Then for $i \neq j$, $N_i \cap N_j = (e)$, and if $a \in N_i$, $b \in N_j$ then $ab = ba$.*

Proof. Suppose that $x \in N_i \cap N_j$. Then we can write x as

$$x = e_1 \cdots e_{i-1} x e_{i+1} \cdots e_j \cdots e_n,$$

where $e_t = e$, viewing x as an element in N_i. Similarly, we can write x as

$$x = e_1 \cdots e_i \cdots e_{j-1} x e_{j+1} \cdots e_n,$$

where $e_t = e$, viewing x as an element of N_j. But every element—and so, in particular x—has a unique representation in the form $m_1 m_2 \cdots m_n$, where $m_i \in N_1, \ldots, m_n \in N_n$. Since the two decompositions in this form for x must coincide, the entry from N_i in each must be equal. In our first decomposition this entry is x, in the other it is e; hence $x = e$. Thus $N_i \cap N_j = (e)$ for $i \neq j$.

Suppose $a \in N_i$, $b \in N_j$, and $i \neq j$. Then $aba^{-1} \in N_j$ since N_j is normal; thus $aba^{-1}b^{-1} \in N_j$. Similarly, since $a^{-1} \in N_i$, $ba^{-1}b^{-1} \in N_i$, whence $aba^{-1}b^{-1} \in N_i$. But then $aba^{-1}b^{-1} \in N_i \cap N_j = (e)$. Thus $aba^{-1}b^{-1} = e$; this gives the desired result $ab = ba$.

One should point out that if $K_1, \ldots, K_n$ are normal subgroups of G such that $G = K_1 K_2 \cdots K_n$ and $K_i \cap K_j = (e)$ for $i \neq j$ it *need not be true* that G is the internal direct product of $K_1, \ldots, K_n$. A more stringent condition is needed (see Problems 8 and 9).

We now can prove the desired isomorphism between the external and internal direct products that was stated earlier.

THEOREM 2.13.1 *Let G be a group and suppose that G is the internal direct product of $N_1, \ldots, N_n$. Let $T = N_1 \times N_2 \times \cdots \times N_n$. Then G and T are isomorphic.*

Proof. Define the mapping $\psi : T \to G$ by

$$\psi((b_1, b_2, \ldots, b_n)) = b_1 b_2 \cdots b_n,$$

where each $b_i \in N_i$, $i = 1, \ldots, n$. We claim that ψ is an isomorphism of T onto G.

To begin with, ψ is certainly onto; for, since G is the internal direct product of $N_1, \ldots, N_n$, if $x \in G$ then $x = a_1 a_2 \cdots a_n$ for some $a_1 \in N_1, \ldots, a_n \in N_n$. But then $\psi((a_1, a_2, \ldots, a_n)) = a_1 a_2 \cdots a_n = x$. The mapping ψ is one-to-one by the *uniqueness* of the representation of every element as a product of elements from $N_1, \ldots, N_n$. For, if $\psi((a_1, \ldots, a_n)) = \psi((c_1, \ldots, c_n))$, where $a_i \in N_i$, $c_i \in N_i$, for $i = 1, 2, \ldots, n$, then, by the definition of ψ, $a_1 a_2 \cdots a_n = c_1 c_2 \cdots c_n$. The uniqueness in the definition of internal direct product forces $a_1 = c_1$, $a_2 = c_2, \ldots, a_n = c_n$. Thus ψ is one-to-one.

All that remains is to show that ψ is a homomorphism of T onto G. If $X = (a_1, \ldots, a_n)$, $Y = (b_1, \ldots, b_n)$ are elements of T then

$$\begin{aligned}\psi(XY) &= \psi((a_1, \ldots, a_n)(b_1, \ldots, b_n)) \\ &= \psi(a_1 b_1, a_2 b_2, \ldots, a_n b_n) \\ &= a_1 b_1 a_2 b_2 \cdots a_n b_n.\end{aligned}$$

However, by Lemma 2.13.1, $a_i b_j = b_j a_i$ if $i \neq j$. This tells us that $a_1 b_1 a_2 b_2 \cdots a_n b_n = a_1 a_2 \cdots a_n b_1 b_2 \cdots b_n$. Thus $\psi(XY) = a_1 a_2 \cdots a_n b_1 b_2 \cdots b_n$. But we can recognize $a_1 a_2 \cdots a_n$ as $\psi((a_1, a_2, \ldots, a_n)) = \psi(X)$ and $b_1 b_2 \cdots b_n$ as $\psi(Y)$. We therefore have $\psi(XY) = \psi(X)\psi(Y)$. In short, we have shown that ψ is an isomorphism of T onto G. This proves the theorem.

Note one particular thing that the theorem proves. If a group G is isomorphic to an external direct product of certain groups G_i, then G *is*, in fact, the internal direct product of groups $\bar{G}_i$ isomorphic to the G_i. We simply say that G is the direct product of the $\bar{G}_i$ (or G_i).

In the next section we shall see that every finite abelian group is a direct product of cyclic groups. Once we have this, we have the structure of all finite abelian groups pretty well under our control.

One should point out that the analog of the direct product of groups exists in the study of almost all algebraic structures. We shall see this later for vector-spaces, rings, and modules. Theorems that describe such an algebraic object in terms of direct products of more describable algebraic objects of the same kind (for example, the case of abelian groups above) are important theorems in general. Through such theorems we can reduce the study of a fairly complex algebraic situation to a much simpler one.

Problems

1. If A and B are groups, prove that $A \times B$ is isomorphic to $B \times A$.
2. If G_1, G_2, G_3 are groups, prove that $(G_1 \times G_2) \times G_3$ is isomorphic to $G_1 \times G_2 \times G_3$. Care to generalize?
3. If $T = G_1 \times G_2 \times \cdots \times G_n$ prove that for each $i = 1, 2, \ldots, n$ there is a homomorphism ϕ_i of T onto G_i. Find the kernel of ϕ_i.
4. Let G be a group and let $T = G \times G$.
   (a) Show that $D = \{(g, g) \in G \times G \mid g \in G\}$ is a group isomorphic to G.
   (b) Prove that D is normal in T if and only if G is abelian.
5. Let G be a finite abelian group. Prove that G is isomorphic to the direct product of its Sylow subgroups.
6. Let A, B be cyclic groups of order m and n, respectively. Prove that $A \times B$ is cyclic if and only if m and n are relatively prime.
7. Use the result of Problem 6 to prove the Chinese Remainder Theorem; namely, if m and n are relatively prime integers and u, v any two integers, then we can find an integer x such that $x \equiv u \bmod m$ and $x \equiv v \bmod n$.
8. Give an example of a group G and normal subgroups $N_1, \ldots, N_n$ such that $G = N_1 N_2 \cdots N_n$ and $N_i \cap N_j = (e)$ for $i \neq j$ and yet G is *not* the internal direct product of $N_1, \ldots, N_n$.
9. Prove that G is the internal direct product of the normal subgroups $N_1, \ldots, N_n$ if and only if
   1. $G = N_1 \cdots N_n$.
   2. $N_i \cap (N_1 N_2 \cdots N_{i-1} N_{i+1} \cdots N_n) = (e)$ for $i = 1, \ldots, n$.
10. Let G be a group, $K_1, \ldots, K_n$ normal subgroups of G. Suppose that $K_1 \cap K_2 \cap \cdots \cap K_n = (e)$. Let $V_i = G/K_i$. Prove that there is an isomorphism of G into $V_1 \times V_2 \times \cdots \times V_n$.

*11. Let G be a finite abelian group such that it contains a subgroup $H_0 \neq (e)$ which lies in *every* subgroup $H \neq (e)$. Prove that G must be cyclic. What can you say about $o(G)$?

12. Let G be a finite abelian group. Using Problem 11 show that G is isomorphic to a subgroup of a direct product of a finite number of finite cyclic groups.

13. Give an example of a finite non-abelian group G which contains a subgroup $H_0 \neq (e)$ such that $H_0 \subset H$ for *all* subgroups $H \neq (e)$ of G.
14. Show that every group of order p^2, p a prime, is either cyclic or is isomorphic to the direct product of two cyclic groups each of order p.
*15. Let $G = A \times A$ where A is cyclic of order p, p a prime. How many automorphisms does G have?
16. If $G = K_1 \times K_2 \times \cdots \times K_n$ describe the center of G in terms of those of the K_i.
17. If $G = K_1 \times K_2 \times \cdots \times K_n$ and $g \in G$, describe

$$N(g) = \{x \in G \mid xg = gx\}.$$

18. If G is a finite group and $N_1, \ldots, N_n$ are normal subgroups of G such that $G = N_1N_2 \cdots N_n$ and $o(G) = o(N_1)o(N_2) \cdots o(N_n)$, prove that G is the direct product of $N_1, N_2, \ldots, N_n$.

2.14 Finite Abelian Groups

We close this chapter with a discussion (and description) of the structure of an arbitrary finite abelian group. The result which we shall obtain is a famous classical theorem, often referred to as the Fundamental Theorem on Finite Abelian Groups. It is a highly satisfying result because of its decisiveness. Rarely do we come out with so compact, succinct, and crisp a result. In it the structure of a finite abelian group is completely revealed, and by means of it we have a ready tool for attacking any structural problem about finite abelian groups. It even has some arithmetic consequences. For instance, one of its by-products is a precise count of how many non-isomorphic abelian groups there are of a given order.

In all fairness one should add that this description of finite abelian groups is not as general as we can go and still get so sharp a theorem. As you shall see in Section 4.5, we completely describe all abelian groups generated by a finite set of elements—a situation which not only covers the finite abelian group case, but much more.

We now state this very fundamental result.

THEOREM 2.14.1 *Every finite abelian group is the direct product of cyclic groups.*

Proof. Our first step is to reduce the problem to a slightly easier one. We have already indicated in the preceding section (see Problem 5 there) that any finite abelian group G is the direct product of its Sylow subgroups. If we knew that each such Sylow subgroup was a direct product of cyclic groups we could put the results together for these Sylow subgroups to realize G as a direct product of cyclic groups. Thus it suffices to prove the theorem for abelian groups of order p^n where p is a prime.

So suppose that G is an abelian group of order p^n. Our objective is to find elements $a_1, \ldots, a_k$ in G such that every element $x \in G$ can be written in a unique fashion as $x = a_1^{\alpha_1} a_2^{\alpha_2} \cdots a_k^{\alpha_k}$. Note that if this were true and $a_1, \ldots, a_k$ were of order $p^{n_1}, \ldots, p^{n_k}$, where $n_1 \geq n_2 \geq \cdots \geq n_k$, then the maximal order of any element in G would be p^{n_1} (Prove!). This gives us a cue of how to go about finding the elements $a_1, \ldots, a_k$ that we seek.

The procedure suggested by this is: let a_1 be an element of maximal order in G. How shall we pick a_2? Well, if $A_1 = (a_1)$ the subgroup generated by a_1, then a_2 maps into an element of highest order in G/A_1. If we can successfully exploit this to find an appropriate a_2, and if $A_2 = (a_2)$, then a_3 would map into an element of maximal order in G/A_1A_2, and so on. With this as guide we can now get down to the brass tacks of the proof.

Let a_1 be an element in G of highest possible order, p^{n_1}, and let $A_1 = (a_1)$. Pick b_2 in G such that $\bar{b}_2$, the image of b_2 in $\bar{G} = G/A_1$, has maximal order p^{n_2}. Since the order of $\bar{b}_2$ divides that of b_2, and since the order of a_1 is maximal, we must have that $n_1 \geq n_2$. In order to get a direct product of A_1 with (b_2) we would need $A_1 \cap (b_2) = (e)$; this might not be true for the initial choice of b_2, so we may have to adapt the element b_2. Suppose that $A_1 \cap (b_2) \neq (e)$; then, since $b_2^{p^{n_2}} \in A_1$ and is the first power of b_2 to fall in A_1 (by our mechanism of choosing b_2) we have that $b_2^{p^{n_2}} = a_1^{i}$. Therefore $(a_1^{i})^{p^{n_1-n_2}} = (b_2^{p^{n_2}})^{p^{n_1-n_2}} = b_2^{p^{n_1}} = e$, whence $a_1^{ip^{n_1-n_2}} = e$. Since a_1 is of order p^{n_1} we must have that $p^{n_1} \mid ip^{n_1-n_2}$, and so $p^{n_2} \mid i$. Thus, recalling what i is, we have $b_2^{p^{n_2}} = a_1^{i} = a_1^{jp^{n_2}}$. This tells us that if $a_2 = a_1^{-j}b_2$ then $a_2^{p^{n_2}} = e$. The element a_2 is indeed the element we seek. Let $A_2 = (a_2)$. We claim that $A_1 \cap A_2 = (e)$. For, suppose that $a_2^{t} \in A_1$; since $a_2 = a_1^{-j}b_2$, we get $(a_1^{-j}b_2)^{t} \in A_1$ and so $b_2^{t} \in A_1$. By choice of b_2, this last relation forces $p^{n_2} \mid t$, and since $a_2^{p^{n_2}} = e$ we must have that $a_2^{t} = e$. In short $A_1 \cap A_2 = (e)$.

We continue one more step in the program we have outlined. Let $b_3 \in G$ map into an element of maximal order in $G/(A_1A_2)$. If the order of the image of b_3 in $G/(A_1A_2)$ is p^{n_3}, we claim that $n_3 \leq n_2 \leq n_1$. Why? By the choice of n_2, $b_3^{p^{n_2}} \in A_1$ so is certainly in A_1A_2. Thus $n_3 \leq n_2$. Since $b_3^{p^{n_3}} \in A_1A_2$, $b_3^{p^{n_3}} = a_1^{i_1}a_2^{i_2}$. We claim that $p^{n_3} \mid i_1$ and $p^{n_3} \mid i_2$. For, $b_3^{p^{n_2}} \in A_1$ hence $(a_1^{i_1}a_2^{i_2})^{p^{n_2-n_3}} = (b_3^{p^{n_3}})^{p^{n_2-n_3}} \equiv b_3^{p^{n_2}} \in A_1$. This tells us that $a_2^{i_2p^{n_2-n_3}} \in A_1$ and so $p^{n_2} \mid i_2p^{n_2-n_3}$, which is to say, $p^{n_3} \mid i_2$. Also $b_3^{p^{n_1}} = e$, hence $(a_1^{i_1}a_2^{i_2})^{p^{n_1-n_3}} = b_3^{p^{n_1}} = e$; this says that $a_1^{i_1p^{n_1-n_3}} \in A_2 \cap A_1 = (e)$, that is, $a_1^{i_1p^{n_1-n_3}} = e$. This yields that $p^{n_3} \mid i_1$. Let $i_1 = j_1p^{n_3}$, $i_2 = j_2p^{n_3}$; thus $b_3^{p^{n_3}} = a_1^{j_1p^{n_3}}a_2^{j_2p^{n_3}}$. Let $a_3 = a_1^{-j_1}a_2^{-j_2}b_3$, $A_3 = (a_3)$; note that $a_3^{p^{n_3}} = e$. We claim that $A_3 \cap (A_1A_2) = (e)$. For if $a_3^{t} \in A_1A_2$ then $(a_1^{-j_1}a_2^{-j_2}b_3)^{t} \in A_1A_2$, giving us $b_3^{t} \in A_1A_2$. But then $p^{n_3} \mid t$, whence, since $a_3^{p^{n_3}} = e$, we have $a_3^{t} = e$. In other words, $A_3 \cap (A_1A_2) = (e)$.

Continuing this way we get cyclic subgroups $A_1 = (a_1)$, $A_2 = (a_2), \ldots, A_k = (a_k)$ of order $p^{n_1}, p^{n_2}, \ldots, p^{n_k}$, respectively, with $n_1 \geq n_2 \geq \cdots \geq n_k$ such that $G = A_1A_2 \cdots A_k$ and such that, for each i, $A_i \cap (A_1A_2 \cdots A_{i-1}) = (e)$. This tells us that every $x \in G$ has a unique representation as $x = a_1'a_2' \cdots a_k'$ where $a_1' \in A_1, \ldots, a_k' \in A_k$. In other words, G is the direct product of the cyclic subgroups $A_1, A_2, \ldots, A_k$. The theorem is now proved.

DEFINITION If G is an abelian group of order p^n, p a prime, and $G = A_1 \times A_2 \times \cdots \times A_k$ where each A_i is cyclic of order p^{n_i} with $n_1 \geq n_2 \geq \cdots \geq n_k > 0$, then the integers $n_1, n_2, \ldots, n_k$ are called the *invariants* of G.

Just because we called the integers above the invariants of G does not mean that they *are* really *the* invariants of G. That is, it is possible that we can assign different sets of invariants to G. We shall soon show that the invariants of G are indeed unique and completely describe G.

Note one other thing about the invariants of G. If $G = A_1 \times \cdots \times A_k$, where A_i is cyclic of order p^{n_i}, $n_1 \geq n_2 \geq \cdots \geq n_k > 0$, then $o(G) = o(A_1)o(A_2) \cdots o(A_k)$, hence $p^n = p^{n_1}p^{n_2} \cdots p^{n_k} = p^{n_1+n_2+\cdots+n_k}$, whence $n = n_1 + n_2 + \cdots + n_k$. In other words, $n_1, n_2, \ldots, n_k$ give us a *partition* of n. We have already run into this concept earlier in studying the conjugate classes in the symmetric group.

Before discussing the uniqueness of the invariants of G, one thing should be made absolutely clear: the elements $a_1, \ldots, a_k$ and the subgroups $A_1, \ldots, A_k$ which they generate, which arose above to give the decomposition of G into a direct product of cyclic groups, are *not* unique. Let's see this in a very simple example. Let $G = \{e, a, b, ab\}$ be an abelian group of order 4 where $a^2 = b^2 = e$, $ab = ba$. Then $G = A \times B$ where $A = (a)$, $B = (b)$ are cyclic groups of order 2. But we have another decomposition of G as a direct product, namely, $G = C \times B$ where $C = (ab)$ and $B = (b)$. So, even in this group of very small order, we can get distinct decompositions of the group as the direct product of cyclic groups. Our claim—which we now want to substantiate—is that while these cyclic subgroups are not unique, their *orders are*

DEFINITION If G is an abelian group and s is any integer, then $G(s) = \{x \in G \mid x^s = e\}$.

Because G is abelian it is evident that $G(s)$ is a subgroup of G. We now prove

LEMMA 2.14.1 *If G and G' are isomorphic abelian groups, then for every integer s, $G(s)$, and $G'(s)$ are isomorphic.*

Proof. Let ϕ be an isomorphism of G onto G'. We claim that ϕ maps $G(s)$ isomorphically onto $G'(s)$. First we show that $\phi(G(s)) \subset G'(s)$. For, if $x \in G(s)$ then $x^s = e$, hence $\phi(x^s) = \phi(e) = e'$. But $\phi(x^s) = \phi(x)^s$; hence $\phi(x)^s = e'$ and so $\phi(x)$ is in $G'(s)$. Thus $\phi(G(s)) \subset G'(s)$.

On the other hand, if $u' \in G'(s)$ then $(u')^s = e'$. But, since ϕ is onto, $u' = \phi(y)$ for some $y \in G$. Therefore $e' = (u')^s = \phi(y)^s = \phi(y^s)$. Because ϕ is one-to-one, we have $y^s = e$ and so $y \in G(s)$. Thus ϕ maps $G(s)$ *onto* $G'(s)$.

Therefore since ϕ is one-to-one, onto, and a homomorphism from $G(s)$ to $G'(s)$, we have that $G(s)$ and $G'(s)$ are isomorphic.

We continue with

LEMMA 2.14.2 *Let G be an abelian group of order p^n, p a prime. Suppose that $G = A_1 \times A_2 \times \cdots \times A_k$, where each $A_i = (a_i)$ is cyclic of order p^{n_i}, and $n_1 \geq n_2 \geq \cdots \geq n_k > 0$. If m is an integer such that $n_t > m \geq n_{t+1}$ then $G(p^m) = B_1 \times \cdots \times B_t \times A_{t+1} \times \cdots \times A_k$ where B_i is cyclic of order p^m, generated by $a_i^{p^{n_i-m}}$, for $i \leq t$. The order of $G(p^m)$ is p^u, where*

$$u = mt + \sum_{i=t+1}^{k} n_i.$$

Proof. First of all, we claim that $A_{t+1}, \ldots, A_k$ are all in $G(p^m)$. For, since $m \geq n_{t+1} \geq \cdots \geq n_k > 0$, if $j \geq t + 1$, $a_j^{p^m} = (a_j^{p^{n_j}})^{p^{m-n_j}} = e$. Hence A_j, for $j \geq t + 1$ lies in $G(p^m)$.

Secondly, if $i \leq t$ then $n_i > m$ and $(a_i^{p^{n_i-m}})^{p^m} = a_i^{p^{n_i}} = e$, whence each such $a_i^{p^{n_i-m}}$ is in $G(p^m)$ and so the subgroup it generates, B_i, is also in $G(p^m)$.

Since $B_1, \ldots, B_t, A_{t+1}, \ldots, A_k$ are all in $G(p^m)$, their product (which is direct, since the product $A_1A_2 \cdots A_k$ is direct) is in $G(p^m)$. Hence $G(p^m) \supset B_1 \times \cdots \times B_t \times A_{t+1} \times \cdots \times A_k$.

On the other hand, if $x = a_1^{\lambda_1}a_2^{\lambda_2} \cdots a_k^{\lambda_k}$ is in $G(p^m)$, since it then satisfies $x^{p^m} = e$, we set $e = x^{p^m} = a_1^{\lambda_1 p^m} \cdots a_k^{\lambda_k p^m}$. However, the product of the subgroups $A_1, \ldots, A_k$ is direct, so we get

$$a_1^{\lambda_1 p^m} = e, \ldots, a_k^{\lambda_k p^m} = e.$$

Thus the order of a_i, that is, p^{n_i} must divide $\lambda_i p^m$ for $i = 1, 2, \ldots, k$. If $i \geq t + 1$ this is automatically true whatever be the choice of $\lambda_{t+1}, \ldots, \lambda_k$ since $m \geq n_{t+1} \geq \cdots \geq n_k$, hence $p^{n_i} \mid p^m$, $i \geq t + 1$. However, for $i \leq t$, we get from $p^{n_i} \mid \lambda_i p^m$ that $p^{n_i-m} \mid \lambda_i$. Therefore $\lambda_i = v_i p^{n_i-m}$ for some integer v_i. Putting all this information into the values of the λ_i's in the expression for x as $x = a_1^{\lambda_1} \cdots a_k^{\lambda_k}$ we see that

$$x = a_1^{v_1 p^{n_1-m}} \cdots a_t^{v_t p^{n_t-m}} a_{t+1}^{\lambda_{t+1}} \cdots a_k^{\lambda_k}.$$

This says that $x \in B_1 \times \cdots \times B_t \times A_{t+1} \times \cdots \times A_k$.

Now since each B_i is of order p^m and since $o(A_i) = p^{n_i}$ and since $G = B_1 \times \cdots \times B_t \times A_{t+1} \times \cdots \times A_k$,

$$o(G) = o(B_1)o(B_2)\cdots o(B_t)o(A_{t+1})\cdots o(A_k) = \underbrace{p^m p^m \cdots p^m}_{t\text{-times}} p^{n_{t+1}} \cdots p^{n_k}.$$

Thus, if we write $o(G) = p^u$, then

$$u = mt + \sum_{i=t+1}^{k} n_i.$$

The lemma is proved.

COROLLARY *If G is as in Lemma 2.14.2, then $o(G(p)) = p^k$.*

Proof. Apply the lemma to the case $m = 1$. Then $t = k$, hence $u = 1k = k$ and so $o(G) = p^k$.

We now have all the pieces required to prove the uniqueness of the invariants of an abelian group of order p^n.

THEOREM 2.14.2 *Two abelian groups of order p^n are isomorphic if and only if they have the same invariants.*

In other words, if G and G' are abelian groups of order p^n and $G = A_1 \times \cdots \times A_k$, where each A_i is a cyclic group of order p^{n_i}, $n_1 \geq \cdots \geq n_k > 0$, and $G' = B_1' \times \cdots \times B_s'$, where each B_i' is a cyclic group of order p^{h_i}, $h_1 \geq \cdots \geq h_s > 0$, then G and G' are isomorphic if and only if $k = s$ and for each i, $n_i = h_i$.

Proof. One way is very easy, namely, if G and G' have the same invariants then they are isomorphic. For then $G = A_1 \times \cdots \times A_k$ where $A_i = (a_i)$ is cyclic of order p^{n_i}, and $G' = B_1' \times \cdots \times B_k'$ where $B_i' = (b_i')$ is cyclic of order p^{n_i}. Map G onto G' by the map $\phi(a_1^{\alpha_1} \cdots a_k^{\alpha_k}) = (b_1')^{\alpha_1} \cdots (b_k')^{\alpha_k}$. We leave it to the reader to verify that this defines an isomorphism of G onto G'.

Now for the other direction. Suppose that $G = A_1 \times \cdots \times A_k$, $G' = B_1' \times \cdots \times B_s'$, A_i, B_i' as described above, cyclic of orders p^{n_i}, p^{h_i}, respectively, where $n_1 \geq \cdots \geq n_k > 0$ and $h_1 \geq \cdots \geq h_s > 0$. We want to show that if G and G' are isomorphic then $k = s$ and each $n_i = h_i$.

If G and G' are isomorphic then, by Lemma 2.14.1, $G(p^m)$ and $G'(p^m)$ must be isomorphic for any integer $m \geq 0$, hence must have the same order. Let's see what this gives us in the special case $m = 1$; that is, what information can we garner from $o(G(p)) = o(G'(p))$. According to the corollary to Lemma 2.14.2, $o(G(p)) = p^k$ and $o(G'(p)) = p^s$. Hence $p^k = p^s$ and so $k = s$. At least we now know that the *number* of invariants for G and G' is the same.

If $n_i \neq h_i$ for some i, let t be the first i such that $n_t \neq h_t$; we may suppose that $n_t > h_t$. Let $m = h_t$. Consider the subgroups, $H = \{x^{p^m} \mid x \in G\}$ and $H' = \{(x')^{p^m} \mid x' \in G\}$, of G and G', respectively. Since G and G' are isomorphic, it follows easily that H and H' are isomorphic. We now examine the invariants of H and H'.

Because $G = A_1 \times \cdots \times A_k$, where $A_i = (a_i)$ is of order p^{n_i}, we get that

$$H = C_1 \times \cdots \times C_t \times \cdots \times C_r,$$

where $C_i = (a_i^{p^m})$ is of order $p^{n_i - m}$, and where r is such that $n_r > m = h_t \geq n_{r-1}$. Thus the invariants of H are $n_1 - m, n_2 - m, \ldots, n_r - m$ and the number of invariants of H is $r \geq t$.

Because $G' = B_1' \times \cdots \times B_k'$, where $B_i = (b_i')$ is cyclic of order p^{h_i}, we get that $H' = D_1' \times \cdots \times D_{t-1}'$, where $D_i' = ((b_i')^{p^m})$ is cyclic of order $p^{h_i - m}$. Thus the invariants of H' are $h_1 - m, \ldots, h_{t-1} - m$ and so the number of invariants of H' is $t - 1$.

But H and H' are isomorphic; as we saw above this forces them to have the same number of invariants. But we saw that assuming that $n_i \neq h_i$ for some i led to a discrepancy in the number of their invariants. In consequence each $n_i = h_i$, and the theorem is proved.

An immediate consequence of this last theorem is that an abelian group of order p^n can be decomposed in only one way—*as far as the orders of the cyclic subgroups is concerned*—as a direct product of cyclic subgroups. Hence the invariants are indeed *the* invariants of G and completely determine G.

If $n_1 \geq \cdots \geq n_k > 0$, $n = n_1 + \cdots + n_k$, is any partition of n, then we can easily construct an abelian group of order p^n whose invariants are $n_1 \geq \cdots \geq n_k > 0$. To do this, let A_i be a cyclic group of order p^{n_i} and let $G = A_1 \times \cdots \times A_k$ be the external direct product of $A_1, \ldots, A_k$. Then, by the very definition, the invariants of G are $n_1 \geq \cdots \geq n_k > 0$. Finally, two different partitions of n give rise to nonisomorphic abelian groups of order p^n. This, too, comes from Theorem 2.14.2. Hence we have

THEOREM 2.14.3 *The number of nonisomorphic abelian groups of order p^n, p a prime, equals the number of partitions of n.*

Note that the answer given in Theorem 2.14.3 does *not* depend on the prime p; it only depends on the exponent n. Hence, for instance, the number of nonisomorphic abelian groups of order 2^4 equals that of orders 3^4, or 5^4, etc. Since there are five partitions of 4, namely: $4 = 4, 3 + 1, 2 + 2, 2 + 1 + 1, 1 + 1 + 1 + 1$, then there are five nonisomorphic abelian groups of order p^4 for any prime p.

Since any finite abelian group is a direct product of its Sylow subgroups, and two abelian groups are isomorphic if and only if their corresponding Sylow subgroups are isomorphic, we have the

COROLLARY *The number of nonisomorphic abelian groups of order $p_1^{\alpha_1}\cdots p_r^{\alpha_r}$, where the p_i are distinct primes and where each $\alpha_i > 0$, is $p(\alpha_1)p(\alpha_2)\cdots p(\alpha_r)$, where $p(u)$ denotes the number of partitions of u.*

Problems

1. If G is an abelian group of order p^n, p a prime and $n_1 \geq n_2 \geq \cdots \geq n_k > 0$, are the invariants of G, show that the maximal order of any element in G is p^{n_1}.
2. If G is a group, $A_1, \ldots, A_k$ normal subgroups of G such that $A_i \cap (A_1A_2\cdots A_{i-1}) = (e)$ for all i, show that G is the direct product of $A_1, \ldots, A_k$ if $G = A_1A_2\cdots A_k$.
3. Using Theorem 2.14.1, prove that if a finite abelian group has subgroups of orders m and n, then it has a subgroup whose order is the least common multiple of m and n.
4. Describe all finite abelian groups of order
   (a) 2^6. (b) 11^6. (c) 7^5. (d) $2^4 \cdot 3^4$.
5. Show how to get all abelian groups of order $2^3 \cdot 3^4 \cdot 5$.
6. If G is an abelian group of order p^n with invariants $n_1 \geq \cdots \geq n_k > 0$ and $H \neq (e)$ is a subgroup of G, show that if $h_1 \geq \cdots \geq h_s > 0$ are the invariants of H, then $k \geq s$ and for each i, $h_i \leq n_i$ for $i = 1, 2, \ldots, s$.

   If G is an abelian group, let $\hat{G}$ be the set of all homomorphisms of G into the group of nonzero complex numbers under multiplication. If $\phi_1, \phi_2 \in \hat{G}$, define $\phi_1 \cdot \phi_2$ by $(\phi_1 \cdot \phi_2)(g) = \phi_1(g)\phi_2(g)$ for all $g \in G$.
7. Show that $\hat{G}$ is an abelian group under the operation defined.
8. If $\phi \in \hat{G}$ and G is finite, show that $\phi(g)$ is a root of unity for every $g \in G$.
9. If G is a finite cyclic group, show that $\hat{G}$ is cyclic and $o(\hat{G}) = o(G)$, hence G and $\hat{G}$ are isomorphic.
10. If $g_1 \neq g_2$ are in G, G a finite abelian group, prove that there is a $\phi \in \hat{G}$ with $\phi(g_1) \neq \phi(g_2)$.
11. If G is a finite abelian group prove that $o(G) = o(\hat{G})$ and G is isomorphic to $\hat{G}$.
12. If $\phi \neq 1 \in \hat{G}$ where G is an abelian group, show that $\sum_{g \in G} \phi(g) = 0$.

Supplementary Problems

There is no relation between the order in which the problems appear and the order of appearance of the sections, in this chapter, which might be relevant to their solutions. No hint is given regarding the difficulty of any problem.

1. (a) If G is a finite abelian group with elements $a_1, a_2, \ldots, a_n$, prove that $a_1 a_2 \cdots a_n$ is an element whose square is the identity.
   (b) If the G in part (a) has no element of order 2 or more than one element of order 2, prove that $a_1 a_2 \cdots a_n = e$.
   (c) If G has one element, y, of order 2, prove that $a_1 a_2 \cdots a_n = y$.
   (d) (*Wilson's theorem*) If p is a prime number show that $(p-1)! \equiv -1(p)$.
2. If p is an odd prime and if

$$1 + \frac{1}{2} + \frac{1}{3} + \cdots + \frac{1}{p-1} = \frac{a}{b},$$

   where a and b are integers, prove that $p \mid a$. If $p > 3$, prove that $p^2 \mid a$.
3. If p is an odd prime, $a \not\equiv 0\ (p)$ is said to be a *quadratic residue of* p if there exists an integer x such that $x^2 \equiv a(p)$. Prove
   (a) The quadratic residues of p form a subgroup Q of the group of nonzero integers mod p under multiplication.
   (b) $o(Q) = (p-1)/2$.
   (c) If $q \in Q$, $n \notin Q$ (n is called a *nonresidue*), then nq is a nonresidue.
   (d) If n_1, n_2 are nonresidues, then $n_1 n_2$ is a residue.
   (e) If a is a quadratic residue of p, then $a^{(p-1)/2} \equiv +1(p)$.
4. Prove that in the integers mod p, p a prime, there are at most n solutions of $x^n \equiv 1(p)$ for every integer n.
5. Prove that the nonzero integers mod p under multiplication form a cyclic group if p is a prime.
6. Give an example of a non-abelian group in which $(xy)^3 = x^3 y^3$ for all x and y.
7. If G is a finite abelian group, prove that the number of solutions of $x^n = e$ in G, where $n \mid o(G)$ is a multiple of n.
8. Same as Problem 7, but do not assume the group to be abelian.
9. Find all automorphisms of S_3 and S_4, the symmetric groups of degree 3 and 4.

DEFINITION *A group G is said to be* solvable *if there exist subgroups $G = N_0 \supset N_1 \supset N_2 \supset \cdots \supset N_r = (e)$ such that each N_i is normal in N_{i-1} and N_{i-1}/N_i is abelian.*

10. Prove that a subgroup of a solvable group and the homomorphic image of a solvable group must be solvable.
11. If G is a group and N is a normal subgroup of G such that both N and G/N are solvable, prove that G is solvable.
12. If G is a group, A a subgroup of G and N a normal subgroup of G, prove that if both A and N are solvable then so is AN.

13. If G is a group, define the sequence of subgroups $G^{(i)}$ of G by
    (1) $G^{(1)}$ = commutator subgroup of G = subgroup of G generated by all $aba^{-1}b^{-1}$ where $a, b \in G$.
    (2) $G^{(i)}$ = commutator subgroup of $G^{(i-1)}$ if $i > 1$.
    Prove
    (a) Each $G^{(i)}$ is a normal subgroup of G.
    (b) G is solvable if and only if $G^{(k)} = (e)$ for some $k \geq 1$.
14. Prove that a solvable group always has an abelian normal subgroup $M \neq (e)$.
    If G is a group, define the sequence of subgroups $G_{(i)}$ by
    (a) $G_{(1)}$ = commutator subgroup of G.
    (b) $G_{(i)}$ = subgroup of G generated by all $aba^{-1}b^{-1}$ where $a \in G$, $b \in G_{(i-1)}$.

G is said to be *nilpotent* if $G_{(k)} = (e)$ for some $k \geq 1$.

15. (a) Show that each $G_{(i)}$ is a normal subgroup of G and $G_{(i)} \supset G^{(i)}$.
    (b) If G is nilpotent, prove it must be solvable.
    (c) Give an example of a group which is solvable but not nilpotent.
16. Show that any subgroup and homomorphic image of a nilpotent group must be nilpotent.
17. Show that every homomorphic image, different from (e), of a nilpotent group has a nontrivial center.
18. (a) Show that any group of order p^n, p a prime, must be nilpotent.
    (b) If G is nilpotent, and $H \neq G$ is a subgroup of G, prove that $N(H) \neq H$ where $N(H) = \{x \in G \mid xHx^{-1} = H\}$.
19. If G is a finite group, prove that G is nilpotent if and only if G is the direct product of its Sylow subgroups.
20. Let G be a finite group and H a subgroup of G. For A, B subgroups of G, define A to be conjugate to B relative to H if $B = x^{-1}Ax$ for some $x \in H$. Prove
    (a) This defines an equivalence relation on the set of subgroups of G.
    (b) The number of subgroups of G conjugate to A relative to H equals the index of $N(A) \cap H$ in H.
21. (a) If G is a finite group and if P is a p-Sylow subgroup of G, prove that P is the only p-Sylow subgroup in $N(P)$.
    (b) If P is a p-Sylow subgroup of G and if $a^{p^k} = e$ then, if $a \in N(P)$, a must be in P.
    (c) Prove that $N(N(P)) = N(P)$.
22. (a) If G is a finite group and P is a p-Sylow subgroup of G, prove that the number of conjugates of P in G is *not* a multiple of p.

(b) Breaking up the conjugate class of P further by using conjugacy relative to P, prove that the conjugate class of P has $1 + kp$ distinct subgroups. (*Hint:* Use part (b) of Problem 20 and Problem 21. Note that together with Problem 23 this gives an alternative proof of Theorem 2.12.3, the third part of Sylow's theorem.)

23. (a) If P is a p-Sylow subgroup of G and B is a subgroup of G of order p^k, prove that if B is not contained in some conjugate of P, then the number of conjugates of P in G is a multiple of p.
    (b) Using part (a) and Problem 22, prove that B must be contained in some conjugate of P.
    (c) Prove that any two p-Sylow subgroups of G are conjugate in G. (This gives another proof of Theorem 2.12.2, the second part of Sylow's theorem.)
24. Combine Problems 22 and 23 to give another proof of all parts of Sylow's theorem.
25. Making a case-by-case discussion using the results developed in this chapter, prove that any group of order less than 60 either is of prime order or has a nontrivial normal subgroup.
26. Using the result of Problem 25, prove that any group of order less than 60 is solvable.
27. Show that the equation $x^2ax = a^{-1}$ is solvable for x in the group G if and only if a is the cube of some element in G.
28. Prove that $(1\ 2\ 3)$ is not a cube of any element in S_n.
29. Prove that $xax = b$ is solvable for x in G if and only if ab is the square of some element in G.
30. If G is a group and $a \in G$ is of finite order and has only a finite number of conjugates in G, prove that these conjugates of a generate a finite normal subgroup of G.
31. Show that a group cannot be written as the set-theoretic union of two proper subgroups.
32. Show that a group G is the set-theoretic union of three proper subgroups if and only if G has, as a homomorphic image, a noncyclic group of order 4.

#33. Let p be a prime and let Z_p be the integers mod p under addition and multiplication. Let G be the group $\begin{pmatrix} a & b \\ c & d \end{pmatrix}$ where $a, b, c, d \in Z_p$ are such that $ad - bc = 1$. Let

$$C = \left\{ \begin{pmatrix} 1 & 0 \\ 0 & 1 \end{pmatrix}, \begin{pmatrix} -1 & 0 \\ 0 & -1 \end{pmatrix} \right\}$$

and let $LF(2, p) = G/C$.

(a) Find the order of $LF(2, p)$.
(b) Prove that $LF(2, p)$ is simple if $p \geq 5$.

#34. Prove that $LF(2, 5)$ is isomorphic to A_5, the alternating group of degree 5.

#35. Let $G = LF(2, 7)$; according to Problem 33, G is a simple group of order 168. Determine exactly how many 2-Sylow, 3-Sylow, and 7-Sylow subgroups there are in G.

Supplementary Reading

BURNSIDE, W., *Theory of Groups of Finite Order*, 2nd ed. Cambridge, England: Cambridge University Press, 1911; New York: Dover Publications, 1955.
HALL, MARSHALL, *Theory of Groups*. New York: The Macmillan Company, 1961.

Topics for Class Discussion

ALPERIN, J. L., "A classification of n-abelian groups," *Canadian Journal of Mathematics*, Vol. XXI (1969), pages 1238–1244.
McKAY, JAMES, H., "Another proof of Cauchy's group theorem," *American Mathematical Monthly*, Vol. 66 (1959), page 119.
SEGAL, I. E., "The automorphisms of the symmetric group," *Bulletin of the American Mathematical Society*, Vol. 46 (1940), page 565.

3

Ring Theory

3.1 Definition and Examples of Rings

As we indicated in Chapter 2, there are certain algebraic systems which serve as the building blocks for the structures comprising the subject which is today called modern algebra. At this stage of the development we have learned something about one of these, namely groups. It is our purpose now to introduce and to study a second such, namely rings. The abstract concept of a group has its origins in the set of mappings, or permutations, of a set onto itself. In contrast, rings stem from another and more familiar source, the set of integers. We shall see that they are patterned after, and are generalizations of, the algebraic aspects of the ordinary integers.

In the next paragraph it will become clear that a ring is quite different from a group in that it is a two-operational system; these operations are usually called addition and multiplication. Yet, despite the differences, the analysis of rings will follow the pattern already laid out for groups. We shall require the appropriate analogs of homomorphism, normal subgroups, factor groups, etc. With the experience gained in our study of groups we shall be able to make the requisite definitions, intertwine them with meaningful theorems, and end up proving results which are both interesting and important about mathematical objects with which we have had long acquaintance. To cite merely one instance, later on in the book, using the tools developed here, we shall prove that it is impossible to trisect an angle of 60° using only a straight-edge and compass.

DEFINITION A nonempty set R is said to be an *associative ring* if in R there are defined two operations, denoted by $+$ and $\cdot$ respectively, such that for all a, b, c in R:

1. $a + b$ is in R.
2. $a + b = b + a$.
3. $(a + b) + c = a + (b + c)$.
4. There is an element 0 in R such that $a + 0 = a$ (for every a in R).
5. There exists an element $-a$ in R such that $a + (-a) = 0$.
6. $a \cdot b$ is in R.
7. $a \cdot (b \cdot c) = (a \cdot b) \cdot c$.
8. $a \cdot (b + c) = a \cdot b + a \cdot c$ and $(b + c) \cdot a = b \cdot a + c \cdot a$ (the two distributive laws).

Axioms 1 through 5 merely state that R is an abelian group under the operation $+$, which we call addition. Axioms 6 and 7 insist that R be closed under an associative operation $\cdot$, which we call multiplication. Axiom 8 serves to interrelate the two operations of R.

Whenever we speak of ring it will be understood we mean associative ring. Nonassociative rings, that is, those in which axiom 7 may fail to hold, do occur in mathematics and are studied, but we shall have no occasion to consider them.

It may very well happen, or not happen, that there is an element 1 in R such that $a \cdot 1 = 1 \cdot a = a$ for every a in R; if there is such we shall describe R as a *ring with unit element.*

If the multiplication of R is such that $a \cdot b = b \cdot a$ for every a, b in R, then we call R a *commutative ring.*

Before going on to work out some properties of rings, we pause to examine some examples. Motivated by these examples we shall define various special types of rings which are of importance.

Example 3.1.1 R is the set of integers, positive, negative, and 0; $+$ is the usual addition and $\cdot$ the usual multiplication of integers. R is a commutative ring with unit element.

Example 3.1.2 R is the set of even integers under the usual operations of addition and multiplication. R is a commutative ring but has no unit element.

Example 3.1.3 R is the set of rational numbers under the usual addition and multiplication of rational numbers. R is a commutative ring with unit element. But even more than that, note that the elements of R different from 0 form an abelian group under multiplication. A ring with this latter property is called a *field.*

Example 3.1.4 R is the set of integers mod 7 under the addition and multiplication mod 7. That is, the elements of R are the seven symbols $\bar{0}, \bar{1}, \bar{2}, \bar{3}, \bar{4}, \bar{5}, \bar{6}$, where

1. $\bar{\imath} + \bar{\jmath} = \bar{k}$ where k is the remainder of $i + j$ on division by 7 (thus, for instance, $\bar{4} + \bar{5} = \bar{2}$ since $4 + 5 = 9$, which, when divided by 7, leaves a remainder of 2).
2. $\bar{\imath} \cdot \bar{\jmath} = \bar{m}$ where m is the remainder of ij on division by 7 (thus, $\bar{5} \cdot \bar{3} = \bar{1}$ since $5 \cdot 3 = 15$ has 1 as a remainder on division by 7).

The student should verify that R is a commutative ring with unit element. However, much more can be shown; namely, since

$$\bar{1} \cdot \bar{1} = \bar{1} = \bar{6} \cdot \bar{6},$$
$$\bar{2} \cdot \bar{4} = \bar{1} = \bar{4} \cdot \bar{2},$$
$$\bar{3} \cdot \bar{5} = \bar{1} = \bar{5} \cdot \bar{3},$$

the nonzero elements of R form an abelian group under multiplication. R is thus a field. Since it only has a finite number of elements it is called a *finite field.*

Example 3.1.5 R is the set of integers mod 6 under addition and multiplication mod 6. If we denote the elements in R by $\bar{0}, \bar{1}, \bar{2}, \ldots, \bar{5}$, one sees that $\bar{2} \cdot \bar{3} = \bar{0}$, yet $\bar{2} \neq \bar{0}$ and $\bar{3} \neq \bar{0}$. *Thus it is possible in a ring R that $a \cdot b = 0$ with neither $a = 0$ nor $b = 0$.* This cannot happen in a field (see Problem 10, end of Section 3.2), thus the ring R in this example is certainly not a field.

Every example given so far has been a commutative ring. We now present a noncommutative ring.

Example 3.1.6 R will be the set of all symbols

$$\alpha_{11}e_{11} + \alpha_{12}e_{12} + \alpha_{21}e_{21} + \alpha_{22}e_{22} = \sum_{i,j=1}^{2} \alpha_{ij}e_{ij},$$

where all the α_{ij} are rational numbers and where we decree

$$\sum_{i,j=1}^{2} \alpha_{ij}e_{ij} = \sum_{i,j=1}^{2} \beta_{ij}e_{ij} \tag{1}$$

if and only if for all $i, j = 1, 2$, $\alpha_{ij} = \beta_{ij}$,

$$\sum_{i,j=1}^{2} \alpha_{ij}e_{ij} + \sum_{i,j=1}^{2} \beta_{ij}e_{ij} = \sum_{i,j=1}^{2} (\alpha_{ij} + \beta_{ij})e_{ij}. \tag{2}$$

$$\left(\sum_{i,j=1}^{2} \alpha_{ij}e_{ij}\right) \cdot \left(\sum_{i,j=1}^{2} \beta_{ij}e_{ij}\right) = \sum_{i,j=1}^{2} \gamma_{ij}e_{ij}, \tag{3}$$

where

$$\gamma_{ij} = \sum_{\nu=1}^{2} \alpha_{i\nu}\beta_{\nu j} = \alpha_{i1}\beta_{1j} + \alpha_{i2}\beta_{2j}.$$

This multiplication, when first seen, looks rather complicated. However, it is founded on relatively simple rules, namely, multiply $\sum\alpha_{ij}e_{ij}$ by $\sum\beta_{ij}e_{ij}$ formally, multiplying out term by term, and collecting terms, and using the relations $e_{ij} \cdot e_{kl} = 0$ for $j \neq k$, $e_{ij} \cdot e_{jl} = e_{il}$ in this term-by-term collecting. (Of course those of the readers who have already encountered some linear algebra will recognize this example as the ring of all 2×2 matrices over the field of rational numbers.)

To illustrate the multiplication, if $a = e_{11} - e_{21} + e_{22}$ and $b = e_{22} + 3e_{12}$, then

$$\begin{aligned} a \cdot b &= (e_{11} - e_{21} + e_{22}) \cdot (e_{22} + 3e_{12}) \\ &= e_{11} \cdot e_{22} + 3e_{11} \cdot e_{12} - e_{21} \cdot e_{22} - 3e_{21} \cdot e_{12} + e_{22} \cdot e_{22} + 3e_{22} \cdot e_{12} \\ &= 0 + 3e_{12} - 0 - 3e_{22} + e_{22} + 0 \\ &= 3e_{12} - 3e_{22} + e_{22} = 3e_{12} - 2e_{22}. \end{aligned}$$

Note that $e_{11} \cdot e_{12} = e_{12}$ whereas $e_{12} \cdot e_{11} = 0$. Thus the multiplication in R is not commutative. Also it is possible for $u \cdot v = 0$ with $u \neq 0$ and $v \neq 0$.

The student should verify that R is indeed a ring. It is called the ring of 2×2 rational matrices. It, and its relative, will occupy a good deal of our time later on in the book.

Example 3.1.7 Let C be the set of all symbols (α, β) where α, β are real numbers. We define

$$(\alpha, \beta) = (\gamma, \delta) \text{ if and only if } \alpha = \gamma \text{ and } \beta = \delta. \tag{1}$$

In C we introduce an addition by defining for $x = (\alpha, \beta)$, $y = (\gamma, \delta)$

$$x + y = (\alpha, \beta) + (\gamma, \delta) = (\alpha + \gamma, \beta + \delta). \tag{2}$$

Note that $x + y$ is again in C. We assert that C is an abelian group under this operation with $(0, 0)$ serving as the identity element for addition, and $(-\alpha, -\beta)$ as the inverse, under addition, of (α, β).

Now that C is endowed with an addition, in order to make of C a ring we still need a multiplication. We achieve this by defining

$$\text{for } X = (\alpha, \beta), \qquad Y = (\gamma, \delta) \text{ in } C,$$

$$X \cdot Y = (\alpha, \beta) \cdot (\gamma, \delta) = (\alpha\gamma - \beta\delta, \alpha\delta + \beta\gamma). \tag{3}$$

Note that $X \cdot Y = Y \cdot X$. Also $X \cdot (1, 0) = (1, 0) \cdot X = X$ so that $(1, 0)$ is a unit element for C.

Again we notice that $X \cdot Y \in C$. Also, if $X = (\alpha, \beta) \neq (0, 0)$ then, since α, β are real and not both 0, $\alpha^2 + \beta^2 \neq 0$; thus

$$Y = \left(\frac{\alpha}{\alpha^2 + \beta^2}, \frac{-\beta}{\alpha^2 + \beta^2}\right)$$

is in C. Finally we see that

$$(\alpha, \beta) \cdot \left(\frac{\alpha}{\alpha^2 + \beta^2}, \frac{-\beta}{\alpha^2 + \beta^2}\right) = (1, 0).$$

All in all we have shown that C is a field. If we write (α, β) as $\alpha + \beta i$, the reader may verify that C is merely a disguised form of the familiar complex numbers.

Example 3.1.8 This last example is often called the ring of *real quaternions.* This ring was first described by the Irish mathematician Hamilton. Initially it was extensively used in the study of mechanics; today its primary interest is that of an important example, although it still plays key roles in geometry and number theory.

Let Q be the set of all symbols $\alpha_0 + \alpha_1 i + \alpha_2 j + \alpha_3 k$, where all the numbers α_0, α_1, α_2, and α_3 are real numbers. We declare two such symbols, $\alpha_0 + \alpha_1 i + \alpha_2 j + \alpha_3 k$ and $\beta_0 + \beta_1 i + \beta_2 j + \beta_3 k$, to be equal if and only if $\alpha_t = \beta_t$ for $t = 0, 1, 2, 3$. In order to make Q into a ring we must define a $+$ and a $\cdot$ for its elements. To this end we define

1. For any $X = \alpha_0 + \alpha_1 i + \alpha_2 j + \alpha_3 k$, $Y = \beta_0 + \beta_1 i + \beta_2 j + \beta_3 k$ in Q, $X + Y = (\alpha_0 + \alpha_1 i + \alpha_2 j + \alpha_3 k) + (\beta_0 + \beta_1 i + \beta_2 j + \beta_3 k) = (\alpha_0 + \beta_0) + (\alpha_1 + \beta_1)i + (\alpha_2 + \beta_2)j + (\alpha_3 + \beta_3)k$

and

2. $X \cdot Y = (\alpha_0 + \alpha_1 i + \alpha_2 j + \alpha_3 k) \cdot (\beta_0 + \beta_1 i + \beta_2 j + \beta_3 k) = (\alpha_0\beta_0 - \alpha_1\beta_1 - \alpha_2\beta_2 - \alpha_3\beta_3) + (\alpha_0\beta_1 + \alpha_1\beta_0 + \alpha_2\beta_3 - \alpha_3\beta_2)i + (\alpha_0\beta_2 + \alpha_2\beta_0 + \alpha_3\beta_1 - \alpha_1\beta_3)j + (\alpha_0\beta_3 + \alpha_3\beta_0 + \alpha_1\beta_2 - \alpha_2\beta_1)k$.

Admittedly this formula for the product seems rather formidable; however, it looks much more complicated than it actually is. It comes from multiplying out two such symbols formally and collecting terms using the relations $i^2 = j^2 = k^2 = ijk = -1$, $ij = -ji = k$, $jk = -kj = i$, $ki = -ik = j$. The latter part of these relations, called the multiplication table of the quaternion units, can be remembered by the little diagram on page 125. As you go around clockwise you read off the product, e.g., $ij = k$, $jk = i$, $ki = j$; while going around counterclockwise you read off the negatives.

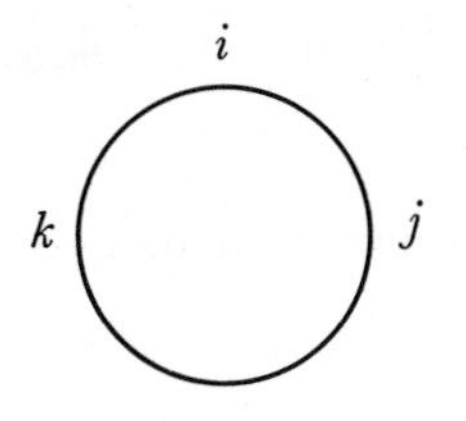

Notice that the elements ± 1, $\pm i$, $\pm j$, $\pm k$ form a non-abelian group of order 8 under this product. In fact, this is the group we called the group of quaternion units in Chapter 2.

The reader may prove that Q is a noncommutative ring in which $0 = 0 + 0i + 0j + 0k$ and $1 = 1 + 0i + 0j + 0k$ serve as the zero and unit elements respectively. Now if $X = \alpha_0 + \alpha_1 i + \alpha_2 j + \alpha_3 k$ is not 0, then not all of $\alpha_0, \alpha_1, \alpha_2, \alpha_3$ are 0; since they are real, $\beta = \alpha_0^2 + \alpha_1^2 + \alpha_2^2 + \alpha_3^2 \neq 0$ follows. Thus

$$Y = \frac{\alpha_0}{\beta} - \frac{\alpha_1}{\beta} i - \frac{\alpha_2}{\beta} j - \frac{\alpha_3}{\beta} k \in Q.$$

A simple computation now shows that $X \cdot Y = 1$. Thus the nonzero elements of Q form a non-abelian group under multiplication. A ring in which the nonzero elements form a group is called a *division ring* or *skew-field*. Of course, a commutative division ring is a field. Q affords us a division ring which is not a field. Many other examples of noncommutative division rings exist, but we would be going too far afield to present one here. The investigation of the nature of division rings and the attempts to classify them form an important part of algebra.

3.2 Some Special Classes of Rings

The examples just discussed in Section 3.1 point out clearly that although rings are a direct generalization of the integers, certain arithmetic facts to which we have become accustomed in the ring of integers need not hold in general rings. For instance, we have seen the possibility of $a \cdot b = 0$ with neither a nor b being zero. Natural examples exist where $a \cdot b \neq b \cdot a$. All these run counter to our experience heretofore.

For simplicity of notation we shall henceforth drop the dot in $a \cdot b$ and merely write this product as ab.

DEFINITION If R is a commutative ring, then $a \neq 0 \in R$ is said to be a *zero-divisor* if there exists a $b \in R$, $b \neq 0$, such that $ab = 0$.

DEFINITION A commutative ring is an *integral domain* if it has no zero-divisors.

The ring of integers, naturally enough, is an example of an integral domain.

DEFINITION A ring is said to be a *division ring* if its nonzero elements form a group under multiplication.

The unit element under multiplication will be written as 1, and the inverse of an element a under multiplication will be denoted by a^{-1}.

Finally we make the definition of the ultra-important object known as a field.

DEFINITION A *field* is a commutative division ring.

In our examples in Section 3.1, we exhibited the noncommutative division ring of real quaternions and the following fields: the rational numbers, complex numbers, and the integers mod 7. Chapter 5 will concern itself with fields and their properties.

We wish to be able to compute in rings in much the same manner in which we compute with real numbers, keeping in mind always that there are differences—it may happen that $ab \neq ba$, or that one cannot divide. To this end we prove the next lemma, which asserts that certain things we should like to be true in rings are indeed true.

LEMMA 3.2.1 *If R is a ring, then for all $a, b \in R$*

1. $a0 = 0a = 0$.
2. $a(-b) = (-a)b = -(ab)$.
3. $(-a)(-b) = ab$.

If, in addition, R has a unit element 1, *then*

4. $(-1)a = -a$.
5. $(-1)(-1) = 1$.

Proof.

1. If $a \in R$, then $a0 = a(0 + 0) = a0 + a0$ (using the right distributive law), and since R is a group under addition, this equation implies that $a0 = 0$.

   Similarly, $0a = (0 + 0)a = 0a + 0a$, using the left distributive law, and so here too, $0a = 0$ follows.
2. In order to show that $a(-b) = -(ab)$ we must demonstrate that $ab + a(-b) = 0$. But $ab + a(-b) = a(b + (-b)) = a0 = 0$ by use of

the distributive law and the result of part 1 of this lemma. Similarly $(-a)b = -(ab)$.

3. That $(-a)(-b) = ab$ is really a special case of part 2; we single it out since its analog in the case of real numbers has been so stressed in our early education. So on with it:

$$\begin{aligned}(-a)(-b) &= -(a(-b)) \quad \text{(by part 2)}\\ &= -(-(ab)) \quad \text{(by part 2)}\\ &= ab\end{aligned}$$

since $-(-x) = x$ is a consequence of the fact that in any group $(u^{-1})^{-1} = u$.

4. Suppose that R has a unit element 1; then $a + (-1)a = 1a + (-1)a = (1 + (-1))a = 0a = 0$, whence $(-1)a = -a$. In particular, if $a = -1$, $(-1)(-1) = -(-1) = 1$, which establishes part 5.

With this lemma out of the way we shall, from now on, feel free to compute with negatives and 0 as we always have in the past. The result of Lemma 3.2.1 is our permit to do so. For convenience, $a + (-b)$ will be written $a - b$.

The lemma just proved, while it is very useful and important, is not very exciting. So let us proceed to results of greater interest. Before we do so, we enunciate a principle which, though completely trivial, provides a mighty weapon when wielded properly. This principle says no more or less than the following: if a postman distributes 101 letters to 100 mailboxes then some mailbox must receive at least two letters. It does not sound very promising as a tool, does it? Yet it will surprise us! Mathematical ideas can often be very difficult and obscure, but no such argument can be made against this very simple-minded principle given above. We formalize it and even give it a name.

THE PIGEONHOLE PRINCIPLE *If n objects are distributed over m places, and if $n > m$, then some place receives at least two objects.*

An equivalent formulation, and one which we shall often use is: If n objects are distributed over n places in such a way that no place receives more than one object, then each place receives *exactly* one object.

We immediately make use of this idea in proving

LEMMA 3.2.2 *A finite integral domain is a field.*

Proof. As we may recall, an integral domain is a commutative ring such that $ab = 0$ if and only if at least one of a or b is itself 0. A field, on the other hand, is a commutative ring with unit element in which every nonzero element has a multiplicative inverse in the ring.

Let D be a finite integral domain. In order to prove that D is a field we must

1. Produce an element $1 \in D$ such that $a1 = a$ for every $a \in D$.
2. For every element $a \neq 0 \in D$ produce an element $b \in D$ such that $ab = 1$.

Let $x_1, x_2, \ldots, x_n$ be all the elements of D, and suppose that $a \neq 0 \in D$. Consider the elements $x_1 a, x_2 a, \ldots, x_n a$; they are all in D. We claim that they are all distinct! For suppose that $x_i a = x_j a$ for $i \neq j$; then $(x_i - x_j)a = 0$. Since D is an integral domain and $a \neq 0$, this forces $x_i - x_j = 0$, and so $x_i = x_j$, contradicting $i \neq j$. Thus $x_1 a, x_2 a, \ldots, x_n a$ are n distinct elements lying in D, which has exactly n elements. By the pigeonhole principle these must account for all the elements of D; stated otherwise, every element $y \in D$ can be written as $x_i a$ for some x_i. In particular, since $a \in D$, $a = x_{i_0} a$ for some $x_{i_0} \in D$. Since D is commutative, $a = x_{i_0} a = a x_{i_0}$. We propose to show that x_{i_0} acts as a unit element for every element of D. For, if $y \in D$, as we have seen, $y = x_i a$ for some $x_i \in D$, and so $y x_{i_0} = (x_i a) x_{i_0} = x_i (a x_{i_0}) = x_i a = y$. Thus x_{i_0} is a unit element for D and we write it as 1. Now $1 \in D$, so by our previous argument, it too is realizable as a multiple of a; that is, there exists a $b \in D$ such that $1 = ba$. The lemma is now completely proved.

COROLLARY *If p is a prime number then J_p, the ring of integers* mod p, *is a field.*

Proof. By the lemma it is enough to prove that J_p is an integral domain, since it only has a finite number of elements. If $a, b \in J_p$ and $ab \equiv 0$, then p must divide the ordinary integer ab, and so p, being a prime, must divide a or b. But then either $a \equiv 0 \bmod p$ or $b \equiv 0 \bmod p$, hence in J_p one of these is 0.

The corollary above assures us that we can find an infinity of fields having a finite number of elements. Such fields are called *finite fields*. The fields J_p do not give all the examples of finite fields; there are others. In fact, in Section 7.1 we give a complete description of all finite fields.

We point out a striking difference between finite fields and fields such as the rational numbers, real numbers, or complex numbers, with which we are more familiar.

Let F be a finite field having q elements (if you wish, think of J_p with its p elements). Viewing F merely as a group under addition, since F has q elements, by Corollary 2 to Theorem 2.4.1,

$$\underbrace{a + a + \cdots + a}_{q\text{-times}} = qa = 0$$

for any $a \in F$. Thus, in F, we have $qa = 0$ for some positive integer q, even if $a \neq 0$. This certainly cannot happen in the field of rational numbers, for instance. We formalize this distinction in the definitions we give below. In these definitions, instead of talking just about fields, we choose to widen the scope a little and talk about integral domains.

DEFINITION An integral domain D is said to be of *characteristic* 0 if the relation $ma = 0$, where $a \neq 0$ is in D, and where m is an integer, can hold only if $m = 0$.

The ring of integers is thus of characteristic 0, as are other familiar rings such as the even integers or the rationals.

DEFINITION An integral domain D is said to be of *finite characteristic* if there exists a *positive* integer m such that $ma = 0$ for all $a \in D$.

If D is of finite characteristic, then we define the *characteristic* of D to be the smallest positive integer p such that $pa = 0$ for all $a \in D$. It is not too hard to prove that if *D is of finite characteristic, then its characteristic is a prime number* (see Problem 6 below).

As we pointed out, any finite field is of finite characteristic. However, an integral domain may very well be infinite yet be of finite characteristic (see Problem 7).

One final remark on this question of characteristic: Why define it for integral domains, why not for arbitrary rings? The question is perfectly reasonable. Perhaps the example we give now points out what can happen if we drop the assumption "integral domain."

Let R be the set of all triples (a, b, c), where $a \in J_2$, the integers mod 2, $b \in J_3$, the integers mod 3, and c is any integer. We introduce a $+$ and a $\cdot$ to make of R a ring. We do so by defining $(a_1, b_1, c_1) + (a_2, b_2, c_2) = (a_1 + a_2, b_1 + b_2, c_1 + c_2)$ and $(a_1, b_1, c_1) \cdot (a_2, b_2, c_2) = (a_1a_2, b_1b_2, c_1c_2)$. It is easy to verify that R is a commutative ring. It is not an integral domain since $(1, 2, 0) \cdot (0, 0, 7) = (0, 0, 0)$, the zero-element of R. Note that in R, $2(1, 0, 0) = (1, 0, 0) + (1, 0, 0) = (2, 0, 0) = (0, 0, 0)$ since addition in the first component is in J_2. Similarly $3(0, 1, 0) = (0, 0, 0)$. Finally, for *no* positive integer m is $m(0, 0, 1) = (0, 0, 0)$.

Thus, from the point of view of the definition we gave above for characteristic, the ring R, which we just looked at, is neither fish nor fowl. The definition just doesn't have any meaning for R. We could generalize the notion of characteristic to arbitrary rings by doing it locally, defining it relative to given elements, rather than globally for the ring itself. We say that R has n-torsion, $n > 0$, if there is an element $a \neq 0$ in R such that $na = 0$, and $ma \neq 0$ for $0 < m < n$. For an integral domain D, it turns out that if D has n-torsion, even for one $n > 0$, then it must be of finite characteristic (see Problem 8).

Problems

R is a ring in all the problems.

1. If $a, b, c, d \in R$, evaluate $(a + b)(c + d)$.
2. Prove that if $a, b \in R$, then $(a + b)^2 = a^2 + ab + ba + b^2$, where by x^2 we mean xx.
3. Find the form of the binomial theorem in a general ring; in other words, find an expression for $(a + b)^n$, where n is a positive integer.
4. If every $x \in R$ satisfies $x^2 = x$, prove that R must be commutative. (A ring in which $x^2 = x$ for all elements is called a *Boolean* ring.)
5. If R is a ring, merely considering it as an abelian group under its addition, we have defined, in Chapter 2, what is meant by na, where $a \in R$ and n is an integer. Prove that if $a, b \in R$ and n, m are integers, then $(na)(mb) = (nm)(ab)$.
6. If D is an integeral domain and D is of finite characteristic, prove that the characteristic of D is a prime number.
7. Give an example of an integral domain which has an infinite number of elements, yet is of finite characteristic.
8. If D is an integral domain and if $na = 0$ for some $a \neq 0$ in D and some integer $n \neq 0$, prove that D is of finite characteristic.
9. If R is a system satisfying all the conditions for a ring with unit element with the possible exception of $a + b = b + a$, prove that the axiom $a + b = b + a$ must hold in R and that R is thus a ring. (*Hint:* Expand $(a + b)(1 + 1)$ in two ways.)
10. Show that the commutative ring D is an integral domain if and only if for $a, b, c \in D$ with $a \neq 0$ the relation $ab = ac$ implies that $b = c$.
11. Prove that Lemma 3.2.2 is false if we drop the assumption that the integral domain is finite.
12. Prove that any field is an integral domain.
13. Useing the pigeonhole principle, prove that if m and n are relatively prime integers and a and b are any integers, there exists an integer x such that $x \equiv a \bmod m$ and $x \equiv b \bmod n$. (*Hint:* Consider the remainders of $a, a + m, a + 2m, \ldots, a + (n - 1)m$ on division by n.)
14. Using the pigeonhole principle, prove that the decimal expansion of a rational number must, after some point, become repeating.

3.3 Homomorphisms

In studying groups we have seen that the concept of a homomorphism turned out to be a fruitful one. This suggests that the appropriate analog for rings could also lead to important ideas. To recall, for groups a homomorphism was defined as a mapping such that $\phi(ab) = \phi(a)\phi(b)$. Since a ring has two operations, what could be a more natural extension of this type of formula than the

DEFINITION A mapping ϕ from the ring R into the ring R' is said to be a *homomorphism* if

1. $\phi(a + b) = \phi(a) + \phi(b)$,
2. $\phi(ab) = \phi(a)\phi(b)$,

for all $a, b \in R$.

As in the case of groups, let us again stress here that the $+$ and $\cdot$ occurring on the left-hand sides of the relations in 1 and 2 are those of R, whereas the $+$ and $\cdot$ occurring on the right-hand sides are those of R'.

A useful observation to make is that a homomorphism of one ring, R, into another, R', is, if we totally ignore the multiplications in both these rings, at least a homomorphism of R into R' when we consider them as abelian groups under their respective additions. Therefore, as far as addition is concerned, all the properties about homomorphisms of groups proved in Chapter 2 carry over. In particular, merely restating Lemma 2.7.2 for the case of the additive group of a ring yields for us

LEMMA 3.3.1 *If ϕ is a homomorphism of R into R', then*

1. $\phi(0) = 0$.
2. $\phi(-a) = -\phi(a)$ *for every* $a \in R$.

A word of caution: if both R and R' have the respective unit elements 1 and $1'$ for their multiplications it need not follow that $\phi(1) = 1'$. However, if R' is an integral domain, or if R' is arbitrary but ϕ is onto, then $\phi(1) = 1'$ is indeed true.

In the case of groups, given a homomorphism we associated with this homomorphism a certain subset of the group which we called the kernel of the homomorphism. What should the appropriate definition of the kernel of a homomorphism be for rings? After all, the ring has two operations, addition and multiplication, and it might be natural to ask which of these should be singled out as the basis for the definition. However, the choice is clear. Built into the definition of an arbitrary ring is the condition that the ring forms an abelian group under addition. The ring multiplication was left much more unrestricted, and so, in a sense, much less under our control than is the addition. For this reason the emphasis is given to the operation of addition in the ring, and we make the

DEFINITION If ϕ is a homomorphism of R into R' then the *kernel of* ϕ, $I(\phi)$, is the set of all elements $a \in R$ such that $\phi(a) = 0$, the zero-element of R'.

LEMMA 3.3.2 *If ϕ is a homomorphism of R into R' with kernel $I(\phi)$, then*

1. *$I(\phi)$ is a subgroup of R under addition.*
2. *If $a \in I(\phi)$ and $r \in R$ then both ar and ra are in $I(\phi)$.*

Proof. Since ϕ is, in particular, a homomorphism of R, as an additive group, into R', as an additive group, (1) follows directly from our results in group theory.

To see (2), suppose that $a \in I(\phi)$, $r \in R$. Then $\phi(a) = 0$ so that $\phi(ar) = \phi(a)\phi(r) = 0\phi(r) = 0$ by Lemma 3.2.1. Similarly $\phi(ra) = 0$. Thus by defining property of $I(\phi)$ both ar and ra are in $I(\phi)$.

Before proceeding we examine these concepts for certain examples.

Example 3.3.1 Let R and R' be two arbitrary rings and define $\phi(a) = 0$ for all $a \in R$. Trivially ϕ is a homomorphism and $I(\phi) = R$. ϕ is called the zero-homomorphism.

Example 3.3.2 Let R be a ring, $R' = R$ and define $\phi(x) = x$ for every $x \in R$. Clearly ϕ is a homomorphism and $I(\phi)$ consists only of 0.

Example 3.3.3 Let $J(\sqrt{2})$ be all real numbers of the form $m + n\sqrt{2}$ where m, n are integers; $J(\sqrt{2})$ forms a ring under the usual addition and multiplication of real numbers. (Verify!) Define $\phi:J(\sqrt{2}) \rightarrow J(\sqrt{2})$ by $\phi(m + n\sqrt{2}) = m - n\sqrt{2}$. ϕ is a homomorphism of $J(\sqrt{2})$ onto $J(\sqrt{2})$ and its kernel $I(\phi)$, consists only of 0. (Verify!)

Example 3.3.4 Let J be the ring of integers, J_n, the ring of integers modulo n. Define $\phi:J \rightarrow J_n$ by $\phi(a) =$ remainder of a on division by n. The student should verify that ϕ is a homomorphism of J onto J_n and that the kernel, $I(\phi)$, of ϕ consists of all multiples of n.

Example 3.3.5 Let R be the set of all continuous, real-valued functions on the closed unit interval. R is made into a ring by the usual addition and multiplication of functions; that it is a ring is a consequence of the fact that the sum and product of two continuous functions are continuous functions. Let F be the ring of real numbers and define $\phi: R \to F$ by $\phi(f(x)) = f(\frac{1}{2})$. ϕ is then a homomorphism of R onto F and its kernel consists of all functions in R vanishing at $x = \frac{1}{2}$.

All the examples given here have used commutative rings. Many beautiful examples exist where the rings are noncommutative but it would be premature to discuss such an example now.

DEFINITION A homomorphism of R into R' is said to be an *isomorphism* if it is a one-to-one mapping.

DEFINITION Two rings are said to be *isomorphic* if there is an isomorphism of one *onto* the other.

The remarks made in Chapter 2 about the meaning of an isomorphism and of the statement that two groups are isomorphic carry over verbatim to rings. Likewise, the criterion given in Lemma 2.7.4 that a homomorphism be an isomorphism translates directly from groups to rings in the form

LEMMA 3.3.3 *The homomorphism ϕ of R into R' is an isomorphism if and only if $I(\phi) = (0)$.*

3.4 Ideals and Quotient Rings

Once the idea of a homomorphism and its kernel have been set up for rings, based on our experience with groups, it should be fruitful to carry over some analog to rings of the concept of normal subgroup. Once this is achieved, one would hope that this analog would lead to a construction in rings like that of the quotient group of a group by a normal subgroup. Finally, if one were an optimist, one would hope that the homomorphism theorems for groups would come over in their entirety to rings.

Fortunately all this can be done, thereby providing us with an incisive technique for analyzing rings.

The first business at hand, then, seems to be to define a suitable "normal subgroup" concept for rings. With a little hindsight this is not difficult. If you recall, normal subgroups eventually turned out to be nothing else than kernels of homomorphisms, even though their primary defining conditions did not involve homomorphisms. Why not use this observation as the keystone to our definition for rings? Lemma 3.3.2 has already provided us with some conditions that a subset of a ring be the kernel of a homomorphism. We now take the point of view that, since no other information is at present available to us, we shall make the conclusions of Lemma 3.3.2 as the starting point of our endeavor, and so we define

DEFINITION A nonempty subset U of R is said to be a (two-sided) *ideal* of R if

1. U is a subgroup of R under addition.
2. For every $u \in U$ and $r \in R$, both ur and ru are in U.

Condition 2 asserts that U "swallows up" multiplication from the right and left by arbitrary ring elements. For this reason U is usually called a two-sided ideal. Since we shall have no occasion, other than in some of the problems, to use any other derivative concept of ideal, we shall merely use the word ideal, rather than two-sided ideal, in all that follows.

Given an ideal U of a ring R, let R/U be the set of all the distinct cosets of U in R which we obtain by considering U as a subgroup of R under addition. We note that we merely say coset, rather than right coset or left coset; this is justified since R is an abelian group under addition. To restate what we have just said, R/U consists of all the cosets, $a + U$, where $a \in R$. By the results of Chapter 2, R/U is automatically a group under addition; this is achieved by the composition law $(a + U) + (b + U) = (a + b) + U$. In order to impose a ring structure on R/U we must define, in it, a multiplication. What is more natural than to define $(a + U)(b + U) = ab + U$? However, we must make sure that this is meaningful. Otherwise put, we are obliged to show that if $a + U = a' + U$ and $b + U = b' + U$, then under our definition of the multiplication, $(a + U)(b + U) = (a' + U)(b' + U)$. Equivalently, it must be established that $ab + U = a'b' + U$. To this end we first note that since $a + U = a' + U$, $a = a' + u_1$, where $u_1 \in U$; similarly $b = b' + u_2$ where $u_2 \in U$. Thus $ab = (a' + u_1)(b + u_2) = a'b' + u_1b' + a'u_2 + u_1u_2$; *since U is an ideal of R*, $u_1b' \in U$, $a'u_2 \in U$, and $u_1u_2 \in U$. Consequently $u_1b' + a'u_2 + u_1u_2 = u_3 \in U$. But then $ab = a'b' + u_3$, from which we deduce that $ab + U = a'b' + u_3 + U$, and since $u_3 \in U$, $u_3 + U = U$. The net consequence of all this is that $ab + U = a'b' + U$. We at least have achieved the principal step on the road to our goal, namely of introducing a well-defined multiplication. The rest now becomes routine. To establish that R/U is a ring we merely have to go through the various axioms which define a ring and check whether they hold in R/U. All these verifications have a certain sameness to them, so we pick one axiom, the right distributive law, and prove it holds in R/U. The rest we leave to the student as informal exercises. If $X = a + U$, $Y = b + U$, $Z = c + U$ are three elements of R/U, where $a, b, c \in R$, then $(X + Y)Z = ((a + U) + (b + U))(c + U) = ((a + b) + U)(c + U) = (a + b)c + U = ac + bc + U = (ac + U) + (bc + U) = (a + U)(c + U) + (b + U)(c + U) = XZ + YZ$.

R/U has now been made into a ring. Clearly, if R is commutative then so is R/U, for $(a + U)(b + U) = ab + U = ba + U = (b + U)(a + U)$. (The converse to this is false.) If R has a unit element 1, then R/U has a unit element $1 + U$. We might ask: In what relation is R/U to R? With the experience we now have in hand this is easy to answer. There is a homomorphism ϕ of R *onto* R/U given by $\phi(a) = a + U$ for every $a \in R$, whose kernel is exactly U. (The reader should verify that ϕ so defined is a homomorphism of R onto R/U with kernel U.)

We summarize these remarks in

LEMMA 3.4.1 *If U is an ideal of the ring R, then R/U is a ring and is a homomorphic image of R.*

With this construction of the *quotient ring* of a ring by an ideal satisfactorily accomplished, we are ready to bring over to rings the homomorphism theorems of groups. Since the proof is an exact verbatim translation of that for groups into the language of rings we merely state the theorem without proof, referring the reader to Chapter 2 for the proof.

THEOREM 3.4.1 *Let R, R' be rings and ϕ a homomorphism of R onto R' with kernel U. Then R' is isomorphic to R/U. Moreover there is a one-to-one correspondence between the set of ideals of R' and the set of ideals of R which contain U. This correspondence can be achieved by associating with an ideal W' in R' the ideal W in R defined by $W = \{x \in R \mid \phi(x) \in W'\}$. With W so defined, R/W is isomorphic to R'/W'.*

Problems

1. If U is an ideal of R and $1 \in U$, prove that $U = R$.
2. If F is a field, prove its only ideals are (0) and F itself.
3. Prove that any homomorphism of a field is either an isomorphism or takes each element into 0.
4. If R is a commutative ring and $a \in R$,
   (a) Show that $aR = \{ar \mid r \in R\}$ is a two-sided ideal of R.
   (b) Show by an example that this may be false if R is not commutative.
5. If U, V are ideals of R, let $U + V = \{u + v \mid u \in U, v \in V\}$. Prove that $U + V$ is also an ideal.
6. If U, V are ideals of R let UV be the set of all elements that can be written as finite sums of elements of the form uv where $u \in U$ and $v \in V$. Prove that UV is an ideal of R.
7. In Problem 6 prove that $UV \subset U \cap V$.
8. If R is the ring of integers, let U be the ideal consisting of all multiples of 17. Prove that if V is an ideal of R and $R \supset V \supset U$ then either $V = R$ or $V = U$. Generalize!

9. If U is an ideal of R, let $r(U) = \{x \in R \mid xu = 0 \text{ for all } u \in U\}$. Prove that $r(U)$ is an ideal of R.
10. If U is an ideal of R let $[R:U] = \{x \in R \mid rx \in U \text{ for every } r \in R\}$. Prove that $[R:U]$ is an ideal of R and that it contains U.
11. Let R be a ring with unit element. Using its elements we define a ring $\tilde{R}$ by defining $a \oplus b = a + b + 1$, and $a \cdot b = ab + a + b$, where $a, b \in R$ and where the addition and multiplication on the right-hand side of these relations are those of R.
    (a) Prove that $\tilde{R}$ is a ring under the operations $\oplus$ and $\cdot$.
    (b) What acts as the zero-element of $\tilde{R}$?
    (c) What acts as the unit-element of $\tilde{R}$?
    (d) Prove that R is isomorphic to $\tilde{R}$.

*12. In Example 3.1.6 we discussed the ring of rational 2×2 matrices. Prove that this ring has no ideals other than (0) and the ring itself.

*13. In Example 3.1.8 we discussed the real quaternions. Using this as a model we define the quaternions over the integers mod p, p an odd prime number, in exactly the same way; however, now considering all symbols of the form $\alpha_0 + \alpha_1 i + \alpha_2 j + \alpha_3 k$, where $\alpha_0, \alpha_1, \alpha_2, \alpha_3$ are integers mod p.
    (a) Prove that this is a ring with p^4 elements whose only ideals are (0) and the ring itself.
    **(b) Prove that this ring is *not* a division ring.

If R is any ring a subset L of R is called a *left-ideal* of R if
1. L is a subgroup of R under addition.
2 $r \in R$, $a \in L$ implies $ra \in L$.

(One can similarly define a *right-ideal*.) An ideal is thus simultaneously a left- and right-ideal of R.

14. For $a \in R$ let $Ra = \{xa \mid x \in R\}$. Prove that Ra is a left-ideal of R.
15. Prove that the intersection of two left-ideals of R is a left-ideal of R.
16. What can you say about the intersection of a left-ideal and right-ideal of R?
17. If R is a ring and $a \in R$ let $r(a) = \{x \in R \mid ax = 0\}$. Prove that $r(a)$ is a right-ideal of R.
18. If R is a ring and L is a left-ideal of R let $\lambda(L) = \{x \in R \mid xa = 0 \text{ for all } a \in L\}$. Prove that $\lambda(L)$ is a two-sided ideal of R.

*19. Let R be a ring in which $x^3 = x$ for every $x \in R$. Prove that R is a commutative ring.

20. If R is a ring with unit element 1 and ϕ is a homomorphism of R *onto* R' prove that $\phi(1)$ is the unit element of R'.

21. If R is a ring with unit element 1 and ϕ is a homomorphism of R into an integral domain R' such that $I(\phi) \neq R$, prove that $\phi(1)$ is the unit element of R'.

3.5 More Ideals and Quotient Rings

We continue the discussion of ideals and quotient rings.

Let us take the point of view, for the moment at least, that a field is the most desirable kind of ring. Why? If for no other reason, we can divide in a field, so operations and results in a field more closely approximate our experience with real and complex numbers. In addition, as was illustrated by Problem 2 in the preceding problem set, a field has no homomorphic images other than itself or the trivial ring consisting of 0. Thus we cannot simplify a field by applying a homomorphism to it. Taking these remarks into consideration it is natural that we try to link a general ring, in some fashion, with fields. What should this linkage involve? We have a machinery whose component parts are homomorphisms, ideals, and quotient rings. With these we will forge the link.

But first we must make precise the rather vague remarks of the preceding paragraph. We now ask the explicit question: Under what conditions is the homomorphic image of a ring a field? For commutative rings we give a complete answer in this section.

Essential to treating this question is the converse to the result of Problem 2 of the problem list at the end of Section 3.4.

LEMMA 3.5.1 *Let R be a commutative ring with unit element whose only ideals are (0) and R itself. Then R is a field.*

Proof. In order to effect a proof of this lemma for any $a \neq 0 \in R$ we must produce an element $b \neq 0 \in R$ such that $ab = 1$.

So, suppose that $a \neq 0$ is in R. Consider the set $Ra = \{xa \mid x \in R\}$. We claim that Ra is an ideal of R. In order to establish this as fact we must show that it is a subgroup of R under addition and that if $u \in Ra$ and $r \in R$ then ru is also in Ra. (We only need to check that ru is in Ra for then ur also is since $ru = ur$.)

Now, if $u, v \in Ra$, then $u = r_1 a$, $v = r_2 a$ for some $r_1, r_2 \in R$. Thus $u + v = r_1 a + r_2 a = (r_1 + r_2)a \in Ra$; similarly $-u = -r_1 a = (-r_1)a \in Ra$. Hence Ra is an additive subgroup of R. Moreover, if $r \in R$, $ru = r(r_1 a) = (rr_1)a \in Ra$. Ra therefore satisfies all the defining conditions for an ideal of R, hence is an ideal of R. (Notice that both the distributive law and associative law of multiplication were used in the proof of this fact.)

By our assumptions on R, $Ra = (0)$ or $Ra = R$. Since $0 \neq a = 1a \in Ra$, $Ra \neq (0)$; thus we are left with the only other possibility, namely that $Ra = R$. This last equation states that every element in R is a multiple of

a by some element of R. In particular, $1 \in R$ and so it can be realized as a multiple of a; that is, there exists an element $b \in R$ such that $ba = 1$. This completes the proof of the lemma.

DEFINITION An ideal $M \neq R$ in a ring R is said to be a *maximal ideal* of R if whenever U is an ideal of R such that $M \subset U \subset R$, then either $R = U$ or $M = U$.

In other words, an ideal of R is a maximal ideal if it is impossible to squeeze an ideal between it and the full ring. Given a ring R there is no guarantee that it has any maximal ideals! If the ring has a unit element this can be proved, assuming a basic axiom of mathematics, the so-called axiom of choice. Also there may be many distinct maximal ideals in a ring R; this will be illustrated for us below in the ring of integers.

As yet we have made acquaintance with very few rings. Only by considering a given concept in many particular cases can one fully appreciate the concept and its motivation. Before proceeding we therefore examine some maximal ideals in two specific rings. When we come to the discussion of polynomial rings we shall exhibit there all the maximal ideals.

Example 3.5.1 Let R be the ring of integers, and let U be an ideal of R. Since U is a subgroup of R under addition, from our results in group theory, we know that U consists of all the multiples of a fixed integer n_0; we write this as $U = (n_0)$. What values of n_0 lead to maximal ideals?

We first assert that if p is a prime number then $P = (p)$ is a maximal ideal of R. For if U is an ideal of R and $U \supset P$, then $U = (n_0)$ for some integer n_0. Since $p \in P \subset U$, $p = mn_0$ for some integer m; because p is a prime this implies that $n_0 = 1$ or $n_0 = p$. If $n_0 = p$, then $P \subset U = (n_0) \subset P$, so that $U = P$ follows; if $n_0 = 1$, then $1 \in U$, hence $r = 1r \in U$ for all $r \in R$ whence $U = R$ follows. Thus no ideal, other than R or P itself, can be put between P and R, from which we deduce that P is maximal.

Suppose, on the other hand, that $M = (n_0)$ is a maximal ideal of R. We claim that n_0 must be a prime number, for if $n_0 = ab$, where a, b are positive integers, then $U = (a) \supset M$, hence $U = R$ or $U = M$. If $U = R$, then $a = 1$ is an easy consequence; if $U = M$, then $a \in M$ and so $a = rn_0$ for some integer r, since every element of M is a multiple of n_0. But then $n_0 = ab = rn_0b$, from which we get that $rb = 1$, so that $b = 1$, $n_0 = a$. Thus n_0 is a prime number.

In this particular example the notion of maximal ideal comes alive—it corresponds exactly to the notion of prime number. One should not, however, jump to any hasty generalizations; this kind of correspondence does not usually hold for more general rings.

Example 3.5.2 Let R be the ring of all the real-valued, continuous functions on the closed unit interval. (See Example 3.3.5.) Let

$$M = \{f(x) \in R \mid f(\tfrac{1}{2}) = 0\}.$$

M is certainly an ideal of R. Moreover, it is a maximal ideal of R, for if the ideal U contains M and $U \neq M$, then there is a function $g(x) \in U$, $g(x) \notin M$. Since $g(x) \notin M$, $g(\frac{1}{2}) = \alpha \neq 0$. Now $h(x) = g(x) - \alpha$ is such that $h(\frac{1}{2}) = g(\frac{1}{2}) - \alpha = 0$, so that $h(x) \in M \subset U$. But $g(x)$ is also in U; therefore $\alpha = g(x) - h(x) \in U$ and so $1 = \alpha\alpha^{-1} \in U$. Thus for any function $t(x) \in R$, $t(x) = 1t(x) \in U$, in consequence of which $U = R$. M is therefore a maximal ideal of R. Similarly if γ is a real number $0 \leq \gamma \leq 1$, then $M_\gamma = \{f(x) \in R \mid f(\gamma) = 0\}$ is a maximal ideal of R. It can be shown (see Problem 4 at the end of this section) that every maximal ideal is of this form. Thus here the maximal ideals correspond to the points on the unit interval.

Having seen some maximal ideals in some concrete rings we are ready to continue the general development with

THEOREM 3.5.1 *If R is a commutative ring with unit element and M is an ideal of R, then M is a maximal ideal of R if and only if R/M is a field.*

Proof. Suppose, first, that M is an ideal of R such that R/M is a field. Since R/M is a field its only ideals are (0) and R/M itself. But by Theorem 3.4.1 there is a one-to-one correspondence between the set of ideals of R/M and the set of ideals of R which contain M. The ideal M of R corresponds to the ideal (0) of R/M whereas the ideal R of R corresponds to the ideal R/M of R/M in this one-to-one mapping. Thus there is no ideal between M and R other than these two, whence M is a maximal ideal.

On the other hand, if M is a maximal ideal of R, by the correspondence mentioned above R/M has only (0) and itself as ideals. Furthermore R/M is commutative and has a unit element since R enjoys both these properties. All the conditions of Lemma 3.5.1 are fulfilled for R/M so we can conclude, by the result of that lemma, that R/M is a field.

We shall have many occasions to refer back to this result in our study of polynomial rings and in the theory of field extensions.

Problems

1. Let R be a ring with unit element, R not necessarily commutative, such that the only right-ideals of R are (0) and R. Prove that R is a division ring.

*2. Let R be a ring such that the only right ideals of R are (0) and R. Prove that either R is a division ring or that R is a ring with a prime number of elements in which $ab = 0$ for every $a, b \in R$.

3. Let J be the ring of integers, p a prime number, and (p) the ideal of J consisting of all multiples of p. Prove
   (a) $J/(p)$ is isomorphic to J_p, the ring of integers mod p.
   (b) Using Theorem 3.5.1 and part (a) of this problem, that J_p is a field.

**4. Let R be the ring of all real-valued continuous functions on the closed unit interval. If M is a maximal ideal of R, prove that there exists a real number γ, $0 \leq \gamma \leq 1$, such that $M = M_\gamma = \{f(x) \in R \mid f(\gamma) = 0\}$.

3.6 The Field of Quotients of an Integral Domain

Let us recall that an integral domain is a commutative ring D with the additional property that it has no zero-divisors, that is, if $ab = 0$ for some $a, b \in D$ then at least one of a or b must be 0. The ring of integers is, of course, a standard example of an integral domain.

The ring of integers has the attractive feature that we can enlarge it to the set of rational numbers, which is a field. Can we perform a similar construction for any integral domain? We will now proceed to show that indeed we can!

DEFINITION A ring R *can be imbedded* in a ring R' if there is an isomorphism of R into R'. (If R and R' have unit elements 1 and $1'$ we insist, in addition, that this isomorphism takes 1 onto $1'$.)

R' will be called an *over-ring* or *extension* of R if R can be imbedded in R'.

With this understanding of imbedding we prove

THEOREM 3.6.1 *Every integral domain can be imbedded in a field.*

Proof. Before becoming explicit in the details of the proof let us take an informal approach to the problem. Let D be our integral domain; roughly speaking the field we seek should be all quotients a/b, where $a, b \in D$ and $b \neq 0$. Of course in D, a/b may very well be meaningless. What should we require of these symbols a/b? Clearly we must have an answer to the following three questions:

1. When is $a/b = c/d$?
2. What is $(a/b) + (c/d)$?
3. What is $(a/b)(c/d)$?

In answer to 1, what could be more natural than to insist that $a/b = c/d$

if and only if $ad = bc$? As for 2 and 3, why not try the obvious, that is, define

$$\frac{a}{b} + \frac{c}{d} = \frac{ad + bc}{bd} \quad \text{and} \quad \frac{a}{b}\frac{c}{d} = \frac{ac}{bd}.$$

In fact in what is to follow we make these considerations our guide. So let us leave the heuristics and enter the domain of mathematics, with precise definitions and rigorous deductions.

Let $\mathscr{M}$ be the set of all ordered pairs (a, b) where $a, b \in D$ and $b \neq 0$. (Think of (a, b) as a/b.) In $\mathscr{M}$ we now define a relation as follows:

$$(a, b) \sim (c, d) \text{ if and only if } ad = bc.$$

We claim that this defines an equivalence relation on $\mathscr{M}$. To establish this we check the three defining conditions for an equivalence relation for this particular relation.

1. If $(a, b) \in \mathscr{M}$, then $(a, b) \sim (a, b)$ since $ab = ba$.
2. If $(a, b), (c, d) \in \mathscr{M}$ and $(a, b) \sim (c, d)$, then $ad = bc$, hence $cb = da$, and so $(c, d) \sim (a, b)$.
3. If $(a, b), (c, d), (e, f)$ are all in $\mathscr{M}$ and $(a, b) \sim (c, d)$ and $(c, d) \sim (e, f)$, then $ad = bc$ and $cf = de$. Thus $bcf = bde$, and since $bc = ad$, it follows that $adf = bde$. Since D is commutative, this relation becomes $afd = bed$; since, moreover, D is an integral domain and $d \neq 0$, this relation further implies that $af = be$. But then $(a, b) \sim (e, f)$ and our relation is transitive.

Let $[a, b]$ be the equivalence class in $\mathscr{M}$ of (a, b), and let F be the set of all such equivalence classes $[a, b]$ where $a, b \in D$ and $b \neq 0$. F is the candidate for the field we are seeking. In order to create out of F a field we must introduce an addition and a multiplication for its elements and then show that under these operations F forms a field.

We first dispose of the addition. Motivated by our heuristic discussion at the beginning of the proof we define

$$[a, b] + [c, d] = [ad + bc, bd].$$

Since D is an integral domain and both $b \neq 0$ and $d \neq 0$ we have that $bd \neq 0$; this, at least, tells us that $[ad + bc, bd] \in F$. We now assert that this addition is well defined, that is, if $[a, b] = [a', b']$ and $[c, d] = [c', d']$, then $[a, b] + [c, d] = [a', b'] + [c', d']$. To see that this is so, from $[a, b] = [a', b']$ we have that $ab' = ba'$; from $[c, d] = [c', d']$ we have that $cd' = dc'$. What we need is that these relations force the equality of $[a, b] + [c, d]$ and $[a', b'] + [c', d']$. From the definition of addition this boils down to showing that $[ad + bc, bd] = [a'd' + b'c', b'd']$, or, in equivalent terms, that $(ad + bc)b'd' = bd(a'd' + b'c')$. Using $ab' = ba'$, $cd' = dc'$

this becomes: $(ad + bc)b'd' = adb'd' + bcb'd' = ab'dd' + bb'cd' = ba'dd' + bb'dc' = bd(a'd' + b'c')$, which is the desired equality.

Clearly $[0, b]$ acts as a zero-element for this addition and $[-a, b]$ as the negative of $[a, b]$. It is a simple matter to verify that F is an abelian group under this addition.

We now turn to the multiplication in F. Again motivated by our preliminary heuristic discussion we define $[a, b][c, d] = [ac, bd]$. As in the case of addition, since $b \neq 0$, $d \neq 0$, $bd \neq 0$ and so $[ac, bd] \in F$. A computation, very much in the spirit of the one just carried out, proves that if $[a, b] = [a', b']$ and $[c, d] = [c', d']$ then $[a, b][c, d] = [a', b'][c', d']$. One can now show that the nonzero elements of F (that is, all the elements $[a, b]$ where $a \neq 0$) form an abelian group under multiplication in which $[d, d]$ acts as the unit element and where

$$[c, d]^{-1} = [d, c] \text{ (since } c \neq 0, [d, c] \text{ is in } F).$$

It is a routine computation to see that the distributive law holds in F. F is thus a field.

All that remains is to show that D can be imbedded in F. We shall exhibit an explicit isomorphism of D into F. Before doing so we first notice that for $x \neq 0$, $y \neq 0$ in D, $[ax, x] = [ay, y]$ because $(ax)y = x(ay)$; let us denote $[ax, x]$ by $[a, 1]$. Define $\phi : D \to F$ by $\phi(a) = [a, 1]$ for every $a \in D$. We leave it to the reader to verify that ϕ is an isomorphism of D into F, and that if D has a unit element 1, then $\phi(1)$ is the unit element of F. The theorem is now proved in its entirety.

F is usually called the *field of quotients* of D. In the special case in which D is the ring of integers, the F so constructed is, of course, the field of rational numbers.

Problems

1. Prove that if $[a, b] = [a', b']$ and $[c, d] = [c', d']$ then $[a, b][c, d] = [a', b'][c', d']$.
2. Prove the distributive law in F.
3. Prove that the mapping $\phi : D \to F$ defined by $\phi(a) = [a, 1]$ is an isomorphism of D into F.
4. Prove that if K is any field which contains D then K contains a subfield isomorphic to F. (*In this sense F is the smallest field containing D.*)

*5. Let R be a commutative ring with unit element. A nonempty subset S of R is called a multiplicative system if
   1. $0 \notin S$.
   2. $s_1, s_2 \in S$ implies that $s_1 s_2 \in S$.

Let $\mathcal{M}$ be the set of all ordered pairs (r, s) where $r \in R$, $s \in S$. In $\mathcal{M}$ define $(r, s) \sim (r', s')$ if there exists an element $s'' \in S$ such that

$$s''(rs' - sr') = 0.$$

(a) Prove that this defines an equivalence relation on $\mathcal{M}$.

Let the equivalence class of (r, s) be denoted by $[r, s]$, and let R_S be the set of all the equivalence classes. In R_S define $[r_1, s_1] + [r_2, s_2] = [r_1s_2 + r_2s_1, s_1s_2]$ and $[r_1, s_1][r_2, s_2] = [r_1r_2, s_1s_2]$.

(b) Prove that the addition and multiplication described above are well defined and that R_S forms a ring under these operations.

(c) Can R be imbedded in R_S?

(d) Prove that the mapping $\phi: R \to R_s$ defined by $\phi(a) = [as, s]$ is a homomorphism of R into R_S and find the kernel of ϕ.

(e) Prove that this kernel has no element of S in it.

(f) Prove that every element of the form $[s_1, s_2]$(where $s_1, s_2 \in S$) in R_S has an inverse in R_S.

6. Let D be an integral domain, $a, b \in D$. Suppose that $a^n = b^n$ and $a^m = b^m$ for two relatively prime positive integers m and n. Prove that $a = b$.

7. Let R be a ring, possibly noncommutative, in which $xy = 0$ implies $x = 0$ or $y = 0$. If $a, b \in R$ and $a^n = b^n$ and $a^m = b^m$ for two relatively prime positive integers m and n, prove that $a = b$.

3.7 Euclidean Rings

The class of rings we propose to study now is motivated by several existing examples—the ring of integers, the Gaussian integers (Section 3.8), and polynomial rings (Section 3.9). The definition of this class is designed to incorporate in it certain outstanding characteristics of the three concrete examples listed above.

DEFINITION An integral domain R is said to be a *Euclidean ring* if for every $a \neq 0$ in R there is defined a nonnegative integer $d(a)$ such that

1. For all $a, b \in R$, both nonzero, $d(a) \leq d(ab)$.
2. For any $a, b \in R$, both nonzero, there exist $t, r \in R$ such that $a = tb + r$ where either $r = 0$ or $d(r) < d(b)$.

We do not assign a value to $d(0)$. The integers serve as an example of a Euclidean ring, where $d(a) =$ absolute value of a acts as the required function. In the next section we shall see that the Gaussian integers also form a Euclidean ring. Out of that observation, and the results developed in this part, we shall prove a classic theorem in number theory due to

Fermat, namely, that every prime number of the form $4n + 1$ can be written as the sum of two squares.

We begin with

THEOREM 3.7.1 *Let R be a Euclidean ring and let A be an ideal of R. Then there exists an element $a_0 \in A$ such that A consists exactly of all a_0x as x ranges over R.*

Proof. If A just consists of the element 0, put $a_0 = 0$ and the conclusion of the theorem holds.

Thus we may assume that $A \neq (0)$; hence there is an $a \neq 0$ in A. Pick an $a_0 \in A$ such that $d(a_0)$ is minimal. (Since d takes on nonnegative integer values this is always possible.)

Suppose that $a \in A$. By the properties of Euclidean rings there exist $t, r \in R$ such that $a = ta_0 + r$ where $r = 0$ or $d(r) < d(a_0)$. Since $a_0 \in A$ and A is an ideal of R, ta_0 is in A. Combined with $a \in A$ this results in $a - ta_0 \in A$; but $r = a - ta_0$, whence $r \in A$. If $r \neq 0$ then $d(r) < d(a_0)$, giving us an element r in A whose d-value is smaller than that of a_0, in contradiction to our choice of a_0 as the element in A of minimal d-value. Consequently $r = 0$ and $a = ta_0$, which proves the theorem.

We introduce the notation $(a) = \{xa \mid x \in R\}$ to represent the ideal of all multiples of a.

DEFINITION An integral domain R with unit element is a *principal ideal ring* if every ideal A in R is of the form $A = (a)$ for some $a \in R$.

Once we establish that a Euclidean ring has a unit element, in virtue of Theorem 3.7.1, we shall know that a Euclidean ring is a principal ideal ring. The converse, however, is false; there are principal ideal rings which are not Euclidean rings. [See the paper by T. Motzkin, *Bulletin of the American Mathematical Society*, Vol. 55 (1949), pages 1142–1146, entitled "The Euclidean algorithm."]

COROLLARY TO THEOREM 3.7.1 *A Euclidean ring possesses a unit element.*

Proof. Let R be a Euclidean ring; then R is certainly an ideal of R, so that by Theorem 3.7.1 we may conclude that $R = (u_0)$ for some $u_0 \in R$. Thus every element in R is a multiple of u_0. Therefore, in particular, $u_0 = u_0c$ for some $c \in R$. If $a \in R$ then $a = xu_0$ for some $x \in R$, hence $ac = (xu_0)c = x(u_0c) = xu_0 = a$. Thus c is seen to be the required unit element.

DEFINITION If $a \neq 0$ and b are in a commutative ring R then a is said to *divide* b if there exists a $c \in R$ such that $b = ac$. We shall use the symbol

$a \mid b$ to represent the fact that a divides b and $a \nmid b$ to mean that a does not divide b.

The proof of the next remark is so simple and straightforward that we omit it.

REMARK 1. *If $a \mid b$ and $b \mid c$ then $a \mid c$.*
2. *If $a \mid b$ and $a \mid c$ then $a \mid (b \pm c)$.*
3. *If $a \mid b$ then $a \mid bx$ for all $x \in R$.*

DEFINITION If $a, b \in R$ then $d \in R$ is said to be a *greatest common divisor* of a and b if

1. $d \mid a$ and $d \mid b$.
2. Whenever $c \mid a$ and $c \mid b$ then $c \mid d$.

We shall use the notation $d = (a, b)$ to denote that d is a greatest common divisor of a and b.

LEMMA 3.7.1 *Let R be a Euclidean ring. Then any two elements a and b in R have a greatest common divisor d. Moreover $d = \lambda a + \mu b$ for some $\lambda, \mu \in R$.*

Proof. Let A be the set of all elements $ra + sb$ where r, s range over R. We claim that A is an ideal of R. For suppose that $x, y \in A$; therefore $x = r_1 a + s_1 b$, $y = r_2 a + s_2 b$, and so $x \pm y = (r_1 \pm r_2)a + (s_1 \pm s_2)b \in A$. Similarly, for any $u \in R$, $ux = u(r_1 a + s_1 b) = (ur_1)a + (us_1)b \in A$.

Since A is an ideal of R, by Theorem 3.7.1 there exists an element $d \in A$ such that every element in A is a mutiple of d. By dint of the fact that $d \in A$ and that every element of A is of the form $ra + sb$, $d = \lambda a + \mu b$ for some $\lambda, \mu \in R$. Now by the corollary to Theorem 3.7.1, R has a unit element 1; thus $a = 1a + 0b \in A$, $b = 0a + 1b \in A$. Being in A, they are both multiples of d, whence $d \mid a$ and $d \mid b$.

Suppose, finally, that $c \mid a$ and $c \mid b$; then $c \mid \lambda a$ and $c \mid \mu b$ so that c certainly divides $\lambda a + \mu b = d$. Therefore d has all the requisite conditions for a greatest common divisor and the lemma is proved.

DEFINITION Let R be a commutative ring with unit element. An element $a \in R$ is a *unit* in R if there exists an element $b \in R$ such that $ab = 1$.

Do not confuse a unit with a unit element! A unit in a ring is an element whose inverse is also in the ring.

LEMMA 3.7.2 *Let R be an integral domain with unit element and suppose that for $a, b \in R$ both $a \mid b$ and $b \mid a$ are true. Then $a = ub$, where u is a unit in R.*

Proof. Since $a \mid b$, $b = xa$ for some $x \in R$; since $b \mid a$, $a = yb$ for some $y \in R$. Thus $b = x(yb) = (xy)b$; but these are elements of an integral domain, so that we can cancel the b and obtain $xy = 1$; y is thus a unit in R and $a = yb$, proving the lemma.

DEFINITION Let R be a commutative ring with unit element. Two elements a and b in R are said to be *associates* if $b = ua$ for some unit u in R.

The relation of being associates is an equivalence relation. (Problem 1 at the end of this section.) Note that in a Euclidean ring any two greatest common divisors of two given elements are associates (Problem 2).

Up to this point we have, as yet, not made use of condition 1 in the definition of a Euclidean ring, namely that $d(a) \leq d(ab)$ for $b \neq 0$. We now make use of it in the proof of

LEMMA 3.7.3 *Let R be a Euclidean ring and $a, b \in R$. If $b \neq 0$ is not a unit in R, then $d(a) < d(ab)$.*

Proof. Consider the ideal $A = (a) = \{xa \mid x \in R\}$ of R. By condition 1 for a Euclidean ring, $d(a) \leq d(xa)$ for $x \neq 0$ in R. Thus the d-value of a is the minimum for the d-value of any element in A. Now $ab \in A$; if $d(ab) = d(a)$, by the proof used in establishing Theorem 3.7.1, since the d-value of ab is minimal in regard to A, every element in A is a multiple of ab. In particular, since $a \in A$, a must be a multiple of ab; whence $a = abx$ for some $x \in R$. Since all this is taking place in an integral domain we obtain $bx = 1$. In this way b is a unit in R, in contradiction to the fact that it was not a unit. The net result of this is that $d(a) < d(ab)$.

DEFINITION In the Euclidean ring R a nonunit π is said to be a *prime element* of R if whenever $\pi = ab$, where a, b are in R, then one of a or b is a unit in R.

A prime element is thus an element in R which cannot be factored in R in a nontrivial way.

LEMMA 3.7.4 *Let R be a Euclidean ring. Then every element in R is either a unit in R or can be written as the product of a finite number of prime elements of R.*

Proof. The proof is by induction on $d(a)$.

If $d(a) = d(1)$ then a is a unit in R (Problem 3), and so in this case, the assertion of the lemma is correct.

We assume that the lemma is true for all elements x in R such that $d(x) < d(a)$. On the basis of this assumption we aim to prove it for a. This would complete the induction and prove the lemma.

If a is a prime element of R there is nothing to prove. So suppose that $a = bc$ where neither b nor c is a unit in R. By Lemma 3.7.3, $d(b) < d(bc) = d(a)$ and $d(c) < d(bc) = d(a)$. Thus by our induction hypothesis b and c can be written as a product of a finite number of prime elements of R; $b = \pi_1\pi_2 \cdots \pi_n$, $c = \pi'_1\pi'_2 \cdots \pi'_m$ where the π's and π''s are prime elements of R. Consequently $a = bc = \pi_1\pi_2 \cdots \pi_n\pi'_1\pi'_2 \cdots \pi'_m$ and in this way a has been factored as a product of a finite number of prime elements. This completes the proof.

DEFINITION In the Euclidean ring R, a and b in R are said to be *relatively prime* if their greatest common divisor is a unit of R.

Since any associate of a greatest common divisor is a greatest common divisor, and since 1 is an associate of any unit, if a and b are relatively prime we may assume that $(a, b) = 1$.

LEMMA 3.7.5 *Let R be a Euclidean ring. Suppose that for $a, b, c \in R$, $a \mid bc$ but $(a, b) = 1$. Then $a \mid c$.*

Proof. As we have seen in Lemma 3.7.1, the greatest common divisor of a and b can be realized in the form $\lambda a + \mu b$. Thus by our assumptions, $\lambda a + \mu b = 1$. Multiplying this relation by c we obtain $\lambda ac + \mu bc = c$. Now $a \mid \lambda ac$, always, and $a \mid \mu bc$ since $a \mid bc$ by assumption; therefore $a \mid (\lambda ac + \mu bc) = c$. This is, of course, the assertion of the lemma.

We wish to show that prime elements in a Euclidean ring play the same role that prime numbers play in the integers. If π in R is a prime element of R and $a \in R$, then either $\pi \mid a$ or $(\pi, a) = 1$, for, in particular, (π, a) is a divisor of π so it must be π or 1 (or any unit). If $(\pi, a) = 1$, one-half our assertion is true; if $(\pi, a) = \pi$, since $(\pi, a) \mid a$ we get $\pi \mid a$, and the other half of our assertion is true.

LEMMA 3.7.6 *If π is a prime element in the Euclidean ring R and $\pi \mid ab$ where $a, b \in R$ then π divides at least one of a or b.*

Proof. Suppose that π does not divide a; then $(\pi, a) = 1$. Applying Lemma 3.7.5 we are led to $\pi \mid b$.

COROLLARY *If π is a prime element in the Euclidean ring R and $\pi \mid a_1a_2 \cdots a_n$ then π divides at least one $a_1, a_2, \ldots, a_n$.*

We carry the analogy between prime elements and prime numbers further and prove

THEOREM 3.7.2 (UNIQUE FACTORIZATION THEOREM) *Let R be a Euclidean ring and $a \neq 0$ a nonunit in R. Suppose that $a = \pi_1\pi_2\cdots\pi_n = \pi'_1\pi'_2\cdots\pi'_m$ where the π_i and π'_j are prime elements of R. Then $n = m$ and each π_i, $1 \leq i \leq n$ is an associate of some π'_j, $1 \leq j \leq m$ and conversely each π'_k is an associate of some π_q.*

Proof. Look at the relation $a = \pi_1\pi_2\cdots\pi_n = \pi'_1\pi'_2\cdots\pi'_m$. But $\pi_1 \mid \pi_1\pi_2\cdots\pi_n$, hence $\pi_1 \mid \pi'_1\pi'_2\cdots\pi'_m$. By Lemma 3.7.6, π_1 must divide some π'_i; since π_1 and π'_i are both prime elements of R and $\pi_1 \mid \pi'_i$ they must be associates and $\pi'_i = u_1\pi_1$, where u_1 is a unit in R. Thus $\pi_1\pi_2\cdots\pi_n = \pi'_1\pi'_2\cdots\pi'_m = u_1\pi_1\pi'_2\cdots\pi'_{i-1}\pi'_{i+1}\cdots\pi'_m$; cancel off π_1 and we are left with $\pi_2\cdots\pi_n = u_1\pi'_2\cdots\pi'_{i-1}\pi'_{i+1}\cdots\pi'_m$. Repeat the argument on this relation with π_2. After n steps, the left side becomes 1, the right side a product of a certain number of π' (the excess of m over n). This would force $n \leq m$ since the π' are not units. Similarly, $m \leq n$, so that $n = m$. In the process we have also showed that every π_i has some π'_i as an associate and conversely.

Combining Lemma 3.7.4 and Theorem 3.7.2 we have that *every nonzero element in a Euclidean ring R can be uniquely written (up to associates) as a product of prime elements or is a unit in R.*

We finish the section by determining all the maximal ideals in a Euclidean ring.

In Theorem 3.7.1 we proved that any ideal A in the Euclidean ring R is of the form $A = (a_0)$ where $(a_0) = \{xa_0 \mid x \in R\}$. We now ask: What conditions imposed on a_0 insure that A is a maximal ideal of R? For this question we have a simple, precise answer, namely

LEMMA 3.7.7 *The ideal $A = (a_0)$ is a maximal ideal of the Euclidean ring R if and only if a_0 is a prime element of R.*

Proof. We first prove that if a_0 is not a prime element, then $A = (a_0)$ is not a maximal ideal. For, suppose that $a_0 = bc$ where $b, c \in R$ and neither b nor c is a unit. Let $B = (b)$; then certainly $a_0 \in B$ so that $A \subset B$. We claim that $A \neq B$ and that $B \neq R$.

If $B = R$ then $1 \in B$ so that $1 = xb$ for some $x \in R$, forcing b to be a unit in R, which it is not. On the other hand, if $A = B$ then $b \in B = A$ whence $b = xa_0$ for some $x \in R$. Combined with $a_0 = bc$ this results in $a_0 = xca_0$, in consequence of which $xc = 1$. But this forces c to be a unit in R, again contradicting our assumption. Therefore B is neither A nor R and since $A \subset B$, A cannot be a maximal ideal of R.

Conversely, suppose that a_0 is a prime element of R and that U is an ideal of R such that $A = (a_0) \subset U \subset R$. By Theorem 3.7.1, $U = (u_0)$. Since $a_0 \in A \subset U = (u_0)$, $a_0 = xu_0$ for some $x \in R$. But a_0 is a prime element of R, from which it follows that either x or u_0 is a unit in R. If u_0

is a unit in R then $U = R$ (see Problem 5). If, on the other hand, x is a unit in R, then $x^{-1} \in R$ and the relation $a_0 = xu_0$ becomes $u_0 = x^{-1}a_0 \in A$ since A is an ideal of R. This implies that $U \subset A$; together with $A \subset U$ we conclude that $U = A$. Therefore there is no ideal of R which fits strictly between A and R. This means that A is a maximal ideal of R.

Problems

1. In a commutative ring with unit element prove that the relation a is an associate of b is an equivalence relation.
2. In a Euclidean ring prove that any two greatest common divisors of a and b are associates.
3. Prove that a necessary and sufficient condition that the element a in the Euclidean ring be a unit is that $d(a) = d(1)$.
4. Prove that in a Euclidean ring (a, b) can be found as follows:

$$\begin{aligned} b &= q_0a + r_1, \quad \text{where} \quad d(r_1) < d(a) \\ a &= q_1r_1 + r_2, \quad \text{where} \quad d(r_2) < d(r_1) \\ r_1 &= q_2r_2 + r_3, \quad \text{where} \quad d(r_3) < d(r_2) \\ &\vdots \\ r_{n-1} &= q_nr_n \end{aligned}$$

and

$$r_n = (a, b).$$

5. Prove that if an ideal U of a ring R contains a unit of R, then $U = R$.
6. Prove that the units in a commutative ring with a unit element form an abelian group.
7. Given two elements a, b in the Euclidean ring R their *least common multiple* $c \in R$ is an element in R such that $a \mid c$ and $b \mid c$ and such that whenever $a \mid x$ and $b \mid x$ for $x \in R$ then $c \mid x$. Prove that any two elements in the Euclidean ring R have a least common multiple in R.
8. In Problem 7, if the least common multiple of a and b is denoted by $[a, b]$, prove that $[a, b] = ab/(a, b)$.

3.8 A Particular Euclidean Ring

An abstraction in mathematics gains in substance and importance when, particularized to a specific example, it sheds new light on this example. We are about to particularize the notion of a Euclidean ring to a concrete ring, the ring of Gaussian integers. Applying the general results obtained about Euclidean rings to the Gaussian integers we shall obtain a highly nontrivial theorem about prime numbers due to Fermat.

Let $J[i]$ denote the set of all complex numbers of the form $a + bi$ where a and b are integers. Under the usual addition and multiplication of complex numbers $J[i]$ forms an integral domain called the domain of *Gaussian integers*.

Our first objective is to exhibit $J[i]$ as a Euclidean ring. In order to do this we must first introduce a function $d(x)$ defined for every nonzero element in $J[i]$ which satisfies

1. $d(x)$ is a nonnegative integer for every $x \neq 0 \in J[i]$.
2. $d(x) \leq d(xy)$ for every $y \neq 0$ in $J[i]$.
3. Given $u, v \in J[i]$ there exist $t, r \in J[i]$ such that $v = tu + r$ where $r = 0$ or $d(r) < d(u)$.

Our candidate for this function d is the following: if $x = a + bi \in J[i]$, then $d(x) = a^2 + b^2$. The $d(x)$ so defined certainly satisfies property 1; in fact, if $x \neq 0 \in J[i]$ then $d(x) \geq 1$. As is well known, for any two complex numbers (not necessarily in $J[i]$) x, y, $d(xy) = d(x)d(y)$; thus if x and y are in addition in $J[i]$ and $y \neq 0$, then since $d(y) \geq 1$, $d(x) = d(x)1 \leq d(x)d(y) = d(xy)$, showing that condition 2 is satisfied. All our effort now will be to show that condition 3 also holds for this function d in $J[i]$. This is done in the proof of

THEOREM 3.8.1 *$J[i]$ is a Euclidean ring.*

Proof. As was remarked in the discussion above, to prove Theorem 3.8.1 we merely must show that, given $x, y \in J[i]$ there exists $t, r \in J[i]$ such that $y = tx + r$ where $r = 0$ or $d(r) < d(x)$.

We first establish this for a very special case, namely, where y is arbitrary in $J[i]$ but where x is an (ordinary) positive integer n. Suppose that $y = a + bi$; by the division algorithm for the ring of integers we can find integers u, v such that $a = un + u_1$ and $b = vn + v_1$ where u_1 and v_1 are integers satisfying $|u_1| \leq \frac{1}{2}n$ and $|v_1| \leq \frac{1}{2}n$. Let $t = u + vi$ and $r = u_1 + v_1 i$; then $y = a + bi = un + u_1 + (vn + v_1)i = (u + vi)n + u_1 + v_1 i = tn + r$. Since $d(r) = d(u_1 + v_1 i) = u_1{}^2 + v_1{}^2 \leq n^2/4 + n^2/4 < n^2 = d(n)$, we see that in this special case we have shown that $y = tn + r$ with $r = 0$ or $d(r) < d(n)$.

We now go to the general case; let $x \neq 0$ and y be arbitrary elements in $J[i]$. Thus $x\bar{x}$ is a positive integer n where $\bar{x}$ is the complex conjugate of x. Applying the result of the paragraph above to the elements $y\bar{x}$ and n we see that there are elements $t, r \in J[i]$ such that $y\bar{x} = tn + r$ with $r = 0$ or $d(r) < d(n)$. Putting into this relation $n = x\bar{x}$ we obtain $d(y\bar{x} - tx\bar{x}) < d(n) = d(x\bar{x})$; applying to this the fact that $d(y\bar{x} - tx\bar{x}) = d(y - tx)d(\bar{x})$ and $d(x\bar{x}) = d(x)d(\bar{x})$ we obtain that $d(y - tx)d(\bar{x}) < d(x)d(\bar{x})$. Since $x \neq 0$, $d(\bar{x})$ is a positive integer, so this inequality simplifies to $d(y - tx) < d(x)$. We represent $y = tx + r_0$, where $r_0 = y - tx$; thus t and r_0 are in

$J[i]$ and as we saw above, $r_0 = 0$ or $d(r_0) = d(y - tx) < d(x)$. This proves the theorem.

Since $J[i]$ has been proved to be a Euclidean ring, we are free to use the results established about this class of rings in the previous section to the Euclidean ring we have at hand, $J[i]$.

LEMMA 3.8.1 *Let p be a prime integer and suppose that for some integer c relatively prime to p we can find integers x and y such that $x^2 + y^2 = cp$. Then p can be written as the sum of squares of two integers, that is, there exist integers a and b such that $p = a^2 + b^2$.*

Proof. The ring of integers is a subring of $J[i]$. Suppose that the integer p is also a prime element of $J[i]$. Since $cp = x^2 + y^2 = (x + yi)(x - yi)$, by Lemma 3.7.6, $p \mid (x + yi)$ or $p \mid (x - yi)$ in $J[i]$. But if $p \mid (x + yi)$ then $x + yi = p(u + vi)$ which would say that $x = pu$ and $y = pv$ so that p also would divide $x - yi$. But then $p^2 \mid (x + yi)(x - yi) = cp$ from which we would conclude that $p \mid c$ contrary to assumption. Similarly if $p \mid (x - yi)$. Thus p is *not* a prime element in $J[i]$! In consequence of this,

$$p = (a + bi)(g + di)$$

where $a + bi$ and $g + di$ are in $J[i]$ and where neither $a + bi$ nor $g + di$ is a unit in $J[i]$. But this means that neither $a^2 + b^2 = 1$ nor $g^2 + d^2 = 1$. (See Problem 2.) From $p = (a + bi)(g + di)$ it follows easily that $p = (a - bi)(g - di)$. Thus

$$p^2 = (a + bi)(g + di)(a - bi)(g - di) = (a^2 + b^2)(g^2 + d^2).$$

Therefore $(a^2 + b^2) \mid p^2$ so $a^2 + b^2 = 1$, p or p^2; $a^2 + b^2 \neq 1$ since $a + bi$ is not a unit, in $J[i]$; $a^2 + b^2 \neq p^2$, otherwise $g^2 + d^2 = 1$, contrary to the fact that $g + di$ is not a unit in $J[i]$. Thus the only feasibility left is that $a^2 + b^2 = p$ and the lemma is thereby established.

The odd prime numbers divide into two classes, those which have a remainder of 1 on division by 4 and those which have a remainder of 3 on division by 4. We aim to show that every prime number of the first kind can be written as the sum of two squares, whereas no prime in the second class can be so represented.

LEMMA 3.8.2 *If p is a prime number of the form $4n + 1$, then we can solve the congruence $x^2 \equiv -1 \bmod p$.*

Proof. Let $x = 1 \cdot 2 \cdot 3 \cdots (p - 1)/2$. Since $p - 1 = 4n$, in this product for x there are an even number of terms, in consequence of which

$$x = (-1)(-2)(-3)\cdots\left(-\left(\frac{p-1}{2}\right)\right).$$

But $p - k \equiv -k \bmod p$, so that

$$x^2 \equiv \left(1 \cdot 2 \cdots \frac{p-1}{2}\right)(-1)(-2) \cdots \left(-\left(\frac{p-1}{2}\right)\right)$$

$$\equiv 1 \cdot 2 \cdots \frac{p-1}{2} \frac{p+1}{2} \cdots (p-1)$$

$$\equiv (p-1)! = -1 \bmod p.$$

We are using here Wilson's theorem, proved earlier, namely that if p is a prime number $(p - 1)! \equiv -1(p)$.

To illustrate this result, if $p = 13$,

$$x = 1 \cdot 2 \cdot 3 \cdot 4 \cdot 5 \cdot 6 = 720 = 5 \bmod 13 \text{ and } 5^2 = -1 \bmod 13.$$

THEOREM 3.8.2 (FERMAT) *If p is a prime number of the form $4n + 1$, then $p = a^2 + b^2$ for some integers a, b.*

Proof. By Lemma 3.8.2 there exists an x such that $x^2 \equiv -1 \bmod p$. The x can be chosen so that $0 \leq x \leq p - 1$ since we only need to use the remainder of x on division by p. We can restrict the size of x even further, namely to satisfy $|x| \leq p/2$. For if $x > p/2$, then $y = p - x$ satisfies $y^2 \equiv -1 \bmod p$ but $|y| \leq p/2$. Thus we may assume that we have an integer x such that $|x| \leq p/2$ and $x^2 + 1$ is a multiple of p, say cp. Now $cp = x^2 + 1 \leq p^2/4 + 1 < p^2$, hence $c < p$ and so $p \nmid c$. Invoking Lemma 3.8.1 we obtain that $p = a^2 + b^2$ for some integers a and b, proving the theorem.

Problems

1. Find all the units in $J[i]$.
2. If $a + bi$ is not a unit of $J[i]$ prove that $a^2 + b^2 > 1$.
3. Find the greatest common divisor in $J[i]$ of
   (a) $3 + 4i$ and $4 - 3i$. (b) $11 + 7i$ and $18 - i$.
4. Prove that if p is a prime number of the form $4n + 3$, then there is no x such that $x^2 \equiv -1 \bmod p$.
5. Prove that no prime of the form $4n + 3$ can be written as $a^2 + b^2$ where a and b are integers.
6. Prove that there is an infinite number of primes of the form $4n + 3$.

*7. Prove there exists an infinite number of primes of the form $4n + 1$.

*8. Determine all the prime elements in $J[i]$.

*9. Determine all positive integers which can be written as a sum of two squares (of integers).

3.9 Polynomial Rings

Very early in our mathematical education—in fact in junior high school or early in high school itself—we are introduced to polynomials. For a seemingly endless amount of time we are drilled, to the point of utter boredom, in factoring them, multiplying them, dividing them, simplifying them. Facility in factoring a quadratic becomes confused with genuine mathematical talent.

Later, at the beginning college level, polynomials make their appearance in a somewhat different setting. Now they are functions, taking on values, and we become concerned with their continuity, their derivatives, their integrals, their maxima and minima.

We too shall be interested in polynomials but from neither of the above viewpoints. To us polynomials will simply be elements of a certain ring and we shall be concerned with algebraic properties of this ring. Our primary interest in them will be that they give us a Euclidean ring whose properties will be decisive in discussing fields and extensions of fields.

Let F be a field. By the *ring of polynomials* in the indeterminate, x, written as $F[x]$, we mean the set of all symbols $a_0 + a_1x + \cdots + a_nx^n$, where n can be any nonnegative integer and where the coefficients $a_1, a_2, \ldots, a_n$ are all in F. In order to make a ring out of $F[x]$ we must be able to recognize when two elements in it are equal, we must be able to add and multiply elements of $F[x]$ so that the axioms defining a ring hold true for $F[x]$. This will be our initial goal.

We could avoid the phrase "the set of all symbols" used above by introducing an appropriate apparatus of sequences but it seems more desirable to follow a path which is somewhat familiar to most readers.

DEFINITION If $p(x) = a_0 + a_1x + \cdots + a_mx^m$ and $q(x) = b_0 + b_1x + \cdots + b_nx^n$ are in $F[x]$, then $p(x) = q(x)$ if and only if for every integer $i \geq 0$, $a_i = b_i$.

Thus two polynomials are declared to be equal if and only if their corresponding coefficients are equal.

DEFINITION If $p(x) = a_0 + a_1x + \cdots + a_mx^m$ and $q(x) = b_0 + b_1x + \cdots + b_nx^n$ are both in $F[x]$, then $p(x) + q(x) = c_0 + c_1x + \cdots + c_tx^t$ where for each i, $c_i = a_i + b_i$.

In other words, add two polynomials by adding their coefficients and collecting terms. To add $1 + x$ and $3 - 2x + x^2$ we consider $1 + x$ as $1 + x + 0x^2$ and add, according to the recipe given in the definition, to obtain as their sum $4 - x + x^2$.

The most complicated item, and the only one left for us to define for $F[x]$, is the multiplication.

DEFINITION If $p(x) = a_0 + a_1x + \cdots + a_mx^m$ and $q(x) = b_0 + b_1x + \cdots + b_nx^n$, then $p(x)q(x) = c_0 + c_1x + \cdots + c_kx^k$ where $c_t = a_tb_0 + a_{t-1}b_1 + a_{t-2}b_2 + \cdots + a_0b_t$.

This definition says nothing more than: multiply the two polynomials by multiplying out the symbols formally, use the relation $x^\alpha x^\beta = x^{\alpha+\beta}$, and collect terms. Let us illustrate the definition with an example:

$$p(x) = 1 + x - x^2, \qquad q(x) = 2 + x^2 + x^3.$$

Here $a_0 = 1$, $a_1 = 1$, $a_2 = -1$, $a_3 = a_4 = \cdots = 0$, and $b_0 = 2$, $b_1 = 0$, $b_2 = 1$, $b_3 = 1$, $b_4 = b_5 = \cdots = 0$. Thus

$$c_0 = a_0b_0 = 1.2 = 2,$$

$$c_1 = a_1b_0 + a_0b_1 = 1.2 + 1.0 = 2,$$

$$c_2 = a_2b_0 + a_1b_1 + a_0b_2 = (-1)(2) + 1.0 + 1.1 = -1,$$

$$c_3 = a_3b_0 + a_2b_1 + a_1b_2 + a_0b_3 = (0)(2) + (-1)(0) + 1.1 + 1.1 = 2,$$

$$\begin{aligned} c_4 &= a_4b_0 + a_3b_1 + a_2b_2 + a_1b_3 + a_0b_4 \\ &= (0)(2) + (0)(0) + (-1)(1) + (1)(1) + 1(0) = 0, \end{aligned}$$

$$\begin{aligned} c_5 &= a_5b_0 + a_4b_1 + a_3b_2 + a_2b_3 + a_1b_4 + a_0b_5 \\ &= (0)(2) + (0)(0) + (0)(1) + (-1)(1) + (1)(0) + (0)(0) = -1, \end{aligned}$$

$$\begin{aligned} c_6 &= a_6b_0 + a_5b_1 + a_4b_2 + a_3b_3 + a_2b_4 + a_1b_5 + a_0b_6 \\ &= (0)(2) + (0)(0) + (0)(1) + (0)(1) + (-1)(0) + (1)(0) + (1)(0) = 0, \end{aligned}$$

$$c_7 = c_8 = \cdots = 0.$$

Therefore according to our definition,

$$(1 + x - x^2)(2 + x^2 + x^3) = c_0 + c_1x + \cdots = 2 + 2x - x^2 + 2x^3 - x^5.$$

If you multiply these together high-school style you will see that you get the same answer. Our definition of product is the one the reader has always known.

Without further ado we assert that $F[x]$ is a ring with these operations, its multiplication is commutative, and it has a unit element. We leave the verification of the ring axioms to the reader.

DEFINITION If $f(x) = a_0 + a_1x + \cdots + a_nx^n \neq 0$ and $a_n \neq 0$ then the *degree* of $f(x)$, written as $\deg f(x)$, is n.

That is, the degree of $f(x)$ is the largest integer i for which the ith coefficient of $f(x)$ is not 0. We do not define the degree of the zero polynomial. We say a polynomial is a *constant* if its degree is 0. The degree function defined on the nonzero elements of $F[x]$ will provide us with the function $d(x)$ needed in order that $F[x]$ be a Euclidean ring.

LEMMA 3.9.1 *If $f(x)$, $g(x)$ are two nonzero elements of $F[x]$, then*

$$\deg (f(x)g(x)) = \deg f(x) + \deg g(x).$$

Proof. Suppose that $f(x) = a_0 + a_1x + \cdots + a_mx^m$ and $g(x) = b_0 + b_1x + \cdots + b_nx^n$ and that $a_m \neq 0$ and $b_n \neq 0$. Therefore $\deg f(x) = m$ and $\deg g(x) = n$. By definition, $f(x)g(x) = c_0 + c_1x + \cdots + c_kx^k$ where $c_t = a_tb_0 + a_{t-1}b_1 + \cdots + a_1b_{t-1} + a_0b_t$. We claim that $c_{m+n} = a_mb_n \neq 0$ and $c_i = 0$ for $i > m + n$. That $c_{m+n} = a_mb_n$ can be seen at a glance by its definition. What about c_i for $i > m + n$? c_i is the sum of terms of the form a_jb_{i-j}; since $i = j + (i - j) > m + n$ then either $j > m$ or $(i - j) > n$. But then one of a_j or b_{i-j} is 0, so that $a_jb_{i-j} = 0$; since c_i is the sum of a bunch of zeros it itself is 0, and our claim has been established. Thus the highest nonzero coefficient of $f(x)g(x)$ is c_{m+n}, whence $\deg f(x)g(x) = m + n = \deg f(x) + \deg g(x)$.

COROLLARY *If $f(x)$, $g(x)$ are nonzero elements in $F[x]$ then* $\deg f(x) \leq \deg f(x)g(x)$.

Proof. Since $\deg f(x)g(x) = \deg f(x) + \deg g(x)$, and since $\deg g(x) \geq 0$, this result is immediate from the lemma.

COROLLARY *$F[x]$ is an integral domain.*

We leave the proof of this corollary to the reader.

Since $F[x]$ is an integral domain, in light of Theorem 3.6.1 we can construct for it its field of quotients. This field merely consists of all quotients of polynomials and is called the field of *rational functions* in x over F.

The function $\deg f(x)$ defined for all $f(x) \neq 0$ in $F[x]$ satisfies

1. $\deg f(x)$ is a nonnegative integer.
2. $\deg f(x) \leq \deg f(x)g(x)$ for all $g(x) \neq 0$ in $F[x]$.

In order for $F[x]$ to be a Euclidean ring with the degree function acting as the d-function of a Euclidean ring we still need that given $f(x), g(x) \in F[x]$, there exist $t(x)$, $r(x)$ in $F[x]$ such that $f(x) = t(x)g(x) + r(x)$ where either $r(x) = 0$ or $\deg r(x) < \deg g(x)$. This is provided us by

LEMMA 3.9.2 (THE DIVISION ALGORITHM) *Given two polynomials $f(x)$ and $g(x) \neq 0$ in $F[x]$, then there exist two polynomials $t(x)$ and $r(x)$ in $F[x]$ such that $f(x) = t(x)g(x) + r(x)$ where $r(x) = 0$ or $\deg r(x) < \deg g(x)$.*

Proof. The proof is actually nothing more than the "long-division" process we all used in school to divide one polynomial by another.

If the degree of $f(x)$ is smaller than that of $g(x)$ there is nothing to prove, for merely put $t(x) = 0$, $r(x) = f(x)$, and we certainly have that $f(x) = 0g(x) + f(x)$ where $\deg f(x) < \deg g(x)$ or $f(x) = 0$.

So we may assume that $f(x) = a_0 + a_1x + \cdots + a_mx^m$ and $g(x) = b_0 + b_1x + \cdots + b_nx^n$ where $a_m \neq 0$, $b_n \neq 0$ and $m \geq n$.

Let $f_1(x) = f(x) - (a_m/b_n)x^{m-n}g(x)$; thus $\deg f_1(x) \leq m - 1$, so by induction on the degree of $f(x)$ we may assume that $f_1(x) = t_1(x)g(x) + r(x)$ where $r(x) = 0$ or $\deg r(x) < \deg g(x)$. But then $f(x) - (a_m/b_n)x^{m-n}g(x) = t_1(x)g(x) + r(x)$, from which, by transposing, we arrive at $f(x) = ((a_m/b_n)x^{m-n} + t_1(x))g(x) + r(x)$. If we put $t(x) = (a_m/b_n)x^{m-n} + t_1(x)$ we do indeed have that $f(x) = t(x)g(x) + r(x)$ where $t(x), r(x) \in F[x]$ and where $r(x) = 0$ or $\deg r(x) < \deg g(x)$. This proves the lemma.

This last lemma fills the gap needed to exhibit $F[x]$ as a Euclidean ring and we now have the right to say

THEOREM 3.9.1 *$F[x]$ is a Euclidean ring.*

All the results of Section 3.7 now carry over and we list these, for our particular case, as the following lemmas. It could be very instructive for the reader to try to prove these directly, adapting the arguments used in Section 3.7 for our particular ring $F[x]$ and its Euclidean function, the degree.

LEMMA 3.9.3 *$F[x]$ is a principal ideal ring.*

LEMMA 3.9.4 *Given two polynomials $f(x)$, $g(x)$ in $F[x]$ they have a greatest common divisor $d(x)$ which can be realized as $d(x) = \lambda(x)f(x) + \mu(x)g(x)$.*

What corresponds to a prime element?

DEFINITION A polynomial $p(x)$ in $F[x]$ is said to be *irreducible* over F if whenever $p(x) = a(x)b(x)$ with $a(x), b(x) \in F[x]$, then one of $a(x)$ or $b(x)$ has degree 0 (i.e., is a constant).

Irreducibility depends on the field; for instance the polynomial $x^2 + 1$ is irreducible over the real field but not over the complex field, for there $x^2 + 1 = (x + i)(x - i)$ where $i^2 = -1$.

LEMMA 3.9.5 *Any polynomial in $F[x]$ can be written in a unique manner as a product of irreducible polynomials in $F[x]$.*

LEMMA 3.9.6 *The ideal $A = (p(x))$ in $F[x]$ is a maximal ideal if and only if $p(x)$ is irreducible over F.*

In Chapter 5 we shall return to take a much closer look at this field $F[x]/(p(x))$, but for now we should like to compute an example.

Let F be the field of rational numbers and consider the polynomial $p(x) = x^3 - 2$ in $F[x]$. As is easily verified, it is irreducible over F, whence $F[x]/(x^3 - 2)$ is a field. What do its elements look like? Let $A = (x^3 - 2)$, the ideal in $F[x]$ generated by $x^3 - 2$.

Any element in $F[x]/(x^3 - 2)$ is a coset of the form $f(x) + A$ of the ideal A with $f(x)$ in $F[x]$. Now, given any polynomial $f(x) \in F[x]$, by the division algorithm, $f(x) = t(x)(x^3 - 2) + r(x)$, where $r(x) = 0$ or $\deg r(x) < \deg (x^3 - 2) = 3$. Thus $r(x) = a_0 + a_1x + a_2x^2$ where a_0, a_1, a_2 are in F; consequently $f(x) + A = a_0 + a_1x + a_2x^2 + t(x)(x^3 - 2) + A = a_0 + a_1x + a_2x^2 + A$ since $t(x)(x^3 - 2)$ is in A, hence by the addition and multiplication in $F[x]/(x^3 - 2)$, $f(x) + A = (a_0 + A) + a_1(x + A) + a_2(x + A)^2$. If we put $t = x + A$, then every element in $F[x]/(x^3 - 2)$ is of the form $a_0 + a_1t + a_2t^2$ with a_0, a_1, a_2 in F. What about t? Since $t^3 - 2 = (x + A)^3 - 2 = x^3 - 2 + A = A = 0$ (since A is the zero element of $F[x]/(x^3 - 2)$) we see that $t^3 = 2$.

Also, if $a_0 + a_1t + a_2t^2 = b_0 + b_1t + b_2t^2$, then $(a_0 - b_0) + (a_1 - b_1)t + (a_2 - b_2)t^2 = 0$, whence $(a_0 - b_0) + (a_1 - b_1)x + (a_2 - b_2)x^2$ is in $A = (x^3 - 2)$. How can this be, since every element in A has degree at least 3? Only if $a_0 - b_0 + (a_1 - b_1)x + (a_2 - b_2)x^2 = 0$, that is, only if $a_0 = b_0$, $a_1 = b_1$, $a_2 = b_2$. Thus every element in $F[x]/(x^3 - 2)$ has a *unique* representation as $a_0 + a_1t + a_2t^2$ where $a_0, a_1, a_2 \in F$. By Lemma 3.9.6, $F[x]/(x^3 - 2)$ is a field. It would be instructive to see this directly; all that it entails is proving that if $a_0 + a_1t + a_2t^2 \neq 0$ then it has an inverse of the form $\alpha + \beta t + \gamma t^2$. Hence we must solve for α, β, γ in the relation $(a_0 + a_1t + a_2t^2)(\alpha + \beta t + \gamma t^2) = 1$, where not all of a_0 a_1, a_2 are 0. Multiplying the relation out and using $t^3 = 2$ we obtain $(a_0\alpha + 2a_2\beta + 2a_1\gamma) + (a_1\alpha + a_0\beta + 2a_2\gamma)t + (a_2\alpha + a_1\beta + a_0\gamma)t^2 = 1$; thus

$$a_0\alpha + 2a_2\beta + 2a_1\gamma = 1,$$

$$a_1\alpha + a_0\beta + 2a_2\gamma = 0,$$

$$a_2\alpha + a_1\beta + a_0\gamma = 0.$$

We can try to solve these three equations in the three unknowns α, β, γ. When we do so we find that a solution exists if and only if

$$a_0^3 + 2a_1^3 + 4a_2^3 - 6a_0a_1a_2 \neq 0.$$

Therefore the problem of proving directly that $F[x]/(x^3 - 2)$ is a field boils down to proving that the only solution in *rational* numbers of

$$a_0^3 + 2a_1^3 + 4a_2^3 = 6a_0a_1a_2 \tag{1}$$

is the solution $a_0 = a_1 = a_2 = 0$. We now proceed to show this. If a solution exists in rationals, by clearing of denominators we can show that a solution exists where a_0, a_1, a_2 are integers. Thus we may assume that a_0, a_1, a_2 are integers satisfying (1). We now assert that we may assume that a_0, a_1, a_2 have no common divisor other than 1, for if $a_0 = b_0 d$, $a_1 = b_1 d$, and $a_2 = b_2 d$, where d is their greatest common divisor, then substituting in (1) we obtain $d^3(b_0^3 + 2b_1^3 + 4b_2^3) = d^3(6b_0b_1b_2)$, and so $b_0^3 + 2b_1^3 + 4b_2^3 = 6b_0b_1b_2$. The problem has thus been reduced to proving that (1) has no solutions in integers which are relatively prime. But then (1) implies that a_0^3 is even, so that a_0 is even; substituting $a_0 = 2\alpha_0$ in (1) gives us $4\alpha_0^3 + a_1^3 + 2a_2^3 = 6\alpha_0 a_1 a_2$. Thus a_1^3, and so, a_1 is even; $a_1 = 2\alpha_1$. Substituting in (1) we obtain $2\alpha_0^3 + 4\alpha_1^3 + a_2^3 = 6\alpha_0\alpha_1 a_2$. Thus a_2^3, and so a_2, is even! But then a_0, a_1, a_2 have 2 as a common factor! This contradicts that they are relatively prime, and we have proved that the equation $a_0^3 + 2a_1^3 + 4a_2^3 = 6a_0a_1a_2$ has no rational solution other than $a_0 = a_1 = a_2 = 0$. Therefore we can solve for α, β, γ and $F[x]/(x^3 - 2)$ is seen, directly, to be a field.

Problems

1. Find the greatest common divisor of the following polynomials over F, the field of rational numbers:
   (a) $x^3 - 6x^2 + x + 4$ and $x^5 - 6x + 1$.
   (b) $x^2 + 1$ and $x^6 + x^3 + x + 1$.
2. Prove that
   (a) $x^2 + x + 1$ is irreducible over F, the field of integers mod 2.
   (b) $x^2 + 1$ is irreducible over the integers mod 7.
   (c) $x^3 - 9$ is irreducible over the integers mod 31.
   (d) $x^3 - 9$ is reducible over the integers mod 11.
3. Let F, K be two fields $F \subset K$ and suppose $f(x), g(x) \in F[x]$ are relatively prime in $F[x]$. Prove that they are relatively prime in $K[x]$.
4. (a) Prove that $x^2 + 1$ is irreducible over the field F of integers mod 11 and prove directly that $F[x]/(x^2 + 1)$ is a field having 121 elements.
   (b) Prove that $x^2 + x + 4$ is irreducible over F, the field of integers mod 11 and prove directly that $F[x]/(x^2 + x + 4)$ is a field having 121 elements.
   *(c) Prove that the fields of part (a) and part (b) are isomorphic.
5. Let F be the field of real numbers. Prove that $F[x]/(x^2 + 1)$ is a field isomorphic to the field of complex numbers.

*6. Define the *derivative* $f'(x)$ of the polynomial

$$f(x) = a_0 + a_1x + \cdots + a_nx^n$$

as

$$f'(x) = a_1 + 2a_2x + 3a_3x^2 + \cdots + na_nx^{n-1}.$$

Prove that if $f(x) \in F[x]$, where F is the field of rational numbers, then $f(x)$ is divisible by the square of a polynomial if and only if $f(x)$ and $f'(x)$ have a greatest common divisor $d(x)$ of positive degree.

7. If $f(x)$ is in $F[x]$, where F is the field of integers mod p, p a prime, and $f(x)$ is irreducible over F of degree n prove that $F[x]/(f(x))$ is a field with p^n elements.

3.10 Polynomials over the Rational Field

We specialize the general discussion to that of polynomials whose coefficients are rational numbers. Most of the time the coefficients will actually be integers. For such polynomials we shall be concerned with their irreducibility.

DEFINITION The polynomial $f(x) = a_0 + a_1x + \cdots + a_nx^n$, where the $a_0, a_1, a_2, \ldots, a_n$ are integers is said to be *primitive* if the greatest common divisor of $a_0, a_1, \ldots, a_n$ is 1.

LEMMA 3.10.1 *If $f(x)$ and $g(x)$ are primitive polynomials, then $f(x)g(x)$ is a primitive polynomial.*

Proof. Let $f(x) = a_0 + a_1x + \cdots + a_nx^n$ and $g(x) = b_0 + b_1x + \cdots + b_mx^m$. Suppose that the lemma was false; then all the coefficients of $f(x)g(x)$ would be divisible by some integer larger than 1, hence by some prime number p. Since $f(x)$ is primitive, p does not divide some coefficient a_i. Let a_j be the first coefficient of $f(x)$ which p does not divide. Similarly let b_k be the first coefficient of $g(x)$ which p does not divide. In $f(x)g(x)$ the coefficient of x^{j+k}, c_{j+k}, is

$$c_{j+k} = a_jb_k + (a_{j+1}b_{k-1} + a_{j+2}b_{k-2} + \cdots + a_{j+k}b_0) + (a_{j-1}b_{k+1} + a_{j-2}b_{k+2} + \cdots + a_0b_{j+k}). \tag{1}$$

Now by our choice of b_k, $p \mid b_{k-1}, b_{k-2}, \ldots$ so that $p \mid (a_{j+1}b_{k-1} + a_{j+2}b_{k-2} + \cdots + a_{j+k}b_0)$. Similarly, by our choice of a_j, $p \mid a_{j-1}, a_{j-2}, \ldots$ so that $p \mid (a_{j-1}b_{k+1} + a_{j-2}b_{k+2} + \cdots + a_0b_{k+j})$. By assumption, $p \mid c_{j+k}$. Thus by (1), $p \mid a_jb_k$, which is nonsense since $p \nmid a_j$ and $p \nmid b_k$. This proves the lemma.

DEFINITION The *content* of the polynomial $f(x) = a_0 + a_1x + \cdots + a_nx^n$, where the a's are integers, is the greatest common divisor of the integers $a_0, a_1, \ldots, a_n$.

Clearly, given any polynomial $p(x)$ with integer coefficients it can be written as $p(x) = dq(x)$ where d is the content of $p(x)$ and where $q(x)$ is a primitive polynomial.

THEOREM 3.10.1 (Gauss' Lemma) *If the primitive polynomial $f(x)$ can be factored as the product of two polynomials having rational coefficients, it can be factored as the product of two polynomials having integer coefficients.*

Proof. Suppose that $f(x) = u(x)v(x)$ where $u(x)$ and $v(x)$ have rational coefficients. By clearing of denominators and taking out common factors we can then write $f(x) = (a/b)\lambda(x)\mu(x)$ where a and b are integers and where both $\lambda(x)$ and $\mu(x)$ have integer coefficients and are primitive. Thus $bf(x) = a\lambda(x)\mu(x)$. The content of the left-hand side is b, since $f(x)$ is primitive; since both $\lambda(x)$ and $\mu(x)$ are primitive, by Lemma 3.10.1 $\lambda(x)\mu(x)$ is primitive, so that the content of the right-hand side is a. Therefore $a = b$, $(a/b) = 1$, and $f(x) = \lambda(x)\mu(x)$ where $\lambda(x)$ and $\mu(x)$ have integer coefficients. This is the assertion of the theorem.

DEFINITION A polynomial is said to be *integer monic* if all its coefficients are integers and its highest coefficient is 1.

Thus an integer monic polynomial is merely one of the form $x^n + a_1x^{n-1} + \cdots + a_n$ where the a's are integers. Clearly an integer monic polynomial is primitive.

COROLLARY *If an integer monic polynomial factors as the product of two non-constant polynomials having rational coefficients then it factors as the product of two integer monic polynomials.*

We leave the proof of the corollary as an exercise for the reader.

The question of deciding whether a given polynomial is irreducible or not can be a difficult and laborious one. Few criteria exist which declare that a given polynomial is or is not irreducible. One of these few is the following result:

THEOREM 3.10.2 (The Eisenstein Criterion) *Let $f(x) = a_0 + a_1x + a_2x^2 + \cdots + a_nx^n$ be a polynomial with integer coefficients. Suppose that for some prime number p, $p \nmid a_n, p \mid a_1, p \mid a_2, \ldots, p \mid a_0, p^2 \nmid a_0$. Then $f(x)$ is irreducible over the rationals.*

Proof. Without loss of generality we may assume that $f(x)$ is primitive, for taking out the greatest common factor of its coefficients does not disturb the hypotheses, since $p \nmid a_n$. If $f(x)$ factors as a product of two rational polynomials, by Gauss' lemma it factors as the product of two polynomials having integer coefficients. Thus if we assume that $f(x)$ is reducible, then

$$f(x) = (b_0 + b_1x + \cdots + b_rx^r)(c_0 + c_1x + \cdots + c_sx^s),$$

where the b's and c's are integers and where $r > 0$ and $s > 0$. Reading off the coefficients we first get $a_0 = b_0c_0$. Since $p \mid a_0$, p must divide one of b_0 or c_0. Since $p^2 \nmid a_0$, p cannot divide both b_0 and c_0. Suppose that $p \mid b_0$, $p \nmid c_0$. Not all the coefficients $b_0, \ldots, b_r$ can be divisible by p; otherwise all the coefficients of $f(x)$ would be divisible by p, which is manifestly false since $p \nmid a_n$. Let b_k be the first b not divisible by p, $k \leq r < n$. Thus $p \mid b_{k-1}$ and the earlier b's. But $a_k = b_kc_0 + b_{k-1}c_1 + b_{k-2}c_2 + \cdots + b_0c_k$, and $p \mid a_k, p \mid b_{k-1}, b_{k-2}, \ldots, b_0$, so that $p \mid b_kc_0$. However, $p \nmid c_0, p \nmid b_k$, which conflicts with $p \mid b_kc_0$. This contradiction proves that we could not have factored $f(x)$ and so $f(x)$ is indeed irreducible.

Problems

1. Let D be a Euclidean ring, F its field of quotients. Prove the Gauss Lemma for polynomials with coefficients in D factored as products of polynomials with coefficients in F.
2. If p is a prime number, prove that the polynomial $x^n - p$ is irreducible over the rationals.
3. Prove that the polynomial $1 + x + \cdots + x^{p-1}$, where p is a prime number, is irreducible over the field of rational numbers. (*Hint:* Consider the polynomial $1 + (x + 1) + (x + 1)^2 + \cdots + (x + 1)^{p-1}$, and use the Eisenstein criterion.)
4. If m and n are relatively prime integers and if

$$\left(x - \frac{m}{n}\right) \mid (a_0 + a_1x + \cdots + a_rx^r),$$

where the a's are integers, prove that $m \mid a_0$ and $n \mid a_r$.
5. If a is rational and $x - a$ divides an integer monic polynomial, prove that a must be an integer.

3.11 Polynomial Rings over Commutative Rings

In defining the polynomial ring in one variable over a field F, no essential use was made of the fact that F was a field; all that was used was that F was a commutative ring. The field nature of F only made itself felt in proving that $F[x]$ was a Euclidean ring.

Thus we can imitate what we did with fields for more general rings. While some properties may be lost, such as "Euclideanism," we shall see that enough remain to lead us to interesting results. The subject could have been developed in this generality from the outset, and we could have obtained the particular results about $F[x]$ by specializing the ring to be a field. However, we felt that it would be healthier to go from the concrete to the abstract rather than from the abstract to the concrete. The price we pay for this is repetition, but even that serves a purpose, namely, that of consolidating the ideas. Because of the experience gained in treating polynomials over fields, we can afford to be a little sketchier in the proofs here.

Let R be a commutative ring with unit element. By *the polynomial ring in x over R, $R[x]$*, we shall mean the set of formal symbols $a_0 + a_1x + \cdots + a_mx^m$, where $a_0, a_1, \ldots, a_m$ are in R, and where equality, addition, and multiplication are defined exactly as they were in Section 3.9. As in that section, $R[x]$ is a commutative ring with unit element.

We now define the *ring of polynomials in the n-variables $x_1, \ldots, x_n$* over R, $R[x_1, \ldots, x_n]$, as follows: Let $R_1 = R[x_1]$, $R_2 = R_1[x_2]$, the polynomial ring in x_2 over $R_1, \ldots, R_n = R_{n-1}[x_n]$. R_n is called the ring of polynomials in $x_1, \ldots, x_n$ over R. Its elements are of the form $\sum a_{i_1 i_2 \ldots i_n} x_1^{i_1} x_2^{i_2} \cdots x_n^{i_n}$, where equality and addition are defined coefficientwise and where multiplication is defined by use of the distributive law and the rule of exponents $(x_1^{i_1} x_2^{i_2} \cdots x_n^{i_n})(x_1^{j_1} x_2^{j_2} \cdots x_n^{j_n}) = x_1^{i_1+j_1} x_2^{i_2+j_2} \cdots x_n^{i_n+j_n}$. Of particular importance is the case in which $R = F$ is a field; here we obtain the ring of polynomials in n-variables over a field.

Of interest to us will be the influence of the structure of R on that of $R[x_1, \ldots, x_n]$. The first result in this direction is

LEMMA 3.11.1 *If R is an integral domain, then so is $R[x]$.*

Proof. For $0 \neq f(x) = a_0 + a_1x + \cdots + a_mx^m$, where $a_m \neq 0$, in $R[x]$, we define the *degree* of $f(x)$ to be m; thus $\deg f(x)$ is the index of the highest nonzero coefficient of $f(x)$. If R is an integral domain we leave it as an exercise to prove that $\deg(f(x)g(x)) = \deg f(x) + \deg g(x)$. But then, for $f(x) \neq 0$, $g(x) \neq 0$, it is impossible to have $f(x)g(x) = 0$. That is, $R[x]$ is an integral domain.

Making successive use of the lemma immediately yields the

COROLLARY *If R is an integral domain, then so is $R[x_1, \ldots, x_n]$.*

In particular, when F is a field, $F[x_1, \ldots, x_n]$ must be an integral domain. As such, we can construct its field of quotients; *we call this the field of rational functions in $x_1, \ldots, x_n$ over F* and denote it by $F(x_1, \ldots, x_n)$. This field plays a vital role in algebraic geometry. For us it shall be of utmost importance in our discussion, in Chapter 5, of Galois theory.

However, we want deeper interrelations between the structures of R and of $R[x_1, \ldots, x_n]$ than that expressed in Lemma 3.11.1. Our development now turns in that direction.

Exactly in the same way as we did for Euclidean rings, we can speak about divisibility, units, etc., in arbitrary integral domains, R, with unit element. Two elements a, b in R are said to be *associates* if $a = ub$ where u

is a unit in R. An element a which is not a unit in R will be called *irreducible* (or a *prime element*) if, whenever $a = bc$ with b, c both in R, then one of b or c must be a unit in R. An irreducible element is thus an element which cannot be factored in a "nontrivial" way.

DEFINITION An integral domain, R, with unit element is a *unique factorization domain* if

a. Any nonzero element in R is either a unit or can be written as the product of a finite number of irreducible elements of R.
b. The decomposition in part (a) is unique up to the order and associates of the irreducible elements.

Theorem 3.7.2 asserts that a Euclidean ring is a unique factorization domain. The converse, however, is false; for example, the ring $F[x_1, x_2]$, where F is a field, is not even a principal ideal ring (hence is certainly not Euclidean), but as we shall soon see it is a unique factorization domain.

In general commutative rings we may speak about the greatest common divisors of elements; the main difficulty is that these, in general, might not exist. However, in unique factorization domains their existence is assured. This fact is not difficult to prove and we leave it as an exercise; equally easy are the other parts of

LEMMA 3.11.2 *If R is a unique factorization domain and if a, b are in R, then a and b have a greatest common divisor (a, b) in R. Moreover, if a and b are relatively prime (i.e., $(a, b) = 1$), whenever $a \mid bc$ then $a \mid c$.*

COROLLARY *If $a \in R$ is an irreducible element and $a \mid bc$, then $a \mid b$ or $a \mid c$.*

We now wish to transfer the appropriate version of the Gauss lemma (Theorem 3.10.1), which we proved for polynomials with integer coefficients, to the ring $R[x]$, where R is a unique factorization domain.

Given the polynomial $f(x) = a_0 + a_1x + \cdots + a_mx^m$ in $R[x]$, then the *content of $f(x)$* is defined to be the greatest common divisor of $a_0, a_1, \ldots, a_m$. It is unique within units of R. We shall denote the content of $f(x)$ by $c(f)$. A polynomial in $R[x]$ is said to be *primitive* if its content is 1 (that is, is a unit in R). Given any polynomial $f(x) \in R[x]$, we can write $f(x) = af_1(x)$ where $a = c(f)$ and where $f_1(x) \in R[x]$ is primitive. (Prove!) Except for multiplication by units of R this decomposition of $f(x)$, as an element of R by a primitive polynomial in $R[x]$, is unique. (Prove!)

The proof of Lemma 3.10.1 goes over completely to our present situation; the only change that must be made in the proof is to replace the prime number p by an irreducible element of R. Thus we have

LEMMA 3.11.3 *If R is a unique factorization domain, then the product of two primitive polynomials in $R[x]$ is again a primitive polynomial in $R[x]$.*

Given $f(x)$, $g(x)$ in $R[x]$ we can write $f(x) = af_1(x)$, $g(x) = bg_1(x)$, where $a = c(f)$, $b = c(g)$ and where $f_1(x)$ and $g_1(x)$ are primitive. Thus $f(x)g(x) = abf_1(x)g_1(x)$. By Lemma 3.11.3, $f_1(x)g_1(x)$ is primitive. Hence the content of $f(x)g(x)$ is ab, that is, it is $c(f)c(g)$. We have proved the

COROLLARY *If R is a unique factorization domain and if $f(x)$, $g(x)$ are in $R[x]$, then $c(fg) = c(f)c(g)$ (up to units).*

By a simple induction, the corollary extends to the product of a finite number of polynomials to read $c(f_1f_2 \cdots f_k) = c(f_1)c(f_2)\cdots c(f_k)$.

Let R be a unique factorization domain. Being an integral domain, by Theorem 3.6.1, it has a field of quotients F. We can consider $R[x]$ to be a subring of $F[x]$. Given any polynomial $f(x) \in F[x]$, then $f(x) = (f_0(x)/a)$, where $f_0(x) \in R[x]$ and where $a \in R$. (Prove!) It is natural to ask for the relation, in terms of reducibility and irreducibility, of a polynomial in $R[x]$ considered as a polynomial in the larger ring $F[x]$

LEMMA 3.11.4 *If $f(x)$ in $R[x]$ is both primitive and irreducible as an element of $R[x]$, then it is irreducible as an element of $F[x]$. Conversely, if the primitive element $f(x)$ in $R[x]$ is irreducible as an element of $F[x]$, it is also irreducible as an element of $R[x]$.*

Proof. Suppose that the primitive element $f(x)$ in $R[x]$ is irreducible in $R[x]$ but is reducible in $F[x]$. Thus $f(x) = g(x)h(x)$, where $g(x)$, $h(x)$ are in $F[x]$ and are of positive degree. Now $g(x) = (g_0(x)/a)$, $h(x) = (h_0(x)/b)$, where $a, b \in R$ and where $g_0(x), h_0(x) \in R[x]$. Also $g_0(x) = \alpha g_1(x)$, $h_0(x) = \beta h_1(x)$, where $\alpha = c(g_0)$, $\beta = c(h_0)$, and $g_1(x)$, $h_1(x)$ are primitive in $R[x]$. Thus $f(x) = (\alpha\beta/ab)g_1(x)h_1(x)$, whence $abf(x) = \alpha\beta g_1(x)h_1(x)$. By Lemma 3.11.3, $g_1(x)h_1(x)$ is primitive, whence the content of the right-hand side is $\alpha\beta$. Since $f(x)$ is primitive, the content of the left-hand side is ab; but then $ab = \alpha\beta$; the implication of this is that $f(x) = g_1(x)h_1(x)$, and we have obtained a nontrivial factorization of $f(x)$ in $R[x]$, contrary to hypothesis. (Note: this factorization is nontrivial since each of $g_1(x)$, $h_1(x)$ are of the same degree as $g(x)$, $h(x)$, so cannot be units in $R[x]$ (see Problem 4).) We leave the converse half of the lemma as an exercise.

LEMMA 3.11.5 *If R is a unique factorization domain and if $p(x)$ is a primitive polynomial in $R[x]$, then it can be factored in a unique way as the product of irreducible elements in $R[x]$.*

Proof. When we consider $p(x)$ as an element in $F[x]$, by Lemma 3.9.5, we can factor it as $p(x) = p_1(x) \cdots p_k(x)$, where $p_1(x), p_2(x), \ldots, p_k(x)$ are irreducible polynomials in $F[x]$. Each $p_i(x) = (f_i(x)/a_i)$, where $f_i(x) \in R[x]$ and $a_i \in R$; moreover, $f_i(x) = c_i q_i(x)$, where $c_i = c(f_i)$ and where $q_i(x)$ is primitive in $R[x]$. Thus each $p_i(x) = (c_i q_i(x)/a_i)$, where $a_i, c_i \in R$ and where $q_i(x) \in R[x]$ is primitive. Since $p_i(x)$ is irreducible in $F[x]$, $q_i(x)$ must also be irreducible in $F[x]$, hence by Lemma 3.11.4 *it is irreducible in* $R[x]$.

Now

$$p(x) = p_1(x) \cdots p_k(x) = \frac{c_1 c_2 \cdots c_k}{a_1 a_2 \cdots a_k} q_1(x) \cdots q_k(x),$$

whence $a_1 a_2 \cdots a_k p(x) = c_1 c_2 \cdots c_k q_1(x) \cdots q_k(x)$. Using the primitivity of $p(x)$ and of $q_1(x) \cdots q_k(x)$, we can read off the content of the left-hand side as $a_1 a_2 \cdots a_k$ and that of the right-hand side as $c_1 c_2 \cdots c_k$. Thus $a_1 a_2 \cdots a_k = c_1 c_2 \cdots c_k$, hence $p(x) = q_1(x) \cdots q_k(x)$. We have factored $p(x)$, in $R[x]$, as a product of irreducible elements.

Can we factor it in another way? If $p(x) = r_1(x) \cdots r_k(x)$, where the $r_i(x)$ are irreducible in $R[x]$, by the primitivity of $p(x)$, each $r_i(x)$ must be primitive, *hence irreducible in* $F[x]$ by Lemma 3.11.4. But by Lemma 3.9.5 we know unique factorization in $F[x]$; the net result of this is that the $r_i(x)$ and the $q_i(x)$ are equal (up to associates) in some order, hence $p(x)$ has a unique factorization as a product of irreducibles in $R[x]$.

We now have all the necessary information to prove the principal theorem of this section.

THEOREM 3.11.1 *If R is a unique factorization domain, then so is* $R[x]$.

Proof. Let $f(x)$ be an arbitrary element in $R[x]$. We can write $f(x)$ in a unique way as $f(x) = cf_1(x)$ where $c = c(f)$ is in R and where $f_1(x)$, in $R[x]$, is primitive. By Lemma 3.11.5 we can decompose $f_1(x)$ in a unique way as the product of irreducible elements of $R[x]$. What about c? Suppose that $c = a_1(x)a_2(x) \cdots a_m(x)$ in $R[x]$; then $0 = \deg c = \deg(a_1(x)) + \deg(a_2(x)) + \cdots + \deg(a_m(x))$. Therefore, each $a_i(x)$ must be of degree 0, that is, it must be an element of R. In other words, the only factorizations of c as an element of $R[x]$ are those it had as an element of R. In particular, an irreducible element in R is still irreducible in $R[x]$. Since R is a unique factorization domain, c has a unique factorization as a product of irreducible elements of R, hence of $R[x]$.

Putting together the unique factorization of $f(x)$ in the form $cf_1(x)$ where $f_1(x)$ is primitive and where $c \in R$ with the unique factorizability of c and of $f_1(x)$ we have proved the theorem.

Given R as a unique factorization domain, then $R_1 = R[x_1]$ is also a unique factorization domain. Thus $R_2 = R_1[x_2] = R[x_1, x_2]$ is also a unique factorization domain. Continuing in this pattern we obtain

COROLLARY 1 *If R is a unique factorization domain then so is $R[x_1, \ldots, x_n]$.*

A special case of Corollary 1 but of independent interest and importance is

COROLLARY 2 *If F is a field then $F[x_1, \ldots, x_n]$ is a unique factorization domain.*

Problems

1. Prove that $R[x]$ is a commutative ring with unit element whenever R is.
2. Prove that $R[x_1, \ldots, x_n] = R[x_{i_1}, \ldots, x_{i_n}]$, where $(i_1, \ldots, i_n)$ is a permutation of $(1, 2, \ldots, n)$.
3. If R is an integral domain, prove that for $f(x)$, $g(x)$ in $R[x]$, $\deg(f(x)g(x)) = \deg(f(x)) + \deg(g(x))$.
4. If R is an integral domain with unit element, prove that any unit in $R[x]$ must already be a unit in R.
5. Let R be a commutative ring with no nonzero *nilpotent* elements (that is, $a^n = 0$ implies $a = 0$). If $f(x) = a_0 + a_1x + \cdots + a_mx^m$ in $R[x]$ is a zero-divisor, prove that there is an element $b \neq 0$ in R such that $ba_0 = ba_1 = \cdots = ba_m = 0$.

*6. Do Problem 5 dropping the assumption that R has no nonzero nilpotent elements.

*7. If R is a commutative ring with unit element, prove that $a_0 + a_1x + \cdots + a_nx^n$ in $R[x]$ has an inverse in $R[x]$ (i.e., is a unit in $R[x]$) if and only if a_0 is a unit in R and $a_1, \ldots, a_n$ are nilpotent elements in R.

8. Prove that when F is a field, $F[x_1, x_2]$ is not a principal ideal ring.
9. Prove, completely, Lemma 3.11.2 and its corollary.
10. (a) If R is a unique factorization domain, prove that every $f(x) \in R[x]$ can be written as $f(x) = af_1(x)$, where $a \in R$ and where $f_1(x)$ is primitive.
    (b) Prove that the decomposition in part (a) is unique (up to associates).
11. If R is an integral domain, and if F is its field of quotients, prove that any element $f(x)$ in $F[x]$ can be written as $f(x) = (f_0(x)/a)$, where $f_0(x) \in R[x]$ and where $a \in R$.
12. Prove the converse part of Lemma 3.11.4.
13. Prove Corollary 2 to Theorem 3.11.1.
14. Prove that a principal ideal ring is a unique factorization domain.
15. If J is the ring of integers, prove that $J[x_1, \ldots, x_n]$ is a unique factorization domain.

Supplementary Problems

1. Let R be a commutative ring; an ideal P of R is said to be a *prime ideal* of R if $ab \in P$, $a, b \in R$ implies that $a \in P$ or $b \in P$. Prove that P is a prime ideal of R if and only if R/P is an integral domain.
2. Let R be a commutative ring with unit element; prove that every maximal ideal of R is a prime ideal.
3. Give an example of a ring in which some prime ideal is not a maximal ideal.
4. If R is a finite commutative ring (i.e., has only a finite number of elements) with unit element, prove that every prime ideal of R is a maximal ideal of R.
5. If F is a field, prove that $F[x]$ is isomorphic to $F[t]$.
6. Find all the automorphisms σ of $F[x]$ with the property that $\sigma(f) = f$ for every $f \in F$.
7. If R is a commutative ring, let $N = \{x \in R \mid x^n = 0 \text{ for some integer } n\}$. Prove
   (a) N is an ideal of R.
   (b) In $\bar{R} = R/N$ if $\bar{x}^m = 0$ for some m then $\bar{x} = 0$.
8. Let R be a commutative ring and suppose that A is an ideal of R. Let $N(A) = \{x \in R \mid x^n \in A \text{ for some } n\}$. Prove
   (a) $N(A)$ is an ideal of R which contains A.
   (b) $N(N(A)) = N(A)$.
   $N(A)$ is often called the *radical* of A.
9. If n is an integer, let J_n be the ring of integers mod n. Describe N (see Problem 7) for J_n in terms of n.
10. If A and B are ideals in a ring R such that $A \cap B = (0)$, prove that for every $a \in A$, $b \in B$, $ab = 0$.
11. If R is a ring, let $Z(R) = \{x \in R \mid xy = yx \text{ all } y \in R\}$. Prove that $Z(R)$ is a subring of R.
12. If R is a division ring, prove that $Z(R)$ is a field.
13. Find a polynomial of degree 3 irreducible over the ring of integers, J_3, mod 3. Use it to construct a field having 27 elements.
14. Construct a field having 625 elements.
15. If F is a field and $p(x) \in F[x]$, prove that in the ring

$$R = \frac{F[x]}{(p(x))},$$

N (see Problem 7) is (0) if an only if $p(x)$ is not divisible by the square of any polynomial.

16. Prove that the polynomial $f(x) = 1 + x + x^3 + x^4$ is not irreducible over any field F.
17. Prove that the polynomial $f(x) = x^4 + 2x + 2$ is irreducible over the field of rational numbers.
18. Prove that if F is a finite field, its characteristic must be a prime number p and F contains p^n elements for some integer. Prove further that if $a \in F$ then $a^{p^n} = a$.
19. Prove that any nonzero ideal in the Gaussian integers $J[i]$ must contain some positive integer.
20. Prove that if R is a ring in which $a^4 = a$ for every $a \in R$ then R must be commutative.
21. Let R and R' be rings and ϕ a mapping from R into R' satisfying
    (a) $\phi(x + y) = \phi(x) + \phi(y)$ for every $x, y \in R$.
    (b) $\phi(xy) = \phi(x)\phi(y)$ or $\phi(y)\phi(x)$.
    Prove that for all $a, b \in R$, $\phi(ab) = \phi(a)\phi(b)$ or that, for all $a, b \in R$, $\phi(a) = \phi(b)\phi(a)$. (*Hint:* If $a \in R$, let
    $$W_a = \{x \in R \mid \phi(ax) = \phi(a)\phi(x)\}$$
    and
    $$U_a = \{x \in R \mid \phi(ax) = \phi(x)\phi(a)\}.)$$
22. Let R be a ring with a unit element, 1, in which $(ab)^2 = a^2b^2$ for all $a, b \in R$. Prove that R must be commutative.
23. Give an example of a noncommutative ring (of course, without 1) in which $(ab)^2 = a^2b^2$ for all elements a and b.
24. (a) Let R be a ring with unit element 1 such that $(ab)^2 = (ba)^2$ for all $a, b \in R$. If in R, $2x = 0$ implies $x = 0$, prove that R must be commutative.
    (b) Show that the result of (a) may be false if $2x = 0$ for some $x \neq 0$ in R.
    (c) Even if $2x = 0$ implies $x = 0$ in R, show that the result of (a) may be false if R does not have a unit element.
25. Let R be a ring in which $x^n = 0$ implies $x = 0$. If $(ab)^2 = a^2b^2$ for all $a, b \in R$, prove that R is commutative.
26. Let R be a ring in which $x^n = 0$ implies $x = 0$. If $(ab)^2 = (ba)^2$ for all $a, b \in R$, prove that R must be commutative.
27. Let $p_1, p_2, \ldots, p_k$ be distinct primes, and let $n = p_1p_2 \cdots p_k$. If R is the ring of integers modulo n, show that there are exactly 2^k elements a in R such that $a^2 = a$.
28. Construct a polynomial $q(x) \neq 0$ with integer coefficients which has no rational roots but is such that for any prime p we can solve the congruence $q(x) \equiv 0 \bmod p$ in the integers.

Supplementary Reading

ZARISKI, OSCAR, and SAMUEL, PIERRE, *Commutative Algebra*, Vol. 1. Princeton, New Jersey: D. Van Nostrand Company, Inc., 1958.

McCOY, N. H., *Rings and Ideals*, Carus Monograph No. 8. La Salle, Illinois: Open Court Publishing Company, 1948.

Topic for Class Discussion

MOTZKIN, T., "The Euclidean algorithm," *Bulletin of the American Mathematical Society*, Vol. 55 (1949), pages 1142–1146.

4

Vector Spaces and Modules

Up to this point we have been introduced to groups and to rings; the former has its motivation in the set of one-to-one mappings of a set onto itself, the latter, in the set of integers. The third algebraic model which we are about to consider—vector space—can, in large part, trace its origins to topics in geometry and physics.

Its description will be reminiscent of those of groups and rings—in fact, part of its structure is that of an abelian group—but a vector space differs from these previous two structures in that one of the products defined on it uses elements outside of the set itself. These remarks will become clear when we make the definition of a vector space.

Vector spaces owe their importance to the fact that so many models arising in the solutions of specific problems turn out to be vector spaces. For this reason the basic concepts introduced in them have a certain universality and are ones we encounter, and keep encountering, in so many diverse contexts. Among these fundamental notions are those of linear dependence, basis, and dimension which will be developed in this chapter. These are potent and effective tools in all branches of mathematics; we shall make immediate and free use of these in many key places in Chapter 5 which treats the theory of fields.

Intimately intertwined with vector spaces are the homomorphisms of one vector space into another (or into itself). These will make up the bulk of the subject matter to be considered in Chapter 6.

In the last part of the present chapter we generalize from vector spaces to modules; roughly speaking, a module is a vector space over a ring instead of over a field. For finitely generated modules over Euclidean rings we shall prove the fundamental basis theorem. This result allows us to give a complete description and construction of all abelian groups which are generated by a finite number of elements.

4.1 Elementary Basic Concepts

DEFINITION A nonempty set V is said to be a *vector space* over a field F if V is an abelian group under an operation which we denote by $+$, and if for every $\alpha \in F$, $v \in V$ there is defined an element, written αv, in V subject to

1. $\alpha(v + w) = \alpha v + \alpha w$;
2. $(\alpha + \beta)v = \alpha v + \beta v$;
3. $\alpha(\beta v) = (\alpha\beta)v$;
4. $1v = v$;

for all $\alpha, \beta \in F$, $v, w \in V$ (where the 1 represents the unit element of F under multiplication).

Note that in Axiom 1 above the $+$ is that of V, whereas on the left-hand side of Axiom 2 it is that of F and on the right-hand side, that of V.

We shall consistently use the following notations:

a. F will be a field.
b. Lowercase Greek letters will be elements of F; we shall often refer to elements of F as *scalars*.
c. Capital Latin letters will denote vector spaces over F.
d. Lowercase Latin letters will denote elements of vector spaces. We shall often call elements of a vector space *vectors*.

If we ignore the fact that V has two operations defined on it and view it for a moment merely as an abelian group under $+$, Axiom 1 states nothing more than the fact that multiplication of the elements of V by a fixed scalar α defines a homomorphism of the abelian group V into itself. From Lemma 4.1.1 which is to follow, if $\alpha \neq 0$ this homomorphism can be shown to be an isomorphism of V onto V.

This suggests that many aspects of the theory of vector spaces (and of rings, too) could have been developed as a part of the theory of groups, had we generalized the notion of a group to that of a *group with operators*. For students already familiar with a little abstract algebra, this is the preferred point of view; since we assumed no familiarity on the reader's part with any abstract algebra, we felt that such an approach might lead to a too sudden introduction to the ideas of the subject with no experience to act as a guide.

Example 4.1.1 Let F be a field and let K be a field which contains F as a subfield. We consider K as a vector space over F, using as the $+$ of the vector space the addition of elements of K, and by defining, for $\alpha \in F$, $v \in K$, αv to be the products of α and v as elements in the field K. Axioms 1, 2, 3 for a vector space are then consequences of the right-distributive law, left-distributive law, and associative law, respectively, which hold for K as a ring.

Example 4.1.2 Let F be a field and let V be the totality of all ordered n-tuples, $(\alpha_1, \ldots, \alpha_n)$ where the $\alpha_i \in F$. Two elements $(\alpha_1, \ldots, \alpha_n)$ and $(\beta_1, \ldots, \beta_n)$ of V are declared to be equal if and only if $\alpha_i = \beta_i$ for each $i = 1, 2, \ldots, n$. We now introduce the requisite operations in V to make of it a vector space by defining:

1. $(\alpha_1, \ldots, \alpha_n) + (\beta_1, \ldots, \beta_n) = (\alpha_1 + \beta_1, \alpha_2 + \beta_2, \ldots, \alpha_n + \beta_n)$.
2. $\gamma(\alpha_1, \ldots, \alpha_n) = (\gamma\alpha_1, \ldots, \gamma\alpha_n)$ for $\gamma \in F$.

It is easy to verify that with these operations, V is a vector space over F. Since it will keep reappearing, we assign a symbol to it, namely $F^{(n)}$.

Example 4.1.3 Let F be any field and let $V = F[x]$, the set of polynomials in x over F. We choose to ignore, at present, the fact that in $F[x]$ we can multiply any two elements, and merely concentrate on the fact that two polynomials can be added and that a polynomial can always be multiplied by an element of F. With these natural operations $F[x]$ is a vector space over F.

Example 4.1.4 In $F[x]$ let V_n be the set of all polynomials of degree less than n. Using the natural operations for polynomials of addition and multiplication, V_n is a vector space over F.

What is the relation of Example 4.1.4 to Example 4.1.2? Any element of V_n is of the form $\alpha_0 + \alpha_1 x + \cdots + \alpha_{n-1}x^{n-1}$, where $\alpha_i \in F$; if we map this element onto the element $(\alpha_0, \alpha_1, \ldots, \alpha_{n-1})$ in $F^{(n)}$ we could reasonably expect, once homomorphism and isomorphism have been defined, to find that V_n and $F^{(n)}$ are isomorphic as vector spaces.

DEFINITION If V is a vector space over F and if $W \subset V$, then W is a *subspace* of V if under the operations of V, W, itself, forms a vector space over F. Equivalently, W is a subspace of V whenever $w_1, w_2 \in W$, $\alpha, \beta \in F$ implies that $\alpha w_1 + \beta w_2 \in W$.

Note that the vector space defined in Example 4.1.4 is a subspace of that defined in Example 4.1.3. Additional examples of vector spaces and subspaces can be found in the problems at the end of this section.

DEFINITION If U and V are vector spaces over F then the mapping T of U into V is said to be a *homomorphism* if

1. $(u_1 + u_2)T = u_1T + u_2T$;
2. $(\alpha u_1)T = \alpha(u_1T)$;

for all $u_1, u_2 \in U$, and all $\alpha \in F$.

As in our previous models, a homomorphism is a mapping preserving all the algebraic structure of our system.

If T, in addition, is one-to-one, we call it an *isomorphism*. The kernel of T is defined as $\{u \in U \mid uT = 0\}$ where 0 is the identity element of the addition in V. It is an exercise that the kernel of T is a subspace of U and that T is an isomorphism if and only if its kernel is (0). Two vector spaces are said to be *isomorphic* if there is an isomorphism of one *onto* the other.

The set of all homomorphisms of U into V will be written as Hom (U, V). Of particular interest to us will be two special cases, Hom (U, F) and Hom (U, U). We shall study the first of these soon; the second, which can be shown to be a ring, is called the *ring of linear transformations* on U. A great deal of our time, later in this book, will be occupied with a detailed study of Hom (U, U).

We begin the material proper with an operational lemma which, as in the case of rings, will allow us to carry out certain natural and simple computations in vector spaces. In the statement of the lemma, 0 represents the zero of the addition in V, o that of the addition in F, and $-v$ the additive inverse of the element v of V.

LEMMA 4.1.1 *If V is a vector space over F then*

1. $\alpha 0 = 0$ *for* $\alpha \in F$.
2. $ov = 0$ *for* $v \in V$.
3. $(-\alpha)v = -(\alpha v)$ *for* $\alpha \in F$, $v \in V$.
4. If $v \neq 0$, *then* $\alpha v = 0$ *implies that* $\alpha = o$.

Proof. The proof is very easy and follows the lines of the analogous results proved for rings; for this reason we give it briefly and with few explanations.

1. Since $\alpha 0 = \alpha(0 + 0) = \alpha 0 + \alpha 0$, we get $\alpha 0 = 0$.
2. Since $ov = (o + o)v = ov + ov$ we get $ov = 0$.

3. Since $0 = (\alpha + (-\alpha))v = \alpha v + (-\alpha)v$, $(-\alpha)v = -(\alpha v)$.
4. If $\alpha v = 0$ and $\alpha \neq o$ then

$$0 = \alpha^{-1}0 = \alpha^{-1}(\alpha v) = (\alpha^{-1}\alpha)v = 1v = v.$$

The lemma just proved shows that multiplication by the zero of V or of F always leads us to the zero of V. Thus there will be no danger of confusion in using the same symbol for both of these, and we henceforth will merely use the symbol 0 to represent both of them.

Let V be a vector space over F and let W be a subspace of V. Considering these merely as abelian groups construct the quotient group V/W; its elements are the cosets $v + W$ where $v \in V$. The commutativity of the addition, from what we have developed in Chapter 2 on group theory, assures us that V/W is an abelian group. We intend to make of it a vector space. If $\alpha \in F$, $v + W \in V/W$, define $\alpha(v + W) = \alpha v + W$. As is usual, we must first show that this product is well defined; that is, if $v + W = v' + W$ then $\alpha(v + W) = \alpha(v' + W)$. Now, because $v + W = v' + W$, $v - v'$ is in W; since W is a subspace, $\alpha(v - v')$ must also be in W. Using part 3 of Lemma 4.1.1 (see Problem 1) this says that $\alpha v - \alpha v' \in W$ and so $\alpha v + W = \alpha v' + W$. Thus $\alpha(v + W) = \alpha v + W = \alpha v' + W = \alpha(v' + W)$; the product has been shown to be well defined. The verification of the vector-space axioms for V/W is routine and we leave it as an exercise. We have shown

LEMMA 4.1.2 *If V is a vector space over F and if W is a subspace of V, then V/W is a vector space over F, where, for $v_1 + W, v_2 + W \in V/W$ and $\alpha \in F$,*

1. $(v_1 + W) + (v_2 + W) = (v_1 + v_2) + W$.
2. $\alpha(v_1 + W) = \alpha v_1 + W$.

V/W is called the *quotient space* of V by W.

Without further ado we now state the first homomorphism theorem for vector spaces; we give no proofs but refer the reader back to the proof of Theorem 2.7.1.

THEOREM 4.1.1 *If T is a homomorphism of U onto V with kernel W, then V is isomorphic to U/W. Conversely, if U is a vector space and W a subspace of U, then there is a homomorphism of U onto U/W.*

The other homomorphism theorems will be found as exercises at the end of this section.

DEFINITION Let V be a vector space over F and let $U_1, \ldots, U_n$ be subspaces of V. V is said to be the *internal direct sum* of $U_1, \ldots, U_n$ if every element $v \in V$ can be written in one and only one way as $v = u_1 + u_2 + \cdots + u_n$ where $u_i \in U_i$.

Given any finite number of vector spaces over F, $V_1, \ldots, V_n$, consider the set V of all ordered n-tuples $(v_1, \ldots, v_n)$ where $v_i \in V_i$. We declare two elements $(v_1, \ldots, v_n)$ and $(v_1', \ldots, v_n')$ of V to be equal if and only if for each i, $v_i = v_i'$. We add two such elements by defining $(v_1, \ldots, v_n) + (w_1, \ldots, w_n)$ to be $(v_1 + w_1, v_2 + w_2, \ldots, v_n + w_n)$. Finally, if $\alpha \in F$ and $(v_1, \ldots, v_n) \in V$ we define $\alpha(v_1, \ldots, v_n)$ to be $(\alpha v_1, \alpha v_2, \ldots, \alpha v_n)$. To check that the axioms for a vector space hold for V with its operations as defined above is straightforward. Thus V itself is a vector space over F. We call V the *external direct sum* of $V_1, \ldots, V_n$ and denote it by writing $V = V_1 \oplus \cdots \oplus V_n$.

THEOREM 4.1.2 *If V is the internal direct sum of $U_1, \ldots, U_n$, then V is isomorphic to the external direct sum of $U_1, \ldots, U_n$.*

Proof. Given $v \in V$, v can be written, by assumption, in one and only one way as $v = u_1 + u_2 + \cdots + u_n$ where $u_i \in U_i$; define the mapping T of V into $U_1 \oplus \cdots \oplus U_n$ by $vT = (u_1, \ldots, u_n)$. Since v has a unique representation of this form, T is well defined. It clearly is onto, for the arbitrary element $(w_1, \ldots, w_n) \in U_1 \oplus \cdots \oplus U_n$ is wT where $w = w_1 + \cdots + w_n \in V$. We leave the proof of the fact that T is one-to-one and a homomorphism to the reader.

Because of the isomorphism proved in Theorem 4.1.2 we shall henceforth merely refer to a direct sum, not qualifying that it be internal or external.

Problems

1. In a vector space show that $\alpha(v - w) = \alpha v - \alpha w$.
2. Prove that the vector spaces in Example 4.1.4 and Example 4.1.2 are isomorphic.
3. Prove that the kernel of a homomorphism is a subspace.
4. (a) If F is a field of real numbers show that the set of real-valued, continuous functions on the closed interval $[0, 1]$ forms a vector space over F.
   (b) Show that those functions in part (a) for which all nth derivatives exist for $n = 1, 2, \ldots$ form a subspace.
5. (a) Let F be the field of all real numbers and let V be the set of all sequences $(a_1, a_2, \ldots, a_n, \ldots)$, $a_i \in F$, where equality, addition and scalar multiplication are defined componentwise. Prove that V is a vector space over F.
   (b) Let $W = \{(a_1, \ldots, a_n, \ldots) \in V \mid \lim_{n \to \infty} a_n = 0\}$. Prove that W is a subspace of V.

*(c) Let $U = \{(a_1, \ldots, a_n, \ldots) \in V \mid \sum_{i=1}^{\infty} a_i^2 \text{ is finite}\}$. Prove that U is a subspace of V and is contained in W.

6. If U and V are vector spaces over F, define an addition and a multiplication by scalars in Hom (U, V) so as to make Hom (U, V) into a vector space over F.

*7. Using the result of Problem 6 prove that Hom $(F^{(n)}, F^{(m)})$ is isomorphic to F^{nm} as a vector space.

8. If $n > m$ prove that there is a homomorphism of $F^{(n)}$ onto $F^{(m)}$ with a kernel W which is isomorphic to $F^{(n-m)}$.

9. If $v \neq 0 \in F^{(n)}$ prove that there is an element $T \in$ Hom $(F^{(n)}, F)$ such that $vT \neq 0$.

10. Prove that there exists an isomorphism of $F^{(n)}$ into Hom (Hom $(F^{(n)}, F), F)$.

11. If U and W are subspaces of V, prove that $U + W = \{v \in V \mid v = u + w,\ u \in U,\ w \in W\}$ is a subspace of V.

12. Prove that the intersection of two subspaces of V is a subspace of V.

13. If A and B are subspaces of V prove that $(A + B)/B$ is isomorphic to $A/(A \cap B)$.

14. If T is a homomorphism of U onto V with kernel W prove that there is a one-to-one correspondence between the subspaces of V and the subspaces of U which contain W.

15. Let V be a vector space over F and let $V_1, \ldots, V_n$ be subspaces of V. Suppose that $V = V_1 + V_2 + \cdots + V_n$ (see Problem 11), and that $V_i \cap (V_1 + \cdots + V_{i-1} + V_{i+1} + \cdots + V_n) = (0)$ for every $i = 1, 2, \ldots, n$. Prove that V is the internal direct sum of $V_1, \ldots, V_n$.

16. Let $V = V_1 \oplus \cdots \oplus V_n$; prove that in V there are subspaces $\bar{V}_i$ isomorphic to V_i such that V is the internal direct sum of the $\bar{V}_i$.

17. Let T be defined on $F^{(2)}$ by $(x_1, x_2)T = (\alpha x_1 + \beta x_2,\ \gamma x_1 + \delta x_2)$ where $\alpha, \beta, \gamma, \delta$ are some fixed elements in F.
   (a) Prove that T is a homomorphism of $F^{(2)}$ into itself.
   (b) Find necessary and sufficient conditions on $\alpha, \beta, \gamma, \delta$ so that T is an isomorphism.

18. Let T be defined on $F^{(3)}$ by $(x_1, x_2, x_3)T = (\alpha_{11}x_1 + \alpha_{12}x_2 + \alpha_{13}x_3,\ \alpha_{21}x_1 + \alpha_{22}x_2 + \alpha_{23}x_3,\ \alpha_{31}x_1 + \alpha_{32}x_2 + \alpha_{33}x_3)$. Show that T is a homomorphism of $F^{(3)}$ into itself and determine necessary and sufficient conditions on the α_{ij} so that T is an isomorphism.

19. Let T be a homomorphism of V into W. Using T, define a homomorphism T^* of Hom (W, F) into Hom (V, F).
20. (a) Prove that $F^{(1)}$ is not isomorphic to $F^{(n)}$ for $n > 1$.
    (b) Prove that $F^{(2)}$ is not isomorphic to $F^{(3)}$.
21. If V is a vector space over an *infinite* field F, prove that V cannot be written as the set-theoretic union of a finite number of proper subspaces.

4.2 Linear Independence and Bases

If we look somewhat more closely at two of the examples described in the previous section, namely Example 4.1.4 and Example 4.1.3, we notice that although they do have many properties in common there is one striking difference between them. This difference lies in the fact that in the former we can find a finite number of elements, $1, x, x^2, \ldots, x^{n-1}$ such that every element can be written as a combination of these with coefficients from F, whereas in the latter no such finite set of elements exists.

We now intend to examine, in some detail, vector spaces which can be generated, as was the space in Example 4.1.4, by a finite set of elements.

DEFINITION If V is a vector space over F and if $v_1, \ldots, v_n \in V$ then any element of the form $\alpha_1 v_1 + \alpha_2 v_2 + \cdots + \alpha_n v_n$, where the $\alpha_i \in F$, is a *linear combination* over F of $v_1, \ldots, v_n$.

Since we usually are working with some fixed field F we shall often say linear combination rather than linear combination over F. Similarly it will be understood that when we say vector space we mean vector space over F.

DEFINITION If S is a nonempty subset of the vector space V, then $L(S)$, the *linear span* of S, is the set of all linear combinations of finite sets of elements of S.

We put, after all, into $L(S)$ the elements required by the axioms of a vector space, so it is not surprising to find

LEMMA 4.2.1 *$L(S)$ is a subspace of V.*

Proof. If v and w are in $L(S)$, then $v = \lambda_1 s_1 + \cdots + \lambda_n s_n$ and $w = \mu_1 t_1 + \cdots + \mu_m t_m$, where the λ's and μ's are in F and the s_i and t_i are all in S. Thus, for $\alpha, \beta \in F$, $\alpha v + \beta w = \alpha(\lambda_1 s_1 + \cdots + \lambda_n s_n) + \beta(\mu_1 t_1 + \cdots + \mu_m t_m) = (\alpha\lambda_1)s_1 + \cdots + (\alpha\lambda_n)s_n + (\beta\mu_1)t_1 + \cdots + (\beta\mu_m)t_m$ and so is again in $L(S)$. $L(S)$ has been shown to be a subspace of V.

The proof of each part of the next lemma is straightforward and easy and we leave the proofs as exercises to the reader.

LEMMA 4.2.2 *If S, T are subsets of V, then*

1. $S \subset T$ *implies* $L(S) \subset L(T)$.
2. $L(S \cup T) = L(S) + L(T)$.
3. $L(L(S)) = L(S)$.

DEFINITION The vector space V is said to be *finite-dimensional* (over F) if there is a *finite* subset S in V such that $V = L(S)$.

Note that $F^{(n)}$ is finite-dimensional over F, for if S consists of the n vectors $(1, 0, \ldots, 0)$, $(0, 1, 0, \ldots, 0), \ldots, (0, 0, \ldots, 0, 1)$, then $V = L(S)$.

Although we have defined what is meant by a finite-dimensional space we have not, as yet, defined what is meant by the dimension of a space. This will come shortly.

DEFINITION If V is a vector space and if $v_1, \ldots, v_n$ are in V, we say that they are *linearly dependent* over F if there exist elements $\lambda_1, \ldots, \lambda_n$ in F, not all of them 0, such that $\lambda_1 v_1 + \lambda_2 v_2 + \cdots + \lambda_n v_n = 0$.

If the vectors $v_1, \ldots, v_n$ are not linearly dependent over F, they are said to be *linearly independent* over F. Here too we shall often contract the phrase "linearly dependent over F" to "linearly dependent." Note that if $v_1, \ldots, v_n$ are linearly independent then none of them can be 0, for if $v_1 = 0$, say, then $\alpha v_1 + 0v_2 + \cdots + 0v_n = 0$ for any $\alpha \neq 0$ in F.

In $F^{(3)}$ it is easy to verify that $(1, 0, 0)$, $(0, 1, 0)$, and $(0, 0, 1)$ are linearly independent while $(1, 1, 0)$, $(3, 1, 3)$, and $(5, 3, 3)$ are linearly dependent.

We point out that linear dependence is a function not only of the vectors but also of the field. For instance, the field of complex numbers is a vector space over the field of real numbers and it is also a vector space over the field of complex numbers. The elements $v_1 = 1$, $v_2 = i$ in it are linearly independent over the reals but are linearly dependent over the complexes, since $iv_1 + (-1)v_2 = 0$.

The concept of linear dependence is an absolutely basic and ultra-important one. We now look at some of its properties.

LEMMA 4.2.3 *If $v_1, \ldots, v_n \in V$ are linearly independent, then every element in their linear span has a unique representation in the form $\lambda_1 v_1 + \cdots + \lambda_n v_n$ with the $\lambda_i \in F$.*

Proof. By definition, every element in the linear span is of the form $\lambda_1 v_1 + \cdots + \lambda_n v_n$. To show uniqueness we must demonstrate that if $\lambda_1 v_1 + \cdots + \lambda_n v_n = \mu_1 v_1 + \cdots + \mu_n v_n$ then $\lambda_1 = \mu_1, \lambda_2 = \mu_2, \ldots, \lambda_n = \mu_n$. But if $\lambda_1 v_1 + \cdots + \lambda_n v_n = \mu_1 v_1 + \cdots + \mu_n v_n$, then we certainly have

$(\lambda_1 - \mu_1)v_1 + (\lambda_2 - \mu_2)v_2 + \cdots + (\lambda_n - \mu_n)v_n = 0$, which by the linear independence of $v_1, \ldots, v_n$ forces $\lambda_1 - \mu_1 = 0$, $\lambda_2 - \mu_2 = 0, \ldots,$ $\lambda_n - \mu_n = 0$.

The next theorem, although very easy and at first glance of a somewhat technical nature, has as consequences results which form the very foundations of the subject. We shall list some of these as corollaries; the others will appear in the succession of lemmas and theorems that are to follow.

THEOREM 4.2.1 *If $v_1, \ldots, v_n$ are in V then either they are linearly independent or some v_k is a linear combination of the preceding ones, $v_1, \ldots, v_{k-1}$.*

Proof. If $v_1, \ldots, v_n$ are linearly independent there is, of course, nothing to prove. Suppose then that $\alpha_1 v_1 + \cdots + \alpha_n v_n = 0$ where not all the α's are 0. Let k be the largest integer for which $\alpha_k \neq 0$. Since $\alpha_i = 0$ for $i > k$, $\alpha_1 v_1 + \cdots + \alpha_k v_k = 0$ which, since $\alpha_k \neq 0$, implies that $v_k = \alpha_k^{-1}(-\alpha_1 v_1 - \alpha_2 v_2 - \cdots - \alpha_{k-1}v_{k-1}) = (-\alpha_k^{-1}\alpha_1)v_1 + \cdots + (-\alpha_k^{-1}\alpha_{k-1})v_{k-1}$. Thus v_k is a linear combination of its predecessors.

COROLLARY 1 *If $v_1, \ldots, v_n$ in V have W as linear span and if $v_1, \ldots, v_k$ are linearly independent, then we can find a subset of $v_1, \ldots, v_n$ of the form $v_1, v_2, \ldots, v_k, v_{i_1}, \ldots, v_{i_r}$ consisting of linearly independent elements whose linear span is also W.*

Proof. If $v_1, \ldots, v_n$ are linearly independent we are done. If not, weed out from this set the first v_j, which is a linear combination of its predecessors. Since $v_1, \ldots, v_k$ are linearly independent, $j > k$. The subset so constructed, $v_1, \ldots, v_k, \ldots, v_{j-1}, v_{j+1}, \ldots, v_n$ has $n - 1$ elements. Clearly its linear span is contained in W. However, we claim that it is actually equal to W; for, given $w \in W$, w can be written as a linear combination of $v_1, \ldots, v_n$. But in this linear combination we can replace v_j by a linear combination of $v_1, \ldots, v_{j-1}$. That is, w is a linear combination of $v_1, \ldots, v_{j-1}, v_{j+1}, \ldots, v_n$.

Continuing this weeding out process, we reach a subset $v_1, \ldots, v_k, v_{i_1}, \ldots, v_{i_r}$ whose linear span is still W but in which no element is a linear combination of the preceding ones. By Theorem 4.2.1 the elements $v_1, \ldots, v_k, v_{i_1}, \ldots, v_{i_r}$ must be linearly independent.

COROLLARY 2 *If V is a finite-dimensional vector space, then it contains a finite set $v_1, \ldots, v_n$ of linearly independent elements whose linear span is V.*

Proof. Since V is finite-dimensional, it is the linear span of a finite number of elements $u_1, \ldots, u_m$. By Corollary 1 we can find a subset of these, denoted by $v_1, \ldots, v_n$, consisting of linearly independent elements whose linear span must also be V.

DEFINITION A subset S of a vector space V is called a *basis* of V if S consists of linearly independent elements (that is, any finite number of elements in S is linearly independent) and $V = L(S)$.

In this terminology we can rephrase Corollary 2 as

COROLLARY 3 *If V is a finite-dimensional vector space and if $u_1, \ldots, u_m$ span V then some subset of $u_1, \ldots, u_m$ forms a basis of V.*

Corollary 3 asserts that a finite-dimensional vector space has a basis containing a finite number of elements $v_1, \ldots, v_n$. Together with Lemma 4.2.3 this tells us that every element in V has a unique representation in the form $\alpha_1 v_1 + \cdots + \alpha_n v_n$ with $\alpha_1, \ldots, \alpha_n$ in F.

Let us see some of the heuristic implications of these remarks. Suppose that V is a finite-dimensional vector space over F; as we have seen above, V has a basis $v_1, \ldots, v_n$. Thus every element $v \in V$ has a unique representation in the form $v = \alpha_1 v_1 + \cdots + \alpha_n v_n$. Let us map V into $F^{(n)}$ by defining the image of $\alpha_1 v_1 + \cdots + \alpha_n v_n$ to be $(\alpha_1, \ldots, \alpha_n)$. By the uniqueness of representation in this form, the mapping is well defined, one-to-one, and onto; it can be shown to have all the requisite properties of an isomorphism. Thus V is isomorphic to $F^{(n)}$ for some n, where in fact n is the number of elements in some basis of V over F. If some other basis of V should have m elements, by the same token V would be isomorphic to $F^{(m)}$. Since both $F^{(n)}$ and $F^{(m)}$ would now be isomorphic to V, they would be isomorphic to each other.

A natural question then arises! Under what conditions on n and m are $F^{(n)}$ and $F^{(m)}$ isomorphic? Our intuition suggests that this can only happen when $n = m$. Why? For one thing, if F should be a field with a finite number of elements—for instance, if $F = J_p$ the integers modulo the prime number p—then $F^{(n)}$ has p^n elements whereas $F^{(m)}$ has p^m elements. Isomorphism would imply that they have the same number of elements, and so we would have $n = m$. From another point of view, if F were the field of real numbers, then $F^{(n)}$ (in what may be a rather vague geometric way to the reader) represents real n-space, and our geometric feeling tells us that n-space is different from m-space for $n \neq m$. Thus we might expect that if F is any field then $F^{(n)}$ is isomorphic to $F^{(m)}$ only if $n = m$. Equivalently, from our earlier discussion, we should expect that any two bases of V have the same number of elements. It is towards this goal that we prove the next lemma.

LEMMA 4.2.4 *If $v_1, \ldots, v_n$ is a basis of V over F and if $w_1, \ldots, w_m$ in V are linearly independent over F, then $m \leq n$.*

Proof. Every vector in V, so in particular w_m, is a linear combination of $v_1, \ldots, v_n$. Therefore the vectors $w_m, v_1, \ldots, v_n$ are linearly dependent.

Moreover, they span V since $v_1, \ldots, v_n$ already do so. Thus some proper subset of these $w_m, v_{i_1}, \ldots, v_{i_k}$ with $k \leq n - 1$ forms a basis of V. We have "traded off" one w, in forming this new basis, for at least one v_i. Repeat this procedure with the set $w_{m-1}, w_m, v_{i_1}, \ldots, v_{i_k}$. From this linearly dependent set, by Corollary 1 to Theorem 4.2.1, we can extract a basis of the form $w_{m-1}, w_m, v_{j_1}, \ldots, v_{j_s}$, $s \leq n - 2$. Keeping up this procedure we eventually get down to a basis of V of the form $w_2, \ldots, w_{m-1}, w_m, v_\alpha, v_\beta \ldots$; since w_1 is not a linear combination of $w_2, \ldots, w_{m-1}$, the above basis must actually include some v. To get to this basis we have introduced $m - 1$ w's, each such introduction having cost us at least one v, and yet there is a v left. Thus $m - 1 \leq n - 1$ and so $m \leq n$.

This lemma has as consequences (which we list as corollaries) the basic results spelling out the nature of the dimension of a vector space. These corollaries are of the utmost importance in all that follows, not only in this chapter but in the rest of the book, in fact in all of mathematics. The corollaries are all theorems in their own rights.

COROLLARY 1 *If V is finite-dimensional over F then any two bases of V have the same number of elements.*

Proof. Let $v_1, \ldots, v_n$ be one basis of V over F and let $w_1, \ldots, w_m$ be another. In particular, $w_1, \ldots, w_m$ are linearly independent over F whence, by Lemma 4.2.4, $m \leq n$. Now interchange the roles of the v's and w's and we obtain that $n \leq m$. Together these say that $n = m$.

COROLLARY 2 *$F^{(n)}$ is isomorphic $F^{(m)}$ if and only if $n = m$.*

Proof. $F^{(n)}$ has, as one basis, the set of n vectors, $(1, 0, \ldots, 0)$, $(0, 1, 0, \ldots, 0), \ldots, (0, 0, \ldots, 0, 1)$. Likewise $F^{(m)}$ has a basis containing m vectors. An isomorphism maps a basis onto a basis (Problem 4, end of this section), hence, by Corollary 1, $m = n$.

Corollary 2 puts on a firm footing the heuristic remarks made earlier about the possible isomorphism of $F^{(n)}$ and $F^{(m)}$. As we saw in those remarks, V is isomorphic to $F^{(n)}$ for some n. By Corollary 2, this n is unique, thus

COROLLARY 3 *If V is finite-dimensional over F then V is isomorphic to $F^{(n)}$ for a unique integer n; in fact, n is the number of elements in any basis of V over F.*

DEFINITION The integer n in Corollary 3 is called the *dimension* of V over F.

The dimension of V over F is thus the number of elements in any basis of V over F.

We shall write the dimension of V over F as dim V, or, the occasional time in which we shall want to stress the role of the field F, as $\dim_F V$.

COROLLARY 4 *Any two finite-dimensional vector spaces over F of the same dimension are isomorphic.*

Proof. If this dimension is n, then each is isomorphic to $F^{(n)}$, hence they are isomorphic to each other.

How much freedom do we have in constructing bases of V? The next lemma asserts that starting with any linearly independent set of vectors we can "blow it up" to a basis of V.

LEMMA 4.2.5 *If V is finite-dimensional over F and if $u_1, \ldots, u_m \in V$ are linearly independent, then we can find vectors $u_{m+1}, \ldots, u_{m+r}$ in V such that $u_1, \ldots, u_m, u_{m+1}, \ldots, u_{m+r}$ is a basis of V.*

Proof. Since V is finite-dimensional it has a basis; let $v_1, \ldots, v_n$ be a basis of V. Since these span V, the vectors $u_1, \ldots, u_m, v_1, \ldots, v_n$ also span V. By Corollary 1 to Theorem 4.2.1 there is a subset of these of the form $u_1, \ldots, u_m, v_{i_1}, \ldots, v_{i_r}$ which consists of linearly independent elements which span V. To prove the lemma merely put $u_{m+1} = v_{i_1}, \ldots, u_{m+r} = v_{i_r}$.

What is the relation of the dimension of a homomorphic image of V to that of V? The answer is provided us by

LEMMA 4.2.6 *If V is finite-dimensional and if W is a subspace of V, then W is finite-dimensional,* $\dim W \leq \dim V$ *and* $\dim V/W = \dim V - \dim W$.

Proof. By Lemma 4.2.4, if $n = \dim V$ then any $n + 1$ elements in V are linearly dependent; in particular, any $n + 1$ elements in W are linearly dependent. Thus we can find a largest set of linearly independent elements in W, $w_1, \ldots, w_m$ and $m \leq n$. If $w \in W$ then $w_1, \ldots, w_m, w$ is a linearly dependent set, whence $\alpha w + \alpha_1 w_1 + \cdots + \alpha_m w_m = 0$, and not all of the α_i's are 0. If $\alpha = 0$, by the linear independence of the w_i we would get that each $\alpha_i = 0$, a contradiction. Thus $\alpha \neq 0$, and so $w = -\alpha^{-1}(\alpha_1 w_1 + \cdots + \alpha_m w_m)$. Consequently, $w_1, \ldots, w_m$ span W; by this, W is finite-dimensional over F, and furthermore, it has a basis of m elements, where $m \leq n$. From the definition of dimension it then follows that $\dim W \leq \dim V$.

Now, let $w_1, \ldots, w_m$ be a basis of W. By Lemma 4.2.5, we can fill this out to a basis, $w_1, \ldots, w_m, v_1, \ldots, v_r$ of V, where $m + r = \dim V$ and $m = \dim W$.

Let $\bar{v}_1, \ldots, \bar{v}_r$ be the images, in $\bar{V} = V/W$, of $v_1, \ldots, v_r$. Since any vector $v \in V$ is of the form $v = \alpha_1 w_1 + \cdots + \alpha_m w_m + \beta_1 v_1 + \cdots + \beta_r v_r$, then $\bar{v}$, the image of v, is of the form $\bar{v} = \beta_1 \bar{v}_1 + \cdots + \beta_r \bar{v}_r$ (since $\bar{w}_1 = \bar{w}_2 = \cdots = \bar{w}_m = 0$). Thus $\bar{v}_1, \ldots, \bar{v}_r$ span V/W. We claim that they are linearly independent, for if $\gamma_1 \bar{v}_1 + \cdots + \gamma_r \bar{v}_r = 0$ then $\gamma_1 v_1 + \cdots + \gamma_r v_r \in W$, and so $\gamma_1 v_1 + \cdots + \gamma_r v_r = \lambda_1 w_1 + \cdots + \lambda_m w_m$, which, by the linear independence of the set $w_1, \ldots, w_m, v_1, \ldots, v_r$ forces $\gamma_1 = \cdots = \gamma_r = \lambda_1 = \cdots = \lambda_m = 0$. We have shown that V/W has a basis of r elements, and so, $\dim V/W = r = \dim V - m = \dim V - \dim W$.

COROLLARY *If A and B are finite-dimensional subspaces of a vector space V, then $A + B$ is finite-dimensional and* $\dim(A + B) = \dim(A) + \dim(B) - \dim(A \cap B)$.

Proof. By the result of Problem 13 at the end of Section 4.1,

$$\frac{A + B}{B} \approx \frac{A}{A \cap B},$$

and since A and B are finite-dimensional, we get that

$$\begin{aligned} \dim(A + B) - \dim B &= \dim\left(\frac{A + B}{B}\right) = \dim\left(\frac{A}{A \cap B}\right) \\ &= \dim A - \dim(A \cap B). \end{aligned}$$

Transposing yields the result stated in the lemma.

Problems

1. Prove Lemma 4.2.2.
2. (a) If F is the field of real numbers, prove that the vectors $(1, 1, 0, 0)$, $(0, 1, -1, 0)$, and $(0, 0, 0, 3)$ in $F^{(4)}$ are linearly independent over F.
   (b) What conditions on the characteristic of F would make the three vectors in (a) linearly dependent?
3. If V has a basis of n elements, give a detailed proof that V is isomorphic to $F^{(n)}$.
4. If T is an isomorphism of V onto W, prove that T maps a basis of V onto a basis of W.
5. If V is finite-dimensional and T is an isomorphism of V into V, prove that T must map V *onto* V.
6. If V is finite-dimensional and T is a homomorphism of V *onto* V, prove that T must be one-to-one, and so an isomorphism.
7. If V is of dimension n, show that any set of n linearly independent vectors in V forms a basis of V.

8. If V is finite-dimensional and W is a subspace of V such that $\dim V = \dim W$, prove that $V = W$.
9. If V is finite-dimensional and T is a homomorphism of V into itself which is not onto, prove that there is some $v \neq 0$ in V such that $vT = 0$.
10. Let F be a field and let $F[x]$ be the polynomials in x over F. Prove that $F[x]$ is not finite-dimensional over F.
11. Let $V_n = \{p(x) \in F[x] \mid \deg p(x) < n\}$. Define T by

$$(\alpha_0 + \alpha_1 x + \cdots + \alpha_{n-1}x^{n-1})T$$
$$= \alpha_0 + \alpha_1(x + 1) + \alpha_2(x + 1)^2 + \cdots + \alpha_{n-1}(x + 1)^{n-1}.$$

Prove that T is an isomorphism of V_n onto itself.
12. Let $W = \{\alpha_0 + \alpha_1 x + \cdots + \alpha_{n-1}x^{n-1} \in F[x] \mid \alpha_0 + \alpha_1 + \cdots + \alpha_{n-1} = 0\}$. Show that W is a subspace of V_n and find a basis of W over F.
13. Let $v_1, \ldots, v_n$ be a basis of V and let $w_1, \ldots, w_n$ be any n elements in V. Define T on V by $(\lambda_1 v_1 + \cdots + \lambda_n v_n)T = \lambda_1 w_1 + \cdots + \lambda_n w_n$.
(a) Show that R is a homomorphism of V into itself.
(b) When is T an isomorphism?
14. Show that any homomorphism of V into itself, when V is finite-dimensional, can be realized as in Problem 13 by choosing appropriate elements $w_1, \ldots, w_n$.
15. Returning to Problem 13, since $v_1, \ldots, v_n$ is a basis of V, each $w_i = \alpha_{i1}v_1 + \cdots + \alpha_{in}v_n$, $\alpha_{ij} \in F$. Show that the n^2 elements α_{ij} of F determine the homomorphism T.

*16. If $\dim_F V = n$ prove that $\dim_F (\text{Hom}(V,V)) = n^2$.

17. If V is finite-dimensional and W is a subspace of V prove that there is a subspace W_1 of V such that $V = W \oplus W_1$.

4.3 Dual Spaces

Given any two vector spaces, V and W, over a field F, we have defined Hom (V, W) to be the set of all vector space homomorphisms of V into W. As yet Hom (V, W) is merely a set with no structure imposed on it. We shall now proceed to introduce operations in it which will turn it into a vector space over F. Actually we have already indicated how to do so in the descriptions of some of the problems in the earlier sections. However we propose to treat the matter more formally here.

Let S and T be any two elements of Hom (V, W); this means that these are both vector space homomorphisms of V into W. Recalling the definition of such a homomorphism, we must have $(v_1 + v_2)S = v_1S + v_2S$ and $(\alpha v_1)S = \alpha(v_1S)$ for all $v_1, v_2 \in V$ and all $\alpha \in F$. The same conditions also hold for T.

We first want to introduce an addition for these elements S and T in Hom (V, W). What is more natural than to define $S + T$ by declaring $v(S + T) = vS + vT$ for all $v \in V$? We must, of course, verify that $S + T$ is in Hom (V, W). By the very definition of $S + T$, if $v_1, v_2 \in V$, then $(v_1 + v_2)(S + T) = (v_1 + v_2)S + (v_1 + v_2)T$; since $(v_1 + v_2)S = v_1S + v_2S$ and $(v_1 + v_2)T = v_1T + v_2T$ and since addition in W is commutative, we get $(v_1 + v_2)(S + T) = v_1S + v_1T + v_2S + v_2T$. Once again invoking the definition of $S + T$, the right-hand side of this relation becomes $v_1(S + T) + v_2(S + T)$; we have shown that $(v_1 + v_2)(S + T) = v_1(S + T) + v_2(S + T)$. A similar computation shows that $(\alpha v)(S + T) = \alpha(v(S + T))$. Consequently $S + T$ is in Hom (V, W). Let 0 be that homomorphism of V into W which sends every element of V onto the zero-element of W; for $S \in$ Hom (V, W) let $-S$ be defined by $v(-S) = -(vS)$. It is immediate that Hom (V, W) is an abelian group under the addition defined above.

Having succeeded in introducing the structure of an abelian group on Hom (V, W), we now turn our attention to defining λS for $\lambda \in F$ and $S \in$ Hom (V, W), our ultimate goal being that of making Hom (V, W) into a vector space over F. For $\lambda \in F$ and $S \in$ Hom (V, W) we define λS by $v(\lambda S) = \lambda(vS)$ for all $v \in V$. We leave it to the reader to show that λS is in Hom (V, W) and that under the operations we have defined, Hom (V, W) is a vector space over F. But we have no assurance that Hom (V, W) has any elements other than the zero-homomorphism. Be that as it may, we have proved

LEMMA 4.3.1 Hom (V, W) *is a vector space over* F *under the operations described above.*

A result such as that of Lemma 4.3.1 really gives us very little information; rather it confirms for us that the definitions we have made are reasonable. We would prefer some results about Hom (V, W) that have more of a bite to them. Such a result is provided us in

THEOREM 4.3.1 *If* V *and* W *are of dimensions* m *and* n, *respectively, over* F, *then* Hom (V, W) *is of dimension* mn *over* F.

Proof. We shall prove the theorem by explicitly exhibiting a basis of Hom (V, W) over F consisting of mn elements.

Let $v_1, \ldots, v_m$ be a basis of V over F and $w_1, \ldots, w_n$ one for W over F. If $v \in V$ then $v = \lambda_1 v_1 + \cdots + \lambda_m v_m$ where $\lambda_1, \ldots, \lambda_m$ are uniquely defined elements of F; define $T_{ij}:V \to W$ by $vT_{ij} = \lambda_i w_j$. From the point of view of the bases involved we are simply letting $v_k T_{ij} = 0$ for $k \neq i$ and $v_i T_{ij} = w_j$. It is an easy exercise to see that T_{ij} is in Hom (V, W). Since i can be any of $1, 2, \ldots, m$ and j any of $1, 2, \ldots, n$ there are mn such T_{ij}'s.

Our claim is that these mn elements constitute a basis of Hom (V, W) over F. For, let $S \in$ Hom (V, W); since $v_1 S \in W$, and since any element in W is a linear combination over F of $w_1, \ldots, w_n$, $v_1 S = \alpha_{11} w_1 + \alpha_{12} w_2 + \cdots + \alpha_{1n} w_n$, for some $\alpha_{11}, \alpha_{12}, \ldots, \alpha_{1n}$ in F. In fact, $v_i S = \alpha_{i1} w_1 + \cdots + \alpha_{in} w_n$ for $i = 1, 2, \ldots, m$. Consider $S_0 = \alpha_{11} T_{11} + \alpha_{12} T_{12} + \cdots + \alpha_{1n} T_{1n} + \alpha_{21} T_{21} + \cdots + \alpha_{2n} T_{2n} + \cdots + \alpha_{i1} T_{i1} + \cdots + \alpha_{in} T_{in} + \cdots + \alpha_{m1} T_{m1} + \cdots + \alpha_{mn} T_{mn}$. Let us compute $v_k S_0$ for the basis vector v_k. Now $v_k S_0 = v_k(\alpha_{11} T_{11} + \cdots + \alpha_{m1} T_{m1} + \cdots + \alpha_{mn} T_{mn}) = \alpha_{11}(v_k T_{11}) + \alpha_{12}(v_k T_{12}) + \cdots + \alpha_{m1}(v_k T_{m1}) + \cdots + \alpha_{mn}(v_k T_{mn})$. Since $v_k T_{ij} = 0$ for $i \neq k$ and $v_k T_{kj} = w_j$, this sum reduces to $v_k S_0 = \alpha_{k1} w_1 + \cdots + \alpha_{kn} w_n$, which, we see, is nothing but $v_k S$. Thus the homomorphisms S_0 and S agree on a basis of V. We claim this forces $S_0 = S$ (see Problem 3, end of this section). However S_0 is a linear combination of the T_{ij}'s, whence S must be the same linear combination. In short, we have shown that the mn elements $T_{11}, T_{12}, \ldots, T_{1n}, \ldots, T_{m1}, \ldots, T_{mn}$ span Hom (V, W) over F.

In order to prove that they form a basis of Hom (V, W) over F there remains but to show their linear independence over F. Suppose that $\beta_{11} T_{11} + \beta_{12} T_{12} + \cdots + \beta_{1n} T_{1n} + \cdots + \beta_{i1} T_{i1} + \cdots + \beta_{in} T_{in} + \cdots + \beta_{m1} T_{m1} + \cdots + \beta_{mn} T_{mn} = 0$ with β_{ij} all in F. Applying this to v_k we get $0 = v_k(\beta_{11} T_{11} + \cdots + \beta_{ij} T_{ij} + \cdots + \beta_{mn} T_{mn}) = \beta_{k1} w_1 + \beta_{k2} w_2 + \cdots + \beta_{kn} w_n$ since $v_k T_{ij} = 0$ for $i \neq k$ and $v_k T_{kj} = w_j$. However, $w_1, \ldots, w_n$ are linearly independent over F, forcing $\beta_{kj} = 0$ for all k and j. Thus the T_{ij} are linearly independent over F, whence they indeed do form a basis of Hom (V, W) over F.

An immediate consequence of Theorem 4.3.1 is that whenever $V \neq (0)$ and $W \neq (0)$ are finite-dimensional vector spaces, then Hom (V, W) does not just consist of the element 0, for its dimension over F is $nm \geq 1$.

Some special cases of Theorem 4.3.1 are themselves of great interest and we list these as corollaries.

COROLLARY 1 *If* $\dim_F V = m$ *then* $\dim_F$ Hom $(V, V) = m^2$.

Proof. In the theorem put $V = W$, and so $m = n$, whence $mn = m^2$.

COROLLARY 2 *If* $\dim_F V = m$ *then* $\dim_F$ Hom $(V, F) = m$.

Proof. As a vector space F is of dimension 1 over F. Applying the theorem yields $\dim_F$ Hom $(V, F) = m$.

Corollary 2 has the interesting consequence that if V is finite-dimensional over F it is isomorphic to Hom (V, F), for, by the corollary, they are of the same dimension over F, whence by Corollary 4 to Lemma 4.2.4 they must be isomorphic. This isomorphism has many shortcomings! Let us explain. It depends heavily on the finite-dimensionality of V, for if V is not finite-dimensional no such isomorphism exists. There is no nice, formal construction of this isomorphism which holds universally for all vector spaces. It depends strongly on the specialities of the finite-dimensional situation. In a few pages we shall, however, show that a "nice" isomorphism does exist for any vector space V into Hom (Hom (V, F), F).

DEFINITION If V is a vector space then its *dual space* is Hom (V, F).

We shall use the notation $\hat{V}$ for the dual space of V. An element of $\hat{V}$ will be called a *linear functional* on V into F.

If V is not finite-dimensional the $\hat{V}$ is usually too large and wild to be of interest. For such vector spaces we often have other additional structures, such as a topology, imposed and then, as the dual space, one does not generally take all of our $\hat{V}$ but rather a properly restricted subspace. If V is finite-dimensional its dual space $\hat{V}$ is always defined, as we did it, as all of Hom (V, F).

In the proof of Theorem 4.3.1 we constructed a basis of Hom (V, W) using a particular basis of V and one of W. The construction depended crucially on the particular bases we had chosen for V and W, respectively. Had we chosen other bases we would have ended up with a different basis of Hom (V, W). As a general principle, it is preferable to give proofs, whenever possible, which are basis-free. Such proofs are usually referred to as invariant ones. An invariant proof or construction has the advantage, other than the mere aesthetic one, over a proof or construction using a basis, in that one does not have to worry how finely everything depends on a particular choice of bases.

The elements of $\hat{V}$ are functions defined on V and having their values in F. In keeping with the functional notation, we shall usually write elements of $\hat{V}$ as f, g, etc. and denote the value on $v \in V$ as $f(v)$ (rather than as vf).

Let V be a finite-dimensional vector space over F and let $v_1, \ldots, v_n$ be a basis of V; let $\hat{v}_i$ be the element of $\hat{V}$ defined by $\hat{v}_i(v_j) = 0$ for $i \neq j$, $\hat{v}_i(v_i) = 1$, and $\hat{v}_i(\alpha_1 v_1 + \cdots + \alpha_i v_i + \cdots + \alpha_n v_n) = \alpha_i$. In fact the $\hat{v}_i$ are nothing but the T_{ij} introduced in the proof of Theorem 4.3.1, for here $W = F$ is one-dimensional over F. Thus we know that $\hat{v}_1, \ldots, \hat{v}_n$ form a basis of $\hat{V}$. We call this basis the *dual basis* of $v_1, \ldots, v_n$. If $v \neq 0 \in V$, by Lemma 4.2.5 we can find a basis of the form $v_1 = v, v_2, \ldots, v_n$ and so there is an element in $\hat{V}$, namely $\hat{v}_1$, such that $\hat{v}_1(v_1) = \hat{v}_1(v) = 1 \neq 0$. We have proved

LEMMA 4.3.2 *If V is finite-dimensional and $v \neq 0 \in V$, then there is an element $f \in \hat{V}$ such that $f(v) \neq 0$.*

In fact, Lemma 4.3.2 is true if V is infinite-dimensional, but as we have no need for the result, and since its proof would involve logical questions that are not relevant at this time, we omit the proof.

Let $v_0 \in V$, where V is any vector space over F. As f varies over $\hat{V}$, and v_0 is kept fixed, $f(v_0)$ defines a functional on $\hat{V}$ into F; *note that we are merely interchanging the role of function and variable.* Let us denote this function by T_{v_0}; in other words $T_{v_0}(f) = f(v_0)$ for any $f \in \hat{V}$. What can we say about T_{v_0}? To begin with, $T_{v_0}(f + g) = (f + g)(v_0) = f(v_0) + g(v_0) = T_{v_0}(f) + T_{v_0}(g)$; furthermore, $T_{v_0}(\lambda f) = (\lambda f)(v_0) = \lambda f(v_0) = \lambda T_{v_0}(f)$. Thus T_{v_0} is in the dual space of $\hat{V}$! We write this space as $\hat{\hat{V}}$ and refer to it as the *second dual* of V.

Given any element $v \in V$ we can associate with it an element T_v in $\hat{\hat{V}}$. Define the mapping $\psi : V \to \hat{\hat{V}}$ by $v\psi = T_v$ for every $v \in V$. Is ψ a homomorphism of V into $\hat{\hat{V}}$? Indeed it is! For, $T_{v+w}(f) = f(v + w) = f(v) + f(w) = T_v(f) + T_w(f) = (T_v + T_w)(f)$, and so $T_{v+w} = T_v + T_w$, that is, $(v + w)\psi = v\psi + w\psi$. Similarly for $\lambda \in F$, $(\lambda v)\psi = \lambda(v\psi)$. Thus ψ defines a homomorphism of V into $\hat{\hat{V}}$. The construction of ψ used no basis or special properties of V; it is an example of an invariant construction.

When is ψ an isomorphism? To answer this we must know when $v\psi = 0$, or equivalently, when $T_v = 0$. But if $T_v = 0$, then $0 = T_v(f) = f(v)$ for all $f \in \hat{V}$. However as we pointed out, without proof, for a general vector space, given $v \neq 0$ there is an $f \in \hat{V}$ with $f(v) \neq 0$. We actually proved this when V is finite-dimensional. Thus for V finite-dimensional (and, in fact, for arbitrary V) ψ is an isomorphism. However, when V is finite-dimensional ψ is an isomorphism onto $\hat{\hat{V}}$; when V is infinite-dimensional ψ is not onto.

If V is finite-dimensional, by the second corollary to Theorem 4.3.1, V and $\hat{V}$ are of the same dimension; similarly, $\hat{V}$ and $\hat{\hat{V}}$ are of the same dimension; since ψ is an isomorphism of V into $\hat{\hat{V}}$, the equality of the dimensions forces ψ to be onto. We have proved

LEMMA 4.3.3 *If V is finite-dimensional, then ψ is an isomorphism of V onto $\hat{\hat{V}}$.*

We henceforth identify V and $\hat{\hat{V}}$, keeping in mind that this identification is being carried out by the isomorphism ψ.

DEFINITION If W is a subspace of V then the *annihilator* of W, $A(W) = \{f \in \hat{V} \mid f(w) = 0 \text{ all } w \in W\}$.

We leave as an exercise to the reader the verification of the fact that $A(W)$ is a subspace of $\hat{V}$. Clearly if $U \subset W$, then $A(U) \supset A(W)$.

Let W be a subspace of V, where V is finite-dimensional. If $f \in \hat{V}$ let $\tilde{f}$ be the restriction of f to W; thus $\tilde{f}$ is defined on W by $\tilde{f}(w) = f(w)$ for every $w \in W$. Since $f \in \hat{V}$, clearly $\tilde{f} \in \hat{W}$. Consider the mapping $T: \hat{V} \to \hat{W}$ defined by $fT = \tilde{f}$ for $f \in \hat{V}$. It is immediate that $(f + g)T = fT + gT$ and that $(\lambda f)T = \lambda(fT)$. Thus T is a homomorphism of $\hat{V}$ into $\hat{W}$. What is the kernel of T? If f is in the kernel of T then the restriction of f to W must be 0; that is, $f(w) = 0$ for all $w \in W$. Also, conversely, if $f(w) = 0$ for all $w \in W$ then f is in the kernel of T. Therefore the kernel of T is exactly $A(W)$.

We now claim that the mapping T is onto $\hat{W}$. What we must show is that given any element $h \in \hat{W}$, then h is the restriction of some $f \in \hat{V}$, that is $h = \tilde{f}$. By Lemma 4.2.5, if $w_1, \ldots, w_m$ is a basis of W then it can be expanded to a basis of V of the form $w_1, \ldots, w_m, v_1, \ldots, v_r$ where $r + m = \dim V$. Let W_1 be the subspace of V spanned by $v_1, \ldots, v_r$. Thus $V = W \oplus W_1$. If $h \in \hat{W}$ define $f \in \hat{V}$ by: let $v \in V$ be written as $v = w + w_1$, $w \in W$, $w_1 \in W_1$; then $f(v) = h(w)$. It is easy to see that f is in $\hat{V}$ and that $\tilde{f} = h$. Thus $h = fT$ and so T maps $\hat{V}$ onto $\hat{W}$. Since the kernel of T is $A(W)$ by Theorem 4.1.1, $\hat{W}$ is isomorphic to $\hat{V}/A(W)$. In particular they have the same dimension. Let $m = \dim W$, $n = \dim V$, and $r = \dim A(W)$. By Corollary 2 to Theorem 4.3.1, $m = \dim \hat{W}$ and $n = \dim \hat{V}$. However, by Lemma 4.2.6 $\dim \hat{V}/A(W) = \dim \hat{V} - \dim A(W) = n - r$, and so $m = n - r$. Transposing, $r = n - m$. We have proved

THEOREM 4.3.2 *If V is finite-dimensional and W is a subspace of V, then $\hat{W}$ is isomorphic to $\hat{V}/A(W)$ and* $\dim A(W) = \dim V - \dim W$.

COROLLARY $A(A(W)) = W$.

Proof. Remember that in order for the corollary even to make sense, since $W \subset V$ and $A(A(W)) \subset \hat{\hat{V}}$, we have identified V with $\hat{\hat{V}}$. Now $W \subset A(A(W))$, for if $w \in W$ then $w\psi = T_w$ acts on $\hat{V}$ by $T_w(f) = f(w)$ and so is 0 for all $f \in A(W)$. However, $\dim A(A(W)) = \dim \hat{\hat{V}} - \dim A(W)$ (applying the theorem to the vector space $\hat{V}$ and its subspace $A(W)$) so that $\dim A(A(W)) = \dim \hat{V} - \dim A(W) = \dim V - (\dim V - \dim W) = \dim W$. Since $W \subset A(A(W))$ and they are of the same dimension, it follows that $W = A(A(W))$.

Theorem 4.3.2 has application to the study of systems of *linear homogeneous equations*. Consider the system of m equations in n unknowns

$$
\begin{aligned}
a_{11}x_1 + a_{12}x_2 + \cdots + a_{1n}x_n &= 0, \\
a_{21}x_1 + a_{22}x_2 + \cdots + a_{2n}x_n &= 0, \\
&\vdots \\
a_{m1}x_1 + a_{m2}x_2 + \cdots + a_{mn}x_n &= 0,
\end{aligned}
$$

where the a_{ij} are in F. We ask for the number of linearly independent solutions $(x_1, \ldots, x_n)$ there are in $F^{(n)}$ to this system.

In $F^{(n)}$ let U be the subspace generated by the m vectors $(a_{11}, a_{12}, \ldots, a_{1n})$, $(a_{21}, a_{22}, \ldots, a_{2n}), \ldots, (a_{m1}, a_{m2}, \ldots, a_{mn})$ and suppose that U is of dimension r. In that case we say the system of equations is of *rank* r.

Let $v_1 = (1, 0, \ldots, 0)$, $v_2 = (0, 1, 0, \ldots, 0), \ldots, v_n = (0, 0, \ldots, 0, 1)$ be used as a basis of $F^{(n)}$ and let $\hat{v}_1, \hat{v}_2, \ldots, \hat{v}_n$ be its dual basis in $\hat{F}^{(n)}$. Any $f \in \hat{F}^{(n)}$ is of the form $f = x_1\hat{v}_1 + x_2\hat{v}_2 + \cdots + x_n\hat{v}_n$, where the $x_i \in F$. When is $f \in A(U)$? In that case, since $(a_{11}, \ldots, a_{1n}) \in U$,

$$\begin{aligned} 0 &= f(a_{11}, a_{12}, \ldots, a_{1n}) \\ &= f(a_{11}v_1 + \cdots + a_{1n}v_n) \\ &= (x_1\hat{v}_1 + x_2\hat{v}_2 + \cdots + x_n\hat{v}_n)(a_{11}v_1 + \cdots + a_{1n}v_n) \\ &= x_1a_{11} + x_2a_{12} + \cdots + x_na_{1n} \end{aligned}$$

since $\hat{v}_i(v_j) = 0$ for $i \neq j$ and $\hat{v}_i(v_i) = 1$. Similarly the other equations of the system are satisfied. Conversely, every solution $(x_1, \ldots, x_n)$ of the system of homogeneous equations yields an element, $x_1\hat{v}_1 + \cdots + x_n\hat{v}_n$, in $A(U)$. Thereby we see that the number of linearly independent solutions of the system of equations is the dimension of $A(U)$, which, by Theorem 4.3.2 is $n - r$. We have proved the following:

THEOREM 4.3.3 *If the system of homogeneous linear equations:*

$$\begin{aligned} a_{11}x_1 + \cdots + a_{1n}x_n &= 0, \\ a_{21}x_1 + \cdots + a_{2n}x_n &= 0, \\ &\vdots \\ a_{m1}x_1 + \cdots + a_{mn}x_n &= 0, \end{aligned}$$

where $a_{ij} \in F$ is of rank r, then there are $n - r$ linearly independent solutions in $F^{(n)}$.

COROLLARY *If $n > m$, that is, if the number of unknowns exceeds the number of equations, then there is a solution $(x_1, \ldots, x_n)$ where not all of $x_1, \ldots, x_n$ are 0.*

Proof. Since U is generated by m vectors, and $m < n$, $r = \dim U \leq m < n$; applying Theorem 4.3.3 yields the corollary.

Problems

1. Prove that $A(W)$ is a subspace of $\hat{V}$.
2. If S is a subset of V let $A(S) = \{f \in \hat{V} \mid f(s) = 0 \text{ all } s \in S\}$. Prove that $A(S) = A(L(S))$, where $L(S)$ is the linear span of S.

3. If $S, T \in \text{Hom}(V, W)$ and $v_i S = v_i T$ for all elements v_i of a basis of V, prove that $S = T$.
4. Complete the proof, with all details, that $\text{Hom}(V, W)$ is a vector space over F.
5. If ψ denotes the mapping used in the text of V into $\hat{\hat{V}}$, give a complete proof that ψ is a vector space homomorphism of V into $\hat{\hat{V}}$.
6. If V is finite-dimensional and $v_1 \neq v_2$ are in V, prove that there is an $f \in \hat{V}$ such that $f(v_1) \neq f(v_2)$.
7. If W_1 and W_2 are subspaces of V, which is finite-dimensional, describe $A(W_1 + W_2)$ in terms of $A(W_1)$ and $A(W_2)$.
8. If V is a finite-dimensional and W_1 and W_2 are subspaces of V, describe $A(W_1 \cap W_2)$ in terms of $A(W_1)$ and $A(W_2)$.
9. If F is the field of real numbers, find $A(W)$ where
   (a) W is spanned by $(1, 2, 3)$ and $(0, 4, -1)$.
   (b) W is spanned by $(0, 0, 1, -1)$, $(2, 1, 1, 0)$, and $(2, 1, 1, -1)$.
10. Find the ranks of the following systems of homogeneous linear equations over F, the field of real numbers, and find all the solutions.
    (a) $x_1 + 2x_2 - 3x_3 + 4x_4 = 0,$
    $x_1 + 3x_2 - x_3 = 0,$
    $6x_1 + x_3 + 2x_4 = 0.$
    (b) $x_1 + 3x_2 + x_3 = 0,$
    $x_1 + 4x_2 + x_3 = 0.$
    (c) $x_1 + x_2 + x_3 + x_4 + x_5 = 0,$
    $x_1 + 2x_2 = 0,$
    $4x_1 + 7x_2 + x_3 + x_4 + x_5 = 0,$
    $x_2 - x_3 - x_4 - x_5 = 0.$
11. If f and g are in $\hat{V}$ such that $f(v) = 0$ implies $g(v) = 0$, prove that $g = \lambda f$ for some $\lambda \in F$.

4.4 Inner Product Spaces

In our discussion of vector spaces the specific nature of F as a field, other than the fact that it is a field, has played virtually no role. In this section we no longer consider vector spaces V over arbitrary fields F; rather, we restrict F to be the field of real or complex numbers. In the first case V is called a *real vector space*, in the second, a *complex vector space*.

We all have had some experience with real vector spaces—in fact both analytic geometry and the subject matter of vector analysis deal with these. What concepts used there can we carry over to a more abstract setting? To begin with, we had in these concrete examples the idea of length; secondly we had the idea of perpendicularity, or, more generally, that of angle. These became special cases of the notion of a dot product (often called a scalar or inner product.)

Let us recall some properties of dot product as it pertained to the special case of the three-dimensional real vectors. Given the vectors $v = (x_1, x_2, x_3)$ and $w = (y_1, y_2, y_3)$, where the x's and y's are real numbers, the dot product of v and w, denoted by $v \cdot w$, was defined as $v \cdot w = x_1 y_1 + x_2 y_2 + x_3 y_3$. Note that the length of v is given by $\sqrt{v \cdot v}$ and the angle θ between v and w is determined by

$$\cos \theta = \frac{v \cdot w}{\sqrt{v \cdot v}\,\sqrt{w \cdot w}}.$$

What formal properties does this dot product enjoy? We list a few:

1. $v \cdot v \geq 0$ and $v \cdot v = 0$ if and only if $v = 0$;
2. $v \cdot w = w \cdot v$;
3. $u \cdot (\alpha v + \beta w) = \alpha(u \cdot v) + \beta(u \cdot w)$;

for any vectors u, v, w and real numbers α, β.

Everything that has been said can be carried over to complex vector spaces. However, to get geometrically reasonable definitions we must make some modifications. If we simply define $v \cdot w = x_1 y_1 + x_2 y_2 + x_3 y_3$ for $v = (x_1, x_2, x_3)$ and $w = (y_1, y_2, y_3)$, where the x's and y's are complex numbers, then it is quite possible that $v \cdot v = 0$ with $v \neq 0$; this is illustrated by the vector $v = (1, i, 0)$. In fact, $v \cdot v$ need not even be real. If, as in the real case, we should want $v \cdot v$ to represent somehow the length of v, we should like that this length be real and that a nonzero vector should not have zero length.

We can achieve this much by altering the definition of dot product slightly. If $\bar{\alpha}$ denotes the complex conjugate of the complex number α, returning to the v and w of the paragraph above let us define $v \cdot w = x_1 \bar{y}_1 + x_2 \bar{y}_2 + x_3 \bar{y}_3$. For real vectors this new definition coincides with the old one; on the other hand, for arbitrary complex vectors $v \neq 0$, not only is $v \cdot v$ real, it is in fact positive. Thus we have the possibility of introducing, in a natural way, a nonnegative length. However, we do lose something; for instance it is no longer true that $v \cdot w = w \cdot v$. In fact the exact relationship between these is $v \cdot w = \overline{w \cdot v}$. Let us list a few properties of this dot product:

1. $v \cdot w = \overline{w \cdot v}$;
2. $v \cdot v \geq 0$, and $v \cdot v = 0$ if and only if $v = 0$;
3. $(\alpha u + \beta v) \cdot w = \alpha(u \cdot w) + \beta(v \cdot w)$;
4. $u \cdot (\alpha v + \beta w) = \bar{\alpha}(u \cdot v) + \bar{\beta}(u \cdot w)$;

for all complex numbers α, β and all complex vectors u, v, w.

We reiterate that in what follows F is either the field of real or complex numbers.

DEFINITION The vector space V over F is said to be an *inner product space* if there is defined for any two vectors $u, v \in V$ an element (u, v) in F such that

1. $(u, v) = \overline{(v, u)}$;
2. $(u, u) \geq 0$ and $(u, u) = 0$ if and only if $u = 0$;
3. $(\alpha u + \beta v, w) = \alpha(u, w) + \beta(v, w)$;

for any $u, v, w \in V$ and $\alpha, \beta \in F$.

A few observations about properties 1, 2, and 3 are in order. A function satisfying them is called an *inner product*. If F is the field of complex numbers, property 1 implies that (u, u) is real, and so property 2 makes sense. Using 1 and 3, we see that $(u, \alpha v + \beta w) = \overline{(\alpha v + \beta w, u)} = \overline{\alpha(v, u) + \beta(w, u)} = \bar{\alpha}\overline{(v, u)} + \bar{\beta}\overline{(w, u)} = \bar{\alpha}(u, v) + \bar{\beta}(u, w)$.

We pause to look at some examples of inner product spaces.

Example 4.4.1 In $F^{(n)}$ define, for $u = (\alpha_1, \ldots, \alpha_n)$ and $v = (\beta_1, \ldots, \beta_n)$, $(u, v) = \alpha_1\bar{\beta}_1 + \alpha_2\bar{\beta}_2 + \cdots + \alpha_n\bar{\beta}_n$. This defines an inner product on $F^{(n)}$.

Example 4.4.2 In $F^{(2)}$ define for $u = (\alpha_1, \alpha_2)$ and $v = (\beta_1, \beta_2)$, $(u, v) = 2\alpha_1\bar{\beta}_1 + \alpha_1\bar{\beta}_2 + \alpha_2\bar{\beta}_1 + \alpha_2\bar{\beta}_2$. It is easy to verify that this defines an inner product on $F^{(2)}$.

Example 4.4.3 Let V be the set of all continuous complex-valued functions on the closed unit interval $[0, 1]$. If $f(t), g(t) \in V$, define

$$(f(t), g(t)) = \int_0^1 f(t)\,\overline{g(t)}\,dt.$$

We leave it to the reader to verify that this defines an inner product on V.

For the remainder of this section V will denote an inner product space.

DEFINITION If $v \in V$ then the *length* of v (or *norm* of v), written $\|v\|$, is defined by $\|v\| = \sqrt{(v, v)}$.

LEMMA 4.4.1 *If $u, v \in V$ and $\alpha, \beta \in F$ then $(\alpha u + \beta v, \alpha u + \beta v) = \alpha\bar{\alpha}(u, u) + \alpha\bar{\beta}(u, v) + \bar{\alpha}\beta(v, u) + \beta\bar{\beta}(v, v)$.*

Proof. By property 3 defining an inner product space, $(\alpha u + \beta v, \alpha u + \beta v) = \alpha(u, \alpha u + \beta v) + \beta(v, \alpha u + \beta v)$; but $(u, \alpha u + \beta v) = \bar{\alpha}(u, u) + \bar{\beta}(u, v)$ and $(v, \alpha u + \beta v) = \bar{\alpha}(v, u) + \bar{\beta}(v, v)$. Substituting these in the expression for $(\alpha u + \beta v, \alpha u + \beta v)$ we get the desired result.

COROLLARY $\|\alpha u\| = |\alpha|\,\|u\|$.

Proof. $\|\alpha u\|^2 = (\alpha u, \alpha u) = \alpha\bar{\alpha}(u, u)$ by Lemma 4.4.1 (with $v = 0$). Since $\alpha\bar{\alpha} = |\alpha|^2$ and $(u, u) = \|u\|^2$, taking square roots yields $\|\alpha u\| = |\alpha|\,\|u\|$.

We digress for a moment, and prove a very elementary and familiar result about real quadratic equations.

LEMMA 4.4.2 *If a, b, c are real numbers such that $a > 0$ and $a\lambda^2 + 2b\lambda + c \geq 0$ for all real numbers λ, then $b^2 \leq ac$.*

Proof. Completing the squares,

$$a\lambda^2 + 2b\lambda + c = \frac{1}{a}(a\lambda + b)^2 + \left(c - \frac{b^2}{a}\right).$$

Since it is greater than or equal to 0 for all λ, in particular this must be true for $\lambda = -b/a$. Thus $c - (b^2/a) \geq 0$, and since $a > 0$ we get $b^2 \leq ac$.

We now proceed to an extremely important inequality, usually known as the *Schwarz inequality:*

THEOREM 4.4.1 *If $u, v \in V$ then $|(u, v)| \leq \|u\|\,\|v\|$.*

Proof. If $u = 0$ then both $(u, v) = 0$ and $\|u\|\,\|v\| = 0$, so that the result is true there.

Suppose, for the moment, that (u, v) is real and $u \neq 0$. By Lemma 4.4.1, for any real number λ, $0 \leq (\lambda u + v, \lambda u + v) = \lambda^2(u, u) + 2(u, v)\lambda + (v, v)$ Let $a = (u, u)$, $b = (u, v)$, and $c = (v, v)$; for these the hypothesis of Lemma 4.4.2 is satisfied, so that $b^2 \leq ac$. That is, $(u, v)^2 \leq (u, u)(v, v)$; from this it is immediate that $|(u,v)| \leq \|u\|\,\|v\|$.

If $\alpha = (u, v)$ is not real, then it certainly is not 0, so that u/α is meaningful. Now,

$$\left(\frac{u}{\alpha}, v\right) = \frac{1}{\alpha}(u, v) = \frac{1}{(u, v)}(u, v) = 1,$$

and so it is certainly real. By the case of the Schwarz inequality discussed in the paragraph above,

$$1 = \left|\left(\frac{u}{\alpha}, v\right)\right| \leq \left\|\frac{u}{\alpha}\right\|\,\|v\|;$$

since

$$\left\|\frac{u}{\alpha}\right\| = \frac{1}{|\alpha|}\|u\|,$$

we get

$$1 \leq \frac{\|u\| \, \|v\|}{|\alpha|},$$

whence $|\alpha| \leq \|u\| \, \|v\|$. Putting in that $\alpha = (u, v)$ we obtain $|(u, v)| \leq \|u\| \, \|v\|$, the desired result.

Specific cases of the Schwarz inequality are themselves of great interest. We point out two of them.

1. If $V = F^{(n)}$ with $(u, v) = \alpha_1\bar{\beta}_1 + \cdots + \alpha_n\bar{\beta}_n$, where $u = (\alpha_1, \ldots, \alpha_n)$ and $v = (\beta_1, \ldots, \beta_n)$, then Theorem 4.4.1 implies that

$$|\alpha_1\bar{\beta}_1 + \cdots + \alpha_n\bar{\beta}_n|^2 \leq (|\alpha_1|^2 + \cdots + |\alpha_n|^2)(|\beta_1|^2 + \cdots + |\beta_n|^2).$$

2. If V is the set of all continuous, complex-valued functions on $[0,1]$ with inner product defined by

$$(f(t), g(t)) = \int_0^1 f(t)\,\overline{g(t)}\,dt,$$

then Theorem 4.4.1 implies that

$$\left|\int_0^1 f(t)\,\overline{g(t)}\,dt\right|^2 \leq \int_0^1 |f(t)|^2\,dt \int_0^1 |g(t)|^2\,dt.$$

The concept of perpendicularity is an extremely useful and important one in geometry. We introduce its analog in general inner product spaces.

DEFINITION If $u, v \in V$ then u is said to be *orthogonal* to v if $(u, v) = 0$.

Note that if u is orthogonal to v then v is orthogonal to u, for $(v, u) = \overline{(u, v)} = \bar{0} = 0$.

DEFINITION If W is a subspace of V, the *orthogonal complement* of W, $W^\perp$, is defined by $W^\perp = \{x \in V \mid (x, w) = 0 \text{ for all } w \in W\}$.

LEMMA 4.4.3 *$W^\perp$ is a subspace of V.*

Proof. If $a, b \in W^\perp$ then for all $\alpha, \beta \in F$ and all $w \in W$, $(\alpha a + \beta b, w) = \alpha(a, w) + \beta(b, w) = 0$ since $a, b \in W^\perp$.

Note that $W \cap W^\perp = (0)$, for if $w \in W \cap W^\perp$ it must be self-orthogonal, that is $(w, w) = 0$. The defining properties of an inner product space rule out this possibility unless $w = 0$.

One of our goals is to show that $V = W + W^\perp$. Once this is done, the remark made above will become of some interest, for it will imply that V is the direct sum of W and $W^\perp$.

DEFINITION The set of vectors $\{v_i\}$ in V is an *orthonormal set* if

1. Each v_i is of length 1 (i.e., $(v_i, v_i) = 1$).
2. For $i \neq j$, $(v_i, v_j) = 0$.

LEMMA 4.4.4 *If $\{v_i\}$ is an orthonormal set, then the vectors in $\{v_i\}$ are linearly independent. If $w = \alpha_1 v_1 + \cdots + \alpha_n v_n$, then $\alpha_i = (w, v_i)$ for $i = 1, 2, \ldots, n$.*

Proof. Suppose that $\alpha_1 v_1 + \alpha_2 v_2 + \cdots + \alpha_n v_n = 0$. Therefore $0 = (\alpha_1 v_1 + \cdots + \alpha_n v_n, v_i) = \alpha_1 (v_1, v_i) + \cdots + \alpha_n (v_n, v_i)$. Since $(v_j, v_i) = 0$ for $j \neq i$ while $(v_i, v_i) = 1$, this equation reduces to $\alpha_i = 0$. Thus the v_j's are linearly independent.

If $w = \alpha_1 v_1 + \cdots + \alpha_n v_n$ then computing as above yields $(w, v_i) = \alpha_i$.

Similar in spirit and in proof to Lemma 4.4.4 is

LEMMA 4.4.5 *If $\{v_1, \ldots, v_n\}$ is an orthonormal set in V and if $w \in V$, then $u = w - (w, v_1)v_1 - (w, v_2)v_2 - \cdots - (w, v_i)v_i - \cdots - (w, v_n)v_n$ is orthogonal to each of $v_1, v_2, \ldots, v_n$.*

Proof. Computing (u, v_i) for any $i \leq n$, using the orthonormality of $v_1, \ldots, v_n$ yields the result.

The construction carried out in the proof of the next theorem is one which appears and reappears in many parts of mathematics. It is a basic procedure and is known as the *Gram-Schmidt orthogonalization process*. Although we shall be working in a finite-dimensional inner product space, the Gram-Schmidt process works equally well in infinite-dimensional situations.

THEOREM 4.4.2 *Let V be a finite-dimensional inner product space; then V has an orthonormal set as a basis.*

Proof. Let V be of dimension n over F and let $v_1, \ldots, v_n$ be a basis of V. From this basis we shall construct an orthonormal set of n vectors; by Lemma 4.4.4 this set is linearly independent so must form a basis of V.

We proceed with the construction. We seek n vectors $w_1, \ldots, w_n$ each of length 1 such that for $i \neq j$, $(w_i, w_j) = 0$. In fact we shall finally produce them in the following form: w_1 will be a multiple of v_1, w_2 will be in the linear span of w_1 and v_2, w_3 in the linear span of w_1, w_2, and v_3, and more generally, w_i in the linear span of $w_1, w_2, \ldots, w_{i-1}, v_i$.

Let

$$w_1 = \frac{v_1}{\|v_1\|};$$

then

$$(w_1, w_1) = \left(\frac{v_1}{\|v_1\|}, \frac{v_1}{\|v_1\|}\right) = \frac{1}{\|v_1\|^2}(v_1, v_1) = 1,$$

whence $\|w_1\| = 1$. We now ask: for what value of α is $\alpha w_1 + v_2$ orthogonal to w_1? All we need is that $(\alpha w_1 + v_2, w_1) = 0$, that is $\alpha(w_1, w_1) + (v_2, w_1) = 0$. Since $(w_1, w_1) = 1$, $\alpha = -(v_2, w_1)$ will do the trick. Let $u_2 = -(v_2, w_1)w_1 + v_2$; u_2 is orthogonal to w_1; since v_1 and v_2 are linearly independent, w_1 and v_2 must be linearly independent, and so $u_2 \neq 0$. Let $w_2 = (u_2/\|u_2\|)$; then $\{w_1, w_2\}$ is an orthonormal set. We continue. Let $u_3 = -(v_3, w_1)w_1 - (v_3, w_2)w_2 + v_3$; a simple check verifies that $(u_3, w_1) = (u_3, w_2) = 0$. Since w_1, w_2, and v_3 are linearly independent (for w_1, w_2 are in the linear span of v_1 and v_2), $u_3 \neq 0$. Let $w_3 = (u_3/\|u_3\|)$; then $\{w_1, w_2, w_3\}$ is an orthonormal set. The road ahead is now clear. Suppose that we have constructed $w_1, w_2, \ldots, w_i$, in the linear span of $v_1, \ldots, v_i$, which form an orthonormal set. How do we construct the next one, w_{i+1}? Merely put $u_{i+1} = -(v_{i+1}, w_1)w_1 - (v_{i+1}, w_2)w_2 - \cdots - (v_{i+1}, w_i)w_i + v_{i+1}$. That $u_{i+1} \neq 0$ and that it is orthogonal to each of $w_1, \ldots, w_i$ we leave to the reader. Put $w_{i+1} = (u_{i+1}/\|u_{i+1}\|)$!

In this way, given r linearly independent elements in V, we can construct an orthonormal set having r elements. If particular, when $\dim V = n$, from any basis of V we can construct an orthonormal set having n elements. This provides us with the required basis for V.

We illustrate the construction used in the last proof in a concrete case. Let F be the real field and let V be the set of polynomials, in a variable x, over F of degree 2 or less. In V we define an inner product by: if $p(x)$, $q(x) \in V$, then

$$(p(x), q(x)) = \int_{-1}^{1} p(x)q(x)\,dx.$$

Let us start with the basis $v_1 = 1$, $v_2 = x$, $v_3 = x^2$ of V. Following the construction used,

$$w_1 = \frac{v_1}{\|v_1\|} = \frac{1}{\sqrt{\int_{-1}^{1} 1\,dx}} = \frac{1}{\sqrt{2}};$$

$$u_2 = -(v_2, w_1)w_1 + v_2,$$

which after the computations reduces to $u_2 = x$, and so

$$w_2 = \frac{u_2}{\|u_2\|} = \frac{x}{\sqrt{\int_{-1}^{1} x^2\,dx}} = \frac{\sqrt{3}}{\sqrt{2}}x;$$

finally,

$$u_3 = -(v_3, w_1)w_1 - (v_3, w_2)w_2 + v_3 = \frac{-1}{3} + x^2,$$

and so

$$w_3 = \frac{u_3}{\|u_3\|} = \frac{\frac{-1}{3} + x^2}{\sqrt{\int_{-1}^{1}\left(\frac{-1}{3} + x^2\right)^2 dx}} = \frac{\sqrt{10}}{4}(-1 + 3x^2).$$

We mentioned the next theorem earlier as one of our goals. We are now able to prove it.

THEOREM 4.4.3 *If V is a finite-dimensional inner product space and if W is a subspace of V, then $V = W + W^{\perp}$. More particularly, V is the direct sum of W and $W^{\perp}$.*

Proof. Because of the highly geometric nature of the result, and because it is so basic, we give several proofs. The first will make use of Theorem 4.4.2 and some of the earlier lemmas. The second will be motivated geometrically.

First Proof. As a subspace of the inner product space V, W is itself an inner product space (its inner product being that of V restricted to W). Thus we can find an orthonormal set $w_1, \ldots, w_r$ in W which is a basis of W. If $v \in V$, by Lemma 4.4.5, $v_0 = v - (v, w_1)w_1 - (v, w_2)w_2 - \cdots - (v, w_r)w_r$ is orthogonal to each of $w_1, \ldots, w_r$ and so is orthogonal to W. Thus $v_0 \in W^{\perp}$, and since $v = v_0 + ((v, w_1)w_1 + \cdots + (v, w_r)w_r)$, $v \in W + W^{\perp}$. Therefore $V = W + W^{\perp}$. Since $W \cap W^{\perp} = (0)$, this sum is direct.

Second Proof. In this proof we shall assume that F is the field of real numbers. The proof works, in almost the same way, for the complex numbers; however, it entails a few extra details which might tend to obscure the essential ideas used.

Let $v \in V$; suppose that we could find a vector $w_0 \in W$ such that $\|v - w_0\| \leq \|v - w\|$ for *all* $w \in W$. We claim that then $(v - w_0, w) = 0$ for all $w \in W$, that is, $v - w_0 \in W^{\perp}$.

If $w \in W$, then $w_0 + w \in W$, in consequence of which

$$(v - w_0, v - w_0) \leq (v - (w_0 + w), v - (w_0 + w)).$$

However, the right-hand side is $(w, w) + (v - w_0, v - w_0) - 2(v - w_0, w)$, leading to $2(v - w_0, w) \leq (w, w)$ for all $w \in W$. If m is any positive integer, since $w/m \in W$ we have that

$$\frac{2}{m}(v - w_0, w) = 2\left(v - w_0, \frac{w}{m}\right) \leq \left(\frac{w}{m}, \frac{w}{m}\right) = \frac{1}{m^2}(w, w),$$

and so $2(v - w_0, w) \leq (1/m)(w, w)$ for any positive integer m. However,

$(1/m)(w, w) \to 0$ as $m \to \infty$, whence $2(v - w_0, w) \leq 0$. Similarly, $-w \in W$, and so $0 \leq -2(v - w_0, w) = 2(v - w_0, -w) \leq 0$, yielding $(v - w_0, w) = 0$ for all $w \in W$. Thus $v - w_0 \in W^\perp$; hence $v \in w_0 + W^\perp \subset W + W^\perp$.

To finish the second proof we must prove the existence of a $w_0 \in W$ such that $\|v - w_0\| \leq \|v - w\|$ for all $w \in W$. We indicate sketchily two ways of proving the existence of such a w_0.

Let $u_1, \ldots, u_k$ be a basis of W; thus any $w \in W$ is of the form $w = \lambda_1 u_1 + \cdots + \lambda_k u_k$. Let $\beta_{ij} = (u_i, u_j)$ and let $\gamma_i = (v, u_i)$ for $v \in V$. Thus $(v - w, v - w) = (v - \lambda_1 u_1 - \cdots - \lambda_k u_k, v - \lambda_1 w_1 - \cdots - \lambda_k w_k) = (v, v) - \sum \lambda_i \lambda_j \beta_{ij} - 2\sum \lambda_i \gamma_i$. This *quadratic* function in the λ's is *nonnegative* and so, by results from the calculus, has a minimum. The λ's for this minimum, $\lambda_1^{(0)}, \lambda_2^{(0)}, \ldots, \lambda_k^{(0)}$ give us the desired vector $w_0 = \lambda_1^{(0)} u_1 + \cdots + \lambda_k^{(0)} u_k$ in W.

A second way of exhibiting such a minimizing w is as follows. In V define a metric ζ by $\zeta(x, y) = \|x - y\|$; one shows that ζ is a proper metric on V, and V is now a metric space. Let $S = \{w \in W \mid \|v - w\| \leq \|v\|\}$; in this metric S is a compact set (prove!) and so the continuous function $f(w) = \|v - w\|$ defined for $w \in S$ takes on a minimum at some point $w_0 \in S$. We leave it to the reader to verify that w_0 is the desired vector satisfying $\|v - w_0\| \leq \|v - w\|$ for all $w \in W$.

COROLLARY *If V is a finite-dimensional inner product space and W is a subspace of V then $(W^\perp)^\perp = W$.*

Proof. If $w \in W$ then for any $u \in W^\perp$, $(w, u) = 0$, whence $W \subset (W^\perp)^\perp$. Now $V = W + W^\perp$ and $V = W^\perp + (W^\perp)^\perp$; from these we get, since the sums are direct, $\dim(W) = \dim((W^\perp)^\perp)$. Since $W \subset (W^\perp)^\perp$ and is of the same dimension as $(W^\perp)^\perp$, it follows that $W = (W^\perp)^\perp$.

Problems

In all the problems V is an inner product space over F.

1. If F is the real field and V is $F^{(3)}$, show that the Schwarz inequality implies that the cosine of an angle is of absolute value at most 1.
2. If F is the real field, find all 4-tuples of real numbers (a, b, c, d) such that for $u = (\alpha_1, \alpha_2)$, $v = (\beta_1, \beta_2) \in F^{(2)}$, $(u, v) = a\alpha_1\beta_1 + b\alpha_2\beta_2 + c\alpha_1\beta_2 + d\alpha_2\beta_1$ defines an inner product on $F^{(2)}$.
3. In V define the *distance* $\zeta(u, v)$ from u to v by $\zeta(u, v) = \|u - v\|$. Prove that
   (a) $\zeta(u, v) \geq 0$ and $\zeta(u, v) = 0$ if and only if $u = v$.
   (b) $\zeta(u, v) = \zeta(v, u)$.
   (c) $\zeta(u, v) \leq \zeta(u, w) + \zeta(w, v)$ (triangle inequality).

4. If $\{w_1, \ldots, w_m\}$ is an orthonormal set in V, prove that

$$\sum_{i=1}^{m} |(w_i, v)|^2 \leq \|v\|^2 \text{ for any } v \in V.$$

(Bessel inequality)

5. If V is finite-dimensional and if $\{w_1, \ldots, w_m\}$ is an orthonormal set in V such that

$$\sum_{i=1}^{m} |(w_i, v)|^2 = \|v\|^2$$

for every $v \in V$, prove that $\{w_1, \ldots, w_m\}$ must be a basis of V.

6. If $\dim V = n$ and if $\{w_1, \ldots, w_m\}$ is an orthonormal set in V, prove that there exist vectors $w_{m+1}, \ldots, w_n$ such that $\{w_1, \ldots, w_m, w_{m+1}, \ldots, w_n\}$ is an orthonormal set (and basis of V).
7. Use the result of Problem 6 to give another proof of Theorem 4.4.3.
8. In V prove the parallelogram law:

$$\|u + v\|^2 + \|u - v\|^2 = 2(\|u\|^2 + \|v\|^2).$$

Explain what this means geometrically in the special case $V = F^{(3)}$, where F is the real field, and where the inner product is the usual dot product.

9. Let V be the real functions $y = f(x)$ satisfying $d^2y/dx^2 + 9y = 0$.
   (a) Prove that V is a two-dimensional real vector space.
   (b) In V define $(y, z) = \int_0^\pi yz\, dx$. Find an orthonormal basis in V.
10. Let V be the set of real functions $y = f(x)$ satisfying

$$\frac{d^3y}{dx^3} - 6\frac{d^2y}{dx^2} + 11\frac{dy}{dx} - 6y = 0.$$

   (a) Prove that V is a three-dimensional real vector space.
   (b) In V define

$$(u, v) = \int_{-\infty}^{0} uv\, dx.$$

   Show that this defines an inner product on V and find an orthonormal basis for V.

11. If W is a subspace of V and if $v \in V$ satisfies $(v, w) + (w, v) \leq (w, w)$ for every $w \in W$, prove that $(v, w) = 0$ for every $w \in W$.
12. If V is a finite-dimensional inner product space and if f is a linear functional on V (i.e., $f \in \hat{V}$), prove that there is a $u_0 \in V$ such that $f(v) = (v, u_0)$ for all $v \in V$.

4.5 Modules

The notion of a module will be a generalization of that of a vector space; instead of restricting the scalars to lie in a field we shall allow them to be elements of an arbitrary ring.

This section has many definitions but only one main theorem. However the definitions are so close in spirit to ones already made for vector spaces that the main ideas to be developed here should not be buried in a sea of definitions.

DEFINITION Let R be any ring; a nonempty set M is said to be an *R-module* (or, a module over R) if M is an abelian group under an operation $+$ such that for every $r \in R$ and $m \in M$ there exists an element rm in M subject to

1. $r(a + b) = ra + rb$;
2. $r(sa) = (rs)a$;
3. $(r + s)a = ra + sa$

for all $a, b \in M$ and $r, s \in R$.

If R has a unit element, 1, and if $1m = m$ for every element m in M, then M is called a *unital* R-module. Note that if R is a field, a unital R-module is nothing more than a vector space over R. *All our modules shall be unital ones.*

Properly speaking, we should call the object we have defined a *left R-module* for we allow multiplication by the elements of R from the left. Similarly we could define a *right R-module.* We shall make no such left-right distinction, it being understood that by the term R-module we mean a left R-module.

Example 4.5.1 Every abelian group G is a module over the ring of integers!

For, write the operation of G as $+$ and let na, for $a \in G$ and n an integer, have the meaning it had in Chapter 2. The usual rules of exponents in abelian groups translate into the requisite properties needed to make of G a module over the integers. Note that it is a unital module.

Example 4.5.2 Let R be any ring and let M be a left-ideal of R. For $r \in R$, $m \in M$, let rm be the product of these elements as elements in R. The definition of left-ideal implies that $rm \in M$, while the axioms defining a ring insure us that M is an R-module. (In this example, by a ring we mean an associative ring, in order to make sure that $r(sm) = (rs)m$.)

Example 4.5.3 The special case in which $M = R$; any ring R is an R-module over itself.

Example 4.5.4 Let R be any ring and let λ be a left-ideal of R. Let M consist of all the cosets, $a + \lambda$, where $a \in R$, of λ in R.

In M define $(a + \lambda) + (b + \lambda) = (a + b) + \lambda$ and $r(a + \lambda) = ra + \lambda$. M can be shown to be an R-module. (See Problem 2, end of this section.) M is usually written as $R - \lambda$ (or, sometimes, as R/λ) and is called the *difference* (or *quotient*) *module of R by λ*.

An additive subgroup A of the R-module M is called a *submodule* of M if whenever $r \in R$ and $a \in A$, then $ra \in A$.

Given an R-module M and a submodule A we could construct the quotient module M/A in a manner similar to the way we constructed quotient groups, quotient rings, and quotient spaces. One could also talk about homomorphisms of one R-module into another one, and prove the appropriate homomorphism theorems. These occur in the problems at the end of this section.

Our interest in modules is in a somewhat different direction; we shall attempt to find a nice decomposition for modules over certain rings.

DEFINITION If M is an R-module and if $M_1, \ldots, M_s$ are submodules of M, then M is said to be the *direct sum* of $M_1, \ldots, M_s$ if every element $m \in M$ can be written in a *unique* manner as $m = m_1 + m_2 + \cdots + m_s$ where $m_1 \in M_1$, $m_2 \in M_2, \ldots, m_s \in M_s$.

As in the case of vector spaces, if M is the direct sum of $M_1, \ldots, M_s$ then M will be isomorphic, as a module, to the set of all s-tuples, $(m_1, \ldots, m_s)$ where the ith component m_i is any element of M_i, where addition is componentwise, and where $r(m_1, \ldots, m_s) = (rm_1, rm_2, \ldots, rm_s)$ for $r \in R$. Thus, knowing the structure of each M_i would enable us to know the structure of M.

Of particular interest and simplicity are modules generated by one element; such modules are called *cyclic*. To be precise:

DEFINITION An R-module M is said to be *cyclic* if there is an element $m_0 \in M$ such that every $m \in M$ is of the form $m = rm_0$ where $r \in R$.

For R, the ring of integers, a cyclic R-module is nothing more than a cyclic group.

We still need one more definition, namely,

DEFINITION An R-module M is said to be *finitely generated* if there exist elements $a_1, \cdots, a_n \in M$ such that every m in M is of the form $m = r_1a_1 + r_2a_2 + \cdots + r_na_n$.

With all the needed definitions finally made, we now come to the theorem which is the primary reason for which this section exists. It is often called the *fundamental theorem on finitely generated modules* over Euclidean rings. In it we shall restrict R to be a Euclidean ring (see Chapter 3, Section 3.7); however the theorem holds in the more general context in which R is any principal ideal domain.

THEOREM 4.5.1 *Let R be a Euclidean ring; then any finitely generated R-module, M, is the direct sum of a finite number of cyclic submodules.*

Proof. Before becoming involved with the machinery of the proof, let us see what the theorem states. The assumption that M is finitely generated tells us that there is a set of elements $a_1, \ldots, a_n \in M$ such that every element in M can be expressed in the form $r_1a_1 + r_2a_2 + \cdots + r_na_n$, where the $r_i \in R$. The conclusion of the theorem states that when R is properly conditioned we can, in fact, find some other set of elements $b_1, \ldots, b_q$ in M such that every element $m \in M$ can be expressed in a unique fashion as $m = s_1b_1 + \cdots + s_qb_q$ with $s_i \in R$. A remark about this uniqueness; it does not mean that the s_i are unique, in fact this may be false; it merely states that the elements s_ib_i are. That is, if $m = s_1b_1 + \cdots + s_qb_q$ and $m = s'_1b_1 + \cdots + s'_qb_q$ we cannot draw the conclusion that $s_1 = s'_1$, $s_2 = s'_2, \ldots, s_q = s'_q$, but rather, we can infer from this that $s_1b_1 = s'_1b_1, \ldots, s_qb_q = s'_qb_q$.

Another remark before we start with the technical argument. Although the theorem is stated for a general Euclidean ring, we shall give the proof in all its detail only for the special case of the ring of integers. At the end we shall indicate the slight modifications needed to make the proof go through for the more general setting. We have chosen this path to avoid cluttering up the essential ideas, which are the same in the general case, with some technical niceties which are of no importance.

Thus we are simply assuming that M is an abelian group which has a finite-generating set. Let us call those generating sets having as few elements as possible *minimal generating* sets and the number of elements in such a minimal generating set the *rank* of M.

Our proof now proceeds by induction on the rank of M.

If the rank of M is 1 then M is generated by a single element, hence it is cyclic; in this case the theorem is true. Suppose that the result is true for all abelian groups of rank $q - 1$, and that M is of rank q.

Given any minimal generating set $a_1, \ldots, a_q$ of M, if any relation of the form $n_1a_1 + n_2a_2 + \cdots + n_qa_q = 0$ ($n_1, \ldots, n_q$ integers) implies that $n_1a_1 = n_2a_2 = \cdots = n_qa_q = 0$, then M is the direct sum of $M_1, M_2, \ldots, M_q$ where each M_i is the cyclic module (i.e., subgroup) generated by a_i, and so we would be done. Consequently, given any minimal generating set

$b_1, \ldots, b_q$ of M, there must be integers $r_1, \ldots, r_q$ such that $r_1 b_1 + \cdots + r_q b_q = 0$ and in which not all of $r_1 b_1, r_2 b_2, \ldots, r_q b_q$ are 0. Among all possible such relations for all minimal generating sets there is a smallest possible positive integer occurring as a coefficient. Let this integer be s_1 and let the generating set for which it occurs be $a_1, \ldots, a_q$. Thus

$$s_1 a_1 + s_2 a_2 + \cdots + s_q a_q = 0. \tag{1}$$

We claim that if $r_1 a_1 + \cdots + r_q a_q = 0$, then $s_1 \mid r_1$; for $r_1 = m s_1 + t$, $0 \leq t < s_1$, and so multiplying Equation (1) by m and subtracting from $r_1 a_1 + \cdots + r_q a_q = 0$ leads to $t a_1 + (r_2 - m s_2) a_2 + \cdots + (r_q - m s_q) a_q = 0$; since $t < s_1$ and s_1 is the minimal possible positive integer in such a relation, we must have that $t = 0$.

We now further claim that $s_1 \mid s_i$ for $i = 2, \ldots, q$. Suppose not; then $s_1 \nmid s_2$, say, so $s_2 = m_2 s_1 + t$, $0 < t < s_1$. Now $a_1' = a_1 + m_2 a_2, a_2, \ldots, a_q$ also generate M, yet $s_1 a_1' + t a_2 + s_3 q_3 + \cdots + s_q a_q = 0$; thus t occurs as a coefficient in some relation among elements of a minimal generating set. But this forces, by the very choice of s_1, that either $t = 0$ or $t \geq s_1$. We are left with $t = 0$ and so $s_1 \mid s_2$. Similarly for the other s_i. Let us write $s_i = m_i s_1$.

Consider the elements $a_1^* = a_1 + m_2 a_2 + m_3 a_3 + \cdots + m_q a_q, a_2, \ldots, a_q$. They generate M; moreover, $s_1 a_1^* = s_1 a_1 + m_2 s_1 a_2 + \cdots + m_q s_1 a_q = s_1 a_1 + s_2 a_2 + \cdots + s_q a_q = 0$. If $r_1 a_1^* + r_2 a_2 + \cdots + r_q a_q = 0$, substituting for a_1^*, we get a relation between $a_1, \ldots, a_q$ in which the coefficient of a_1 is r_1; thus $s_1 \mid r_1$ and so $r_1 a_1^* = 0$. If M_1 is the cyclic module generated by a_1^* and if M_2 is the submodule of M generated by $a_2, \ldots, a_q$, we have just shown that $M_1 \cap M_2 = (0)$. But $M_1 + M_2 = M$ since $a_1^*, a_2, \ldots, a_q$ generate M. *Thus M is the direct sum of M_1 and M_2.* Since M_2 is generated by $a_2, \ldots, a_q$, its rank is at most $q - 1$ (in fact, it is $q - 1$), so by the induction M_2 is the direct sum of cyclic modules. Putting the pieces together we have decomposed M into a direct sum of cyclic modules.

COROLLARY *Any finite abelian group is the direct product (sum) of cyclic groups.*

Proof. The finite abelian group G is certainly finitely generated; in fact it is generated by the finite set consisting of all its elements. Therefore applying Theorem 4.5.1 yields the corollary. This is, of course, the result proved in Theorem 2.14.1.

Suppose that R is a Euclidean ring with Euclidean function d. We modify the proof given for the integers to one for R as follows:

1. Instead of choosing s_1 as the smallest possible positive integer occurring in any relation among elements of a generating set, pick it as that element of R occurring in any relation whose d-value is minimal.

2. In the proof that $s_1 \mid r_1$ for any relation $r_1 a_1 + \cdots + r_q a_q = 0$, the only change needed is that $r_1 = ms_1 + t$ where either

$$t = 0 \quad \text{or} \quad d(t) < d(s_1);$$

the rest goes through. Similarly for the proof that $s_1 \mid s_i$.

Thus with these minor changes the proof holds for general Euclidean rings, whereby Theorem 4.5.1 is completely proved.

Problems

1. Verify that the statement made in Example 4.5.1 that every abelian group is a module over the ring of integers is true.
2. Verify that the set in Example 4.5.4 is an R-module.
3. Suppose that R is a ring with a unit element and that M is a module over R but is not unital. Prove that there exists an $m \neq 0$ in M such that $rm = 0$ for all $r \in R$.

Given two R-modules M and N then the mapping T from M into N is called a *homomorphism* (or *R-homomorphism* or *module homomorphism*) if

1. $(m_1 + m_2)T = m_1 T + m_2 T$;
2. $(rm_1)T = r(m_1 T)$;

for all $m_1, m_2 \in M$ and all $r \in R$.

4. If T is a homomorphism of M into N let $K(T) = \{x \in M \mid xT = 0\}$. Prove that $K(T)$ is a submodule of M and that $I(T) = \{xT \mid x \in M\}$ is a submodule of N.
5. The homomorphism T is said to be an *isomorphism* if it is one-to-one. Prove that T is an isomorphism if and only if $K(T) = (0)$.
6. Let M, N, Q be three R-modules, and let T be a homomorphism of M into N and S a homomorphism of N into Q. Define $TS: M \to Q$ by $m(TS) = (mT)S$ for any $m \in M$. Prove that TS is an R-homomorphism of M into Q and determine its kernel, $K(TS)$.
7. If M is an R-module and A is a submodule of M, define the quotient module M/A (use the analogs in group, rings, and vector spaces as a guide) so that it is an R-module and prove that there is an R-homomorphism of M onto M/A.
8. If T is a homomorphism of M onto N with $K(T) = A$, prove that N is isomorphic (as a module) to M/A.
9. If A and B are submodules of M prove
   (a) $A \cap B$ is a submodule of M.
   (b) $A + B = \{a + b \mid a \in A, b \in B\}$ is a submodule of M.
   (c) $(A + B)/B$ is isomorphic to $A/(A \cap B)$.

10. An R-module M is said to be *irreducible* if its only submodules are (0) and M. Prove that any unital, irreducible R-module is cyclic.
11. If M is an irreducible R-module, prove that either M is cyclic or that for every $m \in M$ and $r \in R$, $rm = 0$.

*12. If M is an irreducible R-module such that $rm \neq 0$ for some $r \in R$ and $m \in M$, prove that any R-homomorphism T of M into M is either an *isomorphism* of M *onto* M or that $mT = 0$ for every $m \in M$.

13. Let M be an R-module and let $E(M)$ be the set of all R-homomorphisms of M into M. Make appropriate definitions of addition and multiplication of elements of $E(M)$ so that $E(M)$ becomes a ring. (*Hint:* imitate what has been done for Hom (V, V), V a vector space.)

*14. If M is an irreducible R-module such that $rm \neq 0$ for some $r \in R$ and $m \in M$, prove that $E(M)$ is a division ring. (This result is known as *Schur's lemma.*)

15. Give a complete proof of Theorem 4.5.1 for finitely generated modules over Euclidean rings.
16. Let M be an R-module; if $m \in M$ let $\lambda(m) = \{x \in R \mid xm = 0\}$. Show that $\lambda(m)$ is a left-ideal of R. It is called the *order* of m.
17. If λ is a left-ideal of R and if M is an R-module, show that for $m \in M$, $\lambda m = \{xm \mid x \in \lambda\}$ is a submodule of M.

*18. Let M be an irreducible R-module in which $rm \neq 0$ for some $r \in R$ and $m \in M$. Let $m_0 \neq 0 \in M$ and let $\lambda(m_0) = \{x \in R \mid xm_0 = 0\}$.
   (a) Prove that $\lambda(m_0)$ is a maximal left-ideal of R (that is, if λ is a left-ideal of R such that $R \supset \lambda \supset \lambda(m_0)$, then $\lambda = R$ or $\lambda = \lambda(m_0)$).
   (b) As R-modules, prove that M is isomorphic to $R - \lambda(m_0)$ (see Example 4.5.4).

Supplementary Reading

HALMOS, PAUL R., *Finite-Dimensional Vector Spaces*, 2nd ed. Princeton, N.J.: D. Van Nostrand Company, Inc., 1958.

5

Fields

In our discussion of rings we have already singled out a special class which we called fields. A field, let us recall, is a commutative ring with unit element in which every nonzero element has a multiplicative inverse. Put another way, a field is a commutative ring in which we can divide by any nonzero element.

Fields play a central role in algebra. For one thing, results about them find important applications in the theory of numbers. For another, their theory encompasses the subject matter of the theory of equations which treats questions about the roots of polynomials.

In our development we shall touch only lightly on the field of algebraic numbers. Instead, our greatest emphasis will be on aspects of field theory which impinge on the theory of equations. Although we shall not treat the material in its fullest or most general form, we shall go far enough to introduce some of the beautiful ideas, due to the brilliant French mathematician Evariste Galois, which have served as a guiding inspiration for algebra as it is today.

5.1 Extension Fields

In this section we shall be concerned with the relation of one field to another. Let F be a field; a field K is said to be an *extension* of F if K contains F. Equivalently, K is an extension of F if F is a subfield of K. *Throughout this chapter F will denote a given field and K an extension of F.*

As was pointed out earlier, in the chapter on vector spaces, if K is an extension of F, then, under the ordinary field operations in K, K is a vector space over F. As a vector space we may talk about linear dependence, dimension, bases, etc., in K relative to F.

DEFINITION The *degree* of K over F is the dimension of K as a vector space over F.

We shall always denote the degree of K over F by $[K:F]$. Of particular interest to us is the case in which $[K:F]$ is finite, that is, when K is finite-dimensional as a vector space over F. This situation is described by saying that K is a *finite extension* of F.

We start off with a relatively simple but, at the same time, highly effective result about finite extensions, namely,

THEOREM 5.1.1 *If L is a finite extension of K and if K is a finite extension of F, then L is a finite extension of F. Moreover, $[L:F] = [L:K][K:F]$.*

Proof. The strategy we employ in the proof is to write down explicitly a basis of L over F. In this way not only do we show that L is a finite extension of F, but we actually prove the sharper result and the one which is really the heart of the theorem, namely that $[L:F] = [L:K][K:F]$.

Suppose, then, that $[L:K] = m$ and that $[K:F] = n$. Let $v_1, \ldots, v_m$ be a basis of L over K and let $w_1, \ldots, w_n$ be a basis of K over F. What could possibly be nicer or more natural than to have the elements v_iw_j, where $i = 1, 2, \ldots, m$, $j = 1, 2, \ldots, n$, serve as a basis of L over F? Whatever else, they do at least provide us with the right number of elements. We now proceed to show that they do in fact form a basis of L over F. What do we need to establish this? First we must show that every element in L is a linear combination of them with coefficients in F, and then we must demonstrate that these mn elements are linearly independent over F.

Let t be any element in L. Since every element in L is a linear combination of $v_1, \ldots, v_m$ with coefficients in K, in particular, t must be of this form. Thus $t = k_1v_1 + \cdots + k_mv_m$, where the elements $k_1, \ldots, k_m$ are all in K. However, every element in K is a linear combination of $w_1, \ldots, w_n$ with coefficients in F. Thus $k_1 = f_{11}w_1 + \cdots + f_{1n}w_n, \ldots, k_i = f_{i1}w_1 + \cdots + f_{in}w_n, \ldots, k_m = f_{m1}w_1 + \cdots + f_{mn}w_n$, where every f_{ij} is in F.

Substituting these expressions for $k_1, \ldots, k_m$ into $t = k_1v_1 + \cdots + k_mv_m$, we obtain $t = (f_{11}w_1 + \cdots + f_{1n}w_n)v_1 + \cdots + (f_{m1}w_1 + \cdots + f_{mn}w_n)v_m$ Multiplying this out, using the distributive and associative laws, we finally arrive at $t = f_{11}v_1w_1 + \cdots + f_{1n}v_1w_n + \cdots + f_{ij}v_iw_j + \cdots + f_{mn}v_mw_n$. Since the f_{ij} are in F, we have realized t as a linear combination over F of the elements v_iw_j. Therefore, the elements v_iw_j do indeed span all of L over F, and so they fulfill the first requisite property of a basis.

We still must show that the elements v_iw_j are linearly independent over F. Suppose that $f_{11}v_1w_1 + \cdots + f_{1n}v_1w_n + \cdots + f_{ij}v_iw_j + \cdots + f_{mn}v_mw_n = 0$, where the f_{ij} are in F. Our objective is to prove that each $f_{ij} = 0$. Regrouping the above expression yields $(f_{11}w_1 + \cdots + f_{1n}w_n)v_1 + \cdots + (f_{i1}w_1 + \cdots + f_{in}w_n)v_i + \cdots + (f_{m1}w_1 + \cdots + f_{mn}w_n)v_m = 0$.

Since the w_i are in K, and since $K \supset F$, all the elements $k_i = f_{i1}w_1 + \cdots + f_{in}w_n$ are in K. Now $k_1v_1 + \cdots + k_mv_m = 0$ with $k_1, \ldots, k_m \in K$. But, by assumption, $v_1, \ldots, v_m$ form a basis of L over K, so, in particular they must be linearly independent over K. The net result of this is that $k_1 = k_2 = \cdots = k_m = 0$. Using the explicit values of the k_i, we get

$$f_{i1}w_1 + \cdots + f_{in}w_n = 0 \qquad \text{for } i = 1, 2, \ldots, m.$$

But now we invoke the fact that the w_i are linearly independent over F; this yields that each $f_{ij} = 0$. In other words, we have proved that the v_iw_j are linearly independent over F. In this way they satisfy the other requisite property for a basis.

We have now succeeded in proving that the mn elements v_iw_j form a basis of L over F. Thus $[L:F] = mn$; since $m = [L:K]$ and $n = [K:F]$ we have obtained the desired result $[L:F] = [L:K][K:F]$.

Suppose that L, K, F are three fields in the relation $L \supset K \supset F$ and, suppose further that $[L:F]$ is finite. Clearly, any elements in L linearly independent over K are, all the more so, linearly independent over F. Thus the assumption that $[L:F]$ is finite forces the conclusion that $[L:K]$ is finite. Also, since K is a subspace of L, $[K:F]$ is finite. By the theorem, $[L:F] = [L:K][K:F]$, whence $[K:F] \mid [L:F]$. We have proved the

COROLLARY *If L is a finite extension of F and K is a subfield of L which contains F, then* $[K:F] \mid [L:F]$.

Thus, for instance, if $[L:F]$ is a prime number, then there can be no fields properly between F and L. A little later, in Section 5.4, when we discuss the construction of certain geometric figures by straightedge and compass, this corollary will be of great significance.

DEFINITION An element $a \in K$ is said to be *algebraic over* F if there exist elements $\alpha_0, \alpha_1, \ldots, \alpha_n$ in F, not all 0, such that $\alpha_0a^n + \alpha_1a^{n-1} + \cdots + \alpha_n = 0$.

If the polynomial $q(x) \in F[x]$, the ring of polynomials in x over F, and if $q(x) = \beta_0x^m + \beta_1x^{m-1} + \cdots + \beta_m$, then for any element $b \in K$, by $q(b)$ we shall mean the element $\beta_0b^m + \beta_1b^{m-1} + \cdots + \beta_m$ in K. In the expression commonly used, $q(b)$ is the *value* of the polynomial $q(x)$ obtained by substituting b for x. The element b is said to *satisfy* $q(x)$ if $q(b) = 0$.

In these terms, $a \in K$ is algebraic over F if there is a nonzero polynomial $p(x) \in F[x]$ which a satisfies, that is, for which $p(a) = 0$.

Let K be an extension of F and let a be in K. Let $\mathcal{M}$ be the collection of all subfields of K which contain both F and a. $\mathcal{M}$ is not empty, for K itself is an element of $\mathcal{M}$. Now, as is easily proved, the intersection of any number of subfields of K is again a subfield of K. Thus the intersection of all those subfields of K which are members of $\mathcal{M}$ is a subfield of K. We denote this subfield by $F(a)$. What are its properties? Certainly it contains both F and a, since this is true for every subfield of K which is a member of $\mathcal{M}$. Moreover, by the very definition of intersection, every subfield of K in $\mathcal{M}$ contains $F(a)$, yet $F(a)$ itself is in $\mathcal{M}$. *Thus $F(a)$ is the smallest subfield of K containing both F and a.* We call $F(a)$ the subfield obtained by *adjoining a* to F.

Our description of $F(a)$, so far, has been purely an external one. We now give an alternative and more constructive description of $F(a)$. Consider all these elements in K which can be expressed in the form $\beta_0 + \beta_1 a + \cdots + \beta_s a^s$; here the β's can range freely over F and s can be any nonnegative integer. As elements in K, one such element can be divided by another, provided the latter is not 0. Let U be the set of all such quotients. We leave it as an exercise to prove that U is a subfield of K.

On one hand, U certainly contains F and a, whence $U \supset F(a)$. On the other hand, any subfield of K which contains both F and a, by virtue of closure under addition and multiplication, must contain all the elements $\beta_0 + \beta_1 a + \cdots + \beta_s a^s$ where each $\beta_i \in F$. Thus $F(a)$ must contain all these elements; being a subfield of K, $F(a)$ must also contain all quotients of such elements. Therefore, $F(a) \supset U$. The two relations $U \subset F(a)$, $U \supset F(a)$ of course imply that $U = F(a)$. In this way we have obtained an internal construction of $F(a)$, namely as U.

We now intertwine the property that $a \in K$ is algebraic over F with macroscopic properties of the field $F(a)$ itself. This is

THEOREM 5.1.2 *The element $a \in K$ is algebraic over F if and only if $F(a)$ is a finite extension of F.*

Proof. As is so very common with so many such "if and only if" propositions, one-half of the proof will be quite straightforward and easy, whereas the other half will be deeper and more complicated.

Suppose that $F(a)$ is a finite extension of F and that $[F(a):F] = m$. Consider the elements $1, a, a^2, \ldots, a^m$; they are all in $F(a)$ and are $m + 1$ in number. By Lemma 4.2.4, these elements are linearly dependent over F. Therefore, there are elements $\alpha_0, \alpha_1, \ldots, \alpha_m$ in F, not all 0, such that $\alpha_0 1 + \alpha_1 a + \alpha_2 a^2 + \cdots + \alpha_m a^m = 0$. Hence a is algebraic over F and satisfies the nonzero polynomial $p(x) = \alpha_0 + \alpha_1 x + \cdots + \alpha_m x^m$ in $F[x]$ of degree at most $m = [F(a):F]$. This proves the "if" part of the theorem.

Now to the "only if" part. Suppose that a in K is algebraic over F. By assumption, a satisfies some nonzero polynomial in $F[x]$; let $p(x)$ be a polynomial in $F[x]$ of smallest positive degree such that $p(a) = 0$. We claim that $p(x)$ is irreducible over F. For, suppose that $p(x) = f(x)g(x)$, where $f(x), g(x) \in F[x]$; then $0 = p(a) = f(a)g(a)$ (see Problem 1) and, since $f(a)$ and $g(a)$ are elements of the field K, the fact that their product is 0 forces $f(a) = 0$ or $g(a) = 0$. Since $p(x)$ is of lowest positive degree with $p(a) = 0$, we must conclude that one of $\deg f(x) \geq \deg p(x)$ or $\deg g(x) \geq \deg p(x)$ must hold. But this proves the irreducibility of $p(x)$.

We define the mapping ψ from $F[x]$ into $F(a)$ as follows. For any $h(x) \in F[x]$, $h(x)\psi = h(a)$. We leave it to the reader to verify that ψ is a ring homomorphism of the ring $F[x]$ into the field $F(a)$ (see Problem 1). What is V, the kernel of ψ? By the very definition of ψ, $V = \{h(x) \in F[x] \mid h(a) = 0\}$. Also, $p(x)$ is an element of lowest degree in the ideal V of $F[x]$. By the results of Section 3.9, every element in V is a multiple of $p(x)$, and since $p(x)$ is irreducible, by Lemma 3.9.6, V is a maximal ideal of $F[x]$. By Theorem 3.5.1, $F[x]/V$ is a field. Now by the general homomorphism theorem for rings (Theorem 3.4.1), $F[x]/V$ is isomorphic to the image of $F[x]$ under ψ. Summarizing, we have shown that the image of $F[x]$ under ψ is a subfield of $F(a)$. This image contains $x\psi = a$ and, for every $\alpha \in F$, $\alpha\psi = \alpha$. Thus the image of $F[x]$ under ψ is a subfield of $F[a]$ which contains both F and a; by the very definition of $F(a)$ we are forced to conclude that the image of $F[x]$ under ψ is *all* of $F(a)$. Put more succinctly, $F[x]/V$ is isomorphic to $F(a)$.

Now, $V = (p(x))$, the ideal generated by $p(x)$; from this we claim that the dimension of $F[x]/V$, as a vector space over F, is precisely equal to $\deg p(x)$ (see Problem 2). In view of the isomorphism between $F[x]/V$ and $F(a)$ we obtain the fact that $[F(a):F] = \deg p(x)$. Therefore, $[F(a):F]$ is certainly finite; this is the contention of the "only if" part of the theorem. Note that we have actually proved more, namely that $[F(a):F]$ is equal to the degree of the polynomial of least degree satisfied by a over F.

The proof we have just given has been somewhat long-winded, but deliberately so. The route followed contains important ideas and ties in results and concepts developed earlier with the current exposition. No part of mathematics is an island unto itself.

We now redo the "only if" part, working more on the inside of $F(a)$. This reworking is, in fact, really identical with the proof already given; the constituent pieces are merely somewhat differently garbed.

Again let $p(x)$ be a polynomial over F of lowest positive degree satisfied by a. Such a polynomial is called a *minimal polynomial* for a over F. We may assume that its coefficient of the highest power of x is 1, that is, it is monic; in that case we can speak of *the* minimal polynomial for a over F for any two minimal, monic polynomials for a over F are equal. (Prove!)

Suppose that $p(x)$ is of degree n; thus $p(x) = x^n + \alpha_1 x^{n-1} + \cdots + \alpha_n$ where the α_i are in F. By assumption, $a^n + \alpha_1 a^{n-1} + \cdots + \alpha_n = 0$, whence $a^n = -\alpha_1 a^{n-1} - \alpha_2 a^{n-2} - \cdots - \alpha_n$. What about a^{n+1}? From the above, $a^{n+1} = -\alpha_1 a^n - \alpha_2 a^{n-1} - \cdots - \alpha_n a$; if we substitute the expression for a^n into the right-hand side of this relation, we realize a^{n+1} as a linear combination of the elements $1, a, \ldots, a^{n-1}$ over F. Continuing this way, we get that a^{n+k}, for $k \geq 0$, is a linear combination over F of $1, a, a^2, \ldots, a^{n-1}$.

Now consider $T = \{\beta_0 + \beta_1 a + \cdots + \beta_{n-1}a^{n-1} \mid \beta_0, \beta_1, \ldots, \beta_{n-1} \in F\}$. Clearly, T is closed under addition; in view of the remarks made in the paragraph above, it is also closed under multiplication. Whatever further it may be, T has at least been shown to be a ring. Moreover, T contains both F and a. We now wish to show that T is more than just a ring, that it is, in fact, a field.

Let $0 \neq u = \beta_0 + \beta_1 a + \cdots + \beta_{n-1}a^{n-1}$ be in T and let $h(x) = \beta_0 + \beta_1 x + \cdots + \beta_{n-1}x^{n-1} \in F[x]$. Since $u \neq 0$, and $u = h(a)$, we have that $h(a) \neq 0$, whence $p(x) \nmid h(x)$. By the irreducibility of $p(x)$, $p(x)$ and $h(x)$ must therefore be relatively prime. Hence we can find polynomials $s(x)$ and $t(x)$ in $F[x]$ such that $p(x)s(x) + h(x)t(x) = 1$. But then $1 = p(a)s(a) + h(a)t(a) = h(a)t(a)$, since $p(a) = 0$; putting into this that $u = h(a)$, we obtain $ut(a) = 1$. The inverse of u is thus $t(a)$; in $t(a)$ all powers of a higher than $n - 1$ can be replaced by linear combinations of $1, a, \ldots, a^{n-1}$ over F, whence $t(a) \in T$. We have shown that every nonzero element of T has its inverse in T; consequently, T is a field. However, $T \subset F(a)$, yet F and a are both contained in T, which results in $T = F(a)$. We have identified $F(a)$ as the set of all expressions $\beta_0 + \beta_1 a + \cdots + \beta_{n-1}a^{n-1}$.

Now T is spanned over F by the elements $1, a, \ldots, a^{n-1}$ in consequence of which $[T:F] \leq n$. However, the elements $1, a, a^2, \ldots, a^{n-1}$ are linearly independent over F, for any relation of the form $\gamma_0 + \gamma_1 a + \cdots + \gamma_{n-1}a^{n-1}$, with the elements $\gamma_i \in F$, leads to the conclusion that a satisfies the polynomial $\gamma_0 + \gamma_1 x + \cdots + \gamma_{n-1}x^{n-1}$ over F of degree less than n. This contradiction proves the linear independence of $1, a, \ldots, a^{n-1}$, and so these elements actually form a basis of T over F, whence, in fact, we now know that $[T:F] = n$. Since $T = F(a)$, the result $[F(a):F] = n$ follows.

DEFINITION The element $a \in K$ is said to be *algebraic of degree n* over F if it satisfies a nonzero polynomial over F of degree n but no nonzero polynomial of lower degree.

In the course of proving Theorem 5.1.2 (in each proof we gave), we proved a somewhat sharper result than that stated in that theorem, namely,

THEOREM 5.1.3 *If $a \in K$ is algebraic of degree n over F, then $[F(a):F] = n$.*

This result adapts itself to many uses. We give now, as an immediate consequence thereof, the very interesting

THEOREM 5.1.4 *If a, b in K are algebraic over F then $a \pm b$, ab, and a/b (if $b \neq 0$) are all algebraic over F. In other words, the elements in K which are algebraic over F form a subfield of K.*

Proof. Suppose that a is algebraic of degree m over F while b is algebraic of degree n over F. By Theorem 5.1.3 the subfield $T = F(a)$ of K is of degree m over F. Now b is algebraic of degree n over F, *a fortiori* it is algebraic of degree at most n over T which contains F. Thus the subfield $W = T(b)$ of K, again by Theorem 5.1.3, is of degree at most n over T. But $[W:F] = [W:T][T:F]$ by Theorem 5.1.1; therefore, $[W:F] \leq mn$ and so W is a finite extension of F. However, a and b are both in W, whence all of $a \pm b$, ab, and a/b are in W. By Theorem 5.1.2, since $[W:F]$ is finite, these elements must be algebraic over F, thereby proving the theorem.

Here, too, we have proved somewhat more. Since $[W:F] \leq mn$, every element in W satisfies a polynomial of degree at most mn over F, whence the

COROLLARY *If a and b in K are algebraic over F of degrees m and n, respectively, then $a \pm b$, ab, and a/b (if $b \neq 0$) are algebraic over F of degree at most mn.*

In the proof of the last theorem we made two extensions of the field F. The first we called T; it was merely the field $F(a)$. The second we called W and it was $T(b)$. Thus $W = (F(a))(b)$; it is customary to write it as $F(a, b)$. Similarly, we could speak about $F(b, a)$; it is not too difficult to prove that $F(a, b) = F(b, a)$. Continuing this pattern, we can define $F(a_1, a_2, \ldots, a_n)$ for elements $a_1, \ldots, a_n$ in K.

DEFINITION The extension K of F is called an *algebraic extension* of F if every element in K is algebraic over F.

We prove one more result along the lines of the theorems we have proved so far.

THEOREM 5.1.5 *If L is an algebraic extension of K and if K is an algebraic extension of F, then L is an algebraic extension of F.*

Proof. Let u be any arbitrary element of L; our objective is to show that u satisfies some nontrivial polynomial with coefficients in F. What information do we have at present? We certainly do know that u satisfies some polynomial $x^n + \sigma_1 x^{n-1} + \cdots + \sigma_n$, where $\sigma_1, \ldots, \sigma_n$ are in K. But K is algebraic over F; therefore, by several uses of Theorem 5.1.3, $M = F(\sigma_1, \ldots, \sigma_n)$ is a finite extension of F. Since u satisfies the polynomial $x^n + \sigma_1 x^{n-1} + \cdots + \sigma_n$ whose coefficients are in M, u is algebraic over M. Invoking Theorem 5.1.2 yields that $M(u)$ is a finite extension of M. However, by Theorem 5.1.1, $[M(u):F] = [M(u):M][M:F]$, whence $M(u)$ is a finite extension of F. But this implies that u is algebraic over F, completing proof of the theorem.

A quick description of Theorem 5.1.5: algebraic over algebraic is algebraic.

The preceding results are of special interest in the particular case in which F is the field of rational numbers and K the field of complex numbers.

DEFINITION A complex number is said to be an *algebraic number* if it is algebraic over the field of rational numbers.

A complex number which is not algebraic is called *transcendental*. At the present stage we have no reason to suppose that there are any transcendental numbers. In the next section we shall prove that the familiar real number e is transcendental. This will, of course, establish the existence of transcendental numbers. In actual fact, they exist in great abundance; in a very well-defined way there are more of them than there are algebraic numbers.

Theorem 5.1.4 applied to algebraic numbers proves the interesting fact that *the algebraic numbers form a field*; that is, the sum, products, and quotients of algebraic numbers are again algebraic numbers.

Theorem 5.1.5 when used in conjunction with the so-called "fundamental theorem of algebra," has the implication that the roots of a polynomial whose coefficients are algebraic numbers are themselves algebraic numbers.

Problems

1. Prove that the mapping $\psi : F[x] \to F(a)$ defined by $h(x)\psi = h(a)$ is a homomorphism.
2. Let F be a field and let $F[x]$ be the ring of polynomials in x over F. Let $g(x)$, of degree n, be in $F[x]$ and let $V = (g(x))$ be the ideal generated by $g(x)$ in $F[x]$. Prove that $F[x]/V$ is an n-dimensional vector space over F.
3. (a) If V is a finite-dimensional vector space over the field K, and if F is a subfield of K such that $[K:F]$ is finite, show that V is a finite-dimensional vector space over F and that moreover $\dim_F (V) = (\dim_K (V))([K:F])$.
   (b) Show that Theorem 5.1.1 is a special case of the result of part (a).

4. (a) Let R be the field of real numbers and Q the field of rational numbers. In R, $\sqrt{2}$ and $\sqrt{3}$ are both algebraic over Q. Exhibit a polynomial of degree 4 over Q satisfied by $\sqrt{2} + \sqrt{3}$.
   (b) What is the degree of $\sqrt{2} + \sqrt{3}$ over Q? Prove your answer.
   (c) What is the degree of $\sqrt{2}\sqrt{3}$ over Q?
5. With the same notation as in Problem 4, show that $\sqrt{2} + \sqrt[3]{5}$ is algebraic over Q of degree 6.

*6. (a) Find an element $u \in R$ such that $Q(\sqrt{2}, \sqrt[3]{5}) = Q(u)$.
   (b) In $Q(\sqrt{2}, \sqrt[3]{5})$ characterize all the elements w such that $Q(w) \neq Q(\sqrt{2}, \sqrt[3]{5})$.

7. (a) Prove that $F(a, b) = F(b, a)$.
   (b) If $(i_1, i_2, \ldots, i_n)$ is any permutation of $(1, 2, \ldots, n)$, prove that

$$F(a_1, \ldots, a_n) = F(a_{i_1}, a_{i_2}, \ldots, a_{i_n}).$$

8. If $a, b \in K$ are algebraic over F of degrees m and n, respectively, and if m and n are relatively prime, prove that $F(a, b)$ is of degree mn over F.
9. Suppose that F is a field having a finite number of elements, q.
   (a) Prove that there is a prime number p such that $\underbrace{a + a + \cdots + a}_{p\text{-times}} = 0$ for all $a \in F$.
   (b) Prove that $q = p^n$ for some integer n.
   (c) If $a \in F$, prove that $a^q = a$.
   (d) If $b \in K$ is algebraic over F, prove $b^{q^m} = b$ for some $m > 0$.

An algebraic number a is said to be an *algebraic integer* if it satisfies an equation of the form $a^m + \alpha_1 a^{m-1} + \cdots + \alpha_m = 0$, where $\alpha_1, \ldots, \alpha_m$ are integers.

10. If a is any algebraic number, prove that there is a positive integer n such that na is an algebraic integer.
11. If the rational number r is also an algebraic integer, prove that r must be an ordinary integer.
12. If a is an algebraic integer and m is an ordinary integer, prove
    (a) $a + m$ is an algebraic integer.
    (b) ma is an algebraic integer.
13. If α is an algebraic integer satisfying $\alpha^3 + \alpha + 1 = 0$ and β is an algebraic integer satisfying $\beta^2 + \beta - 3 = 0$, prove that both $\alpha + \beta$ and $\alpha\beta$ are algebraic integers.

**14. (a) Prove that the sum of two algebraic integers is an algebraic integer.

(b) Prove that the product of two algebraic integers is an algebraic integer.

15. (a) Prove that sin 1° is an algebraic number.
(b) From part (a) prove that sin $m°$ is an algebraic number for any integer m.

5.2 The Transcendence of e

In defining algebraic and transcendental numbers we pointed out that it could be shown that transcendental numbers exist. One way of achieving this would be the demonstration that some specific number is transcendental.

In 1851 Liouville gave a criterion that a complex number be algebraic; using this, he was able to write down a large collection of transcendental numbers. For instance, it follows from his work that the number .101001000000100 ... 10 ... is transcendental; here the number of zeros between successive ones goes as $1!, 2!, \ldots \ldots, n!, \ldots$.

This certainly settled the question of existence. However, the question whether some given, familiar numbers were transcendental still persisted. The first success in this direction was by Hermite, who in 1873 gave a proof that e is transcendental. His proof was greatly simplified by Hilbert. The proof that we shall give here is a variation, due to Hurwitz, of Hilbert's proof.

The number π offered greater difficulties. These were finally overcome by Lindemann, who in 1882 produced a proof that π is transcendental. One immediate consequence of this is the fact that it is impossible, by straightedge and compass, to square the circle, for such a construction would lead to an algebraic number θ such that $\theta^2 = \pi$. But if θ is algebraic then so is θ^2, in virtue of which π would be algebraic, in contradiction to Lindemann's result.

In 1934, working independently, Gelfond and Schneider proved that if a and b are algebraic numbers and if b is irrational, then a^b is transcendental. This answered in the affirmative the question raised by Hilbert whether $2^{\sqrt{2}}$ was transcendental.

For those interested in pursuing the subject of transcendental numbers further, we would strongly recommend the charming books by C. L. Siegel, entitled *Transcendental Numbers*, and by I. Niven, *Irrational Numbers.*

To prove that e is irrational is easy; to prove that π is irrational is much more difficult. For a very clever and neat proof of the latter, see the paper by Niven entitled "A simple proof that π is irrational," *Bulletin of the American Mathematical Society*, Vol. 53 (1947), page 509.

Now to the transcendence of e. Aside from its intrinsic interest, its proof offers us a change of pace. Up to this point all our arguments have been of an algebraic nature; now, for a short while, we return to the more familiar grounds of the calculus. The proof itself will use only elementary calculus; the deepest result needed, therefrom, will be the mean value theorem.

THEOREM 5.2.1 *The number e is transcendental.*

Proof. In the proof we shall use the standard notation $f^{(i)}(x)$ to denote the ith derivative of $f(x)$ with respect to x.

Suppose that $f(x)$ is a polynomial of degree r with real coefficients. Let $F(x) = f(x) + f^{(1)}(x) + f^{(2)}(x) + \cdots + f^{(r)}(x)$. We compute $(d/dx)(e^{-x}F(x))$; using the fact that $f^{(r+1)}(x) = 0$ (since $f(x)$ is of degree r) and the basic property of e, namely that $(d/dx)e^x = e^x$, we obtain $(d/dx)(e^{-x}F(x)) = -e^{-x}f(x)$.

The mean value theorem asserts that if $g(x)$ is a continuously differentiable, single-valued function on the closed interval $[x_1, x_2]$ then

$$\frac{g(x_1) - g(x_2)}{x_1 - x_2} = g^{(1)}(x_1 + \theta(x_2 - x_1)), \qquad \text{where} \quad 0 < \theta < 1.$$

We apply this to our function $e^{-x}F(x)$, which certainly satisfies all the required conditions for the mean value theorem on the closed interval $[x_1, x_2]$ where $x_1 = 0$ and $x_2 = k$, where k is any positive integer. We then obtain that $e^{-k}F(k) - F(0) = -e^{-\theta_k k}f(\theta_k k)k$, where θ_k depends on k and is some real number between 0 and 1. Multiplying this relation through by e^k yields $F(k) - F(0)e^k = -e^{(1-\theta_k)k}f(\theta_k k)k$. We write this out explicitly:

$$\begin{aligned} F(1) - eF(0) &= -e^{(1-\theta_1)}f(\theta_1) = \varepsilon_1, \\ F(2) - e^2F(0) &= -2e^{2(1-\theta_2)}f(2\theta_2) = \varepsilon_2, \\ &\vdots \\ F(n) - e^nF(0) &= -ne^{n(1-\theta_n)}f(n\theta_n) = \varepsilon_n. \end{aligned} \tag{1}$$

Suppose now that e is an algebraic number; then it satisfies some relation of the form

$$c_ne^n + c_{n-1}e^{n-1} + \cdots + c_1e + c_0 = 0, \tag{2}$$

where $c_0, c_1, \ldots, c_n$ are integers and where $c_0 > 0$.

In the relations (1) let us multiply the first equation by c_1, the second by c_2, and so on; adding these up we get $c_1F(1) + c_2F(2) + \cdots + c_nF(n) - F(0)(c_1e + c_2e^2 + \cdots + c_ne^n) = c_1\varepsilon_1 + c_2\varepsilon_2 + \cdots + c_n\varepsilon_n$.

In view of relation (2), $c_1e + c_2e^2 + \cdots + c_ne^n = -c_0$, whence the above equation simplifies to

$$c_0F(0) + c_1F(1) + \cdots + c_nF(n) = c_1\varepsilon_1 + \cdots + c_n\varepsilon_n. \tag{3}$$

All this discussion has held for the $F(x)$ constructed from an arbitrary polynomial $f(x)$. We now see what all this implies for a very specific polynomial, one first used by Hermite, namely,

$$f(x) = \frac{1}{(p-1)!} x^{p-1}(1-x)^p(2-x)^p \cdots (n-x)^p.$$

Here p can be any prime number chosen so that $p > n$ and $p > c_0$. For this polynomial we shall take a very close look at $F(0), F(1), \ldots, F(n)$ and we shall carry out an estimate on the size of $\varepsilon_1, \varepsilon_2, \ldots, \varepsilon_n$.

When expanded, $f(x)$ is a polynomial of the form

$$\frac{(n!)^p}{(p-1)!} x^{p-1} + \frac{a_0 x^p}{(p-1)!} + \frac{a_1 x^{p+1}}{(p-1)!} + \cdots,$$

where $a_0, a_1, \ldots,$ are integers.

When $i \geq p$ we claim that $f^{(i)}(x)$ is a polynomial, with coefficients which are integers all of which are multiples of p. (Prove! See Problem 2.) *Thus for any integer j, $f^{(i)}(j)$, for $i \geq p$, is an integer and is a multiple of p.*

Now, from its very definition, $f(x)$ has a root of multiplicity p at $x = 1, 2, \ldots, n$. Thus for $j = 1, 2, \ldots, n$, $f(j) = 0, f^{(1)}(j) = 0, \ldots, f^{(p-1)}(j) = 0$. However, $F(j) = f(j) + f^{(1)}(j) + \cdots + f^{(p-1)}(j) + f^{(p)}(j) + \cdots + f^{(r)}(j)$; by the discussion above, for $j = 1, 2, \ldots, n$, *$F(j)$ is an integer and is a multiple of p.*

What about $F(0)$? Since $f(x)$ has a root of multiplicity $p-1$ at $x = 0$, $f(0) = f^{(1)}(0) = \cdots = f^{(p-2)}(0) = 0$. For $i \geq p$, $f^{(i)}(0)$ is an integer which is a multiple of p. But $f^{(p-1)}(0) = (n!)^p$ and since $p > n$ and is a prime number, $p \nmid (n!)^p$ so that *$f^{(p-1)}(0)$ is an integer not divisible by p.* Since $F(0) = f(0) + f^{(1)}(0) + \cdots + f^{(p-2)}(0) + f^{(p-1)}(0) + f^{(p)}(0) + \cdots + f^{(r)}(0)$, we conclude that $F(0)$ *is an integer not divisible by p.* Because $c_0 > 0$ and $p > c_0$ and because $p \nmid F(0)$ whereas $p \mid F(1), p \mid F(2), \ldots, p \mid F(n)$, we can assert that $c_0F(0) + c_1F(1) + \cdots + c_nF(n)$ *is an integer and is not divisible by p.*

However, by (3), $c_0F(0) + c_1F(1) + \cdots + c_nF(n) = c_1\varepsilon_1 + \cdots + c_n\varepsilon_n$. What can we say about ε_i? Let us recall that

$$\varepsilon_i = \frac{-e^{i(1-\theta_i)}(1-i\theta_i)^p \cdots (n-i\theta_i)^p(i\theta_i)^{p-1}i}{(p-1)!},$$

where $0 < \theta_i < 1$. Thus

$$|\varepsilon_i| \leq e^n \frac{n^p(n!)^p}{(p-1)!}.$$

As $p \to \infty$,

$$\frac{e^n n^p (n!)^p}{(p-1)!} \to 0,$$

(Prove!) whence we can find a prime number larger than both c_0 and n and large enough to force $|c_1\varepsilon_1 + \cdots + c_n\varepsilon_n| < 1$. But $c_1\varepsilon_1 + \cdots + c_n\varepsilon_n = c_0F(0) + \cdots + c_nF(n)$, *so must be an integer*; since it is smaller than 1 in size our only possible conclusion is that $c_1\varepsilon_1 + \cdots + c_n\varepsilon_n = 0$. Consequently, $c_0F(0) + \cdots + c_nF(n) = 0$; this however is sheer nonsense, since we know that $p \nmid (c_0F(0) + \cdots + c_nF(n))$, whereas $p \mid 0$. This contradiction, stemming from the assumption that e is algebraic, proves that e must be transcendental.

Problems

1. Using the infinite series for e,

$$e = 1 + \frac{1}{1!} + \frac{1}{2!} + \frac{1}{3!} + \cdots + \frac{1}{m!} + \cdots,$$

prove that e is irrational.

2. If $g(x)$ is a polynomial with integer coefficients, prove that if p is a prime number then for $i \geq p$,

$$\frac{d^i}{dx^i}\left(\frac{g(x)}{(p-1)!}\right)$$

is a polynomial with integer coefficients each of which is divisible by p.

3. If a is any real number, prove that $(a^m/m!) \to 0$ as $m \to \infty$.
4. If $m > 0$ and n are integers, prove that $e^{m/n}$ is transcendental.

5.3 Roots of Polynomials

In Section 5.1 we discussed elements in a given extension K of F which were algebraic over F, that is, elements which satisfied polynomials in $F[x]$. We now turn the problem around; given a polynomial $p(x)$ in $F[x]$ we wish to find a field K which is an extension of F in which $p(x)$ has a root. No longer is the field K available to us; in fact it is our prime objective to construct it. Once it is constructed, we shall examine it more closely and see what consequences we can derive.

DEFINITION If $p(x) \in F[x]$, then an element a lying in some extension field of F is called a *root* of $p(x)$ if $p(a) = 0$.

We begin with the familiar result known as the *Remainder Theorem.*

LEMMA 5.3.1 *If $p(x) \in F[x]$ and if K is an extension of F, then for any element $b \in K$, $p(x) = (x - b)q(x) + p(b)$ where $q(x) \in K[x]$ and where* $\deg q(x) = \deg p(x) - 1$.

Proof. Since $F \subset K$, $F[x]$ is contained in $K[x]$, whence we can consider $p(x)$ to be lying in $K[x]$. By the division algorithm for polynomials in $K[x]$, $p(x) = (x - b)q(x) + r$, where $q(x) \in K[x]$ and where $r = 0$ or $\deg r < \deg (x - b) = 1$. Thus either $r = 0$ or $\deg r = 0$; in either case r must be an element of K. But exactly what element of K is it? Since $p(x) = (x - b)q(x) + r$, $p(b) = (b - b)q(b) + r = r$. Therefore, $p(x) = (x - b)q(x) + p(b)$. That the degree of $q(x)$ is one less than that of $p(x)$ is easy to verify and is left to the reader.

COROLLARY *If $a \in K$ is a root of $p(x) \in F[x]$, where $F \subset K$, then in $K[x]$, $(x - a) \mid p(x)$.*

Proof. From Lemma 5.3.1, in $K[x]$, $p(x) = (x - a)q(x) + p(a) = (x - a)q(x)$ since $p(a) = 0$. Thus $(x - a) \mid p(x)$ in $K[x]$.

DEFINITION The element $a \in K$ is a root of $p(x) \in F[x]$ of *multiplicity* m if $(x - a)^m \mid p(x)$, whereas $(x - a)^{m+1} \nmid p(x)$.

A reasonable question to ask is, How many roots can a polynomial have in a given field? Before answering we must decide how to count a root of multiplicity m. *We shall always count it as m roots.* Even with this convention we can prove

LEMMA 5.3.2 *A polynomial of degree n over a field can have at most n roots in any extension field.*

Proof. We proceed by induction on n, the degree of the polynomial $p(x)$. If $p(x)$ is of degree 1, then it must be of the form $\alpha x + \beta$ where α, β are in a field F and where $\alpha \neq 0$. Any a such that $p(a) = 0$ must then imply that $\alpha a + \beta = 0$, from which we conclude that $a = (-\beta/\alpha)$. That is, $p(x)$ has the unique root $-\beta/\alpha$, whence the conclusion of the lemma certainly holds in this case.

Assuming the result to be true in any field for all polynomials of degree less than n, let us suppose that $p(x)$ is of degree n over F. Let K be any extension of F. If $p(x)$ has no roots in K, then we are certainly done, for the number of roots in K, namely zero, is definitely at most n. So, suppose that $p(x)$ has at least one root $a \in K$ and that a is a root of multiplicity m. Since $(x - a)^m \mid p(x)$, $m \leq n$ follows. Now $p(x) = (x - a)^m q(x)$, where $q(x) \in K[x]$ is of degree $n - m$. From the fact that $(x - a)^{m+1} \nmid p(x)$, we get that $(x - a) \nmid q(x)$, whence, by the corollary to Lemma 5.3.1, a is not a root of $q(x)$. If $b \neq a$ is a root, in K, of $p(x)$, then $0 = p(b) = (b - a)^m q(b)$; however, since $b - a \neq 0$ and since we are in a field, we conclude that $q(b) = 0$. That is, any root of $p(x)$, in K, other than a, must be a root of

$q(x)$. Since $q(x)$ is of degree $n - m < n$, by our induction hypothesis, $q(x)$ has at most $n - m$ roots in K, which, together with the other root a, counted m times, tells us that $p(x)$ has at most $m + (n - m) = n$ roots in K. This completes the induction and proves the lemma.

One should point out that commutativity is essential in Lemma 5.3.2. If we consider the ring of real quaternions, which falls short of being a field only in that it fails to be commutative, then the polynomial $x^2 + 1$ has at least 3 roots, i, j, k (in fact, it has an infinite number of roots). In a somewhat different direction we need, even when the ring is commutative, that it be an integral domain, for if $ab = 0$ with $a \neq 0$ and $b \neq 0$ in the commutative ring R, then the polynomial ax of degree 1 over R has at least two distinct roots $x = 0$ and $x = b$ in R.

The previous two lemmas, while interesting, are of subsidiary interest. We now set ourselves to our prime task, that of providing ourselves with suitable extensions of F in which a given polynomial has roots. Once this is done, we shall be able to analyze such extensions to a reasonable enough degree of accuracy to get results. The most important step in the construction is accomplished for us in the next theorem. The argument used will be very reminiscent of some used in Section 5.1.

THEOREM 5.3.1 *If $p(x)$ is a polynomial in $F[x]$ of degree $n \geq 1$ and is irreducible over F, then there is an extension E of F, such that $[E:F] = n$, in which $p(x)$ has a root.*

Proof. Let $F[x]$ be the ring of polynomials in x over F and let $V = (p(x))$ be the ideal of $F[x]$ generated by $p(x)$. By Lemma 3.9.6, V is a maximal ideal of $F[x]$, whence by Theorem 3.5.1, $E = F[x]/V$ is a field. This E will be shown to satisfy the conclusions of the theorem.

First we want to show that E is an extension of F; however, in fact, it is not! But let $\bar{F}$ be the image of F in E; that is, $\bar{F} = \{\alpha + V \mid \alpha \in F\}$. We assert that $\bar{F}$ is a field isomorphic to F; in fact, if ψ is the mapping from $F[x]$ into $F[x]/V = E$ defined by $f(x)\psi = f(x) + V$, then the restriction of ψ to F induces an isomorphism of F onto $\bar{F}$. (Prove!) Using this isomorphism, we identify F and $\bar{F}$; *in this way we can consider E to be an extension of F.*

We claim that E is a finite extension of F of degree $n = \deg p(x)$, for the elements $1 + V, x + V, (x + V)^2 = x^2 + V, \ldots, (x + V)^i = x^i + V, \ldots, (x + V)^{n-1} = x^{n-1} + V$ form a basis of E over F. (Prove!) For convenience of notation let us denote the element $x\psi = x + V$ in the field E as a. Given $f(x) \in F[x]$, what is $f(x)\psi$? We claim that it is merely $f(a)$, for, since ψ is a homomorphism, if $f(x) = \beta_0 + \beta_1 x + \cdots + \beta_k x^k$, then $f(x)\psi = \beta_0\psi + (\beta_1\psi)(x\psi) + \cdots + (\beta_k\psi)(x\psi)^k$, and using the identification indicated above of $\beta\psi$ with β, we see that $f(x)\psi = f(a)$.

In particular, since $p(x) \in V$, $p(x)\psi = 0$; however, $p(x)\psi = p(a)$. *Thus the element $a = x\psi$ in E is a root of $p(x)$.* The field E has been shown to satisfy all the properties required in the conclusion of Theorem 5.3.1, and so this theorem is now proved.

An immediate consequence of this theorem is the

COROLLARY *If $f(x) \in F[x]$, then there is a finite extension E of F in which $f(x)$ has a root. Moreover, $[E:F] \leq \deg f(x)$.*

Proof. Let $p(x)$ be an irreducible factor of $f(x)$; any root of $p(x)$ is a root of $f(x)$. By the theorem there is an extension E of F with $[E:F] = \deg p(x) \leq \deg f(x)$ in which $p(x)$, and so, $f(x)$ has a root.

Although it is, in actuality, a corollary to the above corollary, the next theorem is of such great importance that we single it out as a theorem.

THEOREM 5.3.2 *Let $f(x) \in F[x]$ be of degree $n \geq 1$. Then there is an extension E of F of degree at most $n!$ in which $f(x)$ has n roots (and so, a full complement of roots).*

Proof. In the statement of the theorem, a root of multiplicity m is, of course, counted as m roots.

By the above corollary there is an extension E_0 of F with $[E_0:F] \leq n$ in which $f(x)$ has a root α. Thus in $E_0[x]$, $f(x)$ factors as $f(x) = (x - \alpha)q(x)$, where $q(x)$ is of degree $n - 1$. Using induction (or continuing the above process), there is an extension E of E_0 of degree at most $(n - 1)!$ in which $q(x)$ has $n - 1$ roots. Since any root of $f(x)$ is either α or a root of $q(x)$, we obtain in E all n roots of $f(x)$. Now, $[E:F] = [E:E_0][E_0:F] \leq (n-1)!n = n!$ All the pieces of the theorem are now established.

Theorem 5.3.2 asserts the existence of a finite extension E in which the given polynomial $f(x)$, of degree n, over F has n roots. If $f(x) = a_0x^n + a_1x^{n-1} + \cdots + a_n$, $a_0 \neq 0$ and if the n roots in E are $\alpha_1, \ldots, \alpha_n$, making use of the corollary to Lemma 5.3.1, $f(x)$ can be factored over E as $f(x) = a_0(x - \alpha_1)(x - \alpha_2)\cdots(x - \alpha_n)$. Thus $f(x)$ splits up completely over E as a product of *linear* (first degree) factors. Since a finite extension of F exists with this property, a *finite extension of F of minimal degree* exists which also enjoys this property of decomposing $f(x)$ as a product of linear factors. For such a minimal extension, no proper subfield has the property that $f(x)$ factors over it into the product of linear factors. This prompts the

DEFINITION If $f(x) \in F[x]$, a finite extension E of F is said to be a *splitting field* over F for $f(x)$ if over E (that is, in $E[x]$), but not over any proper subfield of E, $f(x)$ can be factored as a product of linear factors.

We reiterate: *Theorem 5.3.2 guarantees for us the existence of splitting fields.* In fact, it says even more, for it assures that given a polynomial of degree n over F there is a splitting field of this polynomial which is an extension of F of degree at most $n!$ over F. We shall see later that this upper bound of $n!$ is actually taken on; that is, given n, we can find a field F and a polynomial of degree n in $F[x]$ such that the splitting field of $f(x)$ over F has degree $n!$.

Equivalent to the definition we gave of a splitting field for $f(x)$ over F is the statement: *E is a splitting field of $f(x)$ over F if E is a minimal extension of F in which $f(x)$ has n roots, where $n = \deg f(x)$.*

An immediate question arises: given two splitting fields E_1 and E_2 of the same polynomial $f(x)$ in $F[x]$, what is their relation to each other? At first glance, we have no right to assume that they are at all related. Our next objective is to show that they are indeed intimately related; in fact, that they are isomorphic by an isomorphism leaving every element of F fixed. It is in this direction that we now turn.

Let F and F' be two fields and let τ be an isomorphism of F onto F'. For convenience let us denote the image of any $\alpha \in F$ under τ by α'; that is, $\alpha\tau = \alpha'$. We shall maintain this notation for the next few pages.

Can we make use of τ to set up an isomorphism between $F[x]$ and $F'[t]$, the respective polynomial rings over F and F'? Why not try the obvious? For an arbitrary polynomial $f(x) = \alpha_0 x^n + \alpha_1 x^{n-1} + \cdots + \alpha_n \in F[x]$ we define τ^* by $f(x)\tau^* = (\alpha_0 x^n + \alpha_1 x^{n-1} + \cdots + \alpha_n)\tau^* = \alpha_0' t^n + \alpha_1' t^{n-1} + \cdots + \alpha_n'$.

It is an easy and straightforward matter, which we leave to the reader, to verify.

LEMMA 5.3.3 *τ^* defines an isomorphism of $F[x]$ onto $F'[t]$ with the property that $\alpha\tau^* = \alpha'$ for every $\alpha \in F$.*

If $f(x)$ is in $F[x]$ we shall write $f(x)\tau^*$ as $f'(t)$. Lemma 5.3.3 immediately implies that factorizations of $f(x)$ in $F[x]$ result in like factorizations of $f'(t)$ in $F'[t]$, and vice versa. In particular, $f(x)$ is irreducible in $F[x]$ if and only if $f'(t)$ is irreducible in $F'[t]$.

However, at the moment, we are not particularly interested in polynomial rings, but rather, in extensions of F. Let us recall that in the proof of Theorem 5.1.2 we employed quotient rings of polynomial rings to obtain suitable extensions of F. In consequence it should be natural for us to study the relationship between $F[x]/(f(x))$ and $F'[t]/(f'(t))$, where $(f(x))$ denotes the ideal generated by $f(x)$ in $F[x]$ and $(f'(t))$ that generated by $f'(t)$ in $F'[t]$. The next lemma, which is relevant to this question, is actually part of a more general, purely ring-theoretic result, but we shall content ourselves with it as applied in our very special setting.

LEMMA 5.3.4 *There is an isomorphism τ^{**} of $F[x]/(f(x))$ onto $F'[t]/(f'(t))$ with the property that for every $\alpha \in F$, $\alpha\tau^{**} = \alpha'$, $(x + (f(x)))\tau^{**} = t + (f'(t))$.*

Proof. Before starting with the proof proper, we should make clear what is meant by the last part of the statement of the lemma. As we have already done several times, we can consider F as imbedded in $F[x]/(f(x))$ by identifying the element $\alpha \in F$ with the coset $\alpha + (f(x))$ in $F[x]/(f(x))$. Similarly, we can consider F' to be contained in $F'[t]/(f'(t))$. The isomorphism τ^{**} is then supposed to satisfy $[\alpha + (f(x))]\tau^{**} = \alpha' + (f'(t))$.

We seek an isomorphism τ^{**} of $F[x]/(f(x))$ onto $F'[t]/(f'(t))$. What could be simpler or more natural than to try the τ^{**} defined by $[g(x) + (f(x))]\tau^{**} = g'(t) + (f'(t))$ for every $g(x) \in F[x]$? We leave it as an exercise to fill in the necessary details that the τ^{**} so defined is well defined and is an isomorphism of $F[x]/(f(x))$ onto $F'[t]/(f'(t))$ with the properties needed to fulfill the statement of Lemma 5.3.4.

For our purpose—that of proving the uniqueness of splitting fields—Lemma 5.3.4 provides us with the entering wedge, for we can now prove

THEOREM 5.3.3 *If $p(x)$ is irreducible in $F[x]$ and if v is a root of $p(x)$, then $F(v)$ is isomorphic to $F'(w)$ where w is a root of $p'(t)$; moreover, this isomorphism σ can so be chosen that*

1. $v\sigma = w$.
2. $\alpha\sigma = \alpha'$ *for every* $\alpha \in F$.

Proof. Let v be a root of the irreducible polynomial $p(x)$ lying in some extension K of F. Let $M = \{f(x) \in F[x] \mid f(v) = 0\}$. Trivially M is an ideal of $F[x]$, and $M \neq F[x]$. Since $p(x) \in M$ and is an irreducible polynomial, we have that $M = (p(x))$. As in the proof of Theorem 5.1.2, map $F[x]$ into $F(v) \subset K$ by the mapping ψ defined by $q(x)\psi = q(v)$ for every $q(x) \in F[x]$. We saw earlier (in the proof of Theorem 5.1.2) that ψ maps $F[x]$ onto $F(v)$. The kernel of ψ is precisely M, so must be $(p(x))$. By the fundamental homomorphism theorem for rings there is an isomorphism ψ^* of $F[x]/(p(x))$ onto $F(v)$. Note further that $\alpha\psi^* = \alpha$ for every $\alpha \in F$. Summing up: ψ^* is an isomorphism of $F[x]/(p(x))$ onto $F(v)$ leaving every element of F fixed and with the property that $v = [x + (p(x))]\psi^*$.

Since $p(x)$ is irreducible in $F[x]$, $p'(t)$ is irreducible in $F'[t]$ (by Lemma 5.3.3), and so there is an isomorphism θ^* of $F'[t]/(p'(t))$ onto $F'(w)$ where w is a root of $p'(t)$ such that θ^* leaves every element of F' fixed and such that $[t + (p'(t)]\theta^* = w$.

We now stitch the pieces together to prove Theorem 5.3.3. By Lemma 5.3.4 there is an isomorphism τ^{**} of $F[x]/(p(x))$ onto $F'[t]/(p'(t))$ which coincides with τ on F and which takes $x + (p(x))$ onto $t + (p'(t))$. Consider the mapping $\sigma = (\psi^*)^{-1}\tau^{**}\theta^*$ (motivated by

$$F(v) \xrightarrow{(\psi^*)^{-1}} \frac{F[x]}{(p(x))} \xrightarrow{\tau^{**}} \frac{F'[t]}{(p'(t))} \xrightarrow{\theta^*} F'(w))$$

of $F(v)$ onto $F'(w)$. It is an isomorphism of $F(v)$ onto $F'(w)$ since all the mapping ψ^*, τ^{**}, and θ^* are isomorphisms and onto. Moreover, since $v = [x + (p(x))]\psi^*$, $v\sigma = (v(\psi^*)^{-1})\tau^{**}\theta^* = ([x + (p(x)]\tau^{**})\theta^* = [t + (p'(t))]\theta^* = w$. Also, for $\alpha \in F$, $\alpha\sigma = (\alpha(\psi^*)^{-1})\tau^{**}\theta^* = (\alpha\tau^{**})\theta^* = \alpha'\theta^* = \alpha'$. We have shown that σ is an isomorphism satisfying all the requirements of the isomorphism in the statement of the theorem. Thus Theorem 5.3.3 has been proved.

A special case, but itself of interest, is the

COROLLARY *If $p(x) \in F[x]$ is irreducible and if a, b are two roots of $p(x)$, then $F(a)$ is isomorphic to $F(b)$ by an isomorphism which takes a onto b and which leaves every element of F fixed.*

We now come to the theorem which is, as we indicated earlier, the foundation stone on which the whole Galois theory rests. For us it is the focal point of this whole section.

THEOREM 5.3.4 *Any splitting fields E and E' of the polynomials $f(x) \in F[x]$ and $f'(t) \in F'[t]$, respectively, are isomorphic by an isomorphism ϕ with the property that $\alpha\phi = \alpha'$ for every $\alpha \in F$. (In particular, any two splitting fields of the same polynomial over a given field F are isomorphic by an isomorphism leaving every element of F fixed.)*

Proof. We should like to use an argument by induction; in order to do so, we need an integer-valued indicator of size which we can decrease by some technique or other. We shall use as our indicator the degree of some splitting field over the initial field. It may seem artificial (in fact, it may even be artificial), but we use it because, as we shall soon see, Theorem 5.3.3 provides us with the mechanism for decreasing it.

If $[E:F] = 1$, then $E = F$, whence $f(x)$ splits into a product of linear factors over F itself. By Lemma 5.3.3 $f'(t)$ splits over F' into a product of linear factors, hence $E' = F'$. But then $\phi = \tau$ provides us with an isomorphism of E onto E' coinciding with τ on F.

Assume the result to be true for any field F_0 and any polynomial $f(x) \in F_0[x]$ provided the degree of some splitting field E_0 of $f(x)$ has degree less than n over F_0, that is, $[E_0:F_0] < n$.

Suppose that $[E:F] = n > 1$, where E is a splitting field of $f(x)$ over F. Since $n > 1$, $f(x)$ has an irreducible factor $p(x)$ of degree $r > 1$. Let $p'(t)$ be the corresponding irreducible factor of $f'(t)$. Since E splits $f(x)$, a full complement of roots of $f(x)$, and so, *a priori*, of roots of $p(x)$, are in E. Thus there is a $v \in E$ such that $p(v) = 0$; by Theorem 5.1.3, $[F(v):F] = r$. Similarly, there is a $w \in E'$ such that $p'(w) = 0$. By Theorem 5.3.4 there is an isomorphism σ of $F(v)$ onto $F'(w)$ with the property that $\alpha\sigma = \alpha'$ for every $\alpha \in F$.

Since $[F(v):F] = r > 1$,

$$[E:F(v)] = \frac{[E:F]}{[F(v):F]} = \frac{n}{r} < n.$$

We claim that E is a splitting field for $f(x)$ considered as a polynomial over $F_0 = F(v)$, for no subfield of E, containing F_0 and hence F, can split $f(x)$, since E is assumed to be a splitting field of $f(x)$ over F. Similarly E' is a splitting field for $f'(t)$ over $F_0' = F'(w)$. By our induction hypothesis there is an isomorphism ϕ of E onto E' such that $a\phi = a\sigma$ for all $a \in F_0$. But for every $\alpha \in F$, $\alpha\sigma = \alpha'$ hence for every $\alpha \in F \subset F_0$, $\alpha\phi = \alpha\sigma = \alpha'$. This completes the induction and proves the theorem.

To see the truth of the "(in particular . . .)" part, let $F = F'$ and let τ be the identity map $\alpha\tau = \alpha$ for every $\alpha \in F$. Suppose that E_1 and E_2 are two splitting fields of $f(x) \in F[x]$. Considering $E_1 = E \supset F$ and $E_2 = E' \supset F' = F$, and applying the theorem just proved, yields that E_1 and E_2 are isomorphic by an isomorphism leaving every element of F fixed.

In view of the fact that any two splitting fields of the same polynomial over F are isomorphic and by an isomorphism leaving every element of F fixed, we are justified in speaking about *the* splitting field, rather than *a* splitting field, for it is essentially unique.

Examples

1. Let F be any field and let $p(x) = x^2 + \alpha x + \beta$, $\alpha, \beta \in F$, be in $F[x]$. If K is any extension of F in which $p(x)$ has a root, a, then the element $b = -\alpha - a$ also in K is also a root of $p(x)$. If $b = a$ it is easy to check that $p(x)$ must then be $p(x) = (x - a)^2$, and so both roots of $p(x)$ are in K. If $b \neq a$ then again both roots of $p(x)$ are in K. Consequently, $p(x)$ can be split by an extension of degree 2 of F. We could also get this result directly by invoking Theorem 5.3.2.

2. Let F be the field of rational numbers and let $f(x) = x^3 - 2$. In the field of complex numbers the three roots of $f(x)$ are $\sqrt[3]{2}$, $\omega\sqrt[3]{2}$, $\omega^2\sqrt[3]{2}$, where $\omega = (-1 + \sqrt{3}\,i)/2$ and where $\sqrt[3]{2}$ is a real cube root of 2. Now $F(\sqrt[3]{2})$ cannot split $x^3 - 2$, for, as a subfield of the real field, it cannot contain the complex, but not real, number $\omega\sqrt[3]{2}$. Without explicitly determining it, what can we say about E, the splitting field of $x^3 - 2$ over

F? By Theorem 5.3.2, $[E:F] \leq 3! = 6$; by the above remark, since $x^3 - 2$ is irreducible over F and since $[F(\sqrt[3]{2}):F] = 3$, by the corollary to Theorem 5.1.1, $3 = [F(\sqrt[3]{2}):F] \mid [E:F]$. Finally, $[E:F] > [F(\sqrt[3]{2}):F] = 3$. The only way out is $[E:F] = 6$. We could, of course, get this result by making two extensions $F_1 = F(\sqrt[3]{2})$ and $E = F_1(\omega)$ and showing that ω satisfies an irreducible quadratic equation over F_1.

3. Let F be the field of rational numbers and let

$$f(x) = x^4 + x^2 + 1 \in F[x].$$

We claim that $E = F(\omega)$, where $\omega = (-1 + \sqrt{3}\,i)/2$, is a splitting field of $f(x)$. Thus $[E:F] = 2$, far short of the maximum possible $4! = 24$.

Problems

1. In the proof of Lemma 5.3.1, prove that the degree of $q(x)$ is one less than that of $p(x)$.
2. In the proof of Theorem 5.3.1, prove in all detail that the elements $1 + V, x + V, \ldots, x^{n-1} + V$ form a basis of E over F.
3. Prove Lemma 5.3.3 in all detail.
4. Show that τ^{**} in Lemma 5.3.4 is well defined and is an isomorphism of $F[x]/(f(x))$ onto $F[t]/(f'(t))$.
5. In Example 3 at the end of this section prove that $F(\omega)$ is the splitting field of $x^4 + x^2 + 1$.
6. Let F be the field of rational numbers. Determine the degrees of the splitting fields of the following polynomials over F.
   (a) $x^4 + 1$. (b) $x^6 + 1$.
   (c) $x^4 - 2$. (d) $x^5 - 1$.
   (e) $x^6 + x^3 + 1$.
7. If p is a prime number, prove that the splitting field over F, the field of rational numbers, of the polynomial $x^p - 1$ is of degree $p - 1$.

**8. If $n > 1$, prove that the splitting field of $x^n - 1$ over the field of rational numbers is of degree $\Phi(n)$ where Φ is the Euler Φ-function.
   (This is a well-known theorem. I know of no easy solution, so don't be disappointed if you fail to get it. If you get an easy proof, I would like to see it. This problem occurs in an equivalent form as Problem 15, Section 5.6.)

*9. If F is the field of rational numbers, find necessary and sufficient conditions on a and b so that the splitting field of $x^3 + ax + b$ has degree exactly 3 over F.

10. Let p be a prime number and let $F = J_p$, the field of integers mod p.
   (a) Prove that there is an irreducible polynomial of degree 2 over F.

(b) Use this polynomial to construct a field with p^2 elements.

*(c) Prove that any two irreducible polynomials of degree 2 over F lead to isomorphic fields with p^2 elements.

11. If E is an extension of F and if $f(x) \in F[x]$ and if ϕ is an automorphism of E leaving every element of F fixed, prove that ϕ must take a root of $f(x)$ lying in E into a root of $f(x)$ in E.
12. Prove that $F(\sqrt[3]{2})$, where F is the field of rational numbers, has no automorphisms other than the identity automorphism.
13. Using the result of Problem 11, prove that if the complex number α is a root of the polynomial $p(x)$ having *real* coefficients then $\bar{\alpha}$, the complex conjugate of α, is also a root of $p(x)$.
14. Using the result of Problem 11, prove that if m is an integer which is not a perfect square and if $\alpha + \beta\sqrt{m}$ (α, β rational) is the root of a polynomial $p(x)$ having *rational coefficients*, then $\alpha - \beta\sqrt{m}$ is also a root of $p(x)$.

*15. If F is the field of real numbers, prove that if ϕ is an automorphism of F, then ϕ leaves every element of F fixed.

16 (a) Find *all* real quaternions $t = a_0 + a_1 i + a_2 j + a_3 k$ satisfying $t^2 = -1$

*(b) For a t as in part (a) prove we can find a real quaternion s such that $sts^{-1} = i$.

5.4 Construction with Straightedge and Compass

We pause in our general development to examine some implications of the results obtained so far in some familiar, geometric situations.

A real number α is said to be a *constructible number* if by the use of straightedge and compass alone we can construct a line segment of length α. We assume that we are given some fundamental unit length. Recall that from high-school geometry we can construct with a straightedge and compass a line perpendicular to and a line parallel to a given line through a given point. From this it is an easy exercise (see Problem 1) to prove that if α and β are constructible numbers then so are $\alpha \pm \beta$, $\alpha\beta$, and when $\beta \neq 0$, α/β. Therefore, the set of constructible numbers form a subfield, W, of the field of real numbers.

In particular, since $1 \in W$, W must contain F_0, the field of rational numbers. We wish to study the relation of W to the rational field.

Since we shall have many occasions to use the phrase "construct by straightedge and compass" (and variants thereof) *the words construct, constructible, construction, will always mean by straightedge and compass.*

If $w \in W$, we can reach w from the rational field by a *finite* number of constructions.

Let F be any subfield of the field of real numbers. Consider all the points (x, y) in the real Euclidean plane both of whose coordinates x and y are in F; we call the set of these points the *plane of F*. Any straight line joining two points in the plane of F has an equation of the form $ax + by + c = 0$ where a, b, c are all in F (see Problem 2). Moreover, any circle having as center a point in the plane of F and having as radius an element of F has an equation of the form $x^2 + y^2 + ax + by + c = 0$, where all of a, b, c are in F (see Problem 3). We call such lines and circles *lines and circles in F*.

Given two lines in F which intersect in the real plane, then their intersection point is a point in the plane of F (see Problem 4). On the other hand, the intersection of a line in F and a circle in F need not yield a point in the plane of F. But, using the fact that the equation of a line in F is of the form $ax + by + c = 0$ and that of a circle in F is of the form $x^2 + y^2 + dx + ey + f = 0$, where a, b, c, d, e, f are all in F, we can show that when a line and circle of F intersect in the real plane, they intersect either in a point in the plane of F or in the plane of $F(\sqrt{\gamma})$ for some positive γ in F (see Problem 5). Finally, the intersection of two circles in F can be realized as that of a line in F and a circle in F, for if these two circles are $x^2 + y^2 + a_1x + b_1y + c_1 = 0$ and $x^2 + y^2 + a_2x + b_2y + c_2 = 0$, then their intersection is the intersection of either of these with the line $(a_1 - a_2)x + (b_1 - b_2)y + (c_1 - c_2) = 0$, so also yields a point either in the plane of F or of $F(\sqrt{\gamma})$ for some positive γ in F.

Thus lines and circles of F lead us to points either in F or in quadratic extensions of F. If we now are in $F(\sqrt{\gamma_1})$ for some quadratic extension of F, then lines and circles in $F(\sqrt{\gamma_1})$ intersect in points in the plane of $F(\sqrt{\gamma_1}, \sqrt{\gamma_2})$ where γ_2 is a positive number in $F(\sqrt{\gamma_1})$. A point is constructible from F if we can find real numbers $\lambda_1, \ldots, \lambda_n$, such that $\lambda_1^2 \in F$, $\lambda_2^2 \in F(\lambda_1)$, $\lambda_3^2 \in F(\lambda_1, \lambda_2), \ldots, \lambda_n^2 \in F(\lambda_1, \ldots, \lambda_{n-1})$, such that the point is in the plane of $F(\lambda_1, \ldots, \lambda_n)$. Conversely, if $\gamma \in F$ is such that $\sqrt{\gamma}$ is real then we can realize γ as an intersection of lines and circles in F (see Problem 6). Thus a point is constructible from F if and only if we can find a finite number of real numbers $\lambda_1, \ldots, \lambda_n$, such that

1. $[F(\lambda_1):F] = 1$ or 2;
2. $[F(\lambda_1, \ldots, \lambda_i):F(\lambda_1, \ldots, \lambda_{i-1})] = 1$ or 2 for $i = 1, 2, \ldots, n$;

and such that our point lies in the plane of $F(\lambda_1, \ldots, \lambda_n)$.

We have defined a real number α to be constructible if by use of straightedge and compass we can construct a line segment of length α. But this translates, in terms of the discussion above, into: α is constructible if starting from the plane of the rational numbers, F_0, we can imbed α in a field obtained from F_0 by a finite number of quadratic extensions. This is

THEOREM 5.4.1 *The real number α is constructible if and only if we can find a finite number of real numbers $\lambda_1, \ldots, \lambda_n$ such that*

1. $\lambda_1{}^2 \in F_0$,
2. $\lambda_i{}^2 \in F_0(\lambda_1, \ldots, \lambda_{i-1})$ *for* $i = 1, 2, \ldots, n$,

such that $\alpha \in F_0(\lambda_1, \ldots, \lambda_n)$.

However, we can compute the degree of $F_0(\lambda_1, \ldots, \lambda_n)$ over F_0, for by Theorem 5.1.1

$$\begin{aligned}[F_0(\lambda_1, \ldots, \lambda_n):F_0] &= [F_0(\lambda_1, \ldots, \lambda_n):F_0(\lambda_1, \ldots, \lambda_{n-1})] \cdots \\ &\quad \times [F_0(\lambda_1, \ldots, \lambda_i):F_0(\lambda_1, \ldots, \lambda_{i-1})] \cdots \\ &\quad \times [F_0(\lambda_1):F_0].\end{aligned}$$

Since each term in the product is either 1 or 2, we get that

$$[F_0(\lambda_1, \ldots, \lambda_n):F_0] = 2^r,$$

and thus the

COROLLARY 1 *If α is constructible then α lies in some extension of the rationals of degree a power of* 2.

If α is constructible, by Corollary 1 above, there is a subfield K of the real field such that $\alpha \in K$ and such that $[K:F_0] = 2^r$. However, $F_0(\alpha) \subset K$, whence by the corollary to Theorem 5.1.1 $[F_0(\alpha):F_0] \mid [K:F_0] = 2^r$; thereby $[F_0(\alpha):F_0]$ is also a power of 2. However, if α satisfies an irreducible polynomial of degree k over F_0, we have proved in Theorem 5.1.3 that $[F_0(\alpha):F_0] = k$. Thus we get the important criterion for nonconstructibility

COROLLARY 2 *If the real number α satisfies an irreducible polynomial over the field of rational numbers of degree k, and if k is not a power of* 2, *then α is not constructible.*

This last corollary enables us to settle the ancient problem of trisecting an angle by straightedge and compass, for we prove

THEOREM 5.4.2 *It is impossible, by straightedge and compass alone, to trisect* 60°.

Proof. If we could trisect 60° by straightedge and compass, then the length $\alpha = \cos 20°$ would be constructible. At this point, let us recall the identity $\cos 3\theta = 4\cos^3\theta - 3\cos\theta$. Putting $\theta = 20°$ and remembering that $\cos 60° = \frac{1}{2}$, we obtain $4\alpha^3 - 3\alpha = \frac{1}{2}$, whence $8\alpha^3 - 6\alpha - 1 = 0$. Thus α is a root of the polynomial $8x^3 - 6x - 1$ over the rational field.

However, this polynomial is irreducible over the rational field (Problem 7(a)), and since its degree is 3, which certainly is not a power of 2, by Corollary 2 to Theorem 5.4.1, α is not constructible. Thus 60° cannot be trisected by straightedge and compass.

Another ancient problem is that of duplicating the cube, that is, of constructing a cube whose volume is twice that of a given cube. If the original cube is the unit cube, this entails constructing a length α such that $\alpha^3 = 2$. Since the polynomial $x^3 - 2$ is irreducible over the rationals (Problem 7(b)), by Corollary 2 to Theorem 5.4.1, α is not constructible. Thus

THEOREM 5.4.3 *By straightedge and compass it is impossible to duplicate the cube.*

We wish to exhibit yet another geometric figure which cannot be constructed by straightedge and compass, namely, the regular septagon. To carry out such a construction would require the constructibility of $\alpha = 2\cos(2\pi/7)$. However, we claim that α satisfies $x^3 + x^2 - 2x - 1$ (Problem 8) and that this polynomial is irreducible over the field of rational numbers (Problem 7(c)). Thus again using Corollary 2 to Theorem 5.4.1 we obtain

THEOREM 5.4.4 *It is impossible to construct a regular septagon by straightedge and compass.*

Problems

1. Prove that if α, β are constructible, then so are $\alpha \pm \beta$, $\alpha\beta$, and α/β (when $\beta \neq 0$).
2. Prove that a line in F has an equation of the form $ax + by + c = 0$ with a, b, c in F.
3. Prove that a circle in F has an equation of the form
$$x^2 + y^2 + ax + by + c = 0,$$
with a, b, c in F.
4. Prove that two lines in F, which intersect in the real plane, intersect at a point in the plane of F.
5. Prove that a line in F and a circle in F which intersect in the real plane do so at a point either in the plane of F or in the plane of $F(\sqrt{\gamma})$ where γ is a positive number in F.
6. If $\gamma \in F$ is positive, prove that $\sqrt{\gamma}$ is realizable as an intersection of lines and circles in F.

7. Prove that the following polynomials are irreducible over the field of rational numbers.
   (a) $8x^3 - 6x - 1$.
   (b) $x^3 - 2$.
   (c) $x^3 + x^2 - 2x - 1$.
8. Prove that $2 \cos (2\pi/7)$ satisfies $x^3 + x^2 - 2x - 1$. (*Hint:* Use $2 \cos (2\pi/7) = e^{2\pi i/7} + e^{-2\pi i/7}$.)
9. Prove that the regular pentagon is constructible.
10. Prove that the regular hexagon is constructible.
11. Prove that the regular 15-gon is constructible.
12. Prove that it is possible to trisect 72°.
13. Prove that a regular 9-gon is not constructible.

*14. Prove a regular 17-gon is constructible.

5.5 More about Roots

We return to the general exposition. Let F be any field and, as usual, let $F[x]$ be the ring of polynomials in x over F.

DEFINITION If $f(x) = \alpha_0 x^n + \alpha_1 x^{n-1} + \cdots + \alpha_i x^{n-i} + \cdots + \alpha_{n-1}x + \alpha_n$ in $F[x]$, then the *derivative* of $f(x)$, written as $f'(x)$, is the polynomial $f'(x) = n\alpha_0 x^{n-1} + (n-1)\alpha_1 x^{n-2} + \cdots + (n-i)\alpha_i x^{n-i-1} + \cdots + \alpha_{n-1}$ in $F[x]$.

To make this definition or to prove the basic formal properties of the derivatives, as applied to polynomials, does not require the concept of a limit. However, since the field F is arbitrary, we might expect some strange things to happen.

At the end of Section 5.2, we defined what is meant by the characteristic of a field. Let us recall it now. A field F is said to be of characteristic 0 if $ma \neq 0$ for $a \neq 0$ in F and $m > 0$, an integer. If $ma = 0$ for some $m > 0$ and some $a \neq 0 \in F$, then F is said to be of finite characteristic. In this second case, the characteristic of F is defined to be the smallest positive integer p such that $pa = 0$ for all $a \in F$. It turned out that if F is of finite characteristic then its characteristic p is a prime number.

We return to the question of the derivative. Let F be a field of characteristic $p \neq 0$. In this case, the derivative of the polynomial x^p is $px^{p-1} = 0$. Thus the usual result from the calculus that a polynomial whose derivative is 0 must be a constant no longer need hold true. However, if the characteristic of F is 0 and if $f'(x) = 0$ for $f(x) \in F[x]$, it is indeed true that $f(x) = \alpha \in F$ (see Problem 1). Even when the characteristic of F is $p \neq 0$, we can still describe the polynomials with zero derivative; if $f'(x) = 0$, then $f(x)$ is a polynomial in x^p (see Problem 2).

We now prove the analogs of the formal rules of differentiation that we know so well.

LEMMA 5.5.1 *For any* $f(x), g(x) \in F[x]$ *and any* $\alpha \in F$,

1. $(f(x) + g(x))' = f'(x) + g'(x)$.
2. $(\alpha f(x))' = \alpha f'(x)$.
3. $(f(x)g(x))' = f'(x)g(x) + f(x)g'(x)$.

Proof. The proofs of parts 1 and 2 are extremely easy and are left as exercises. To prove part 3, note that from parts 1 and 2 it is enough to prove it in the highly special case $f(x) = x^i$ and $g(x) = x^j$ where both i and j are positive. But then $f(x)g(x) = x^{i+j}$, whence $(f(x)g(x))' = (i+j)x^{i+j-1}$; however, $f'(x)g(x) = ix^{i-1}x^j = ix^{i+j-1}$ and $f(x)g'(x) = jx^ix^{j-1} = jx^{i+j-1}$; consequently, $f'(x)g(x) + f(x)g'(x) = (i+j)x^{i+j-1} = (f(x)g(x))'$.

Recall that in elementary calculus the equivalence is shown between the existence of a multiple root of a function and the simultaneous vanishing of the function and its derivative at a given point. Even in our setting, where F is an arbitrary field, such an interrelation exists.

LEMMA 5.5.2 *The polynomial* $f(x) \in F[x]$ *has a multiple root if and only if* $f(x)$ *and* $f'(x)$ *have a nontrivial (that is, of positive degree) common factor.*

Proof. Before proving the lemma proper, a related remark is in order, namely, if $f(x)$ and $g(x)$ in $F[x]$ have a nontrivial common factor in $K[x]$, for K an extension of F, then they have a nontrivial common factor in $F[x]$. For, were they relatively prime as elements in $F[x]$, then we would be able to find two polynomials $a(x)$ and $b(x)$ in $F[x]$ such that $a(x)f(x) + b(x)g(x) = 1$. Since this relation also holds for those elements viewed as elements of $K[x]$, in $K[x]$ they would have to be relatively prime.

Now to the lemma itself. From the remark just made, we may assume, without loss of generality, that the roots of $f(x)$ all lie in F (otherwise extend F to K, the splitting field of $f(x)$). If $f(x)$ has a multiple root α, then $f(x) = (x-\alpha)^m q(x)$, where $m > 1$. However, as is easily computed, $((x-\alpha)^m)' = m(x-\alpha)^{m-1}$ whence, by Lemma 5.5.1, $f'(x) = (x-\alpha)^m q'(x) + m(x-\alpha)^{m-1}q(x) = (x-\alpha)r(x)$, since $m > 1$. But this says that $f(x)$ and $f'(x)$ have the common factor $x - \alpha$, thereby proving the lemma in one direction.

On the other hand, if $f(x)$ has no multiple root then $f(x) = (x-\alpha_1)(x-\alpha_2)\cdots(x-\alpha_n)$ where the α_i's are all distinct (we are supposing $f(x)$ to be monic). But then

$$f'(x) = \sum_{i=1}^{n} (x-\alpha_1)\cdots\widehat{(x-\alpha_i)}\cdots(x-\alpha_n)$$

where the $\wedge$ denotes the term is omitted. We claim no root of $f(x)$ is a root of $f'(x)$, for

$$f'(\alpha_i) = \prod_{j \neq i} (\alpha_i - \alpha_j) \neq 0,$$

since the roots are all distinct. However, if $f(x)$ and $f'(x)$ have a nontrivial common factor, they have a common root, namely, any root of this common factor. The net result is that $f(x)$ and $f'(x)$ have no nontrivial common factor, and so the lemma has been proved in the other direction.

COROLLARY 1 *If $f(x) \in F[x]$ is irreducible, then*

1. *If the characteristic of F is 0, $f(x)$ has no multiple roots.*
2. *If the characteristic of F is $p \neq 0$, $f(x)$ has a multiple root only if it is of the form $f(x) = g(x^p)$.*

Proof. Since $f(x)$ is irreducible, its only factors in $F[x]$ are 1 and $f(x)$. If $f(x)$ has a multiple root, then $f(x)$ and $f'(x)$ have a nontrivial common factor by the lemma, hence $f(x) \mid f'(x)$. However, since the degree of $f'(x)$ is less than that of $f(x)$, the only possible way that this can happen is for $f'(x)$ to be 0. In characteristic 0 this implies that $f(x)$ is a constant, which has no roots; in characteristic $p \neq 0$, this forces $f(x) = g(x^p)$.

We shall return in a moment to discuss the implications of Corollary 1 more fully. But first, for later use in Chapter 7 in our treatment of finite fields, we prove the rather special

COROLLARY 2 *If F is a field of characteristic $p \neq 0$, then the polynomial $x^{p^n} - x \in F[x]$, for $n \geq 1$, has distinct roots.*

Proof. The derivative of $x^{p^n} - x$ is $p^n x^{p^n - 1} - 1 = -1$, since F is of characteristic p. Therefore, $x^{p^n} - x$ and its derivative are certainly relatively prime, which, by the lemma, implies that $x^{p^n} - x$ has no multiple roots.

Corollary 1 does not rule out the possibility that in characteristic $p \neq 0$ an irreducible polynomial might have multiple roots. To clinch matters, we exhibit an example where this actually happens. Let F_0 be a field of characteristic 2 and let $F = F_0(x)$ be the field of rational functions in x over F_0. We claim that the polynomial $t^2 - x$ in $F[t]$ is irreducible over F and that its roots are equal. To prove irreducibility we must show that there is no rational function in $F_0(x)$ whose square is x; this is the content of Problem 4. To see that $t^2 - x$ has a multiple root, notice that its derivative (the derivative is with respect to t; for x, being in F, is considered as a constant) is $2t = 0$. Of course, the analogous example works for any prime characteristic.

Now that the possibility has been seen to be an actuality, it points out a sharp difference between the case of characteristic 0 and that of characteristic p. The presence of irreducible polynomials with multiple roots in the latter case leads to many interesting, but at the same time complicating, subtleties. These require a more elaborate and sophisticated treatment which we prefer to avoid at this stage of the game. *Therefore, we make the flat assumption for the rest of this chapter that all fields occurring in the text material proper are fields of characteristic* 0.

DEFINITION The extension K of F is a *simple extension of* F if $K = F(\alpha)$ for some α in K.

In characteristic 0 (or in properly conditioned extensions in characteristic $p \neq 0$; see Problem 14) all finite extensions are realizable as simple extensions. This result is

THEOREM 5.5.1 *If F is of characteristic* 0 *and if a, b, are algebraic over F, then there exists an element $c \in F(a, b)$ such that $F(a, b) = F(c)$.*

Proof. Let $f(x)$ and $g(x)$, of degrees m and n, be the irreducible polynomials over F satisfied by a and b, respectively. Let K be an extension of F in which both $f(x)$ and $g(x)$ split completely. Since the characteristic of F is 0, all the roots of $f(x)$ are distinct, as are all those of $g(x)$. Let the roots of $f(x)$ be $a = a_1, a_2, \ldots, a_m$ and those of $g(x)$, $b = b_1, b_2, \ldots, b_n$.

If $j \neq 1$, then $b_j \neq b_1 = b$, hence the equation $a_i + \lambda b_j = a_1 + \lambda b_1 = a + \lambda b$ has only one solution λ in K, namely,

$$\lambda = \frac{a_i - a}{b - b_j}$$

Since F is of characteristic 0 it has an infinite number of elements, so we can find an element $\gamma \in F$ such that $a_i + \gamma b_j \neq a + \gamma b$ for all i and for all $j \neq 1$. Let $c = a + \gamma b$; our contention is that $F(c) = F(a, b)$. Since $c \in F(a, b)$, we certainly do have that $F(c) \subset F(a, b)$. We will now show that both a and b are in $F(c)$ from which it will follow that $F(a, b) \subset F(c)$.

Now b satisfies the polynomial $g(x)$ over F, hence satisfies $g(x)$ considered as a polynomial over $K = F(c)$. Moreover, if $h(x) = f(c - \gamma x)$ then $h(x) \in K[x]$ and $h(b) = f(c - \gamma b) = f(a) = 0$, since $a = c - \gamma b$. Thus in some extension of K, $h(x)$ and $g(x)$ have $x - b$ as a common factor. We assert that $x - b$ is in fact their greatest common divisor. For, if $b_j \neq b$ is another root of $g(x)$, then $h(b_j) = f(c - \gamma b_j) \neq 0$, since by our choice of γ, $c - \gamma b_j$ for $j \neq 1$ avoids all roots a_i of $f(x)$. Also, since $(x - b)^2 \nmid g(x)$, $(x - b)^2$ cannot divide the greatest common divisor of $h(x)$ and $g(x)$. Thus $x - b$ is the greatest common divisor of $h(x)$ and $g(x)$ over some extension of K. But then they have a nontrivial greatest common divisor over K, which must be a divisor of $x - b$. Since the degree of $x - b$ is 1, we see that the greatest common divisor of $g(x)$ and $h(x)$ in $K[x]$ is exactly $x - b$. Thus $x - b \in K[x]$, whence $b \in K$; remembering that $K = F(c)$, we obtain that $b \in F(c)$. Since $a = c - \gamma b$, and since $b, c \in F(c)$, $\gamma \in F \subset F(c)$, we get that $a \in F(c)$, whence $F(a, b) \subset F(c)$. The two opposite containing relations combine to yield $F(a, b) = F(c)$.

A simple induction argument extends the result from 2 elements to any finite number, that is, if $\alpha_1, \ldots, \alpha_n$ are algebraic over F, then there is an element $c \in F(\alpha_1, \ldots, \alpha_n)$ such that $F(c) = F(\alpha_1, \ldots, \alpha_n)$. Thus the

COROLLARY *Any finite extension of a field of characteristic* 0 *is a simple extension.*

Problems

1. If F is of characteristic 0 and $f(x) \in F[x]$ is such that $f'(x) = 0$, prove that $f(x) = \alpha \in F$.
2. If F is of characteristic $p \neq 0$ and if $f(x) \in F[x]$ is such that $f'(x) = 0$, prove that $f(x) = g(x^p)$ for some polynomial $g(x) \in F[x]$.
3. Prove that $(f(x) + g(x))' = f'(x) + g'(x)$ and that $(\alpha f(x))' = \alpha f'(x)$ for $f(x), g(x) \in F[x]$ and $\alpha \in F$.
4. Prove that there is no rational function in $F(x)$ such that its square is x.
5. Complete the induction needed to establish the corollary to Theorem 5.5.1.

An element a in an extension K of F is called *separable over* F if it satisfies a polynomial over F having no multiple roots. An extension K of F is called *separable* over F if all its elements are separable over F. A field F is called *perfect* if all finite extensions of F are separable.

6. Show that any field of characteristic 0 is perfect.
7. (a) If F is of characteristic $p \neq 0$ show that for $a, b \in F$, $(a + b)^{p^m} = a^{p^m} + b^{p^m}$.
   (b) If F is of characteristic $p \neq 0$ and if K is an extension of F let $T = \{a \in K \mid a^{p^n} \in F \text{ for some } n\}$. Prove that T is a subfield of K.
8. If K, T, F are as in Problem 7(b) show that any automorphism of K leaving every element of F fixed also leaves every element of T fixed.

*9. Show that a field F of characteristic $p \neq 0$ is perfect if and only if for every $a \in F$ we can find a $b \in F$ such that $b^p = a$.

10. Using the result of Problem 9, prove that any finite field is perfect.

**11. If K is an extension of F prove that the set of elements in K which are separable over F forms a subfield of K.

12. If F is of characteristic $p \neq 0$ and if K is a finite extension of F, prove that given $a \in K$ either $a^{p^n} \in F$ for some n or we can find an integer m such that $a^{p^m} \notin F$ and is separable over F.

13. If K and F are as in Problem 12, and if no element which is in K but not in F is separable over F, prove that given $a \in K$ we can find an integer n, depending on a, such that $a^{p^n} \in F$.

14. If K is a finite, separable extension of F prove that K is a simple extension of F.

15. If one of a or b is separable over F, prove that $F(a, b)$ is a simple extension of F.

5.6 The Elements of Galois Theory

Given a polynomial $p(x)$ in $F[x]$, the polynomial ring in x over F, we shall associate with $p(x)$ a group, called the *Galois group* of $p(x)$. There is a very close relationship between the roots of a polynomial and its Galois group; in fact, the Galois group will turn out to be a certain permutation group of the roots of the polynomial. We shall make a study of these ideas in this, and in the next, section.

The means of introducing this group will be through the splitting field of $p(x)$ over F, the Galois group of $p(x)$ being defined as a certain group of automorphisms of this splitting field. This accounts for our concern, in so many of the theorems to come, with the automorphisms of a field. A beautiful duality, expressed in the fundamental theorem of the Galois theory (Theorem 5.6.6), exists between the subgroups of the Galois group and the subfields of the splitting field. From this we shall eventually derive a condition for the solvability by means of radicals of the roots of a polynomial in terms of the algebraic structure of its Galois group. From this will follow the classical result of Abel that the general polynomial of degree 5 is not solvable by radicals. Along the way we shall also derive, as side results, theorems of great interest in their own right. One such will be the fundamental theorem on symmetric functions. Our approach to the subject is founded on the treatment given it by Artin.

Recall that we are assuming that all our fields are of characteristic 0, hence we can (and shall) make free use of Theorem 5.5.1 and its corollary.

By an *automorphism of the field K* we shall mean, as usual, a mapping σ of K onto itself such that $\sigma(a + b) = \sigma(a) + \sigma(b)$ and $\sigma(ab) = \sigma(a)\sigma(b)$ for all $a, b \in K$. Two automorphisms σ and τ of K are said to be distinct if $\sigma(a) \neq \tau(a)$ for some element a in K.

We begin the material with

THEOREM 5.6.1 *If K is a field and if $\sigma_1, \ldots, \sigma_n$ are distinct automorphisms of K, then it is impossible to find elements $a_1, \ldots, a_n$, not all 0, in K such that $a_1\sigma_1(u) + a_2\sigma_2(u) + \cdots + a_n\sigma_n(u) = 0$ for all $u \in K$.*

Proof. Suppose we could find a set of elements $a_1, \ldots, a_n$ in K, not all 0, such that $a_1\sigma_1(u) + \cdots + a_n\sigma_n(u) = 0$ for all $u \in K$. Then we could find such a relation having as few nonzero terms as possible; on renumbering we can assume that this minimal relation is

$$a_1\sigma_1(u) + \cdots + a_m\sigma_m(u) = 0 \tag{1}$$

where $a_1, \ldots, a_m$ are all different from 0.

If m were equal to 1 then $a_1\sigma_1(u) = 0$ for all $u \in K$, leading to $a_1 = 0$, contrary to assumption. Thus we may assume that $m > 1$. Since the automorphisms are distinct there is an element $c \in K$ such that $\sigma_1(c) \neq \sigma_m(c)$. Since $cu \in K$ for all $u \in K$, relation (1) must also hold for cu, that is, $a_1\sigma_1(cu) + a_2\sigma_2(cu) + \cdots + a_m\sigma_m(cu) = 0$ for all $u \in K$. Using the hypothesis that the σ's are automorphisms of K, this relation becomes

$$a_1\sigma_1(c)\sigma_1(u) + a_2\sigma_2(c)\sigma_2(u) + \cdots + a_m\sigma_m(c)\sigma_m(u) = 0. \tag{2}$$

Multiplying relation (1) by $\sigma_1(c)$ and subtracting the result from (2) yields

$$a_2(\sigma_2(c) - \sigma_1(c))\sigma_2(u) + \cdots + a_m(\sigma_m(c) - \sigma_1(c))\sigma_m(u) = 0. \tag{3}$$

If we put $b_i = a_i(\sigma_i(c) - \sigma_1(c))$ for $i = 2, \ldots, m$, then the b_i are in K, $b_m = a_m(\sigma_m(c) - \sigma_1(c)) \neq 0$, since $a_m \neq 0$, and $\sigma_m(c) - \sigma_1(c) \neq 0$ yet $b_2\sigma_2(u) + \cdots + b_m\sigma_m(u) = 0$ for all $u \in K$. This produces a shorter relation, contrary to the choice made; thus the theorem is proved.

DEFINITION If G is a group of automorphisms of K, then the *fixed field* of G is the set of all elements $a \in K$ such that $\sigma(a) = a$ for all $\sigma \in G$.

Note that this definition makes perfectly good sense even if G is not a group but is merely a set of automorphisms of K. However, the fixed field of a set of automorphisms and that of the group of automorphisms generated by this set (in the group of all automorphisms of K) are equal (Problem 1), hence we lose nothing by defining the concept just for groups of automorphisms. Besides, we shall only be interested in the fixed fields of groups of automorphisms.

Having called the set, in the definition above, the fixed *field* of G, it would be nice if this terminology were accurate. That it is we see in

LEMMA 5.6.1 *The fixed field of G is a subfield of K.*

Proof. Let a, b be in the fixed field of G. Thus for all $\sigma \in G$, $\sigma(a) = a$ and $\sigma(b) = b$. But then $\sigma(a \pm b) = \sigma(a) \pm \sigma(b) = a \pm b$ and $\sigma(ab) = \sigma(a)\sigma(b) = ab$; hence $a \pm b$ and ab are again in the fixed field of G. If $b \neq 0$, then $\sigma(b^{-1}) = \sigma(b)^{-1} = b^{-1}$, hence b^{-1} also falls in the fixed field of G. Thus we have verified that the fixed field of G is indeed a subfield of K.

We shall be concerned with the automorphisms of a field which behave in a prescribed manner on a given subfield.

DEFINITION Let K be a field and let F be a subfield of K. Then the *group of automorphisms of K relative to F*, written $G(K, F)$, is the set of *all* automorphisms of K leaving every element of F fixed; that is, the automorphism σ of K is in $G(K, F)$ if and only if $\sigma(\alpha) = \alpha$ for every $\alpha \in F$.

It is not surprising, and is quite easy to prove

LEMMA 5.6.2 *$G(K, F)$ is a subgroup of the group of all automorphisms of K.*

We leave the proof of this lemma to the reader. One remark: K contains the field of rational numbers F_0, since K is of characteristic 0, and it is easy to see that the fixed field of any group of automorphisms of K, being a field, must contain F_0. Hence, every rational number is left fixed by every automorphism of K.

We pause to examine a few examples of the concepts just introduced.

Example 5.6.1 Let K be the field of complex numbers and let F be the field of real numbers. We compute $G(K, F)$. If σ is any automorphism of K, since $i^2 = -1$, $\sigma(i)^2 = \sigma(i^2) = \sigma(-1) = -1$, hence $\sigma(i) = \pm i$. If, in addition, σ leaves every real number fixed, then for any $a + bi$ where a, b are real, $\sigma(a + bi) = \sigma(a) + \sigma(b)\sigma(i) = a \pm bi$. Each of these possibilities, namely the mapping $\sigma_1(a + bi) = a + bi$ and $\sigma_2(a + bi) = a - bi$ defines an automorphism of K, σ_1 being the identity automorphism and σ_2 complex-conjugation. Thus $G(K, F)$ is a group of order 2.

What is the fixed field of $G(K, F)$? It certainly must contain F, but does it contain more? If $a + bi$ is in the fixed field of $G(K, F)$ then $a + bi = \sigma_2(a + bi) = a - bi$, whence $b = 0$ and $a = a + bi \in F$. In this case we see that the fixed field of $G(K, F)$ is precisely F itself.

Example 5.6.2 Let F_0 be the field of rational numbers and let $K = F_0(\sqrt[3]{2})$ where $\sqrt[3]{2}$ is the real cube root of 2. Every element in K is of the form $\alpha_0 + \alpha_1\sqrt[3]{2} + \alpha_2(\sqrt[3]{2})^2$, where $\alpha_0, \alpha_1, \alpha_2$ are rational numbers. If

σ is an automorphism of K, then $\sigma(\sqrt[3]{2})^3 = \sigma((\sqrt[3]{2})^3) = \sigma(2) = 2$, hence $\sigma(\sqrt[3]{2})$ must also be a cube root of 2 lying in K. However, there is only one real cube root of 2, and since K is a subfield of the real field, we must have that $\sigma(\sqrt[3]{2}) = \sqrt[3]{2}$. But then $\sigma(\alpha_0 + \alpha_1\sqrt[3]{2} + \alpha_2(\sqrt[3]{2})^2) = \alpha_0 + \alpha_1\sqrt[3]{2} + \alpha_2(\sqrt[3]{2})^2$, that is, σ is the identity automorphism of K. We thus see that $G(K, F_0)$ consists only of the identity map, and in this case the *fixed field of* $G(K, F_0)$ *is not* F_0 *but is, in fact, larger, being all of* K.

Example 5.6.3 Let F_0 be the field of rational numbers and let $\omega = e^{2\pi i/5}$; thus $\omega^5 = 1$ and ω satisfies the polynomial $x^4 + x^3 + x^2 + x + 1$ over F_0. By the Eisenstein criterion one can show that $x^4 + x^3 + x^2 + x + 1$ is irreducible over F_0 (see Problem 3). Thus $K = F_0(\omega)$ is of degree 4 over F_0 and every element in K is of the form $\alpha_0 + \alpha_1\omega + \alpha_2\omega^2 + \alpha_3\omega^3$ where all of $\alpha_0, \alpha_1, \alpha_2$, and α_3 are in F_0. Now, for any automorphism σ of K, $\sigma(\omega) \neq 1$, since $\sigma(1) = 1$, and $\sigma(\omega)^5 = \sigma(\omega^5) = \sigma(1) = 1$, whence $\sigma(\omega)$ is also a 5th root of unity. In consequence, $\sigma(\omega)$ can only be one of $\omega, \omega^2, \omega^3$, or ω^4. We claim that each of these possibilities actually occurs, for let us define the four mappings $\sigma_1, \sigma_2, \sigma_3$, and σ_4 by $\sigma_i(\alpha_0 + \alpha_1\omega + \alpha_2\omega^2 + \alpha_3\omega^3) = \alpha_0 + \alpha_1(\omega^i) + \alpha_2(\omega^i)^2 + \alpha_3(\omega^i)^3$, for $i = 1, 2, 3$, and 4. Each of these defines an automorphism of K (Problem 4). Therefore, since $\sigma \in G(K, F_0)$ is completely determined by $\sigma(\omega)$, $G(K, F_0)$ is a group of order 4, with σ_1 as its unit element. In light of $\sigma_2{}^2 = \sigma_4$, $\sigma_2{}^3 = \sigma_3$, $\sigma_2{}^4 = \sigma_1$, $G(K, F_0)$ is a cyclic group of order 4. One can easily prove that the fixed field of $G(K, F_0)$ is F_0 itself (Problem 5). The subgroup $A = \{\sigma_1, \sigma_4\}$ of $G(K, F_0)$ has as its fixed field the set of all elements $\alpha_0 + \alpha_2(\omega^2 + \omega^3)$, which is an extension of F_0 of degree 2.

The examples, although illustrative, are still too special, for note that in each of them $G(K, F)$ turned out to be a cyclic group. This is highly atypical for, in general, $G(K, F)$ need not even be abelian (see Theorem 5.6.3). However, despite their speciality, they do bring certain important things to light. For one thing they show that we must study the effect of the automorphisms on the roots of polynomials and, for another, they point out that F *need not* be equal to all of the fixed field of $G(K, F)$. The cases in which this does happen are highly desirable ones and are situations with which we shall soon spend much time and effort.

We now compute an important bound on the size of $G(K, F)$.

THEOREM 5.6.2 *If K is a finite extension of F, then $G(K, F)$ is a finite group and its order, $o(G(K, F))$ satisfies $o(G(K, F)) \leq [K:F]$.*

Proof. Let $[K:F] = n$ and suppose that $u_1, \ldots, u_n$ is a basis of K over F. Suppose we can find $n + 1$ distinct automorphisms $\sigma_1, \sigma_2, \ldots, \sigma_{n+1}$

in $G(K, F)$. By the corollary to Theorem 4.3.3 the system of n homogeneous linear equations in the $n + 1$ unknowns $x_1, \ldots, x_{n+1}$:

$$\begin{aligned}
\sigma_1(u_1)x_1 + \sigma_2(u_1)x_2 + \cdots + \sigma_{n+1}(u_1)x_{n+1} &= 0 \\
&\vdots \\
\sigma_1(u_i)x_1 + \sigma_2(u_i)x_2 + \cdots + \sigma_{n+1}(u_i)x_{n+1} &= 0 \\
&\vdots \\
\sigma_1(u_n)x_1 + \sigma_2(u_n)x_2 + \cdots + \sigma_{n+1}(u_n)x_{n+1} &= 0
\end{aligned}$$

has a nontrivial solution (not all 0) $x_1 = a_1, \ldots, x_{n+1} = a_{n+1}$ in K. Thus

$$a_1\sigma_1(u_i) + a_2\sigma_2(u_i) + \cdots + a_{n+1}\sigma_{n+1}(u_i) = 0 \tag{1}$$

for $i = 1, 2, \ldots, n$.

Since every element in F is left fixed by each σ_i and since an arbitrary element t in K is of the form $t = \alpha_1 u_1 + \cdots + \alpha_n u_n$ with $\alpha_1, \ldots, \alpha_n$ in F, then from the system of equations (1) we get $a_1\sigma_1(t) + \cdots + a_{n+1}\sigma_{n+1}(t) = 0$ for all $t \in K$. But this contradicts the result of Theorem 5.6.1. Thus Theorem 5.6.2 has been proved.

Theorem 5.6.2 is of central importance in the Galois theory. However, aside from its key role there, it serves us well in proving a classic result concerned with symmetric rational functions. This result on symmetric functions in its turn will play an important part in the Galois theory.

First a few remarks on the field of rational functions in n-variables over a field F. Let us recall that in Section 3.11 we defined the ring of polynomials in the n-variables $x_1, \ldots, x_n$ over F and from this defined the field of rational functions in $x_1, \ldots, x_n$, $F(x_1, \ldots, x_n)$, over F as the ring of all quotients of such polynomials.

Let S_n be the symmetric group of degree n considered to be acting on the set $[1, 2, \ldots, n]$; for $\sigma \in S_n$ and i an integer with $1 \leq i \leq n$, let $\sigma(i)$ be the image of i under σ. We can make S_n act on $F(x_1, \ldots, x_n)$ in the following natural way: for $\sigma \in S_n$ and $r(x_1, \ldots, x_n) \in F(x_1, \ldots, x_n)$, define the mapping which takes $r(x_1, \ldots, x_n)$ onto $r(x_{\sigma(1)}, \ldots, x_{\sigma(n)})$. We shall write this mapping of $F(x_1, \ldots, x_n)$ onto itself also as σ. It is obvious that these mappings define automorphisms of $F(x_1, \ldots, x_n)$. What is the fixed field of $F(x_1, \ldots, x_n)$ with respect to S_n? It consists of all rational functions $r(x_1, \ldots, x_n)$ such that $r(x_1, \ldots, x_n) = r(x_{\sigma(1)}, \ldots, x_{\sigma(n)})$ for all $\sigma \in S_n$. But these are precisely those elements in $F(x_1, \ldots, x_n)$ which are known as the *symmetric rational functions*. Being the fixed field of S_n they form a subfield of $F(x_1, \ldots, x_n)$, called the field of symmetric rational functions which we shall denote by S. We shall be concerned with three questions:

1. What is $[F(x_1, \ldots, x_n):S]$?
2. What is $G(F(x_1, \ldots, x_n), S)$?
3. Can we describe S in terms of some particularly easy extension of F?

We shall answer these three questions simultaneously.

We can explicitly produce in S some particularly simple functions constructed from $x_1, \ldots, x_n$ known as the *elementary symmetric functions* in $x_1, \ldots, x_n$. These are defined as follows:

$$a_1 = x_1 + x_2 + \cdots + x_n = \sum_{i=1}^{n} x_i$$

$$a_2 = \sum_{i<j} x_i x_j$$

$$a_3 = \sum_{i<j<k} x_i x_j x_k$$

$$\vdots$$

$$a_n = x_1 x_2 \cdots x_n.$$

That these are symmetric functions is left as an exercise. For $n = 2$, 3 and 4 we write them out explicitly below.

$n = 2$

$$a_1 = x_1 + x_2.$$

$$a_2 = x_1 x_2.$$

$n = 3$

$$a_1 = x_1 + x_2 + x_3.$$

$$a_2 = x_1 x_2 + x_1 x_3 + x_2 x_3.$$

$$a_3 = x_1 x_2 x_3.$$

$n = 4$

$$a_1 = x_1 + x_2 + x_3 + x_4.$$

$$a_2 = x_1 x_2 + x_1 x_3 + x_1 x_4 + x_2 x_3 + x_2 x_4 + x_3 x_4.$$

$$a_3 = x_1 x_2 x_3 + x_1 x_2 x_4 + x_1 x_3 x_4 + x_2 x_3 x_4.$$

$$a_4 = x_1 x_2 x_3 x_4.$$

Note that when $n = 2$, x_1 and x_2 are the roots of the polynomial $t^2 - a_1 t + a_2$, that when $n = 3$, x_1, x_2, and x_3 are roots of $t^3 - a_1 t^2 + a_2 t - a_3$ and that when $n = 4$, x_1, x_2, x_3, and x_4 are all roots of $t^4 - a_1 t^3 + a_2 t^2 - a_3 t + a_4$.

Since $a_1, \ldots, a_n$ are all in S, the field $F(a_1, \ldots, a_n)$ obtained by adjoining $a_1, \ldots, a_n$ to F must lie in S. Our objective is now twofold, namely, to prove

1. $[F(x_1, \ldots, x_n):S] = n!$.
2. $S = F(a_1, \ldots, a_n)$.

Since the group S_n is a group of automorphisms of $F(x_1, \ldots, x_n)$ leaving S fixed, $S_n \subset G(F(x_1, \ldots, x_n), S)$. Thus, by Theorem 5.6.2,

$[F(x_1, \ldots, x_n):S] \geq o(G(F(x_n, \ldots, x_n), S)) \geq o(S_n) = n!$. If we could show that $[F(x_1,\ldots,x_n):F(a_1,\ldots,a_n)] \leq n!$, well then, since $F(a_1,\ldots,a_n)$ is a subfield of S, we would have $n! \geq [F(x_1, \ldots, x_n):F(a_1, \ldots, a_n)] = [F(x_1, \ldots, x_n):S][S:F(a_1, \ldots, a_n)] \geq n!$. But then we would get that $[F(x_1, \ldots, x_n):S] = n!$, $[S:F(a_1, \ldots, a_n)] = 1$ and so $S = F(a_1, \ldots, a_n)$, and, finally, $S_n = G(F(x_1, \ldots, x_n), S)$ (this latter from the second sentence of this paragraph). These are precisely the conclusions we seek.

Thus we merely must prove that $[F(x_1,\ldots,x_n):F(a_1, \ldots, a_n)] \leq n!$. To see how this settles the whole affair, note that the polynomial $p(t) = t^n - a_1t^{n-1} + a_2t^{n-2} \cdots + (-1)^n a_n$, which has coefficients in $F(a_1,\ldots,a_n)$, factors over $F(x_1, \ldots, x_n)$ as $p(t) = (t - x_1)(t - x_2) \cdots (t - x_n)$. (This is in fact the origin of the elementary symmetric functions.) Thus $p(t)$, of degree n over $F(a_1, \ldots, a_n)$, splits as a product of linear factors over $F(x_1, \ldots, x_n)$. It cannot split over a proper subfield of $F(x_1, \ldots, x_n)$ which contains $F(a_1, \ldots, a_n)$ for this subfield would then have to contain both F and each of the roots of $p(t)$, namely, $x_1, x_2, \ldots, x_n$; but then this subfield would be all of $F(x_1, \ldots, x_n)$. *Thus we see that $F(x_1, \ldots, x_n)$ is the splitting field of the polynomial $p(t) = t^n - a_1t^{n-1} + \cdots + (-1)^n a_n$ over $F(a_1, \ldots, a_n)$.* Since $p(t)$ is of degree n, by Theorem 5.3.2 we get $[F(x_1, \ldots, x_n):F(a_1, \ldots, a_n)] \leq n!$. Thus all our claims are established. We summarize the whole discussion in the basic and important result

THEOREM 5.6.3 *Let F be a field and let $F(x_1, \ldots, x_n)$ be the field of rational functions in $x_1, \ldots, x_n$ over F. Suppose that S is the field of symmetric rational functions; then*

1. $[F(x_1, \ldots, x_n):S] = n!$.
2. $G(F(x_1, \ldots, x_n), S) = S_n$, *the symmetric group of degree n.*
3. *If $a_1, \ldots, a_n$ are the elementary symmetric functions in $x_1, \ldots, x_n$, then $S = F(a_1, a_2, \ldots, a_n)$.*
4. *$F(x_1, \ldots, x_n)$ is the splitting field over $F(a_1, \ldots, a_n) = S$ of the polynomial $t^n - a_1t^{n-1} + a_2t^{n-2} \cdots + (-1)^n a_n$.*

We mentioned earlier that given any integer n it is possible to construct a field and a polynomial of degree n over this field whose splitting field is of maximal possible degree, $n!$, over this field. Theorem 5.6.3 explicitly provides us with such an example for if we put $S = F(a_1, \ldots, a_n)$, the rational function field in n variables $a_1, \ldots, a_n$ and consider the splitting field of the polynomial $t^n - a_1t^{n-1} + a_2t^{n-2} \cdots + (-1)^n a_n$ over S then it is of degree $n!$ over S.

Part 3 of Theorem 5.6.3 is a very classical theorem. *It asserts that a symmetric rational function in n variables is a rational function in the elementary symmetric functions of these variables.* This result can even be sharpened to: A symmetric polynomial in n variables is a *polynomial* in their elementary symmetric functions (see Problem 7). This result is known as the *theorem on symmetric polynomials.*

In the examples we discussed of groups of automorphisms of fields and of fixed fields under such groups, we saw that it might very well happen that F is actually smaller than the whole fixed field of $G(K, F)$. Certainly F is always contained in this field but need not fill it out. Thus to impose the condition on an extension K of F that F be precisely the fixed field of $G(K, F)$ is a genuine limitation on the type of extension of F that we are considering. It is in this kind of extension that we shall be most interested.

DEFINITION K is a *normal extension* of F if K is a finite extension of F such that F is the fixed field of $G(K, F)$.

Another way of saying the same thing: If K is a normal extension of F, then every element in K which is outside F is moved by some element in $G(K, F)$. In the examples discussed, Examples 5.6.1 and 5.6.3 were normal extensions whereas Example 5.6.2 was not.

An immediate consequence of the assumption of normality is that it allows us to calculate with great accuracy the size of the fixed field of any subgroup of $G(K, F)$ and, in particular, to sharpen Theorem 5.6.2 from an inequality to an equality.

THEOREM 5.6.4 *Let K be a normal extension of F and let H be a subgroup of $G(K, F)$; let $K_H = \{x \in K \mid \sigma(x) = x \text{ for all } \sigma \in H\}$ be the fixed field of H. Then*

1. $[K:K_H] = o(H)$.
2. $H = G(K, K_H)$.

(In particular, when $H = G(K, F)$, $[K:F] = o(G(K, F))$.)

Proof. Since very element in H leaves K_H elementwise fixed, certainly $H \subset G(K, K_H)$. By Theorem 5.6.2 we know that $[K:K_H] \geq o(G(K, K_H))$; and since $o(G(K, K_H)) \geq o(H)$ we have the inequalities $[K:K_H] \geq o(G(K, K_H)) \geq o(H)$. If we could show that $[K:K_H] = o(H)$, it would immediately follow that $o(H) = o(G(K, K_H))$ and as a subgroup of $G(K, K_H)$ having order that of $G(K, K_H)$, we would obtain that $H = G(K, K_H)$. So we must merely show that $[K:K_H] = o(H)$ to prove everything.

By Theorem 5.5.1 there exists an $a \in K$ such that $K = K_H(a)$; this a must therefore satisfy an irreducible polynomial over K_H of degree $m = [K:K_H]$ and no nontrivial polynomial of lower degree (Theorem 5.1.3). Let the elements of H be $\sigma_1, \sigma_2, \ldots, \sigma_h$, where σ_1 is the identity of $G(K, F)$

and where $h = o(H)$. Consider the elementary symmetric functions of $a = \sigma_1(a), \sigma_2(a), \ldots, \sigma_h(a)$, namely,

$$\alpha_1 = \sigma_1(a) + \sigma_2(a) + \cdots + \sigma_h(a) = \sum_{i=1}^{h} \sigma_i(a)$$

$$\alpha_2 = \sum_{i<j} \sigma_i(a)\sigma_j(a)$$

$$\vdots$$

$$\alpha_h = \sigma_1(a)\sigma_2(a)\cdots\sigma_h(a).$$

Each α_i is invariant under every $\sigma \in H$. (Prove!) Thus, by the definition of K_H, $\alpha_1, \alpha_2, \ldots, \alpha_h$ are all elements of K_H. However, a (as well as $\sigma_2(a), \ldots, \sigma_h(a)$) is a root of the polynomial $p(x) = (x - \sigma_1(a))(x - \sigma_2(a))\cdots(x - \sigma_h(a)) = x^h - \alpha_1 x^{h-1} + \alpha_2 x^{h-2} + \cdots + (-1)^h \alpha_h$ having coefficients in K_H. By the nature of a, this forces $h \geq m = [K:K_H]$, whence $o(H) \geq [K:K_H]$. Since we already know that $o(H) \leq [K:K_H]$ we obtain $o(H) = [K:K_H]$, the desired conclusion.

When $H = G(K, F)$, *by the normality of K over F*, $K_H = F$; consequently for this particular case we read off the result $[K:F] = o(G(K, F))$.

We are rapidly nearing the central theorem of the Galois theory. What we still lack is the relationship between splitting fields and normal extensions. This gap is filled by

THEOREM 5.6.5 *K is a normal extension of F if and only if K is the splitting field of some polynomial over F.*

Proof. In one direction the proof will be highly reminiscent of that of Theorem 5.6.4.

Suppose that K is a normal extension of F; by Theorem 5.5.1, $K = F(a)$. Consider the polynomial $p(x) = (x - \sigma_1(a))(x - \sigma_2(a))\cdots(x - \sigma_n(a))$ over K, where $\sigma_1, \sigma_2, \ldots, \sigma_n$ are all the elements of $G(K, F)$. Expanding $p(x)$ we see that $p(x) = x^n - \alpha_1 x^{n-1} + \alpha_2 x^{n-2} + \cdots + (-1)^n \alpha_n$ where $\alpha_1, \ldots, \alpha_n$ are the elementary symmetric functions in $a = \sigma_1(a), \sigma_2(a), \ldots, \sigma_n(a)$. But then $\alpha_1, \ldots, \alpha_n$ are each invariant with respect to every $\sigma \in G(K, F)$, whence by the normality of K over F, must all be in F. Therefore, K splits the polynomial $p(x) \in F[x]$ into a product of linear factors. Since a is a root of $p(x)$ and since a generates K over F, a can be in no proper subfield of K which contains F. Thus K is the splitting field of $p(x)$ over F.

Now for the other direction; it is a little more complicated. We separate off one piece of its proof in

LEMMA 5.6.3 *Let K be the splitting field of $f(x)$ in $F[x]$ and let $p(x)$ be an*

irreducible factor of $f(x)$ in $F[x]$. If the roots of $p(x)$ are $\alpha_1, \ldots, \alpha_r$, then for each i there exists an automorphism σ_i in $G(K, F)$ such that $\sigma_i(\alpha_1) = \alpha_i$.

Proof. Since every root of $p(x)$ is a root of $f(x)$, it must lie in K. Let α_1, α_i be any two roots of $p(x)$. By Theorem 5.3.3, there is an isomorphism τ of $F_1 = F(\alpha_1)$ onto $F_1' = F(\alpha_i)$ taking α_1 onto α_i and leaving every element of F fixed. Now K is the splitting field of $f(x)$ considered as a polynomial over F_1; likewise, K is the splitting field of $f(x)$ considered as a polynomial over F_1'. By Theorem 5.3.4 there is an isomorphism σ_i of K onto K (thus an automorphism of K) coinciding with τ on F_1. But then $\sigma_i(\alpha_1) = \tau(\alpha_1) = \alpha_i$ and σ_i leaves every element of F fixed. This is, of course, exactly what Lemma 5.6.3 claims.

We return to the completion of the proof of Theorem 5.6.5. Assume that K is the splitting field of the polynomial $f(x)$ in $F[x]$. We want to show that K is normal over F. We proceed by induction on $[K:F]$, assuming that for any pair of fields K_1, F_1 of degree less than $[K:F]$ that whenever K_1 is the splitting field over F_1 of a polynomial in $F_1[x]$, then K_1 is normal over F_1.

If $f(x) \in F[x]$ splits into linear factors over F, then $K = F$, which is certainly a normal extension of F. So, assume that $f(x)$ has an irreducible factor $p(x) \in F[x]$ of degree $r > 1$. The r *distinct* roots $\alpha_1, \alpha_2, \ldots, \alpha_r$ of $p(x)$ all lie in K and K is the splitting field of $f(x)$ considered as a polynomial over $F(\alpha_1)$. Since

$$[K:F(\alpha_1)] = \frac{[K:F]}{[F(\alpha_1):F]} = \frac{n}{r} < n,$$

by our induction hypothesis K is a normal extension of $F(\alpha_1)$.

Let $\theta \in K$ be left fixed by every automorphism $\sigma \in G(K, F)$; we would like to show that θ is in F. Now, any automorphism in $G(K, F(\alpha_1))$ certainly leaves F fixed, hence leaves θ fixed; by the normality of K over $F(\alpha_1)$, this implies that θ is in $F(\alpha_1)$. Thus

$$\theta = \lambda_0 + \lambda_1\alpha_1 + \lambda_2\alpha_1^{\,2} + \cdots + \lambda_{r-1}\alpha_1^{\,r-1} \quad \text{where } \lambda_0, \ldots, \lambda_{r-1} \in F. \tag{1}$$

By Lemma 5.6.3 there is an automorphism σ_i of K, $\sigma_i \in G(K, F)$, such that $\sigma_i(\alpha_1) = \alpha_i$; since this σ_i leaves θ and each λ_j fixed, applying it to (1) we obtain

$$\theta = \lambda_0 + \lambda_1\alpha_i + \lambda_2\alpha_i^{\,2} + \cdots + \lambda_{r-1}\alpha_i^{\,r-1} \quad \text{for } i = 1, 2, \ldots, r. \tag{2}$$

Thus the polynomial

$$q(x) = \lambda_{r-1}x^{r-1} + \lambda_{r-2}x^{r-2} + \cdots + \lambda_1 x + (\lambda_0 - \theta)$$

in $K[x]$, of degree at most $r - 1$, has the r distinct roots $\alpha_1, \alpha_2, \ldots, \alpha_r$.

This can only happen if all its coefficients are 0; in particular, $\lambda_0 - \theta = 0$ whence $\theta = \lambda_0$ so is in F. This completes the induction and proves that K is a normal extension of F. Theorem 5.6.5 is now completely proved.

DEFINITION Let $f(x)$ be a polynomial in $F[x]$ and let K be its splitting field over F. The *Galois group* of $f(x)$ is the group $G(K, F)$ of all the automorphisms of K, leaving every element of F fixed.

Note that the Galois group of $f(x)$ can be considered as a group of permutations of its roots, for if α is a root of $f(x)$ and if $\sigma \in G(K, F)$, then $\sigma(\alpha)$ is also a root of $f(x)$.

We now come to the result known as the *fundamental theorem of Galois theory*. It sets up a one-to-one correspondence between the subfields of the splitting field of $f(x)$ and the subgroups of its Galois group. Moreover, it gives a criterion that a subfield of a normal extension itself be a normal extension of F. This fundamental theorem will be used in the next section to derive conditions for the solvability by radicals of the roots of a polynomial.

THEOREM 5.6.6 *Let $f(x)$ be a polynomial in $F[x]$, K its splitting field over F, and $G(K, F)$ its Galois group. For any subfield T of K which contains F let $G(K, T) = \{\sigma \in G(K, F) \mid \sigma(t) = t \text{ for every } t \in T\}$ and for any subgroup H of $G(K, F)$ let $K_H = \{x \in K \mid \sigma(x) = x \text{ for every } \sigma \in H\}$. Then the association of T with $G(K, T)$ sets up a one-to-one correspondence of the set of subfields of K which contain F onto the set of subgroups of $G(K, F)$ such that*

1. $T = K_{G(K,T)}$.
2. $H = G(K, K_H)$.
3. $[K:T] = o(G(K, T))$, $[T:F]$ = *index of* $G(K, T)$ *in* $G(K, F)$.
4. *T is a normal extension of F if and only if $G(K, T)$ is a normal subgroup of $G(K, F)$.*
5. *When T is a normal extension of F, then $G(T, F)$ is isomorphic to $G(K, F)/G(K, T)$.*

Proof. Since K is the splitting field of $f(x)$ over F it is also the splitting field of $f(x)$ over any subfield T which contains F, therefore, by Theorem 5.6.5, K is a normal extension of T. Thus, by the definition of normality, T is the fixed field of $G(K, T)$, that is, $T = K_{G(K,T)}$, proving part 1.

Since K is a normal extension of F, by Theorem 5.6.4, given a subgroup H of $G(K, F)$, then $H = G(K, K_H)$, which is the assertion of part 2. Moreover, this shows that any subgroup of $G(K, F)$ arises in the form $G(K, T)$, whence the association of T with $G(K, T)$ maps the set of all subfields of K containing F *onto* the set of all subgroups of $G(K, F)$. That it is one-to-one is clear, for, if $G(K, T_1) = G(K, T_2)$ then, by part 1, $T_1 = K_{G(K,T_1)} = K_{G(K,T_2)} = T_2$.

Since K is normal over T, again using Theorem 5.6.4, $[K:T] = o(G(K, T))$; but then we have $o(G(K, F)) = [K:F] = [K:T][T:F] = o(G(K, T))[T:F]$, whence

$$[T:F] = \frac{o(G(K, F))}{o(G(K, T))} = \text{index of } G(K, T)$$

in $G(K, F)$. This is part 3.

The only parts which remain to be proved are those which pertain to normality. We first make the following observation. T is a normal extension of F if and only if for every $\sigma \in G(K, F)$, $\sigma(T) \subset T$. Why? We know by Theorem 5.5.1 that $T = F(a)$; thus if $\sigma(T) \subset T$, then $\sigma(a) \in T$ for all $\sigma \in G(K, F)$. But, as we saw in the proof of Theorem 5.6.5, this implies that T is the splitting field of

$$p(x) = \prod_{\sigma \in G(K,F)} (x - \sigma(a))$$

which has coefficients in F. As a splitting field, T, by Theorem 5.6.5, is a normal extension of F. Conversely, if T is a normal extension of F, then $T = F(a)$, where the minimal polynomial of a, $p(x)$, over F has all its roots in T (Theorem 5.6.5). However, for any $\sigma \in G(K, F)$, $\sigma(a)$ is also a root of $p(x)$, whence $\sigma(a)$ must be in T. Since T is generated by a over F, we get that $\sigma(T) \subset T$ for every $\sigma \in G(K, F)$.

Thus T is a normal extension of F if and only if for any $\sigma \in G(K, F)$, $\tau \in G(K, T)$ and $t \in T$, $\sigma(t) \in T$ and so $\tau(\sigma(t)) = \sigma(t)$; that is, if and only if $\sigma^{-1}\tau\sigma(t) = t$. But this says that T is normal over F if and only if $\sigma^{-1}G(K, T)\sigma \subset G(K, T)$ for every $\sigma \in G(K, F)$. This last condition being precisely that which defines $G(K, T)$ as a normal subgroup of $G(K, F)$, we see that part 4 is proved.

Finally, if T is normal over F, given $\sigma \in G(K, F)$, since $\sigma(T) \subset T$, σ induces an automorphism σ_* of T defined by $\sigma_*(t) = \sigma(t)$ for every $t \in T$. Because σ_* leaves every element of F fixed, σ_* must be in $G(T, F)$. Also, as is evident, for any $\sigma, \psi \in G(K, F)$, $(\sigma\psi)_* = \sigma_*\psi_*$ whence the mapping of $G(K, F)$ into $G(T, F)$ defined by $\sigma \to \sigma_*$ is a homomorphism of $G(K, F)$ into $G(T, F)$. What is the kernel of this homomorphism? It consists of all elements σ in $G(K, F)$ such that σ_* is the identity map on T. That is, the kernel is the set of all $\sigma \in G(K, F)$ such that $t = \sigma_*(t) = \sigma(t)$; by the very definition, we get that the kernel is exactly $G(K, T)$. The image of $G(K, F)$ in $G(T, F)$, by Theorem 2.7.1 is isomorphic to $G(K, F)/G(K, T)$, whose order is $o(G(K, F))/o(G(K, T)) = [T:F]$ (by part 3) $= o(G(T, F))$ (by Theorem 5.6.4). Thus the image of $G(K, F)$ in $G(T, F)$ is all of $G(T, F)$ and so we have $G(T, F)$ isomorphic to

$G(K, F)/G(K, T)$. This finishes the proof of part 5 and thereby completes the proof of Theorem 5.6.6.

Problems

1. If K is a field and S a set of automorphisms of K, prove that the fixed field of S and that of $\bar{S}$ (the subgroup of the group of all automorphisms of K generated by S) are identical.
2. Prove Lemma 5.6.2.
3. Using the Eisenstein criterion, prove that $x^4 + x^3 + x^2 + x + 1$ is irreducible over the field of rational numbers.
4. In Example 5.6.3, prove that each mapping σ_i defined is an automorphism of $F_0(\omega)$.
5. In Example 5.6.3, prove that the fixed field of $F_0(\omega)$ under σ_1, σ_2, σ_3, σ_4 is precisely F_0.
6. Prove directly that any automorphism of K must leave every rational number fixed.

*7. Prove that a symmetric polynomial in $x_1, \ldots, x_n$ is a polynomial in the elementary symmetric functions in $x_1, \ldots, x_n$.

8. Express the following as polynomials in the elementary symmetric functions in x_1, x_2, x_3:
   (a) $x_1^2 + x_2^2 + x_3^2$.
   (b) $x_1^3 + x_2^3 + x_3^3$.
   (c) $(x_1 - x_2)^2(x_1 - x_3)^2(x_2 - x_3)^2$.
9. If $\alpha_1, \alpha_2, \alpha_3$ are the roots of the cubic polynomial $x^3 + 7x^2 - 8x + 3$, find the cubic polynomial whose roots are

   (a) $\alpha_1^2, \alpha_2^2, \alpha_3^2$. (b) $\dfrac{1}{\alpha_1}, \dfrac{1}{\alpha_2}, \dfrac{1}{\alpha_3}$. (c) $\alpha_1^3, \alpha_2^3, \alpha_3^3$.

*10. Prove *Newton's identities*, namely, if $\alpha_1, \alpha_2, \ldots, \alpha_n$ are the roots of $f(x) = x^n + a_1 x^{n-1} + a_2 x^{n-2} + \cdots + a_n$ and if $s_k = \alpha_1^k + \alpha_2^k + \cdots + \alpha_n^k$ then
   (a) $s_k + a_1 s_{k-1} + a_2 s_{k-2} + \cdots + a_{k-1} s_1 + k a_k = 0$ if $k = 1, 2, \ldots, n$.
   (b) $s_k + a_1 s_{k-1} + \cdots + a_n s_{k-n} = 0$ for $k > n$.
   (c) For $n = 5$, apply part (a) to determine s_2, s_3, s_4, and s_5.

11. Prove that the elementary symmetric functions in $x_1, \ldots, x_n$ are indeed symmetric functions in $x_1, \ldots, x_n$.
12. If $p(x) = x^n - 1$ prove that the Galois group of $p(x)$ over the field of rational numbers is abelian.

The complex number ω is a *primitive nth root of unity* if $\omega^n = 1$ but $\omega^m \neq 1$ for $0 < m < n$. F_0 will denote the field of rational numbers.

13. (a) Prove that there are $\phi(n)$ primitive nth roots of unity where $\phi(n)$ is the Euler ϕ-function.
    (b) If ω is a primitive nth root of unity prove that $F_0(\omega)$ is the splitting field of $x^n - 1$ over F_0 (and so is a normal extension of F_0).
    (c) If $\omega_1, \ldots, \omega_{\phi(n)}$ are the $\phi(n)$ primitive nth roots of unity, prove that any automorphism of $F_0(\omega_1)$ takes ω_1 into some ω_i.
    (d) Prove that $[F_0(\omega_1):F_0] \leq \phi(n)$.

14. The notation is as in Problem 13.
    *(a) Prove that there is an automorphism σ_i of $F_0(\omega_1)$ which takes ω_1 into ω_i.
    (b) Prove the polynomial $p_n(x) = (x - \omega_1)(x - \omega_2) \cdots (x - \omega_{\phi(n)})$ has rational coefficients. (The polynomial $p_n(x)$ is called the *nth cyclotomic polynomial.*)
    *(c) Prove that, in fact, the coefficients of $p_n(x)$ are integers.

**15. Use the results of Problems 13 and 14 to prove that $p_n(x)$ is irreducible over F_0 for all $n \geq 1$. (See Problem 8, Section 3.)

16. For $n = 3, 4, 6$, and 8, calculate $p_n(x)$ explicitly, show that it has integer coefficients and prove directly that it is irreducible over F_0.

17. (a) Prove that the Galois group of $x^3 - 2$ over F_0 is isomorphic to S_3, the symmetric group of degree 3.
    (b) Find the splitting field, K, of $x^3 - 2$ over F_0.
    (c) For every subgroup H of S_3 find K_H and check the correspondence given in Theorem 5.6.6.
    (d) Find a normal extension in K of degree 2 over F_0.

18. If the field F contains a primitive nth root of unity, prove that the Galois group of $x^n - a$, for $a \in F$, is abelian.

5.7 Solvability by Radicals

Given the specific polynomial $x^2 + 3x + 4$ over the field of rational numbers F_0, from the quadratic formula for its roots we know that its roots are $(-3 \pm \sqrt{-7})/2$; thus the field $F_0(\sqrt{7}\, i)$ is the splitting field of $x^2 + 3x + 4$ over F_0. Consequently there is an element $\gamma = -7$ in F_0 such that the extension field $F_0(\omega)$ where $\omega^2 = \gamma$ is such that it contains all the roots of $x^2 + 3x + 4$.

From a slightly different point of view, given the *general* quadratic polynomial $p(x) = x^2 + a_1 x + a_2$ over F, we can consider it as a *particular* polynomial over the field $F(a_1, a_2)$ of rational functions in the two variables a_1 and a_2 over F; in the extension obtained by adjoining ω to $F(a_1, a_2)$ where $\omega^2 = a_1{}^2 - 4a_2 \in F(a_1, a_2)$, we find all the roots of $p(x)$. There is a formula which expresses the roots of $p(x)$ in terms of a_1, a_2 and square roots of rational functions of these.

For a cubic equation the situation is very similar; given the general cubic equation $p(x) = x^3 + a_1x^2 + a_2x + a_3$ an explicit formula can be given, involving combinations of square roots and cube roots of rational functions in a_1, a_2, a_3. While somewhat messy, they are explicitly given by *Cardan's formulas:* Let $p = a_2 - (a_1{}^2/3)$ and

$$q = \frac{2a_1{}^3}{27} - \frac{a_1a_2}{3} + a_3$$

and let

$$P = \sqrt[3]{-\frac{q}{2} + \sqrt{\frac{p^3}{27} + \frac{q^2}{4}}}$$

and

$$Q = \sqrt[3]{-\frac{q}{2} - \sqrt{\frac{p^3}{27} + \frac{q^2}{4}}}$$

(with cube roots chosen properly); then the roots are $P + Q - (a_1/3)$, $\omega P + \omega^2 Q - (a_1/3)$, and $\omega^2 P + \omega Q - (a_1/3)$, where $\omega \neq 1$ is a cube root of 1. The above formulas only serve to illustrate for us that by adjoining a certain square root and then a cube root to $F(a_1, a_2, a_3)$ we reach a field in which $p(x)$ has its roots.

For fourth-degree polynomials, which we shall not give explicitly, by using rational operations and square roots, we can reduce the problem to that of solving a certain cubic, so here too a formula can be given expressing the roots in terms of combinations of radicals (surds) of rational functions of the coefficients.

For polynomials of degree five and higher, no such universal radical formula can be given, for we shall prove that it is impossible to express their roots, in general, in this way.

Given a field F and a polynomial $p(x) \in F[x]$, we say that *$p(x)$ is solvable by radicals over F* if we can find a finite sentence of fields $F_1 = F(\omega_1)$, $F_2 = F_1(\omega_2), \ldots, F_k = F_{k-1}(\omega_k)$ such that $\omega_1{}^{r_1} \in F$, $\omega_2{}^{r_2} \in F_1, \ldots$, $\omega_k{}^{r_k} \in F_{k-1}$ such that the roots of $p(x)$ all lie in F_k.

If K is the splitting field of $p(x)$ over F, then $p(x)$ is solvable by radicals over F if we can find a sequence of fields as above such that $K \subset F_k$. An important remark, and one we shall use later, in the proof of Theorem 5.7.2, is that if such an F_k can be found, we can, without loss of generality, assume it to be a *normal* extension of F; we leave its proof as a problem (Problem 1).

By the *general polynomial of degree n over F*, $p(x) = x^n + a_1x^{n-1} + \cdots + a_n$, we mean the following: Let $F(a_1, \ldots, a_n)$ be the field of rational functions, in the n variables $a_1, \ldots, a_n$ over F, and consider the particular polynomial $p(x) = x^n + a_1x^{n-1} + \cdots + a_n$ over the field $F(a_1, \ldots, a_n)$. We say that it is solvable by radicals if it is solvable by radicals over $F(a_1, \ldots, a_n)$. This really expresses the intuitive idea of "finding a formula" for the roots of $p(x)$ involving combinations of mth roots, for various m's, of rational functions in $a_1, a_2, \ldots, a_n$. For $n = 2, 3$, and 4, we pointed out that this can always be done. For $n \geq 5$, Abel proved that this cannot be done. However, this does not exclude the possibility that a given polynomial over F may be solvable by radicals. In fact, we shall give a criterion for this in terms of the Galois group of the polynomial. But first we must develop a few purely group-theoretical results. Some of these occurred as problems at the end of Chapter 2, but we nevertheless do them now officially.

DEFINITION A group G is said to be *solvable* if we can find a finite chain of subgroups $G = N_0 \supset N_1 \supset N_2 \supset \cdots \supset N_k = (e)$, where each N_i is a normal subgroup of N_{i-1} and such that every factor group N_{i-1}/N_i is abelian.

Every abelian group is solvable, for merely take $N_0 = G$ and $N_1 = (e)$ to satisfy the above definition. The symmetric group of degree 3, S_3, is solvable for take $N_1 = \{e, (1, 2, 3), (1, 3, 2)\}$; N_1 is a normal subgroup of S_3 and S_3/N_1 and $N_1/(e)$ are both abelian being of orders 2 and 3, respectively. It can be shown that S_4 is solvable (Problem 3). For $n \geq 5$ we show in Theorem 5.7.1 below that S_n is *not* solvable.

We seek an alternative description for solvability. Given the group G and elements a, b in G, then the *commutator* of a and b is the element $a^{-1}b^{-1}ab$. The *commutator subgroup*, G', of G is the subgroup of G generated by all the commutators in G. (It is *not* necessarily true that the set of commutators itself forms a subgroup of G.) It was an exercise before that G' is a normal subgroup of G. Moreover, the group G/G' is abelian, for, given any two elements in it, aG', bG', with $a, b \in G$, then

$$\begin{aligned}(aG')(bG') &= abG' = ba(a^{-1}b^{-1}ab)G' \\ &= (\text{since } a^{-1}b^{-1}ab \in G')\ baG' = (bG')(aG').\end{aligned}$$

On the other hand, if M is a normal subgroup of G such that G/M is abelian, then $M \supset G'$, for, given $a, b \in G$, then $(aM)(bM) = (bM)(aM)$, from which we deduce $abM = baM$ whence $a^{-1}b^{-1}abM = M$ and so $a^{-1}b^{-1}ab \in M$. Since M contains all commutators, it contains the group these generate, namely G'.

G' is a group in its own right, so we can speak of its commutator subgroup $G^{(2)} = (G')'$. This is the subgroup of G generated by all elements $(a')^{-1}(b')^{-1}a'b'$ where $a', b' \in G'$. It is easy to prove that not only is $G^{(2)}$ a normal subgroup of G' but it is also a normal subgroup of G (Problem 4).

We continue this way and define the higher commutator subgroups $G^{(m)}$ by $G^{(m)} = (G^{(m-1)})'$. Each $G^{(m)}$ is a normal subgroup of G (Problem 4) and $G^{(m-1)}/G^{(m)}$ is an abelian group.

In terms of these higher commutator subgroups of G, we have a very succinct criterion for solvability, namely,

LEMMA 5.7.1 *G is solvable if and only if $G^{(k)} = (e)$ for some integer k.*

Proof. If $G^{(k)} = (e)$ let $N_0 = G$, $N_1 = G'$, $N_2 = G^{(2)}, \ldots, N_k = G^{(k)} = (e)$. We have

$$G = N_0 \supset N_1 \supset N_2 \supset \cdots \supset N_k = (e);$$

each N_i being normal in G is certainly normal in N_{i-1}. Finally,

$$\frac{N_{i-1}}{N_i} = \frac{G^{(i-1)}}{G^{(i)}} = \frac{G^{(i-1)}}{(G^{(i-1)})'}$$

hence is abelian. Thus by the definition of solvability G is a solvable group.

Conversely, if G is a solvable group, there is a chain $G = N_0 \supset N_1 \supset N_2 \supset \cdots \supset N_k = (e)$ where each N_i is normal in N_{i-1} and where N_{i-1}/N_i is abelian. But then the commutator subgroup N'_{i-1} of N_{i-1} must be contained in N_i. Thus $N_1 \supset N'_0 = G'$, $N_2 \supset N'_1 \supset (G')' = G^{(2)}$, $N_3 \supset N'_2 \supset (G^{(2)})' = G^{(3)}, \ldots, N_i \supset G^{(i)}$, $(e) = N_k \supset G^{(k)}$. We therefore obtain that $G^{(k)} = (e)$.

COROLLARY *If G is a solvable group and if $\bar{G}$ is a homomorphic image of G, then $\bar{G}$ is solvable.*

Proof. Since $\bar{G}$ is a homomorphic image of G it is immediate that $(\bar{G})^{(k)}$ is the image of $G^{(k)}$. Since $G^{(k)} = (e)$ for some k, $(\bar{G})^{(k)} = (e)$ for the same k, whence by the lemma $\bar{G}$ is solvable.

The next lemma is the key step in proving that the infinite family of groups S_n, with $n \geq 5$, is not solvable; here S_n is the symmetric group of degree n.

LEMMA 5.7.2 *Let $G = S_n$, where $n \geq 5$; then $G^{(k)}$ for $k = 1, 2, \ldots,$ contains every 3-cycle of S_n.*

Proof. We first remark that for an arbitrary group G, if N is a normal subgroup of G, then N' must also be a normal subgroup of G (Problem 5).

We claim that if N is a normal subgroup of $G = S_n$, where $n \geq 5$, which contains every 3-cycle in S_n, then N' must also contain every 3-cycle. For suppose $a = (1, 2, 3)$, $b = (1, 4, 5)$ are in N (we are using here that $n \geq 5$); then $a^{-1}b^{-1}ab = (3, 2, 1)(5, 4, 1)(1, 2, 3)(1, 4, 5) = (1, 4, 2)$, as a commutator of elements of N must be in N'. Since N' is a normal subgroup of G, for any $\pi \in S_n$, $\pi^{-1}(1, 4, 2)\pi$ must also be in N'. Choose a π in S_n such that $\pi(1) = i_1$, $\pi(4) = i_2$, and $\pi(2) = i_3$, where i_1, i_2, i_3 are any three distinct integers in the range from 1 to n; then $\pi^{-1}(1, 4, 2)\pi = (i_1, i_2, i_3)$ is in N'. Thus N' contains all 3-cycles.

Letting $N = G$, which is certainly normal in G and contains all 3-cycles, we get that G' contains all 3-cycles; since G' is normal in G, $G^{(2)}$ contains all 3-cycles; since $G^{(2)}$ is normal in G, $G^{(3)}$ contains all 3-cycles. Continuing this way we obtain that $G^{(k)}$ contains all 3-cycles for arbitrary k.

A direct consequence of this lemma is the interesting group-theoretic result.

THEOREM 5.7.1 *S_n is not solvable for $n \geq 5$.*

Proof. If $G = S_n$, by Lemma 5.7.2, $G^{(k)}$ contains all 3-cycles in S_n for every k. Therefore, $G^{(k)} \neq (e)$ for any k, whence by Lemma 5.7.1, G cannot be solvable.

We now interrelate the solvability by radicals of $p(x)$ with the solvability, as a group, of the Galois group of $p(x)$. The very terminology is highly suggestive that such a relation exists. But first we need a result about the Galois group of a certain type of polynomial.

LEMMA 5.7.3 *Suppose that the field F has all nth roots of unity (for some particular n) and suppose that $a \neq 0$ is in F. Let $x^n - a \in F[x]$ and let K be its splitting field over F. Then*

1. *$K = F(u)$ where u is any root of $x^n - a$.*
2. *The Galois group of $x^n - a$ over F is abelian.*

Proof. Since F contains all nth roots of unity, it contains $\xi = e^{2\pi i/n}$; note that $\xi^n = 1$ but $\xi^m \neq 1$ for $0 < m < n$.

If $u \in K$ is any root of $x^n - a$, then $u, \xi u, \xi^2 u, \ldots, \xi^{n-1}u$ are all the roots of $x^n - a$. That they are roots is clear; that they are distinct follows from: $\xi^i u = \xi^j u$ with $0 \leq i < j < n$, then since $u \neq 0$, and $(\xi^i - \xi^j)u = 0$, we must have $\xi^i = \xi^j$, which is impossible since $\xi^{j-i} = 1$, with $0 < j - i < n$. Since $\xi \in F$, all of $u, \xi u, \ldots, \xi^{n-1}u$ are in $F(u)$, thus $F(u)$ splits $x^n - a$; since no proper subfield of $F(u)$ which contains F also contains u, no proper subfield of $F(u)$ can split $x^n - a$. Thus $F(u)$ is the splitting field of $x^n - a$, and we have proved that $K = F(u)$.

If σ, τ are any two elements in the Galois group of $x^n - a$, that is, if σ, τ are automorphisms of $K = F(u)$ leaving every element of F fixed, then since both $\sigma(u)$ and $\tau(u)$ are roots of $x^n - a$, $\sigma(u) = \xi^i u$ and $\tau(u) = \xi^j u$ for some i and j. Thus $\sigma\tau(u) = \sigma(\xi^j u) = \xi^j \sigma(u)$ (since $\xi^j \in F$) $= \xi^i \xi^j u = \xi^{i+j}u$; similarly, $\tau\sigma(u) = \xi^{i+j}u$. Therefore, $\sigma\tau$ and $\tau\sigma$ agree on u and on

F hence on all of $K = F(u)$. But then $\sigma\tau = \tau\sigma$, whence the Galois group is abelian.

Note that the lemma says that when F has all nth roots of unity, then adjoining one root of $x^n - a$ to F, where $a \in F$, gives us the whole splitting field of $x^n - a$; thus this must be a normal extension of F.

We assume for the rest of the section that F is a field which contains all nth roots of unity for every integer n. We have

THEOREM 5.7.2 *If $p(x) \in F[x]$ is solvable by radicals over F, then the Galois group over F of $p(x)$ is a solvable group.*

Proof. Let K be the splitting field of $p(x)$ over F; the Galois group of $p(x)$ over F is $G(K, F)$. Since $p(x)$ is solvable by radicals, there exists a sequence of fields

$$F \subset F_1 = F(\omega_1) \subset F_2 = F_1(\omega_2) \subset \cdots \subset F_k = F_{k-1}(\omega_k),$$

where $\omega_1^{r_1} \in F$, $\omega_2^{r_2} \in F_1, \ldots, \omega_k^{r_k} \in F_{k-1}$ and where $K \subset F_k$. As we pointed out, without loss of generality we may assume that F_k is a normal extension of F. As a normal extension of F, F_k is also a normal extension of any intermediate field, hence F_k is a normal extension of each F_i.

By Lemma 5.7.3 each F_i is a normal extension of F_{i-1} and since F_k is normal over F_{i-1}, by Theorem 5.6.6, $G(F_k, F_i)$ is a normal subgroup in $G(F_k, F_{i-1})$. Consider the chain

$$G(F_k, F) \supset G(F_k, F_1) \supset G(F_k, F_2) \supset \cdots \supset G(F_k, F_{k-1}) \supset (e). \quad (1)$$

As we just remarked, each subgroup in this chain is a normal subgroup in the one preceding it. Since F_i is a normal extension of F_{i-1}, by the fundamental theorem of Galois theory (Theorem 5.6.6) the group of F_i over F_{i-1}, $G(F_i, F_{i-1})$ is isomorphic to $G(F_k, F_{i-1})/G(F_k, F_i)$. However, by Lemma 5.7.3, $G(F_i, F_{i-1})$ is an abelian group. Thus each quotient group $G(F_k, F_{i-1})/G(F_k, F_i)$ of the chain (1) is abelian.

Thus the group $G(F_k, F)$ is solvable! Since $K \subset F_k$ and is a normal extension of F (being a splitting field), by Theorem 5.6.6, $G(F_k, K)$ is a normal subgroup of $G(F_k, F)$ and $G(K, F)$ is isomorphic to $G(F_k, F)/G(F_k, K)$. Thus $G(K, F)$ is a homomorphic image of $G(F_k, F)$, a solvable group; by the corollary to Lemma 5.7.1, $G(K, F)$ itself must then be a solvable group. Since $G(K, F)$ is the Galois group of $p(x)$ over F the theorem has been proved.

We make two remarks without proof.

1. The converse of Theorem 5.7.2 is also true; that is, if the Galois group of $p(x)$ over F is solvable then $p(x)$ is solvable by radicals over F.

2. Theorem 5.7.2 and its converse are true even if F does not contain roots of unity.

Recalling what is meant by the general polynomial of degree n over F, $p(x) = x^n + a_1x^{n-1} + \cdots + a_n$, and what is meant by solvable by radicals, we close with the great, classic theorem of Abel:

THEOREM 5.7.3 *The general polynomial of degree $n \geq 5$ is not solvable by radicals.*

Proof. In Theorem 5.6.3 we saw that if $F(a_1, \ldots, a_n)$ is the field of rational functions in the n variables $a_1, \ldots, a_n$, then the Galois group of the polynomial $p(t) = t^n + a_1t^{n-1} + \cdots + a_n$ over $F(a_1, \ldots, a_n)$ was S_n, the symmetric group of degree n. By Theorem 5.7.1, S_n is not a solvable group when $n \geq 5$, thus by Theorem 5.7.2, $p(t)$ is not solvable by radicals over $F(a_1, \ldots, a_n)$ when $n \geq 5$.

Problems

*1. If $p(x)$ is solvable by radicals over F, prove that we can find a sequence of fields

$$F \subset F_1 = F(\omega_1) \subset F_2 = F_1(\omega_2) \subset \cdots \subset F_k = F_{k-1}(\omega_k),$$

where $\omega_1^{r_1} \in F$, $\omega_2^{r_2} \in F_1, \ldots, \omega_k^{r_k} \in F_{k-1}$, F_k containing all the roots of $p(x)$, such that F_k is *normal* over F.

2. Prove that a subgroup of a solvable group is solvable.
3. Prove that S_4 is a solvable group.
4. If G is a group, prove that all $G^{(k)}$ are normal subgroups of G.
5. If N is a normal subgroup of G prove that N' must also be a normal subgroup of G.
6. Prove that the alternating group (the group of even permutations in S_n) A_n has no nontrivial normal subgroups for $n \geq 5$.

5.8 Galois Groups over the Rationals

In Theorem 5.3.2 we saw that, given a field F and a polynomial $p(x)$, of degree n, in $F[x]$, then the splitting field of $p(x)$ over F has degree at most $n!$ over F. In the preceding section we saw that this upper limit of $n!$ is, indeed, taken on for some choice of F and some polynomial $p(x)$ of degree n over F. In fact, if F_0 is any field and if F is the field of rational functions in the variables $a_1, \ldots, a_n$ over F_0, it was shown that the splitting field, K, of the polynomial $p(x) = x^n + a_1x^{n-1} + \cdots + a_n$ over F has degree exactly $n!$ over F. Moreover, it was shown that the Galois group of K over

F is S_n, the symmetric group of degree n. This turned out to be the basis for the fact that the general polynomial of degree n, with $n \geq 5$, is not solvable by radicals.

However, it would be nice to know that the phenomenon described above can take place with fields which are more familiar to us than the field of rational functions in n variables. What we shall do will show that for any prime number p, at least, we can find polynomials of degree p over the field of rational numbers whose splitting fields have degree $p!$ over the rationals. This way we will have polynomials with rational coefficients whose Galois group over the rationals is S_p. In light of Theorem 5.7.2, we will conclude from this that the roots of these polynomials cannot be expressed in combinations of radicals involving rational numbers. Although in proving Theorem 5.7.2 we used that roots of unity were in the field, and roots of unity do not lie in the rationals, we make use of remark 2 following the proof of Theorem 5.7.2 here, namely that Theorem 5.7.2 remains valid even in the absence of roots of unity.

We shall make use of the fact that polynomials with rational coefficients have all their roots in the complex field.

We now prove

THEOREM 5.8.1 *Let $q(x)$ be an irreducible polynomial of degree p, p a prime, over the field Q of rational numbers. Suppose that $q(x)$ has exactly two nonreal roots in the field of complex numbers. Then the Galois group of $q(x)$ over Q is S_p, the symmetric group of degree p. Thus the splitting field of $q(x)$ over Q has degree $p!$ over Q.*

Proof. Let K be the splitting field of the polynomial $q(x)$ over Q. If α is a root of $q(x)$ in K, then, since $q(x)$ is irreducible over Q, by Theorem 5.1.3, $[Q(\alpha):Q] = p$. Since $K \supset Q(\alpha) \supset Q$ and, according to Theorem 5.1.1, $[K:Q] = [K:Q(\alpha)][Q(\alpha):Q] = [K:Q(\alpha)]p$, we have that $p \mid [K:Q]$. If G is the Galois group of K over Q, by Theorem 5.6.4, $o(G) = [K:F]$. Thus $p \mid o(G)$. Hence, by Cauchy's theorem (Theorem 2.11.3), G has an element σ of order p.

To this point we have not used our hypothesis that $q(x)$ has exactly two nonreal roots. We use it now. If α_1, α_2 are these nonreal roots, then $\alpha_1 = \bar{\alpha}_2$, $\alpha_2 = \bar{\alpha}_1$ (see Problem 13, Section 5.3), where the bar denotes the complex conjugate. If $\alpha_3, \ldots, \alpha_p$ are the other roots, then, since they are real, $\bar{\alpha}_i = \alpha_i$ for $i \geq 3$. Thus the complex conjugate mapping takes K into itself, is an automorphism τ of K over Q, and interchanges α_1 and α_2, leaving the other roots of $q(x)$ fixed.

Now, the elements of G take roots of $q(x)$ into roots of $q(x)$, so induce permutations of $\alpha_1, \ldots, \alpha_p$. In this way we imbed G in S_p. The automorphism τ described above is the transposition $(1, 2)$ since $\tau(\alpha_1) = \alpha_2$,

$\tau(\alpha_2) = \alpha_1$, and $\tau(\alpha_i) = \alpha_i$ for $i \geq 3$. What about the element $\sigma \in G$, which we mentioned above, which has order p? As an element of S_p, σ has order p. But the only elements of order p in S_p are p-cycles. Thus σ must be a p-cycle.

Therefore G, as a subgroup of S_p, contains a transposition and a p-cycle. It is a relatively easy exercise (see Problem 4) to prove that any transposition and any p-cycle in S_p generate S_p. Thus σ and τ generate S_p. But since they are in G, the group generated by σ and τ must be in G. The net result of this is that $G = S_p$. In other words, the Galois group of $q(x)$ over Q is indeed S_p. This proves the theorem.

The theorem gives us a fairly general criterion to get S_p as a Galois group over Q. Now we must produce polynomials of degree p over the rationals which are irreducible over Q and have exactly two nonreal roots. To produce irreducible polynomials, we use the Eisenstein criterion (Theorem 3.10.2). To get all but two real roots one can play around with the coefficients, but always staying in a context where the Eisenstein criterion is in force.

We do it explicitly for $p = 5$. Let $q(x) = 2x^5 - 10x + 5$. By the Eisenstein criterion, $q(x)$ is irreducible over Q. We graph $y = q(x) = 2x^5 - 10x + 5$. By elementary calculus it has a maximum at $x = -1$ and a minimum at $x = 1$ (see Figure 5.8.1). As the graph clearly indicates,

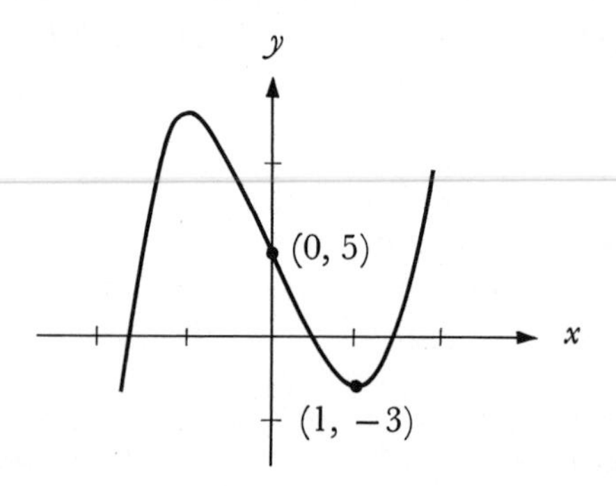

Figure 5.8.1

$y = q(x) = 2x^5 - 10x + 5$ crosses the x-axis exactly three times, so $q(x)$ has exactly three roots which are real. Hence the other two roots must be complex, nonreal numbers. Therefore $q(x)$ satisfies the hypothesis of Theorem 5.8.1, in consequence of which the Galois group of $q(x)$ over Q is S_5. Using Theorem 5.7.2, we know that it is not possible to express the roots of $q(x)$ in a combination of radicals of rational numbers.

Problems

1. In S_5 show that (1 2) and (1 2 3 4 5) generate S_5.
2. In S_5 show that (1 2) and (1 3 2 4 5) generate S_5.
3. If $p > 2$ is a prime, show that (1 2) and $(1\ 2 \cdots p-1\ p)$ generate S_p.
4. Prove that any transposition and p-cycle in S_p, p a prime, generate S_p.
5. Show that the following polynomials over Q are irreducible and have exactly two nonreal roots.
   (a) $p(x) = x^3 - 3x - 3$,
   (b) $p(x) = x^5 - 6x + 3$,
   (c) $p(x) = x^5 + 5x^4 + 10x^3 + 10x^2 - x - 2$.
6. What are the Galois groups over Q of the polynomials in Problem 5?
7. Construct a polynomial of degreee 7 with rational coefficients whose Galois group over Q is S_7.

Supplementary Reading

ARTIN, E., *Galois Theory*, 2nd ed. Notre Dame Mathematical Lectures Number 2. Notre Dame, Ind.: Notre Dame Press, 1966.

KAPLANSKY, IRVING, *Fields and Rings*, 2nd ed. Chicago: University of Chicago Press, 1972.

POLLARD, H., *Theory of Algebraic Numbers*, Carus Monograph, Number 9. New York: John Wiley & Sons, 1950.

VAN DER WAERDEN, B. L., *Modern Algebra*, Vol. 1. New York: Ungar Publishing Company, 1949.

WEISNER, L., *Theory of Equations*. New York: The Macmillan Company, 1938.

SIEGAL, C. L., *Transcendental Numbers*, Annals of Mathematics Studies Number 16. Princeton, N.J.: Princeton University Press, 1949. Milwood, N.Y.: Kraus Reprint Company, 1949.

NIVEN, I., *Irrational Numbers*, Carus Monograph Number 11. New York: John Wiley & Sons, 1956.

Topics for Class Discussion

NIVEN, I., "A simple proof of the irrationality of π," *Bulletin of the American Mathematical Society*, Vol. 53 (1947), page 509.

6

Linear Transformations

In Chapter 4 we defined, for any two vector spaces V and W over the same field F, the set Hom (V, W) of all vector space homomorphisms of V into W. In fact, we introduced into Hom (V, W) the operations of addition and of multiplication by scalars (elements of F) in such a way that Hom (V, W) itself became a vector space over F.

Of much greater interest is the special case $V = W$, for here, in addition to the vector space operations, we can introduce a multiplication for any two elements under which Hom (V, V) becomes a ring. Blessed with this twin nature—that of a vector space and of a ring—Hom (V, V) acquires an extremely rich structure. It is this structure and its consequences that impart so much life and sparkle to the subject and which justify most fully the creation of the abstract concept of a vector space.

Our main concern shall be concentrated on Hom (V, V) where V will not be an arbitrary vector space but rather will be restricted to be a finite-dimensional vector space over a field F. The finite-dimensionality of V imposes on Hom (V, V) the consequence that each of its elements satisfies a polynomial over F. This fact, perhaps more than any other, gives us a ready entry into Hom (V, V) and allows us to probe both deeply and effectively into its structure.

The subject matter to be considered often goes under the name of *linear algebra*. It encompasses the isomorphic *theory of matrices*. The statement that its results are in constant everyday use in every aspect of mathematics (and elsewhere) is not in the least exaggerated.

A popular myth is that mathematicians revel in the inapplicability of their discipline and are disappointed when one of their results is "soiled" by use in the outside world. This is sheer nonsense! It is true that a mathematician does not depend for his value judgments on the applicability of a given result outside of mathematics proper but relies, rather, on some intrinsic, and at times intangible, mathematical criteria. However, it is equally true that the converse is false—the utility of a result has never lowered its mathematical value. A perfect case in point is the subject of linear algebra; it is real mathematics, interesting and exciting on its own, yet it is probably that part of mathematics which finds the widest application—in physics, chemistry, economics, in fact in almost every science and pseudoscience.

6.1 The Algebra of Linear Transformations

Let V be a vector space over a field F and let Hom (V, V), as before, be the set of all vector-space-homomorphisms of V into itself. In Section 4.3 we showed that Hom (V, V) forms a vector space over F, where, for $T_1, T_2 \in \text{Hom}(V, V)$, $T_1 + T_2$ is defined by $v(T_1 + T_2) = vT_1 + vT_2$ for all $v \in V$ and where, for $\alpha \in F$, αT_1 is defined by $v(\alpha T_1) = \alpha(vT_1)$.

For $T_1, T_2 \in \text{Hom}(V, V)$, since $vT_1 \in V$ for any $v \in V$, $(vT_1)T_2$ makes sense. As we have done for mappings of any set into itself, we define T_1T_2 by $v(T_1T_2) = (vT_1)T_2$ for any $v \in V$. We now claim that $T_1T_2 \in$ Hom (V, V). To prove this, we must show that for all $\alpha, \beta \in F$ and all $u, v \in V$, $(\alpha u + \beta v)(T_1T_2) = \alpha(u(T_1T_2)) + \beta(v(T_1T_2))$. We compute

$$\begin{aligned}(\alpha u + \beta v)(T_1T_2) &= ((\alpha u + \beta v)T_1)T_2 \\ &= (\alpha(uT_1) + \beta(vT_1))T_2 \\ &= \alpha(uT_1)T_2 + \beta(vT_1)T_2 \\ &= \alpha(u(T_1T_2)) + \beta(v(T_1T_2)).\end{aligned}$$

We leave as an exercise the following properties of this product in Hom (V, V):

1. $(T_1 + T_2)T_3 = T_1T_3 + T_2T_3$;
2. $T_3(T_1 + T_2) = T_3T_1 + T_3T_2$;
3. $T_1(T_2T_3) = (T_1T_2)T_3$;
4. $\alpha(T_1T_2) = (\alpha T_1)\,T_2 = T_1(\alpha T_2)$;

for all $T_1, T_2, T_3 \in$ Hom (V, V) and all $\alpha \in F$.

Note that properties 1, 2, 3, above, are exactly what are required to make of Hom (V, V) an associative ring. Property 4 intertwines the character of Hom (V, V), as a vector space over F, with its character as a ring.

Note further that there is an element, I, in Hom (V, V), defined by $vI = v$ for all $v \in V$, with the property that $TI = IT = T$ for every $T \in$ Hom (V, V). Thereby, Hom (V, V) is a ring with a unit element. Moreover, if in property 4 above we put $T_2 = I$, we obtain $\alpha T_1 = T_1(\alpha I)$. Since $(\alpha I)T_1 = \alpha(IT_1) = \alpha T_1$, we see that $(\alpha I)T_1 = T_1(\alpha I)$ for all $T_1 \in$ Hom (V, V), and so αI commutes with every element of Hom (V, V). *We shall always write, in the future, αI merely as α.*

DEFINITION An associative ring A is called an *algebra* over F if A is a vector space over F such that for all $a, b \in A$ and $\alpha \in F$, $\alpha(ab) = (\alpha a)b = a(\alpha b)$.

Homomorphisms, isomorphisms, ideals, etc., of algebras are defined as for rings with the additional proviso that these must preserve, or be invariant under, the vector space structure.

Our remarks above indicate that Hom (V, V) is an algebra over F. For convenience of notation we henceforth shall write Hom (V, V) as $A(V)$; whenever we want to emphasize the role of the field F we shall denote it by $A_F(V)$.

DEFINITION A *linear transformation* on V, over F, is an element of $A_F(V)$.

We shall, at times, refer to $A(V)$ as the *ring, or algebra, of linear transformations on* V.

For arbitrary algebras A, with unit element, over a field F, we can prove the analog of Cayley's theorem for groups; namely,

LEMMA 6.1.1 *If A is an algebra, with unit element, over F, then A is isomorphic to a subalgebra of $A(V)$ for some vector space V over F.*

Proof. Since A is an algebra over F, it must be a vector space over F. We shall use $V = A$ to prove the theorem.

If $a \in A$, let $T_a : A \to A$ be defined by $vT_a = va$ for every $v \in A$. We assert that T_a is a linear transformation on $V(=A)$. By the right-distributive law $(v_1 + v_2)T_a = (v_1 + v_2)a = v_1 a + v_2 a = v_1 T_a + v_2 T_a$. Since A is an algebra, $(\alpha v)T_a = (\alpha v)a = \alpha(va) = \alpha(vT_a)$ for $v \in A$, $\alpha \in F$. Thus T_a is indeed a linear transformation on A.

Consider the mapping $\psi : A \to A(V)$ defined by $a\psi = T_a$ for every $a \in A$. We claim that ψ is an isomorphism of A into $A(V)$. To begin with, if $a, b \in A$ and $\alpha, \beta \in F$, then for all $v \in A$, $vT_{\alpha a + \beta b} = v(\alpha a + \beta b) = \alpha(va) + \beta(vb)$ [by the left-distributive law and the fact that A is an algebra over F] $= \alpha(vT_a) + \beta(vT_b) = v(\alpha T_a + \beta T_b)$ since both T_a and T_b are linear transformations. In consequence, $T_{\alpha a + \beta b} = \alpha T_a + \beta T_b$, whence ψ is a vector-space homomorphism of A into $A(V)$. Next, we compute, for

$a, b \in A$, $vT_{ab} = v(ab) = (va)b = (vT_a)T_b = v(T_aT_b)$ (we have used the associative law of A in this computation), which implies that $T_{ab} = T_aT_b$. In this way, ψ is also a ring-homomorphism of A. So far we have proved that ψ is a homomorphism of A, as an algebra, into $A(V)$. All that remains is to determine the kernel of ψ. Let $a \in A$ be in the kernel of ψ; then $a\psi = 0$, whence $T_a = 0$ and so $vT_a = 0$ for all $v \in V$. Now $V = A$, and A has a unit element, e, hence $eT_a = 0$. However, $0 = eT_a = ea = a$, proving that $a = 0$. The kernel of ψ must therefore merely consist of 0, thus implying that ψ is an isomorphism of A into $A(V)$. This completes the proof of the lemma.

The lemma points out the universal role played by the particular algebras, $A(V)$, for in these we can find isomorphic copies of any algebra.

Let A be an algebra, with unit element e, over F, and let $p(x) = \alpha_0 + \alpha_1 x + \cdots + \alpha_n x^n$ be a polynomial in $F[x]$. For $a \in A$, by $p(a)$, we shall mean the element $\alpha_0 e + \alpha_1 a + \cdots + \alpha_n a^n$ in A. If $p(a) = 0$ we shall say *a satisfies $p(x)$*.

LEMMA 6.1.2 *Let A be an algebra, with unit element, over F, and suppose that A is of dimension m over F. Then every element in A satisfies some nontrivial polynomial in $F[x]$ of degree at most m.*

Proof. Let e be the unit element of A; if $a \in A$, consider the $m + 1$ elements $e, a, a^2, \ldots, a^m$ in A. Since A is m-dimensional over F, by Lemma 4.2.4, $e, a, a^2, \ldots, a^m$, being $m + 1$ in number, must be linearly dependent over F. In other words, there are elements $\alpha_0, \alpha_1, \ldots, \alpha_m$ in F, not all 0, such that $\alpha_0 e + \alpha_1 a + \cdots + \alpha_m a^m = 0$. But then a satisfies the nontrivial polynomial $q(x) = \alpha_0 + \alpha_1 x + \cdots + \alpha_m x^m$, of degree at most m, in $F[x]$.

If V is a finite-dimensional vector space over F, of dimension n, by Corollary 1 to Theorem 4.3.1, $A(V)$ is of dimension n^2 over F. Since $A(V)$ is an algebra over F, we can apply Lemma 6.1.2 to it to obtain that every element in $A(V)$ satisfies a polynomial over F of degree at most n^2. This fact will be of central significance in all that follows, so we single it out as

THEOREM 6.1.1 *If V is an n-dimensional vector space over F, then, given any element T in $A(V)$, there exists a nontrivial polynomial $q(x) \in F[x]$ of degree at most n^2, such that $q(T) = 0$.*

We shall see later that we can assert much more about the degree of $q(x)$; in fact, we shall eventually be able to say that we can choose such a $q(x)$ of degree at most n. This fact is a famous theorem in the subject, and is known as the Cayley-Hamilton theorem. For the moment we can get by without any sharp estimate of the degree of $q(x)$; all we need is that a suitable $q(x)$ exists.

Since for finite-dimensional V, given $T \in A(V)$, some polynomial $q(x)$ exists for which $q(T) = 0$, a nontrivial polynomial of lowest degree with this property, $p(x)$, exists in $F[x]$. We call $p(x)$ a *minimal polynomial* for T over F. If T satisfies a polynomial $h(x)$, then $p(x) \mid h(x)$.

DEFINITION An element $T \in A(V)$ is called *right-invertible* if there exists an $S \in A(V)$ such that $TS = 1$. (Here 1 denotes the unit element of $A(V)$.)

Similarly, we can define left-invertible, if there is a $U \in A(V)$ such that $UT = 1$. If T is both right- and left-invertible and if $TS = UT = 1$, it is an easy exercise that $S = U$ and that S is unique.

DEFINITION An element T in $A(V)$ is *invertible* or *regular* if it is both right- and left-invertible; that is, if there is an element $S \in A(V)$ such that $ST = TS = 1$. We write S as T^{-1}.

An element in $A(V)$ which is not regular is called *singular*.

It is quite possible that an element in $A(V)$ is right-invertible but is not invertible. An example of such: Let F be the field of real numbers and let V be $F[x]$, the set of all polynomials in x over F. In V let S be defined by

$$q(x)S = \frac{d}{dx}\, q(x)$$

and T by

$$q(x)\,T = \int_1^x q(x)\, dx.$$

Then $ST \neq 1$, whereas $TS = 1$. As we shall see in a moment, if V is finite-dimensional over F, then an element in $A(V)$ which is right-invertible is invertible.

THEOREM 6.1.2 *If V is finite-dimensional over F, then $T \in A(V)$ is invertible if and only if the constant term of the minimal polynomial for T is not 0.*

Proof. Let $p(x) = \alpha_0 + \alpha_1 x + \cdots + \alpha_k x^k$, $\alpha_k \neq 0$, be the minimal polynomial for T over F.

If $\alpha_0 \neq 0$, since $0 = p(T) = \alpha_k T^k + \alpha_{k-1} T^{k-1} + \cdots + \alpha_1 T + \alpha_0$, we obtain

$$\begin{aligned} 1 &= T\left(-\frac{1}{\alpha_0}\left(\alpha_k T^{k-1} + \alpha_{k-1} T^{k-2} + \cdots + \alpha_1\right)\right) \\ &= \left(-\frac{1}{\alpha_0}\left(\alpha_k T^{k-1} + \cdots + \alpha_1\right)\right) T. \end{aligned}$$

Therefore,

$$S = -\frac{1}{\alpha_0}(\alpha_k T^{k-1} + \cdots + \alpha_1)$$

acts as an inverse for T, whence T is invertible.

Suppose, on the other hand, that T is invertible, yet $\alpha_0 = 0$. Thus $0 = \alpha_1 T + \alpha_2 T^2 + \cdots + \alpha_k T^k = (\alpha_1 + \alpha_2 T + \cdots + \alpha_k T^{k-1})T$. Multiplying this relation from the right by T^{-1} yields $\alpha_1 + \alpha_2 T + \cdots + \alpha_k T^{k-1} = 0$, whereby T satisfies the polynomial $q(x) = \alpha_1 + \alpha_2 x + \cdots + \alpha_k x^{k-1}$ in $F[x]$. Since the degree of $q(x)$ is less than that of $p(x)$, this is impossible. Consequently, $\alpha_0 \neq 0$ and the other half of the theorem is established.

COROLLARY 1 *If V is finite-dimensional over F and if $T \in A(V)$ is invertible, then T^{-1} is a polynomial expression in T over F.*

Proof. Since T is invertible, by the theorem, $\alpha_0 + \alpha_1 T + \cdots + \alpha_k T^k = 0$ with $\alpha_0 \neq 0$. But then

$$T^{-1} = -\frac{1}{\alpha_0}(\alpha_1 + \alpha_2 T + \cdots + \alpha_k T^{k-1}).$$

COROLLARY 2 *If V is finite-dimensional over F and if $T \in A(V)$ is singular, then there exists an $S \neq 0$ in $A(V)$ such that $ST = TS = 0$.*

Proof. Because T is not regular, the constant term of its minimal polynomial must be 0. That is, $p(x) = \alpha_1 x + \cdots + \alpha_k x^k$, whence $0 = \alpha_1 T + \cdots + \alpha_k T^k$. If $S = \alpha_1 + \cdots + \alpha_k T^{k-1}$, then $S \neq 0$ (since $\alpha_1 + \cdots + \alpha_k x^{k-1}$ is of lower degree than $p(x)$) and $ST = TS = 0$.

COROLLARY 3 *If V is finite-dimensional over F and if $T \in A(V)$ is right-invertible, then it is invertible.*

Proof. Let $TU = 1$. If T were singular, there would be an $S \neq 0$ such that $ST = 0$. However, $0 = (ST)U = S(TU) = S1 = S \neq 0$, a contradiction. Thus T is regular.

We wish to transfer the information contained in Theorem 6.1.2 and its corollaries from $A(V)$ to the action of T on V. A most basic result in this vein is

THEOREM 6.1.3 *If V is finite-dimensional over F, then $T \in A(V)$ is singular if and only if there exists a $v \neq 0$ in V such that $vT = 0$.*

Proof. By Corollary 2 to Theorem 6.1.2, T is singular if and only if there is an $S \neq 0$ in $A(V)$ such that $ST = TS = 0$. Since $S \neq 0$ there is an element $w \in V$ such that $wS \neq 0$.

Let $v = wS$; then $vT = (wS)T = w(ST) = w0 = 0$. We have produced a nonzero vector v in V which is annihilated by T. Conversely, if $vT = 0$ with $v \neq 0$, we leave as an exercise the fact that T is not invertible.

We seek still another characterization of the singularity or regularity of a linear transformation in terms of its overall action on V.

DEFINITION If $T \in A(V)$, then the *range* of T, VT, is defined by $VT = \{vT \mid v \in V\}$.

The range of T is easily shown to be a subvector space of V. It merely consists of all the images by T of the elements of V. Note that the range of T is all of V if and only if T is onto.

THEOREM 6.1.4 *If V is finite-dimensional over F, then $T \in A(V)$ is regular if and only if T maps V onto V.*

Proof. As happens so often, one-half of this is almost trivial; namely, if T is regular then, given $v \in V$, $v = (vT^{-1})T$, whence $VT = V$ and T is onto.

On the other hand, suppose that T is not regular. We must show that T is not onto. Since T is singular, by Theorem 6.1.3, there exists a vector $v_1 \neq 0$ in V such that $v_1 T = 0$. By Lemma 4.2.5 we can fill out, from v_1, to a basis $v_1, v_2, \ldots, v_n$ of V. Then every element in VT is a linear combination of the elements $w_1 = v_1 T$, $w_2 = v_2 T, \ldots, w_n = v_n T$. Since $w_1 = 0$, VT is spanned by the $n - 1$ elements $w_2, \ldots, w_n$; therefore $\dim VT \leq n - 1 < n = \dim V$. But then VT must be different from V; that is, T is not onto.

Theorem 6.1.4 points out that we can distinguish regular elements from singular ones, in the finite-dimensional case, according as their ranges are or are not all of V. If $T \in A(V)$ this can be rephrased as: T is regular if and only if $\dim (VT) = \dim V$. This suggests that we could use $\dim (VT)$ not only as a test for regularity, but even as a measure of the degree of singularity (or, lack of regularity) for a given $T \in A(V)$.

DEFINITION If V is finite-dimensional over F, then the *rank* of T is the dimension of VT, the range of T, over F.

We denote the rank of T by $r(T)$. At one end of the spectrum, if $r(T) = \dim V$, T is regular (and so, not at all singular). At the other end, if $r(T) = 0$, then $T = 0$ and so T is as singular as it can possibly be. The rank, as a function on $A(V)$, is an important function, and we now investigate some of its properties.

LEMMA 6.1.3 *If V is finite-dimensional over F then for $S, T \in A(V)$.*

1. $r(ST) \leq r(T)$;
2. $r(TS) \leq r(T)$;

(and so, $r(ST) \leq \min\{r(T), r(S)\}$)

3. $r(ST) = r(TS) = r(T)$ *for S regular in $A(V)$.*

Proof. We go through 1, 2, and 3 in order.

1. Since $VS \subset V$, $V(ST) = (VS)T \subset VT$, whence, by Lemma 4.2.6, $\dim(V(ST)) \leq \dim VT$; that is, $r(ST) \leq r(T)$.

2. Suppose that $r(T) = m$. Therefore, VT has a basis of m elements, $w_1, w_2, \ldots, w_m$. But then $(VT)S$ is spanned by $w_1S, w_2S, \ldots, w_mS$, hence has dimension at most m. Since $r(TS) = \dim(V(TS)) = \dim((VT)S) \leq m = \dim VT = r(T)$, part 2 is proved.

3. If S is invertible then $VS = V$, whence $V(ST) = (VS)T = VT$. Thereby, $r(ST) = \dim(V(ST)) = \dim(VT) = r(T)$. On the other hand, if VT has $w_1, \ldots, w_m$ as a basis, the regularity of S implies that $w_1S, \ldots, w_mS$ are linearly independent. (Prove!) Since these span $V(TS)$ they form a basis of $V(TS)$. But then $r(TS) = \dim(V(TS)) = \dim(VT) = r(T)$.

COROLLARY *If $T \in A(V)$ and if $S \in A(V)$ is regular, then $r(T) = r(STS^{-1})$.*

Proof. By part 3 of the lemma, $r(STS^{-1}) = r(S(TS^{-1})) = r((TS^{-1})S) = r(T)$.

Problems

In all problems, *unless stated otherwise*, V will denote a finite-dimensional vector space over a field F.

1. Prove that $S \in A(V)$ is regular if and only if whenever $v_1, \ldots, v_n \in V$ are linearly independent, then $v_1S, v_2S, \ldots, v_nS$ are also linearly independent.
2. Prove that $T \in A(V)$ is completely determined by its values on a basis of V.
3. Prove Lemma 6.1.1 even when A does not have a unit element.
4. If A is the field of complex numbers and F is the field of real numbers, then A is an algebra over F of dimension 2. For $a = \alpha + \beta i$ in A, compute the action of T_a (see Lemma 6.1.1) on a basis of A over F.
5. If V is two-dimensional over F and $A = A(V)$, write down a basis of A over F and compute T_a for each a in this basis.
6. If $\dim_F V > 1$ prove that $A(V)$ is not commutative.
7. In $A(V)$ let $Z = \{T \in A(V) \mid ST = TS \text{ for all } S \in A(V)\}$. Prove that

Z merely consists of the multiples of the unit element of $A(V)$ by the elements of F.

*8. If $\dim_F(V) > 1$ prove that $A(V)$ has no two-sided ideals other than (0) and $A(V)$.

**9. Prove that the conclusion of Problem 8 is false if V is not finite-dimensional over F.

10. If V is an arbitrary vector space over F and if $T \in A(V)$ is both right- and left-invertible, prove that the right inverse and left inverse must be equal. From this, prove that the inverse of T is unique.

11. If V is an arbitrary vector space over F and if $T \in A(V)$ is right-invertible with a *unique* right inverse, prove that T is invertible.

12. Prove that the regular elements in $A(V)$ form a group.

13. If F is the field of integers modulo 2 and if V is two-dimensional over F, compute the group of regular elements in $A(V)$ and prove that this group is isomorphic to S_3, the symmetric group of degree 3.

*14. If F is a finite field with q elements, compute the order of the group of regular elements in $A(V)$ where V is two-dimensional over F.

*15. Do Problem 14 if V is assumed to be n-dimensional over F.

*16. If V is finite-dimensional, prove that every element in $A(V)$ can be written as a sum of regular elements.

17. An element $E \in A(V)$ is called an *idempotent* if $E^2 = E$. If $E \in A(V)$ is an idempotent, prove that $V = V_0 \oplus V_1$ where $v_0 E = 0$ for all $v_0 \in V_0$ and $v_1 E = v_1$ for all $v_1 \in V_1$.

18. If $T \in A_F(V)$, F of characteristic not 2, satisfies $T^3 = T$, prove that $V = V_0 \oplus V_1 \oplus V_2$ where
(a) $v_0 \in V_0$ implies $v_0 T = 0$.
(b) $v_1 \in V_1$ implies $v_1 T = v_1$.
(c) $v_2 \in V_2$ implies $v_2 T = -v_2$.

*19. If V is finite-dimensional and $T \neq 0 \in A(V)$, prove that there is an $S \in A(V)$ such that $E = TS \neq 0$ is an idempotent.

20. The element $T \in A(V)$ is called *nilpotent* if $T^m = 0$ for some m. If T is *nilpotent* and if $vT = \alpha v$ for some $v \neq 0$ in V, with $\alpha \in F$, prove that $\alpha = 0$.

21. If $T \in A(V)$ is nilpotent, prove that $\alpha_0 + \alpha_1 T + \alpha_2 T^2 + \cdots + \alpha_k T^k$ is regular, provided that $\alpha_0 \neq 0$.

22. If A is a finite-dimensional algebra over F and if $a \in A$, prove that for some integer $k > 0$ and some polynomial $p(x) \in F[x]$, $a^k = a^{k+1}p(a)$.

23. Using the result of Problem 22, prove that for $a \in A$ there is a polynomial $q(x) \in F[x]$ such that $a^k = a^{2k}q(a)$.

24. Using the result of Problem 23, prove that given $a \in A$ either a is nilpotent or there is an element $b \neq 0$ in A of the form $b = ah(a)$, where $h(x) \in F[x]$, such that $b^2 = b$.

25. If A is an algebra over F (not necessarily finite-dimensional) and if for $a \in A$, $a^2 - a$ is nilpotent, prove that either a is nilpotent or there is an element b of the form $b = ah(a) \neq 0$, where $h(x) \in F[x]$, such that $b^2 = b$.

*26. If $T \neq 0 \in A(V)$ is singular, prove that there is an element $S \in A(V)$ such that $TS = 0$ but $ST \neq 0$.

27. Let V be two-dimensional over F with basis v_1, v_2. Suppose that $T \in A(V)$ is such that $v_1T = \alpha v_1 + \beta v_2$, $v_2T = \gamma v_1 + \delta v_2$, where $\alpha, \beta, \gamma, \delta \in F$. Find a nonzero polynomial in $F[x]$ of degree 2 satisfied by T.

28. If V is three-dimensional over F with basis v_1, v_2, v_3 and if $T \in A(V)$ is such that $v_iT = \alpha_{i1}v_1 + \alpha_{i2}v_2 + \alpha_{i3}v_3$ for $i = 1, 2, 3$, with all $\alpha_{ij} \in F$, find a polynomial of degree 3 in $F[x]$ satisfied by T.

29. Let V be n-dimensional over F with a basis $v_1, \ldots, v_n$. Suppose that $T \in A(V)$ is such that

$$v_1T = v_2,\ v_2T = v_3, \ldots, v_{n-1}T = v_n,$$
$$v_nT = -\alpha_n v_1 - \alpha_{n-1}v_2 - \cdots - \alpha_1 v_n,$$

where $\alpha_1, \ldots, \alpha_n \in F$. Prove that T satisfies the polynomial

$$p(x) = x^n + \alpha_1 x^{n-1} + \alpha_2 x^{n-2} + \cdots + \alpha_n \text{ over } F.$$

30. If $T \in A(V)$ satisfies a polynomial $q(x) \in F[x]$, prove that for $S \in A(V)$, S regular, STS^{-1} also satisfies $q(x)$.

31. (a) If F is the field of rational numbers and if V is three-dimensional over F with a basis v_1, v_2, v_3, compute the rank of $T \in A(V)$ defined by

$$v_1T = v_1 - v_2,$$
$$v_2T = v_1 + v_3,$$
$$v_3T = v_2 + v_3.$$

(b) Find a vector $v \in V$, $v \neq 0$. such that $vT = 0$.

32. Prove that the range of T and $U = \{v \in V \mid vT = 0\}$ are subspaces of V.

33. If $T \in A(V)$, let $V_0 = \{v \in V \mid vT^k = 0 \text{ for some } k\}$. Prove that V_0 is a subspace and that if $vT^m \in V_0$, then $v \in V_0$.

34. Prove that the minimal polynomial of T over F divides all polynomials satisfied by T over F.

*35. If $n(T)$ is the dimension of the U of Problem 32 prove that $r(T) + n(T) = \dim V$.

6.2 Characteristic Roots

For the rest of this chapter our interest will be limited to linear transformations on finite-dimensional vector spaces. *Thus, henceforth, V will always denote a finite-dimensional vector space over a field F.*

The algebra $A(V)$ has a unit element; for ease of notation we shall write this as 1, and by the symbol $\lambda - T$, for $\lambda \in F$, $T \in A(V)$ we shall mean $\lambda 1 - T$.

DEFINITION If $T \in A(V)$ then $\lambda \in F$ is called a *characteristic root* (or *eigenvalue*) of T if $\lambda - T$ is singular.

We wish to characterize the property of being a characteristic root in the behavior of T on V. We do this in

THEOREM 6.2.1 *The element $\lambda \in F$ is a characteristic root of $T \in A(V)$ if and only if for some $v \neq 0$ in V, $vT = \lambda v$.*

Proof. If λ is a characteristic root of T then $\lambda - T$ is singular, whence, by Theorem 6.1.3, there is a vector $v \neq 0$ in V such that $v(\lambda - T) = 0$. But then $\lambda v = vT$.

On the other hand, if $vT = \lambda v$ for some $v \neq 0$ in V, then $v(\lambda - T) = 0$, whence, again by Theorem 6.1.3, $\lambda - T$ must be singular, and so, λ is a characteristic root of T.

LEMMA 6.2.1 *If $\lambda \in F$ is a characteristic root of $T \in A(V)$, then for any polynomial $q(x) \in F[x]$, $q(\lambda)$ is a characteristic root of $q(T)$.*

Proof. Suppose that $\lambda \in F$ is a characteristic root of T. By Theorem 6.2.1, there is a nonzero vector v in V such that $vT = \lambda v$. What about vT^2? Now $vT^2 = (\lambda v)T = \lambda(vT) = \lambda(\lambda v) = \lambda^2 v$. Continuing in this way, we obtain that $vT^k = \lambda^k v$ for all positive integers k. If $q(x) = \alpha_0 x^m + \alpha_1 x^{m-1} + \cdots + \alpha_m$, $\alpha_i \in F$, then $q(T) = \alpha_0 T^m + \alpha_1 T^{m-1} + \cdots + \alpha_m$, whence $vq(T) = v(\alpha_0 T^m + \alpha_1 T^{m-1} + \cdots + \alpha_m) = \alpha_0(vT^m) + \alpha_1(vT^{m-1}) + \cdots + \alpha_m v = (\alpha_0 \lambda^m + \alpha_1 \lambda^{m-1} + \cdots + \alpha_m)v = q(\lambda)v$ by the remark made above. Thus $v(q(\lambda) - q(T)) = 0$, hence, by Theorem 6.2.1, $q(\lambda)$ is a characteristic root of $q(T)$.

As immediate consequence of Lemma 6.2.1, in fact as a mere special case (but an extremely important one), we have

THEOREM 6.2.2 *If $\lambda \in F$ is a characteristic root of $T \in A(V)$, then λ is a root of the minimal polynomial of T. In particular, T only has a finite number of characteristic roots in F.*

Proof. Let $p(x)$ be the minimal polynomial over F of T; thus $p(T) = 0$. If $\lambda \in F$ is a characteristic root of T, there is a $v \neq 0$ in V with $vT = \lambda v$. As in the proof of Lemma 6.2.1, $vp(T) = p(\lambda)v$; but $p(T) = 0$, which thus implies that $p(\lambda)v = 0$. Since $v \neq 0$, by the properties of a vector space, we must have that $p(\lambda) = 0$. Therefore, λ is a root of $p(x)$. Since $p(x)$ has only a finite number of roots (in fact, since $\deg p(x) \leq n^2$ where $n = \dim_F V$, $p(x)$ has at most n^2 roots) in F, there can only be a finite number of characteristic roots of T in F.

If $T \in A(V)$ and if $S \in A(V)$ is regular, then $(STS^{-1})^2 = STS^{-1}STS^{-1} = ST^2S^{-1}$, $(STS^{-1})^3 = ST^3S^{-1}, \ldots, (STS^{-1})^i = ST^iS^{-1}$. Consequently, for any $q(x) \in F[x]$, $q(STS^{-1}) = Sq(T)S^{-1}$. In particular, if $q(T) = 0$, then $q(STS^{-1}) = 0$. Thus if $p(x)$ is the minimal polynomial for T, then it follows easily that $p(x)$ is also the minimal polynomial for STS^{-1}. We have proved

LEMMA 6.2.2 *If $T, S \in A(V)$ and if S is regular, then T and STS^{-1} have the same minimal polynomial.*

DEFINITION The element $0 \neq v \in V$ is called a *characteristic vector* of T belonging to the characteristic root $\lambda \in F$ if $vT = \lambda v$.

What relation, if any, must exist between characteristic vectors of T belonging to different characteristic roots? This is answered in

THEOREM 6.2.3 *If $\lambda_1, \ldots, \lambda_k$ in F are distinct characteristic roots of $T \in A(V)$ and if $v_1, \ldots, v_k$ are characteristic vectors of T belonging to $\lambda_1, \ldots, \lambda_k$, respectively, then $v_1, \ldots, v_k$ are linearly independent over F.*

Proof. For the theorem to require any proof, k must be larger than 1; so we suppose that $k > 1$.

If $v_1, \ldots, v_k$ are linearly dependent over F, then there is a relation of the form $\alpha_1 v_1 + \cdots + \alpha_k v_k = 0$, where $\alpha_1, \ldots, \alpha_k$ are all in F and not all of them are 0. In all such relations, there is one having as few nonzero coefficients as possible. By suitably renumbering the vectors, we can assume this shortest relation to be

$$\beta_1 v_1 + \cdots + \beta_j v_j = 0, \qquad \beta_1 \neq 0, \ldots, \beta_j \neq 0. \tag{1}$$

We know that $v_i T = \lambda_i v_i$, so, applying T to equation (1), we obtain

$$\lambda_1 \beta_1 v_1 + \cdots + \lambda_j \beta_j v_j = 0. \tag{2}$$

Multiplying equation (1) by λ_1 and subtracting from equation (2), we obtain

$$(\lambda_2 - \lambda_1)\beta_2 v_2 + \cdots + (\lambda_j - \lambda_1)\beta_j v_j = 0.$$

Now $\lambda_i - \lambda_1 \neq 0$ for $i > 1$, and $\beta_i \neq 0$, whence $(\lambda_i - \lambda_1)\beta_i \neq 0$. But then we have produced a shorter relation than that in (1) between $v_1, v_2, \ldots, v_k$. This contradiction proves the theorem.

COROLLARY 1 *If $T \in A(V)$ and if $\dim_F V = n$ then T can have at most n distinct characteristic roots in F.*

Proof. Any set of linearly independent vectors in V can have at most n elements. Since any set of distinct characteristic roots of T, by Theorem 6.2.3, gives rise to a corresponding set of linearly independent characteristic vectors, the corollary follows.

COROLLARY 2 *If $T \in A(V)$ and if $\dim_F V = n$, and if T has n distinct characteristic roots in F, then there is a basis of V over F which consists of characteristic vectors of T.*

We leave the proof of this corollary to the reader. Corollary 2 is but the first of a whole class of theorems to come which will specify for us that a given linear transformation has a certain desirable basis of the vector space on which its action is easily describable.

Problems

In all the problems V is a vector space over F.

1. If $T \in A(V)$ and if $q(x) \in F[x]$ is such that $q(T) = 0$, is it true that every root of $q(x)$ in F is a characteristic root of T? Either prove that this is true or give an example to show that it is false.
2. If $T \in A(V)$ and if $p(x)$ is the minimal polynomial for T over F, suppose that $p(x)$ has all its roots in F. Prove that every root of $p(x)$ is a characteristic root of T.
3. Let V be two-dimensional over the field F, of real numbers, with a basis v_1, v_2. Find the characteristic roots and corresponding characteristic vectors for T defined by
   (a) $v_1 T = v_1 + v_2, \quad v_2 T = v_1 - v_2$.
   (b) $v_1 T = 5v_1 + 6v_2, \quad v_2 T = -7v_2$.
   (c) $v_1 T = v_1 + 2v_2, \quad v_2 T = 3v_1 + 6v_2$.
4. Let V be as in Problem 3, and suppose that $T \in A(V)$ is such that $v_1 T = \alpha v_1 + \beta v_2$, $v_2 T = \gamma v_1 + \delta v_2$, where $\alpha, \beta, \gamma, \delta$ are in F.
   (a) Find necessary and sufficient conditions that 0 be a characteristic root of T in terms of $\alpha, \beta, \gamma, \delta$.

(b) In terms of $\alpha, \beta, \gamma, \delta$ find necessary and sufficient conditions that T have two distinct characteristic roots in F.

5. If V is two-dimensional over a field F prove that every element in $A(V)$ satisfies a polynomial of degree 2 over F.

*6. If V is two-dimensional over F and if $S, T \in A(V)$, prove that $(ST - TS)^2$ commutes with all elements of $A(V)$.

7. Prove Corollary 2 to Theorem 6.2.3.

8. If V is n-dimensional over F and $T \in A(V)$ is nilpotent (i.e., $T^k = 0$ for some k), prove that $T^n = 0$. (*Hint:* If $v \in V$ use the fact that $v, vT, vT^2, \ldots, vT^n$ must be linearly dependent over F.)

6.3 Matrices

Although we have been discussing linear transformations for some time, it has always been in a detached and impersonal way; to us a linear transformation has been a symbol (very often T) which acts in a certain way on a vector space. When one gets right down to it, outside of the few concrete examples encountered in the problems, we have really never come face to face with specific linear transformations. At the same time it is clear that if one were to pursue the subject further there would often arise the need of making a thorough and detailed study of a given linear transformation. To mention one precise problem, presented with a linear transformation (and suppose, for the moment, that we have a means of recognizing it), how does one go about, in a "practical" and computable way, finding its characteristic roots?

What we seek first is a simple notation, or, perhaps more accurately, representation, for linear transformations. We shall accomplish this by use of a particular basis of the vector space and by use of the action of a linear transformation on this basis. Once this much is achieved, by means of the operations in $A(V)$ we can induce operations for the symbols created, making of them an algebra. This new object, infused with an algebraic life of its own, can be studied as a mathematical entity having an interest by itself. This study is what comprises the subject of *matrix theory*.

However, to ignore the source of these matrices, that is, to investigate the set of symbols independently of what they represent, can be costly, for we would be throwing away a great deal of useful information. Instead we shall always use the interplay between the abstract, $A(V)$, and the concrete, the matrix algebra, to obtain information one about the other.

Let V be an n-dimensional vector space over a field F and let $v_1, \ldots, v_n$ be a basis of V over F. If $T \in A(V)$ then T is determined on any vector as soon as we know its action on a basis of V. Since T maps V into V, $v_1 T$,

$v_2T, \ldots, v_nT$ must all be in V. As elements of V, each of these is realizable in a *unique* way as a linear combination of $v_1, \ldots, v_n$ over F. Thus

$$
\begin{aligned}
v_1T &= \alpha_{11}v_1 + \alpha_{12}v_2 + \cdots + \alpha_{1n}v_n \\
v_2T &= \alpha_{21}v_1 + \alpha_{22}v_2 + \cdots + \alpha_{2n}v_n \\
v_iT &= \alpha_{i1}v_1 + \alpha_{i2}v_2 + \cdots + \alpha_{in}v_n \\
&\vdots \\
v_nT &= \alpha_{n1}v_1 + \alpha_{n2}v_2 + \cdots + \alpha_{nn}v_n,
\end{aligned}
$$

where each $\alpha_{ij} \in F$. This system of equations can be written more compactly as

$$v_iT = \sum_{j=1}^{n} \alpha_{ij}v_j, \qquad \text{for} \quad i = 1, 2, \ldots, n.$$

The ordered set of n^2 numbers α_{ij} in F completely describes T. They will serve as the means of representing T.

DEFINITION Let V be an n-dimensioned vector space over F and let $v_1, \ldots, v_n$ be a basis for V over F. If $T \in A(V)$ then the *matrix of* T *in the basis* $v_1, \ldots, v_n$, written as $m(T)$, is

$$m(T) = \begin{pmatrix} \alpha_{11} & \alpha_{12} & \cdots & \alpha_{1n} \\ \alpha_{21} & \alpha_{22} & \cdots & \alpha_{2n} \\ \vdots & \vdots & & \vdots \\ \alpha_{n1} & \alpha_{n2} & \cdots & \alpha_{nn} \end{pmatrix},$$

where $v_iT = \sum_j \alpha_{ij}v_j$.

A matrix then is an ordered, square array of elements of F, with, as yet, no further properties, which represents the effect of a linear transformation on a given basis.

Let us examine an example. Let F be a field and let V be the set of all polynomials in x of degree $n - 1$ or less over F. On V let D be defined by $(\beta_0 + \beta_1x + \cdots + \beta_{n-1}x^{n-1})D = \beta_1 + 2\beta_2x + \cdots + i\beta_ix^{i-1} + \cdots + (n-1)\beta_{n-1}x^{n-2}$. It is trivial that D is a linear transformation on V; in fact, it is merely the differentiation operator.

What is the matrix of D? The questions is meaningless unless we specify a basis of V. Let us first compute the matrix of D in the basis $v_1 = 1$, $v_2 = x$, $v_3 = x^2, \ldots, v_i = x^{i-1}, \ldots, v_n = x^{n-1}$. Now,

$$
\begin{aligned}
v_1D &= 1D = 0 = 0v_1 + 0v_2 + \cdots + 0v_n \\
v_2D &= xD = 1 = 1v_1 + 0v_2 + \cdots + 0v_n \\
&\vdots \\
v_iD &= x^{i-1}D = (i-1)x^{i-2} \\
&= 0v_1 + 0v_2 + \cdots + 0v_{i-2} + (i-1)v_{i-1} + 0v_i \\
&\quad + \cdots + 0v_n \\
&\vdots \\
v_nD &= x^{n-1}D = (n-1)x^{n-2} \\
&= 0v_1 + 0v_2 + \cdots + 0v_{n-2} + (n-1)v_{n-1} + 0v_n.
\end{aligned}
$$

Going back to the very definition of the matrix of a linear transformation in a given basis, we see the matrix of D in the basis $v_1, \ldots, v_n$, $m_1(D)$, is in fact

$$m_1(D) = \begin{pmatrix} 0 & 0 & 0 & \cdots & 0 & 0 \\ 1 & 0 & 0 & \cdots & 0 & 0 \\ 0 & 2 & 0 & \cdots & 0 & 0 \\ 0 & 0 & 3 & \cdots & 0 & 0 \\ 0 & 0 & 0 & \cdots & (n-1) & 0 \end{pmatrix}$$

However, there is nothing special about the basis we just used, or in how we numbered its elements. Suppose we merely renumber the elements of this basis; we then get an equally good basis $w_1 = x^{n-1}$, $w_2 = x^{n-2}, \ldots, w_i = x^{n-i}, \ldots, w_n = 1$. What is the matrix of the same linear transformation D in this basis? Now,

$$\begin{aligned} w_1 D &= x^{n-1} D = (n-1)x^{n-2} \\ &= 0w_1 + (n-1)w_2 + 0w_3 + \cdots + 0w_n \\ &\vdots \\ w_i D &= x^{n-i} D = (n-i)x^{n-i-1} \\ &= 0w_1 + \cdots + 0w_i + (n-i)w_{i+1} + 0w_{i+2} + \cdots + 0w_n \\ &\vdots \\ w_n D &= 1D = 0 = 0w_1 + 0w_2 + \cdots + 0w_n, \end{aligned}$$

whence $m_2(D)$, the matrix of D in this basis is

$$m_2(D) = \begin{pmatrix} 0 & (n-1) & 0 & 0 & \cdots & 0 & 0 \\ 0 & 0 & (n-2) & 0 & \cdots & 0 & 0 \\ 0 & 0 & 0 & (n-3) & \cdots & 0 & 0 \\ 0 & \cdots & \cdots & \cdots & \cdots & \cdots & \\ \vdots & & & & & & \\ 0 & 0 & 0 & \cdots & \cdots & 0 & 1 \\ 0 & 0 & 0 & \cdots & \cdots & 0 & 0 \end{pmatrix}.$$

Before leaving this example, let us compute the matrix of D in still another basis of V over F. Let $u_1 = 1$, $u_2 = 1 + x$, $u_3 = 1 + x^2, \ldots, u_n = 1 + x^{n-1}$; it is easy to verify that $u_1, \ldots, u_n$ form a basis of V over F. What is the matrix of D in this basis? Since

$$\begin{aligned} u_1 D &= 1D = 0 = 0u_1 + 0u_2 + \cdots + 0u_n \\ u_2 D &= (1 + x)D = 1 = 1u_1 + 0u_2 + \cdots + 0u_n \\ u_3 D &= (1 + x^2)D = 2x = 2(u_2 - u_1) = -2u_1 + 2u_2 + 0u_3 + \cdots + 0u_n \\ &\vdots \\ u_n D &= (1 + x^{n-1})D = (n-1)x^{n-2} = (n-1)(u_n - u_1) \\ &= -(n-1)u_1 + 0u_2 + \cdots + 0u_{n-2} + (n-1)u_{n-1} + 0u_n. \end{aligned}$$

The matrix, $m_3(D)$, of D in this basis is

$$m_3(D) = \begin{pmatrix} 0 & 0 & 0 & \cdots & 0 & 0 \\ 1 & 0 & 0 & \cdots & 0 & 0 \\ -2 & 2 & 0 & \cdots & 0 & 0 \\ -3 & 0 & 3 & \cdots & 0 & 0 \\ \cdot & \cdot & \cdot & \cdots & 0 & 0 \\ \cdot & \cdot & \cdot & \cdots & 0 & 0 \\ \vdots & & & & & \\ -(n-1) & 0 & 0 & \cdots & (n-1) & 0 \end{pmatrix}.$$

By the example worked out we see that the matrices of D, for the three bases used, depended completely on the basis. Although different from each other, they still represent the same linear transformation, D, and we could reconstruct D from any of them if we knew the basis used in their determination. However, although different, we might expect that some relationship must hold between $m_1(D)$, $m_2(D)$, and $m_3(D)$. This exact relationship will be determined later.

Since the basis used at any time is completely at our disposal, given a linear transformation T (whose definition, after all, does not depend on any basis) it is natural for us to seek a basis in which the matrix of T has a particularly nice form. For instance, if T is a linear transformation on V, which is n-dimensional over F, and if T has n distinct characteristic roots $\lambda_1, \ldots, \lambda_n$ in F, then by Corollary 2 to Theorem 6.2.3 we can find a basis $v_1, \ldots, v_n$ of V over F such that $v_iT = \lambda_i v_i$. In this basis T has as matrix the especially simple matrix,

$$m(T) = \begin{pmatrix} \lambda_1 & 0 & 0 & \cdots & 0 \\ 0 & \lambda_2 & 0 & \cdots & 0 \\ \cdot & \cdot & \cdot & & \cdot \\ \cdot & \cdot & & & \cdot \\ 0 & 0 & \cdot & \cdots & \lambda_n \end{pmatrix}.$$

We have seen that once a basis of V is picked, to every linear transformation we can associate a matrix. Conversely, having picked a fixed basis $v_1, \ldots, v_n$ of V over F, a given matrix

$$\begin{pmatrix} \alpha_{11} & \cdots & \alpha_{1n} \\ \vdots & & \vdots \\ \alpha_{n1} & \cdots & \alpha_{nn} \end{pmatrix}, \qquad \alpha_{ij} \in F,$$

gives rise to a linear transformation T defined on V by $v_iT = \sum_j \alpha_{ij}v_j$ on this basis. Notice that the matrix of the linear transformation T, just constructed, in the basis $v_1, \ldots, v_n$ is exactly the matrix with which we started. Thus every possible square array serves as the matrix of some linear transformation in the basis $v_1, \ldots, v_n$.

It is clear what is intended by the phrase the first row, second row, . . . , of a matrix, and likewise by the first column, second column, ... In the matrix

$$\begin{pmatrix} \alpha_{11} & \cdots & \alpha_{1n} \\ \vdots & & \vdots \\ \alpha_{n1} & \cdots & \alpha_{nn} \end{pmatrix},$$

the element α_{ij} is in the ith row and jth column; we refer to it as the (i, j) *entry* of the matrix.

To write out the whole square array of a matrix is somewhat awkward; instead we shall always write a matrix as (α_{ij}); this indicates that the (i, j) entry of the matrix is α_{ij}.

Suppose that V is an n-dimensional vector space over F and $v_1, \ldots, v_n$ is a basis of V over F which will remain fixed in the following discussion. Suppose that S and T are linear transformations on V over F having matrices $m(S) = (\sigma_{ij})$, $m(T) = (\tau_{ij})$, respectively, in the given basis. Our objective is to transfer the algebraic structure of $A(V)$ to the set of matrices having entries in F.

To begin with, $S = T$ if and only if $vS = vT$ for any $v \in V$, hence, if and only if $v_iS = v_iT$ for any $v_1, \ldots, v_n$ forming a basis of V over F. Equivalently, $S = T$ if and only if $\sigma_{ij} = \tau_{ij}$ for each i and j.

Given that $m(S) = (\sigma_{ij})$ and $m(T) = (\tau_{ij})$, can we explicitly write down $m(S + T)$? Because $m(S) = (\sigma_{ij})$, $v_iS = \sum_j \sigma_{ij}v_j$; likewise, $v_iT = \sum_j \tau_{ij}v_j$, whence

$$v_i(S + T) = v_iS + v_iT = \sum_j \sigma_{ij}v_j + \sum_j \tau_{ij}v_j = \sum_j (\sigma_{ij} + \tau_{ij})v_j.$$

But then, by what is meant by the matrix of a linear transformation in a given basis, $m(S + T) = (\lambda_{ij})$ where $\lambda_{ij} = \sigma_{ij} + \tau_{ij}$ for every i and j. A computation of the same kind shows that for $\gamma \in F$, $m(\gamma S) = (\mu_{ij})$ where $\mu_{ij} = \gamma\sigma_{ij}$ for every i and j.

The most interesting, and complicated, computation is that of $m(ST)$. Now

$$v_i(ST) = (v_iS)T = \left(\sum_k \sigma_{ik}v_k\right)T = \sum_k \sigma_{ik}(v_kT).$$

However, $v_kT = \sum_j \tau_{kj}v_j$; substituting in the above formula yields

$$v_i(ST) = \sum_k \sigma_{ik}\left(\sum_j \tau_{kj}v_j\right) = \sum_j \left(\sum_k \sigma_{ik}\tau_{kj}\right)v_j.$$

(Prove!) Therefore, $m(ST) = (\nu_{ij})$, where for each i and j, $\nu_{ij} = \sum_k \sigma_{ik}\tau_{kj}$.

At first glance the rule for computing the matrix of the product of two linear transformations in a given basis seems complicated. However, note that the (i, j) entry of $m(ST)$ is obtained as follows: Consider the rows of S as vectors and the columns of T as vectors; then the (i, j) entry of $m(ST)$ is merely the dot product of the ith row of S with the jth column of T.

Let us illustrate this with an example. Suppose that

$$m(S) = \begin{pmatrix} 1 & 2 \\ 3 & 4 \end{pmatrix}$$

and

$$m(T) = \begin{pmatrix} -1 & 0 \\ 2 & 3 \end{pmatrix};$$

the dot product of the first row of S with the first column of T is $(1)(-1) + (2)(2) = 3$, whence the $(1, 1)$ entry of $m(ST)$ is 3; the dot product of the first row of S with the second column of T is $(1)(0) + (2)(3) = 6$, whence the $(1, 2)$ entry of $m(ST)$ is 6; the dot product of the second row of S with the first column of T is $(3)(-1) + (4)(2) = 5$, whence the $(2, 1)$ entry of $m(ST)$ is 5; and, finally the dot product of the second row of S with the second column of T is $(3)(0) + (4)(3) = 12$, whence the $(2, 2)$ entry of $M(ST)$ is 12. Thus

$$m(ST) = \begin{pmatrix} 3 & 6 \\ 5 & 12 \end{pmatrix}.$$

The previous discussion has been intended to serve primarily as a motivation for the constructions we are about to make.

Let F be a field; an *$n \times n$ matrix* over F will be a square array of elements in F,

$$\begin{pmatrix} \alpha_{11} & \alpha_{12} & \cdots & \alpha_{1n} \\ \vdots & \vdots & & \vdots \\ \alpha_{n1} & \alpha_{n2} & \cdots & \alpha_{nn} \end{pmatrix}$$

(which we write as (α_{ij})). Let $F_n = \{(\alpha_{ij}) \mid \alpha_{ij} \in F\}$; in F_n we want to introduce the notion of equality of its elements, an addition, scalar multiplication by elements of F and a multiplication so that it becomes an algebra over F. We use the properties of $m(T)$ for $T \in A(V)$ as our guide in this.

1. We declare $(\alpha_{ij}) = (\beta_{ij})$, for two matrices in F_n, if and only if $\alpha_{ij} = \beta_{ij}$ for each i and j.
2. We define $(\alpha_{ij}) + (\beta_{ij}) = (\lambda_{ij})$ where $\lambda_{ij} = \alpha_{ij} + \beta_{ij}$ for every i, j.
3. We define, for $\gamma \in F$, $\gamma(\alpha_{ij}) = (\mu_{ij})$ where $\mu_{ij} = \gamma\alpha_{ij}$ for every i and j.
4. We define $(\alpha_{ij})(\beta_{ij}) = (\nu_{ij})$, where, for every i and j, $\nu_{ij} = \sum_k \alpha_{ik}\beta_{kj}$.

Let V be an n-dimensional vector space over F and let $v_1, \ldots, v_n$ be a basis of V over F; the matrix, $m(T)$, in the basis $v_1, \ldots, v_n$ associates with $T \in A(V)$ an element, $m(T)$, in F_n. Without further ado we claim that the mapping from $A(V)$ into F_n defined by mapping T onto $m(T)$ is an algebra isomorphism of $A(V)$ onto F_n. Because of this isomorphism, F_n is an associative algebra over F (as can also be verified directly). We call F_n the *algebra of all $n \times n$ matrices over F.*

Every basis of V provides us with an algebra isomorphism of $A(V)$ onto F_n. It is a theorem that every algebra isomorphism of $A(V)$ onto F_n is so obtainable.

In light of the very specific nature of the isomorphism between $A(V)$ and F_n, we shall often identify a linear transformation with its matrix, in some basis, and $A(V)$ with F_n. In fact, F_n can be considered as $A(V)$ acting on the vector space $V = F^{(n)}$ of all n-tuples over F, where for the basis $v_1 = (1, 0, \ldots, 0)$, $v_2 = (0, 1, 0, \ldots, 0), \ldots,$ $v_n = (0, 0, \ldots, 0, 1)$, $(\alpha_{ij}) \in F_n$ acts as $v_i(\alpha_{ij}) = i$th row of (α_{ij}).

We summarize what has been done in

THEOREM 6.3.1 *The set of all $n \times n$ matrices over F form an associative algebra, F_n, over F. If V is an n-dimensional vector space over F, then $A(V)$ and F_n are isomorphic as algebras over F. Given any basis $v_1, \ldots, v_n$ of V over F, if for $T \in A(V)$, $m(T)$ is the matrix of T in the basis $v_1, \ldots. v_n$, the mapping $T \to m(T)$ provides an algebra isomorphism of $A(V)$ onto F_n.*

The zero under addition in F_n is the *zero-matrix* all of whose entries are 0; we shall often write it merely as 0. The *unit matrix*, which is the unit element of F_n under multiplication, is the matrix whose diagonal entries are 1 and whose entries elsewhere are 0; we shall write it as I, I_n (when we wish to emphasize the size of matrices), or merely as 1. For $\alpha \in F$, the matrices

$$\alpha I = \begin{pmatrix} \alpha & & \\ & \ddots & \\ & & \alpha \end{pmatrix}$$

(*blank spaces indicate only* 0 *entries*) are called *scalar matrices*. Because of the isomorphism between $A(V)$ and F_n, it is clear that $T \in A(V)$ is invertible if and only if $m(T)$, as a matrix, has an inverse in F_n.

Given a linear transformation $T \in A(V)$, if we pick two bases, $v_1, \ldots, v_n$ and $w_1, \ldots, w_n$ of V over F, each gives rise to a matrix, namely, $m_1(T)$ and $m_2(T)$, the matrices of T in the bases $v_1, \ldots, v_n$ and $w_1, \ldots, w_n$, respectively. As matrices, that is, as elements of the matrix algebra F_n, what is the relationship between $m_1(T)$ and $m_2(T)$?

THEOREM 6.3.2 *If V is n-dimensional over F and if $T \in A(V)$ has the matrix $m_1(T)$ in the basis $v_1, \ldots, v_n$ and the matrix $m_2(T)$ in the basis $w_1, \ldots, w_n$ of V over F, then there is an element $C \in F_n$ such that $m_2(T) = Cm_1(T)C^{-1}$.*

In fact, if S is the linear transformation of V defined by $v_iS = w_i$ for $i = 1, 2, \ldots, n$, then C can be chosen to be $m_1(S)$.

Proof. Let $m_1(T) = (\alpha_{ij})$ and $m_2(T) = (\beta_{ij})$; thus $v_iT = \sum_j \alpha_{ij}v_j$, $w_iT = \sum_j \beta_{ij}w_j$.

Let S be the linear transformation on V defined by $v_iS = w_i$. Since $v_1, \ldots, v_n$ and $w_1, \ldots, w_n$ are bases of V over F, S maps V onto V, hence, by Theorem 6.1.4, S is invertible in $A(V)$.

Now $w_iT = \sum_j \beta_{ij}w_j$; since $w_i = v_iS$, on substituting this in the expression for w_iT we obtain $(v_iS)T = \sum_j \beta_{ij}(v_jS)$. But then $v_i(ST) = (\sum_j \beta_{ij}v_j)S$; since S is invertible, this further simplifies to $v_i(STS^{-1}) = \sum_j \beta_{ij}v_j$. By the very definition of the matrix of a linear transformation in a given basis, $m_1(STS^{-1}) = (\beta_{ij}) = m_2(T)$. However, the mapping $T \to m_1(T)$ is an isomorphism of $A(V)$ onto F_n; therefore, $m_1(STS^{-1}) = m_1(S)m_1(T)m_1(S^{-1}) = m_1(S)m_1(T)m_1(S)^{-1}$. Putting the pieces together, we obtain $m_2(T) = m_1(S)m_1(T)m_1(S)^{-1}$, which is exactly what is claimed in the theorem.

We illustrate this last theorem with the example of the matrix of D, in various bases, worked out earlier. To minimize the computation, suppose that V is the vector space of all polynomials over F of degree 3 or less, and let D be the differentiation operator defined by $(\alpha_0 + \alpha_1x + \alpha_2x^2 + \alpha_3x^3)D = \alpha_1 + 2\alpha_2x + 3\alpha_3x^2$.

As we saw earlier, in the basis $v_1 = 1$, $v_2 = x$, $v_3 = x^2$, $v_4 = x^3$, the matrix of D is

$$m_1(D) = \begin{pmatrix} 0 & 0 & 0 & 0 \\ 1 & 0 & 0 & 0 \\ 0 & 2 & 0 & 0 \\ 0 & 0 & 3 & 0 \end{pmatrix}.$$

In the basis $u_1 = 1$, $u_2 = 1 + x$, $u_3 = 1 + x^2$, $u_4 = 1 + x^3$, the matrix of D is

$$m_2(D) = \begin{pmatrix} 0 & 0 & 0 & 0 \\ 1 & 0 & 0 & 0 \\ -2 & 2 & 0 & 0 \\ -3 & 0 & 3 & 0 \end{pmatrix}.$$

Let S be the linear transformation of V defined by $v_1S = w_1(=v_1)$, $v_2S = w_2 = 1 + x = v_1 + v_2$, $v_3S = w_3 = 1 + x^2 = v_1 + v_3$, and also $v_4S = w_4 = 1 + x^3 = v_1 + v_4$. The matrix of S in the basis v_1, v_2, v_3, v_4 is

$$C = \begin{pmatrix} 1 & 0 & 0 & 0 \\ 1 & 1 & 0 & 0 \\ 1 & 0 & 1 & 0 \\ 1 & 0 & 0 & 1 \end{pmatrix}.$$

A simple computation shows that

$$C^{-1} = \begin{pmatrix} 1 & 0 & 0 & 0 \\ -1 & 1 & 0 & 0 \\ -1 & 0 & 1 & 0 \\ -1 & 0 & 0 & 1 \end{pmatrix}.$$

Then

$$Cm_1(D)C^{-1} = \begin{pmatrix} 1 & 0 & 0 & 0 \\ 1 & 1 & 0 & 0 \\ 1 & 0 & 1 & 0 \\ 1 & 0 & 0 & 1 \end{pmatrix} \begin{pmatrix} 0 & 0 & 0 & 0 \\ 1 & 0 & 0 & 0 \\ 0 & 2 & 0 & 0 \\ 0 & 0 & 3 & 0 \end{pmatrix} \begin{pmatrix} 1 & 0 & 0 & 0 \\ -1 & 1 & 0 & 0 \\ -1 & 0 & 1 & 0 \\ -1 & 0 & 0 & 1 \end{pmatrix}$$

$$= \begin{pmatrix} 0 & 0 & 0 & 0 \\ 1 & 0 & 0 & 0 \\ -2 & 2 & 0 & 0 \\ -3 & 0 & 3 & 0 \end{pmatrix} = m_2(D),$$

as it should be, according to the theorem. (Verify all the computations used!)

The theorem asserts that, knowing the matrix of a linear transformation in any one basis allows us to compute it in any other, as long as we know the linear transformation (or matrix) of the change of basis.

We still have not answered the question: Given a linear transformation, how does one compute its characteristic roots? This will come later. From the matrix of a linear transformation we shall show how to construct a polynomial whose roots are precisely the characteristic roots of the linear transformation.

Problems

1. Compute the following matrix products:

(a) $\begin{pmatrix} 1 & 2 & 3 \\ 1 & -1 & 2 \\ 3 & 4 & 5 \end{pmatrix} \begin{pmatrix} 1 & 0 & 1 \\ 0 & 2 & 3 \\ -1 & -1 & -1 \end{pmatrix}.$

(b) $\begin{pmatrix} 1 & 6 \\ -6 & 1 \end{pmatrix} \begin{pmatrix} 3 & -2 \\ 2 & 3 \end{pmatrix}.$

(c) $\begin{pmatrix} \frac{1}{3} & \frac{1}{3} & \frac{1}{3} \\ \frac{1}{3} & \frac{1}{3} & \frac{1}{3} \\ \frac{1}{3} & \frac{1}{3} & \frac{1}{3} \end{pmatrix}^2.$

(d) $\begin{pmatrix} 1 & 1 \\ -1 & -1 \end{pmatrix}^2.$

2. Verify all the computations made in the example illustrating Theorem 6.3.2.

3. In F_n prove directly, using the definitions of sum and product, that
   (a) $A(B + C) = AB + AC$;
   (b) $(AB)C = A(BC)$;
   for $A, B, C \in F_n$.
4. In F_2 prove that for any two elements A and B, $(AB - BA)^2$ is a scalar matrix.
5. Let V be the vector space of polynomials of degree 3 or less over F. In V define T by $(\alpha_0 + \alpha_1 x + \alpha_2 x^2 + \alpha_3 x^3)T = \alpha_0 + \alpha_1(x + 1) + \alpha_2(x + 1)^2 + \alpha_3(x + 1)^3$. Compute the matrix of T in the basis
   (a) $1, x, x^2, x^3$.
   (b) $1, 1 + x, 1 + x^2, 1 + x^3$.
   (c) If the matrix in part (a) is A and that in part (b) is B, find a matrix C so that $B = CAC^{-1}$.
6. Let $V = F^{(3)}$ and suppose that

$$\begin{pmatrix} 1 & 1 & 2 \\ -1 & 2 & 1 \\ 0 & 1 & 3 \end{pmatrix}$$

   is the matrix of $T \in A(V)$ in the basis $v_1 = (1, 0, 0)$, $v_2 = (0, 1, 0)$, $v_3 = (0, 0, 1)$. Find the matrix of T in the basis
   (a) $u_1 = (1, 1, 1)$, $u_2 = (0, 1, 1)$, $u_3 = (0, 0, 1)$.
   (b) $u_1 = (1, 1, 0)$, $u_2 = (1, 2, 0)$, $u_3 = (1, 2, 1)$.
7. Prove that, given the matrix

$$A = \begin{pmatrix} 0 & 1 & 0 \\ 0 & 0 & 1 \\ 6 & -11 & 6 \end{pmatrix} \in F_3$$

   (where the characteristic of F is not 2), then
   (a) $A^3 - 6A^2 + 11A - 6 = 0$.
   (b) There exists a matrix $C \in F_3$ such that

$$CAC^{-1} = \begin{pmatrix} 1 & 0 & 0 \\ 0 & 2 & 0 \\ 0 & 0 & 3 \end{pmatrix}.$$

8. Prove that it is impossible to find a matrix $C \in F_2$ such that

$$C\begin{pmatrix} 1 & 1 \\ 0 & 1 \end{pmatrix}C^{-1} = \begin{pmatrix} \alpha & 0 \\ 0 & \beta \end{pmatrix},$$

   for any $\alpha, \beta \in F$.
9. A matrix $A \in F_n$ is said to be a *diagonal* matrix if all the entries off the main diagonal of A are 0, i.e., if $A = (\alpha_{ij})$ and $\alpha_{ij} = 0$ for $i \neq j$. If A is a diagonal matrix all of whose entries on the main diagonal

are distinct, find all the matrices $B \in F_n$ which commute with A, that is, all matrices B such that $BA = AB$.

10. Using the result of Problem 9, prove that the only matrices in F_n which commute with all matrices in F_n are the scalar matrices.

11. Let $A \in F_n$ be the matrix

$$A = \begin{pmatrix} 0 & 1 & 0 & 0 & \dots & 0 & 0 \\ 0 & 0 & 1 & 0 & \dots & 0 & 0 \\ 0 & 0 & 0 & 1 & \dots & 0 & 0 \\ \vdots & & & & & \vdots & \vdots \\ 0 & 0 & 0 & 0 & \dots & 0 & 1 \\ 0 & 0 & 0 & 0 & \dots & 0 & 0 \end{pmatrix},$$

whose entries everywhere, except on the superdiagonal, are 0, and whose entries on the superdiagonal are 1's. Prove $A^n = 0$ but $A^{n-1} \neq 0$.

*12. If A is as in Problem 11, find all matrices in F_n which commute with A and show that they must be of the form $\alpha_0 + \alpha_1 A + \alpha_2 A^2 + \cdots + \alpha_{n-1}A^{n-1}$ where $\alpha_0, \alpha_1, \ldots, \alpha_{n-1} \in F$.

13. Let $A \in F_2$ and let $C(A) = \{B \in F_2 \mid AB = BA\}$. Let $C(C(A)) = \{G \in F_2 \mid GX = XG \text{ for all } X \in C(A)\}$. Prove that if $G \in C(C(A))$ then G is of the form $\alpha_0 + \alpha_1 A$, $\alpha_0, \alpha_1 \in F$.

14. Do Problem 13 for $A \in F_3$, proving that every $G \in C(C(A))$ is of the form $\alpha_0 + \alpha_1 A + \alpha_2 A^2$.

15. In F_n let the matrices E_{ij} be defined as follows: E_{ij} is the matrix whose only nonzero entry is the (i, j) entry, which is 1. Prove
    (a) The E_{ij} form a basis of F_n over F.
    (b) $E_{ij}E_{kl} = 0$ for $j \neq k$; $E_{ij}E_{jl} = E_{il}$.
    (c) Given i, j, there exists a matrix C such that $CE_{ii}C^{-1} = E_{jj}$.
    (d) If $i \neq j$ there exists a matrix C such that $CE_{ij}C^{-1} = E_{12}$.
    (e) Find all $B \in F_n$ commuting with E_{12}.
    (f) Find all $B \in F_n$ commuting with E_{11}.

16. Let F be the field of real numbers and let C be the field of complex numbers. For $a \in C$ let $T_a: C \to C$ by $xT_a = xa$ for all $x \in C$. Using the basis $1, i$ find the matrix of the linear transformation T_a and so get an isomorphic representation of the complex numbers as 2×2 matrices over the real field.

17. Let Q be the division ring of quaternions over the real field. Using the basis $1, i, j, k$ of Q over F, proceed as in Problem 16 to find an isomorphic representation of Q by 4×4 matrices over the field of real numbers.

*18. Combine the results of Problems 16 and 17 to find an isomorphic representation of Q as 2×2 matrices over the field of complex numbers.

19. Let $\mathcal{M}$ be the set of all $n \times n$ matrices having entries 0 and 1 in such a way that there is one 1 in each row and column. (Such matrices are called *permutation matrices*.)
    (a) If $M \in \mathcal{M}$, describe AM in terms of the rows and columns of A.
    (b) If $M \in \mathcal{M}$, describe MA in terms of the rows and columns of A.
20. Let $\mathcal{M}$ be as in Problem 19. Prove
    (a) $\mathcal{M}$ has $n!$ elements.
    (b) If $M \in \mathcal{M}$, then it is invertible and its inverse is again in $\mathcal{M}$.
    (c) Give the explicit form of the inverse of M.
    (d) Prove that $\mathcal{M}$ is a group under matrix multiplication.
    (e) Prove that $\mathcal{M}$ is isomorphic, as a group, to S_n, the symmetric group of degree n.
21. Let $A = (\alpha_{ij})$ be such that for each i, $\sum_j \alpha_{ij} = 1$. Prove that 1 is a characteristic root of A (that is, $1 - A$ is not invertible).
22. Let $A = (\alpha_{ij})$ be such that for every j, $\sum_i \alpha_{ij} = 1$. Prove that 1 is a characteristic root of A.
23. Find necessary and sufficient conditions on $\alpha, \beta, \gamma, \delta$, so that $A = \begin{pmatrix} \alpha & \beta \\ \gamma & \delta \end{pmatrix}$ is invertible. When it is invertible, write down A^{-1} explicitly.
24. If $E \in F_n$ is such that $E^2 = E \neq 0$ prove that there is a matrix $C \in F_n$ such that
$$CEC^{-1} = \left(\begin{array}{cccc|ccc} 1 & 0 & \cdots & 0 & 0 & \cdots & 0 \\ 0 & 1 & \cdots & 0 & & & \\ \vdots & \vdots & & \vdots & \vdots & & \vdots \\ 0 & 0 & \cdots & 1 & 0 & \cdots & 0 \\ \hline 0 & & \cdots & 0 & 0 & \cdots & 0 \\ 0 & & \cdots & 0 & 0 & \cdots & 0 \end{array}\right),$$
where the unit matrix in the top left corner is $r \times r$, where r is the rank of E.
25. If F is the real field, prove that it is impossible to find matrices $A, B \in F_n$ such that $AB - BA = 1$.
26. If F is of characteristic 2, prove that in F_2 it is possible to find matrices A, B such that $AB - BA = 1$.
27. The matrix A is called *triangular* if all the entries above the main diagonal are 0. (If all the entries below the main diagonal are 0 the matrix is also called triangular).
    (a) If A is triangular and no entry on the main diagonal is 0, prove that A is invertible.
    (b) If A is triangular and an entry on the main diagonal is 0, prove that A is singular.

28. If A is triangular, prove that its characteristic roots are precisely the elements on its main diagonal.
29. If $N^k = 0$, $N \in F_n$, prove that $1 + N$ is invertible and find its inverse as a polynomial in N.
30. If $A \in F_n$ is triangular and all the entries on its main diagonal are 0, prove that $A^n = 0$.
31. If $A \in F_n$ is triangular and all the entries on its main diagonal are equal to $\alpha \neq 0 \in F$, find A^{-1}.
32. Let S, T be linear transformations on V such that the matrix of S in one basis is equal to the matrix of T in another. Prove there exists a linear transformation A on V such that $T = ASA^{-1}$.

6.4 Canonical Forms: Triangular Form

Let V be an n-dimensional vector space over a field F.

DEFINITION The linear transformations $S, T \in A(V)$ are said to be *similar* if there exists an invertible element $C \in A(V)$ such that $T = CSC^{-1}$.

In view of the results of Section 6.3, this definition translates into one about matrices. In fact, since F_n acts as $A(V)$ on $F^{(n)}$, the above definition already defines similarity of matrices. By it, $A, B \in F_n$ are similar if there is an invertible $C \in F_n$ such that $B = CAC^{-1}$.

The relation on $A(V)$ defined by similarity is an equivalence relation; the equivalence class of an element will be called its *similarity* class. Given two linear transformations, how can we determine whether or not they are similar? Of course, we could scan the similarity class of one of these to see if the other is in it, but this procedure is not a feasible one. Instead we try to establish some kind of landmark in each similarity class and a way of going from any element in the class to this landmark. We shall prove the existence of linear transformations in each similarity class whose matrix, in some basis, is of a particularly nice form. These matrices will be called the *canonical forms*. To determine if two linear transformations are similar, we need but compute a particular canonical form for each and check if these are the same.

There are many possible canonical forms; we shall only consider three of these, namely, the triangular form, Jordan form, and the rational canonical form, in this and the next three sections.

DEFINITION The subspace W of V is *invariant under* $T \in A(V)$ if $WT \subset W$.

LEMMA 6.4.1 *If $W \subset V$ is invariant under T, then T induces a linear transformation $\bar{T}$ on V/W, defined by $(v + W)\bar{T} = vT + W$. If T satisfies* *the polynomial $q(x) \in F[x]$, then so does $\bar{T}$. If $p_1(x)$ is the minimal polynomial for $\bar{T}$ over F and if $p(x)$ is that for T, then $p_1(x) \mid p(x)$.*

Proof. Let $\bar{V} = V/W$; the elements of $\bar{V}$ are, of course, the cosets $v + W$ of W in V. Given $\bar{v} = v + W \in \bar{V}$ define $\bar{v}\bar{T} = vT + W$. To verify that $\bar{T}$ has all the formal properties of a linear transformation on $\bar{V}$ is an easy matter *once it has been established that $\bar{T}$ is well defined on $\bar{V}$*. We thus content ourselves with proving this fact.

Suppose that $\bar{v} = v_1 + W = v_2 + W$ where $v_1, v_2 \in V$. We must show that $v_1T + W = v_2T + W$. Since $v_1 + W = v_2 + W$, $v_1 - v_2$ must be in W, and since W is invariant under T, $(v_1 - v_2)T$ must also be in W. Consequently $v_1T - v_2T \in W$, from which it follows that $v_1T + W = v_2T + W$, as desired. We now know that $\bar{T}$ defines a linear transformation on $\bar{V} = V/W$.

If $\bar{v} = v + W \in \bar{V}$, then $\bar{v}(\overline{T^2}) = vT^2 + W = (vT)T + W = (vT + W)\bar{T} = ((v + W)\bar{T})\bar{T} = \bar{v}(\bar{T})^2$; thus $(\overline{T^2}) = (\bar{T})^2$. Similarly, $(\overline{T^k}) = (\bar{T})^k$ for any $k \geq 0$. Consequently, for any polynomial $q(x) \in F[x]$, $\overline{q(T)} = q(\bar{T})$. For any $q(x) \in F[x]$ with $q(T) = 0$, since $\bar{0}$ is the zero transformation on $\bar{V}$, $0 = \overline{q(T)} = q(\bar{T})$.

Let $p_1(x)$ be the minimal polynomial over F satisfied by $\bar{T}$. If $q(\bar{T}) = 0$ for $q(x) \in F[x]$, then $p_1(x) \mid q(x)$. If $p(x)$ is the minimal polynomial for T over F, then $p(T) = 0$, whence $p(\bar{T}) = 0$; in consequence, $p_1(x) \mid p(x)$.

As we saw in Theorem 6.2.2, all the characteristic roots of T which lie in F are roots of the minimal polynomial of T over F. *We say that all the characteristic roots of T are in F if all the roots of the minimal polynomial of T over F lie in F.*

In Problem 27 at the end of the last section, we defined a matrix as being *triangular* if all its entries above the main diagonal were 0. Equivalently, if T is a linear transformation on V over F, the matrix of T in the basis $v_1, \ldots, v_n$ is triangular if

$$\begin{aligned} v_1T &= \alpha_{11}v_1 \\ v_2T &= \alpha_{21}v_1 + \alpha_{22}v_2 \\ &\vdots \\ v_iT &= \alpha_{i1}v_1 + \alpha_{i2}v_2 + \cdots + \alpha_{ii}v_i, \\ v_nT &= \alpha_{n1}v_1 + \cdots + \alpha_{mn}v_n, \end{aligned}$$

i.e., if v_iT is a linear combination only of v_i and its predecessors in the basis.

THEOREM 6.4.1 *If $T \in A(V)$ has all its characteristic roots in F, then there is a basis of V in which the matrix of T is triangular.*

Proof. The proof goes by induction on the dimension of V over F.

If $\dim_F V = 1$, then every element in $A(V)$ is a scalar, and so the theorem is true here.

Suppose that the theorem is true for all vector spaces over F of dimension $n - 1$, and let V be of dimension n over F.

The linear transformation T on V has all its characteristic roots in F; let $\lambda_1 \in F$ be a characteristic root of T. There exists a nonzero vector v_1 in V such that $v_1 T = \lambda_1 v_1$. Let $W = \{\alpha v_1 \mid \alpha \in F\}$; W is a one-dimensional subspace of V, and is invariant under T. Let $\bar{V} = V/W$; by Lemma 4.2.6, $\dim \bar{V} = \dim V - \dim W = n - 1$. By Lemma 6.4.1, T induces a linear transformation $\bar{T}$ on $\bar{V}$ whose minimal polynomial over F divides the minimal polynomial of T over F. Thus all the roots of the minimal polynomial of $\bar{T}$, being roots of the minimal polynomial of T, must lie in F. The linear transformation $\bar{T}$ in its action on $\bar{V}$ satisfies the hypothesis of the theorem; since $\bar{V}$ is $(n-1)$-dimensional over F, by our induction hypothesis, there is a basis $\bar{v}_2, \bar{v}_3, \ldots, \bar{v}_n$ of $\bar{V}$ over F such that

$$\begin{aligned}
\bar{v}_2 \bar{T} &= \alpha_{22}\bar{v}_2 \\
\bar{v}_3 \bar{T} &= \alpha_{32}\bar{v}_2 + \alpha_{33}\bar{v}_3 \\
&\vdots \\
\bar{v}_i \bar{T} &= \alpha_{i2}\bar{v}_2 + \alpha_{i3}\bar{v}_3 + \cdots + \alpha_{ii}\bar{v}_i \\
&\vdots \\
\bar{v}_n \bar{T} &= \alpha_{n2}\bar{v}_2 + \alpha_{n3}\bar{v}_3 + \cdots + \alpha_{nn}\bar{v}_n.
\end{aligned}$$

Let $v_2, \ldots, v_n$ be elements of V mapping into $\bar{v}_2, \ldots, \bar{v}_n$, respectively. Then $v_1, v_2, \ldots, v_n$ form a basis of V (see Problem 3, end of this section). Since $\bar{v}_2\bar{T} = \alpha_{22}\bar{v}_2$, $\bar{v}_2\bar{T} - \alpha_{22}\bar{v}_2 = 0$, whence $v_2 T - \alpha_{22}v_2$ must be in W. Thus $v_2 T - \alpha_{22}v_2$ is a multiple of v_1, say $\alpha_{21}v_1$, yielding, after transposing, $v_2 T = \alpha_{21}v_1 + \alpha_{22}v_2$. Similarly, $v_i T - \alpha_{i2}v_2 - \alpha_{i3}v_3 - \cdots - \alpha_{ii}v_i \in W$, whence $v_i T = \alpha_{i1}v_1 + \alpha_{i2}v_2 + \cdots + \alpha_{ii}v_i$. The basis $v_1, \ldots, v_n$ of V over F provides us with a basis where every $v_i T$ is a linear combination of v_i and its predecessors in the basis. Therefore, the matrix of T in this basis is triangular. This completes the induction and proves the theorem.

We wish to restate Theorem 6.4.1 for matrices. Suppose that the matrix $A \in F_n$ has all its characteristic roots in F. A defines a linear transformation T on $F^{(n)}$ whose matrix in the basis

$$v_1 = (1, 0, \ldots, 0), v_2 = (0, 1, 0, \ldots, 0), \ldots, v_n = (0, 0, \ldots, 0, 1),$$

is precisely A. The characteristic roots of T, being equal to those of A, are all in F, whence by Theorem 6.4.1, there is a basis of $F^{(n)}$ in which the matrix of T is triangular. However, by Theorem 6.3.2, this change of basis merely changes the matrix of T, namely A, in the first basis, into CAC^{-1} for a suitable $C \subset F_n$. Thus

ALTERNATIVE FORM OF THEOREM 6.4.1 *If the matrix $A \in F_n$ has all its characteristic roots in F, then there is a matrix $C \in F_n$ such that CAC^{-1} is a triangular matrix.*

Theorem 6.4.1 (in either form) is usually described by saying that T (or A) can be *brought to triangular form over* F.

If we glance back at Problem 28 at the end of Section 6.3, we see that after T has been brought to triangular form, the elements on the main diagonal of its matrix play the following significant role: *they are precisely the characteristic roots of* T.

We conclude the section with

THEOREM 6.4.2 *If V is n-dimensional over F and if $T \in A(V)$ has all its characteristic roots in F, then T satisfies a polynomial of degree n over F.*

Proof. By Theorem 6.4.1, we can find a basis $v_1, \ldots, v_n$ of V over F such that:

$$\begin{aligned} v_1 T &= \lambda_1 v_1 \\ v_2 T &= \alpha_{21} v_1 + \lambda_2 v_2 \\ &\vdots \\ v_i T &= \alpha_{i1} v_1 + \cdots + \alpha_{i,i-1} v_{i-1} + \lambda_i v_i, \end{aligned}$$

for $i = 1, 2, \ldots, n$.

Equivalently

$$\begin{aligned} v_1(T - \lambda_1) &= 0 \\ v_2(T - \lambda_2) &= \alpha_{21} v_1 \\ &\vdots \\ v_i(T - \lambda_1) &= \alpha_{i1} v_1 + \cdots + \alpha_{i,i-1} v_{i-1}, \end{aligned}$$

for $i = 1, 2, \ldots, n$.

What is $v_2(T - \lambda_2)(T - \lambda_1)$? As a result of $v_2(T - \lambda_2) = \alpha_{21} v_1$ and $v_1(T - \lambda_1) = 0$, we obtain $v_2(T - \lambda_2)(T - \lambda_1) = 0$. Since

$$\begin{aligned} (T - \lambda_2)(T - \lambda_1) &= (T - \lambda_1)(T - \lambda_2), \\ v_1(T - \lambda_2)(T - \lambda_1) &= v_1(T - \lambda_1)(T - \lambda_2) = 0. \end{aligned}$$

Continuing this type of computation yields

$$\begin{aligned} v_1(T - \lambda_i)(T - \lambda_{i-1}) \cdots (T - \lambda_1) &= 0, \\ v_2(T - \lambda_i)(T - \lambda_{i-1}) \cdots (T - \lambda_1) &= 0, \\ &\vdots \\ v_i(T - \lambda_i)(T - \lambda_{i-1}) \cdots (T - \lambda_1) &= 0. \end{aligned}$$

For $i = n$, the matrix $S = (T - \lambda_n)(T - \lambda_{n-1}) \cdots (T - \lambda_1)$ satisfies $v_1 S = v_2 S = \cdots = v_n S = 0$. Then, since S annihilates a basis of V, S must annihilate all of V. Therefore, $S = 0$. Consequently, T satisfies the polynomial $(x - \lambda_1)(x - \lambda_2) \cdots (x - \lambda_n)$ in $F[x]$ of degree n, proving the theorem.

Unfortunately, it is in the nature of things that not every linear transformation on a vector space over every field F has all its characteristic roots in F. This depends totally on the field F. For instance, if F is the field of real numbers, then the minimal equation of

$$\begin{pmatrix} 0 & 1 \\ -1 & 0 \end{pmatrix}$$

over F is $x^2 + 1$, which has no roots in F. Thus we have no right to assume that characteristic roots always lie in the field in question. However, we may ask, can we slightly enlarge F to a new field K so that everything works all right over K?

The discussion will be made for matrices; it could be carried out equally well for linear transformations. What would be needed would be the following: given a vector space V over a field F of dimension n, and given an extension K of F, then we can embed V into a vector space V_K over K of dimension n over K. One way of doing this would be to take a basis $v_1, \ldots, v_n$ of V over F and to consider V_K as the set of all $\alpha_1 v_1 + \cdots + \alpha_n v_n$ with the $\alpha_i \in K$, considering the v_i linearly independent over K. This heavy use of a basis is unaesthetic; the whole thing can be done in a basis-free way by introducing the concept of *tensor product* of vector spaces. We shall not do it here; instead we argue with matrices (which is effectively the route outlined above using a fixed basis of V).

Consider the algebra F_n. If K is any extension field of F, then $F_n \subset K_n$ the set of $n \times n$ matrices over K. Thus any matrix over F can be considered as a matrix over K. If $T \in F_n$ has the minimal polynomial $p(x)$ over F, considered as an element of K_n it might conceivably satisfy a different polynomial $p_0(x)$ over K. But then $p_0(x) \mid p(x)$, since $p_0(x)$ divides all polynomials over K (and hence all polynomials over F) which are satisfied by T. We now specialize K. By Theorem 5.3.2 there is a finite extension, K, of F in which the minimal polynomial, $p(x)$, for T over F has all its roots. As an element of K_n, for this K, does T have all its characteristic roots in K? As an element of K_n, the minimal polynomial for T over K, $p_0(x)$ divides $p(x)$ so all the roots of $p_0(x)$ are roots of $p(x)$ and therefore lie in K. Consequently, as an element in K_n, T has all its characteristic roots in K.

Thus, given T in F_n, by going to the splitting field, K, of its minimal polynomial we achieve the situation where the hypotheses of Theorems 6.4.1 and 6.4.2 are satisfied, not over F, but over K. Therefore, for instance, T can be brought to triangular form over K and satisfies a polynomial of degree n over K. Sometimes, when luck is with us, knowing that a certain result is true over K we can "cut back" to F and know that the result is still true over F. However, going to K is no panacea for there are frequent situations when the result for K implies nothing for F. This is why we have two types of "canonical form" theorems, those which assume that all the characteristic roots of T lie in F and those which do not.

A final word; if $T \in F_n$, by the phrase "*a characteristic root of* T" we shall mean an element λ in the splitting field K of the minimal polynomial $p(x)$ of T over F such that $\lambda - T$ is not invertible in K_n. It is a fact (see Problem 5) that every root of the minimal polynomial of T over F is a characteristic root of T.

Problems

1. Prove that the relation of similarity is an equivalence relation in $A(V)$.
2. If $T \in F_n$ and if $K \supset F$, prove that as an element of K_n, T is invertible if and only if it is already invertible in F_n.
3. In the proof of Theorem 6.4.1 prove that $v_1, \ldots, v_n$ is a basis of V.
4. Give a proof, using matrix computations, that if A is a triangular $n \times n$ matrix with entries $\lambda_1, \ldots, \lambda_n$ on the diagonal, then

$$(A - \lambda_1)(A - \lambda_2) \cdots (A - \lambda_n) = 0.$$

*5. If $T \in F_n$ has minimal polynomial $p(x)$ over F, prove that every root of $p(x)$, in its splitting field K, is a characteristic root of T.

6. If $T \in A(V)$ and if $\lambda \in F$ is a characteristic root of T in F, let $U_\lambda = \{v \in V \mid vT = \lambda v\}$. If $S \in A(V)$ commutes with T, prove that U_λ is invariant under S.

*7. If $\mathcal{M}$ is a commutative set of elements in $A(V)$ such that every $M \in \mathcal{M}$ has all its characteristic roots in F, prove that there is a $C \in A(V)$ such that every CMC^{-1}, for $M \in \mathcal{M}$, is in triangular form.

8. Let W be a subspace of V invariant under $T \in A(V)$. By restricting T to W, T induces a linear transformation $\tilde{T}$ (defined by $w\tilde{T} = wT$ for every $w \in W$). Let $\tilde{p}(x)$ be the minimal polynomial of $\tilde{T}$ over F.
   (a) Prove that $\widetilde{p(x)} \mid p(x)$, the minimal polynomial of T over F.
   (b) If T induces $\bar{T}$ on V/W satisfying the minimal polynomial $\bar{p}(x)$ over F, prove that $p(x) \mid \tilde{p}(x)\bar{p}(x)$.
   *(c) If $\tilde{p}(x)$ and $\bar{p}(x)$ are relatively prime, prove that $p(x) = \tilde{p}(x)\bar{p}(x)$.
   *(d) Give an example of a T for which $p(x) \neq \tilde{p}(x)\bar{p}(x)$.

9. Let $\mathcal{M}$ be a nonempty set of elements in $A(V)$; the subspace $W \subset V$ is said to be *invariant under* $\mathcal{M}$ if for every $M \in \mathcal{M}$, $WM \subset W$. If W is invariant under $\mathcal{M}$ and is of dimension r over F, prove that there exists a basis of V over F such that every $M \in \mathcal{M}$ has a matrix, in this basis, of the form

$$\left(\begin{array}{c|c} M_1 & 0 \\ \hline M_{12} & M_2 \end{array}\right),$$

where M_1 is an $r \times r$ matrix and M_2 is an $(n - r) \times (n - r)$ matrix.

10. In Problem 9 prove that M_1 is the matrix of the linear transformation $\tilde{M}$ induced by M on W, and that M_2 is the matrix of the linear transformation $\bar{M}$ induced by M on V/W.

*11. The nonempty set, $\mathcal{M}$, of linear transformations in $A(V)$ is called an *irreducible* set if the only subspaces of V invariant under $\mathcal{M}$ are (0) and V. If $\mathcal{M}$ is an irreducible set of linear transformations on V and if

$$D = \{T \in A(V) \mid TM = MT \text{ for all } M \in \mathcal{M}\},$$

prove that D is a division ring.

*12. Do Problem 11 by using the result (Schur's lemma) of Problem 14, end of Chapter 4, page 206.

*13. If F is such that all elements in $A(V)$ have all their characteristic roots in F, prove that the D of Problem 11 consists only of scalars.

14. Let F be the field of real numbers and let

$$\begin{pmatrix} 0 & 1 \\ -1 & 0 \end{pmatrix} \in F_2.$$

(a) Prove that the set $\mathcal{M}$ consisting only of

$$\begin{pmatrix} 0 & 1 \\ -1 & 0 \end{pmatrix}$$

is an irreducible set.

(b) Find the set D of all matrices commuting with

$$\begin{pmatrix} 0 & 1 \\ -1 & 0 \end{pmatrix}$$

and prove that D is isomorphic to the field of complex numbers.

15. Let F be the field of real numbers.

(a) Prove that the set

$$\mathcal{M} = \left\{ \begin{pmatrix} 0 & 1 & 0 & 0 \\ -1 & 0 & 0 & 0 \\ 0 & 0 & 0 & 1 \\ 0 & 0 & -1 & 0 \end{pmatrix}, \begin{pmatrix} 0 & 0 & 0 & 1 \\ 0 & 0 & 1 & 0 \\ 0 & -1 & 0 & 0 \\ -1 & 0 & 0 & 0 \end{pmatrix} \right\}$$

is an irreducible set.

(b) Find all $A \in F_4$ such that $AM = MA$ for all $M \in \mathcal{M}$.

(c) Prove that the set of all A in part (b) is a division ring isomorphic to the division ring of quaternions over the real field.

16. A set of linear transformations, $\mathcal{M} \subset A(V)$, is called *decomposable* if there is a subspace $W \subset V$ such that $V = W \oplus W_1$, $W \neq (0)$, $W \neq V$, and each of W and W_1 is invariant under $\mathcal{M}$. If $\mathcal{M}$ is not decomposable, it is called *indecomposable*.

(a) If $\mathcal{M}$ is a decomposable set of linear transformations on V, prove that there is a basis of V in which every $M \in \mathcal{M}$ has a matrix of the form

$$\left(\begin{array}{c|c} M_1 & 0 \\ \hline 0 & M_2 \end{array}\right),$$

where M_1 and M_2 are square matrices.

(b) If V is an n-dimensional vector space over F and if $T \in A(V)$ satisfies $T^n = 0$ but $T^{n-1} \neq 0$, prove that the set $\{T\}$ (consisting of T) is indecomposable.

17. Let $T \in A(V)$ and suppose that $p(x)$ is the minimal polynomial for T over F.
    (a) If $p(x)$ is divisible by two distinct irreducible polynomials $p_1(x)$ and $p_2(x)$ in $F[x]$, prove that $\{T\}$ is decomposable.
    (b) If $\{T\}$, for some $T \in A(V)$ is indecomposable, prove that the minimal polynomial for T over F is the power of an irreducible polynomial.
18. If $T \in A(V)$ is nilpotent, prove that T can be brought to triangular form over F, and in that form all the elements on the diagonal are 0.
19. If $T \in A(V)$ has only 0 as a characteristic root, prove that T is nilpotent.

6.5 Canonical Forms: Nilpotent Transformations

One class of linear transformations which have all their characteristic roots in F is the class of nilpotent ones, for their characteristic roots are all 0, hence are in F. Therefore by the result of the previous section a nilpotent linear transformation can always be brought to triangular form over F. For some purposes this is not sharp enough, and as we shall soon see, a great deal more can be said.

Although the class of nilpotent linear transformations is a rather restricted one, it nevertheless merits study for its own sake. More important for our purposes, once we have found a good canonical form for these we can readily find a good canonical form for all linear transformations which have all their characteristic roots in F.

A word about the line of attack that we shall follow is in order. We could study these matters from a "ground-up" approach or we could invoke results about the decomposition of modules which we obtained in Chapter 4. We have decided on a compromise between the two; we treat the material in this section and the next (on Jordan forms) independently of the notion of a module and the results about modules developed in Chapter 4. However, in the section dealing with the rational canonical form we shall completely change point of view, introducing via a given linear transformation a module structure on the vector spaces under discussion; making use of

Theorem 4.5.1 we shall then get a decomposition of a vector space, and the resulting canonical form, relative to a given linear transformation.

Even though we do not use a module theoretic approach now, the reader should note the similarity between the arguments used in proving Theorem 4.5.1 and those used to prove Lemma 6.5.4.

Before concentrating our efforts on nilpotent linear transformations we prove a result of interest which holds for arbitrary ones.

LEMMA 6.5.1 *If $V = V_1 \oplus V_2 \oplus \cdots \oplus V_k$, where each subspace V_i is of dimension n_i and is invariant under T, an element of $A(V)$, then a basis of V can be found so that the matrix of T in this basis is of the form*

$$\begin{pmatrix} A_1 & 0 & \dots & 0 \\ 0 & A_2 & \dots & 0 \\ \vdots & \vdots & & \vdots \\ 0 & 0 & \dots & A_k \end{pmatrix}$$

where each A_i is an $n_i \times n_i$ matrix and is the matrix of the linear transformation induced by T on V_i.

Proof. Choose a basis of V as follows: $v_1^{(1)}, \dots, v_{n_1}^{(1)}$ is a basis of V_1, $v_1^{(2)}, v_2^{(2)}, \dots, v_{n_2}^{(2)}$ is a basis of V_2, and so on. Since each V_i is invariant under T, $v_j^{(i)}T \in V_i$ so is a linear combination of $v_1^{(i)}, v_2^{(i)}, \dots, v_{n_i}^{(i)}$, and of only these. Thus the matrix of T in the basis so chosen is of the desired form. That each A_i is the matrix of T_i, the linear transformation induced on V_i by T, is clear from the very definition of the matrix of a linear transformation.

We now narrow our attention to nilpotent linear transformations.

LEMMA 6.5.2 *If $T \in A(V)$ is nilpotent, then $\alpha_0 + \alpha_1 T + \cdots + \alpha_m T^m$, where the $\alpha_i \in F$, is invertible if $\alpha_0 \neq 0$.*

Proof. If S is nilpotent and $\alpha_0 \neq 0 \in F$, a simple computation shows that

$$(\alpha_0 + S)\left(\frac{1}{\alpha_0} - \frac{S}{\alpha_0^{\,2}} + \frac{S^2}{\alpha_0^{\,3}} + \cdots + (-1)^{r-1}\frac{S^{r-1}}{\alpha_0^{\,r}}\right) = 1,$$

if $S^r = 0$. Now if $T^r = 0$, $S = \alpha_1 T + \alpha_2 T^2 + \cdots + \alpha_m T^m$ also must satisfy $S^r = 0$. (Prove!) Thus for $\alpha_0 \neq 0$ in F, $\alpha_0 + S$ is invertible.

Notation. M_t will denote the $t \times t$ matrix

$$\begin{pmatrix} 0 & 1 & 0 & \dots & 0 & 0 \\ 0 & 0 & 1 & \dots & 0 & 0 \\ \vdots & & & & & \vdots \\ 0 & 0 & & \dots & 0 & 1 \\ 0 & 0 & & \dots & 0 & 0 \end{pmatrix},$$

all of whose entries are 0 except on the superdiagonal, where they are all 1's.

DEFINITION If $T \in A(V)$ is nilpotent, then k is called the *index of nilpotence* of T if $T^k = 0$ but $T^{k-1} \neq 0$.

The key result about nilpotent linear transformations is

THEOREM 6.5.1 *If $T \in A(V)$ is nilpotent, of index of nilpotence n_1, then a basis of V can be found such that the matrix of T in this basis has the form*

$$\begin{pmatrix} M_{n_1} & 0 & \cdots & 0 \\ 0 & M_{n_2} & \cdots & 0 \\ \vdots & & \ddots & \vdots \\ 0 & 0 & \cdots & M_{n_r} \end{pmatrix},$$

where $n_1 \geq n_2 \geq \cdots \geq n_r$ and where $n_1 + n_2 + \cdots + n_r = \dim_F V$.

Proof. The proof will be a little detailed, so as we proceed we shall separate parts of it out as lemmas.

Since $T^{n_1} = 0$ but $T^{n_1 - 1} \neq 0$, we can find a vector $v \in V$ such that $vT^{n_1-1} \neq 0$. We claim that the vectors $v, vT, \ldots, vT^{n_1-1}$ are linearly independent over F. For, suppose that $\alpha_1 v + \alpha_2 vT + \cdots + \alpha_{n_1} vT^{n_1-1} = 0$ where the $\alpha_i \in F$; let α_s be the first nonzero α, hence

$$vT^{s-1}(\alpha_s + \alpha_{s+1}T + \cdots + \alpha_{n_1}T^{n_1-s}) = 0.$$

Since $\alpha_s \neq 0$, by Lemma 6.5.2, $\alpha_s + \alpha_{s+1}T + \cdots + \alpha_{n_1}T^{n_1-s}$ is invertible, and therefore $vT^{s-1} = 0$. However, $s < n_1$, thus this contradicts that $vT^{n_1-1} \neq 0$. Thus no such nonzero α_s exists and $v, vT, \ldots, vT^{n_1-1}$ have been shown to be linearly independent over F.

Let V_1 be the subspace of V spanned by $v_1 = v$, $v_2 = vT, \ldots, v_{n_1} = vT^{n_1-1}$; V_1 is invariant under T, and, in the basis above, the linear transformation induced by T on V_1 has as matrix M_{n_1}.

So far we have produced the upper left-hand corner of the matrix of the theorem. We must somehow produce the rest of this matrix.

LEMMA 6.5.3 *If $u \in V_1$ is such that $uT^{n_1-k} = 0$, where $0 < k \leq n_1$, then $u = u_0T^k$ for some $u_0 \in V_1$.*

Proof. Since $u \in V_1$, $u = \alpha_1 v + \alpha_2 vT + \cdots + \alpha_k vT^{k-1} + a_{k+1}vT^k + \cdots + \alpha_{n_1}vT^{n_1-1}$. Thus $0 = uT^{n_1-k} = \alpha_1 vT^{n_1-k} + \cdots + \alpha_k vT^{n_1-1}$. However, $vT^{n_1-k}, \ldots, vT^{n_1-1}$ are linearly independent over F, whence $\alpha_1 = \alpha_2 = \cdots = \alpha_k = 0$, and so, $u = \alpha_{k+1}vT^k + \cdots + \alpha_{n_1}vT^{n_1-1} = u_0T^k$, where $u_0 = \alpha_{k+1}v + \cdots + \alpha_{n_1}vT^{n_1-k-1} \in V_1$.

The argument, so far, has been fairly straightforward. Now it becomes a little sticky.

LEMMA 6.5.4 *There exists a subspace W of V, invariant under T, such that $V = V_1 \oplus W$.*

Proof. Let W be a subspace of V, of largest possible dimension, such that

1. $V_1 \cap W = (0)$;
2. W is invariant under T.

We want to show that $V = V_1 + W$. Suppose not; then there exists an element $z \in V$ such that $z \notin V_1 + W$. Since $T^{n_1} = 0$, there exists an integer k, $0 < k \leq n_1$, such that $zT^k \in V_1 + W$ and such that $zT^i \notin V_1 + W$ for $i < k$. Thus $zT^k = u + w$, where $u \in V_1$ and where $w \in W$. But then $0 = zT^{n_1} = (zT^k)T^{n_1-k} = uT^{n_1-k} + wT^{n_1-k}$; however, since both V_1 and W are invariant under T, $uT^{n_1-k} \in V_1$ and $wT^{n_1-k} \in W$. Now, since $V_1 \cap W = (0)$, this leads to $uT^{n_1-k} = -wT^{n_1-k} \in V_1 \cap W = (0)$, resulting in $uT^{n_1-k} = 0$. By Lemma 6.5.3, $u = u_0T^k$ for some $u_0 \in V_1$; therefore, $zT^k = u + w = u_0T^k + w$. Let $z_1 = z - u_0$; then $z_1T^k = zT^k - u_0T^k = w \in W$, and since W is invariant under T this yields $z_1T^m \in W$ for all $m \geq k$. On the other hand, if $i < k$, $z_1T^i = zT^i - u_0T^i \notin V_1 + W$, for otherwise zT^i must fall in $V_1 + W$, contradicting the choice of k.

Let W_1 be the subspace of V spanned by W and $z_1, z_1T, \ldots, z_1T^{k-1}$. Since $z_1 \notin W$, and since $W_1 \supset W$, the dimension of W_1 must be larger than that of W. Moreover, since $z_1T^k \in W$ and since W is invariant under T, W_1 must be invariant under T. By the maximal nature of W there must be an element of the form $w_0 + \alpha_1 z_1 + \alpha_2 z_1 T + \cdots + \alpha_k z_1 T^{k-1} \neq 0$ in $W_1 \cap V_1$, where $w_0 \in W$. Not all of $\alpha_1, \ldots, \alpha_k$ can be 0; otherwise we would have $0 \neq w_0 \in W \cap V_1 = (0)$, a contradiction. Let α_s be the first nonzero α; then $w_0 + z_1T^{s-1}(\alpha_s + \alpha_{s+1}T + \cdots + \alpha_k T^{k-s}) \in V_1$. Since $\alpha_s \neq 0$, by Lemma 6.5.2, $\alpha_s + \alpha_{s+1}T + \cdots + \alpha_k T^{k-s}$ is invertible and its inverse, R, is a polynomial in T. Thus W and V_1 are invariant under R; however, from the above, $w_0R + z_1T^{s-1} \in V_1R \subset V_1$, forcing $z_1T^{s-1} \in V_1 + WR \subset V_1 + W$. Since $s - 1 < k$ this is impossible; therefore $V_1 + W = V$. Because $V_1 \cap W = (0)$, $V = V_1 \oplus W$, and the lemma is proved.

The hard work, for the moment, is over; we now complete the proof of Theorem 6.5.1.

By Lemma 6.5.4, $V = V_1 \oplus W$ where W is invariant under T. Using the basis $v_1, \ldots, v_{n_1}$ of V_1 and any basis of W as a basis of V, by Lemma 6.5.1, the matrix of T in this basis has the form

$$\begin{pmatrix} M_{n_1} & 0 \\ 0 & A_2 \end{pmatrix},$$

where A_2 is the matrix of T_2, the linear transformation induced on W by T. Since $T^{n_1} = 0$, $T_2{}^{n_2} = 0$ for some $n_2 \leq n_1$. Repeating the argument used for T on V for T_2 on W we can decompose W as we did V (or, invoke an induction on the dimension of the vector space involved). Continuing this way, we get a basis of V in which the matrix of T is of the form

$$\begin{pmatrix} M_{n_1} & 0 & \dots & 0 \\ 0 & M_{n_2} & & \\ \vdots & & \ddots & \vdots \\ 0 & \dots & & M_{n_r} \end{pmatrix}.$$

That $n_1 + n_2 + \cdots + n_r = \dim V$ is clear, since the size of the matrix is $n \times n$ where $n = \dim V$.

DEFINITION The integers $n_1, n_2, \ldots, n_r$ are called the *invariants* of T.

DEFINITION If $T \in A(V)$ is nilpotent, the subspace M of V, of dimension m, which is invariant under T, is called *cyclic with respect to* T if

1. $MT^m = (0)$, $MT^{m-1} \neq (0)$;
2. there is an element $z \in M$ such that $z, zT, \ldots, zT^{m-1}$ form a basis of M.

(Note: Condition 1 is actually implied by Condition 2).

LEMMA 6.5.5 *If M, of dimension m, is cyclic with respect to T, then the dimension of MT^k is $m - k$ for all $k \leq m$.*

Proof. A basis of MT^k is provided us by taking the image of any basis of M under T^k. Using the basis $z, zT, \ldots, zT^{m-1}$ of M leads to a basis zT^k, $zT^{k+1}, \ldots, zT^{m-1}$ of MT^k. Since this basis has $m - k$ elements, the lemma is proved.

Theorem 6.5.1 tells us that given a nilpotent T in $A(V)$ we can find integers $n_1 \geq n_2 \geq \cdots \geq n_r$ and subspaces, $V_1, \ldots, V_r$ of V cyclic with respect to T and of dimensions $n_1, n_2, \ldots, n_r$, respectively such that $V = V_1 \oplus \cdots \oplus V_r$.

Is it possible that we can find other integers $m_1 \geq m_2 \geq \cdots \geq m_s$ and subspaces $U_1, \ldots, U_s$ of V, cyclic with respect to T and of dimensions $m_1, \ldots, m_s$, respectively, such that $V = U_1 \oplus \cdots \oplus U_s$? We claim that we cannot, or in other words that $s = r$ and $m_1 = n_1, m_2 = n_2, \ldots, m_r = n_r$. Suppose that this were not the case; then there is a first integer i such that $m_i \neq n_i$. We may assume that $m_i < n_i$.

Consider VT^{m_i}. On one hand, since $V = V_1 \oplus \cdots \oplus V_r$, $VT^{m_i} = V_1T^{m_i} \oplus \cdots \oplus V_iT^{m_i} \oplus \cdots \oplus V_rT^{m_i}$. Since $\dim V_1T^{m_i} = n_1 - m_i$, $\dim V_2T^{m_i} = n_2 - m_i, \ldots, \dim V_iT^{m_i} = n_i - m_i$ (by Lemma 6.5.5), $\dim VT^{m_i} \geq (n_1 - m_i) + (n_2 - m_i) + \cdots + (n_i - m_i)$. On the other hand, since $V = U_1 \oplus \cdots \oplus U_s$ and since $U_jT^{m_i} = (0)$ for $j \geq i$, $VT^{m_i} = U_1T^{m_i} \oplus U_2T^{m_i} + \cdots \oplus U_{i-1}T^{m_i}$. Thus

$$\dim VT^{m_i} = (m_1 - m_i) + (m_2 - m_i) + \cdots + (m_{i-1} - m_i).$$

By our choice of i, $n_1 = m_1, n_2 = m_2, \ldots, n_{i-1} = m_{i-1}$, whence

$$\dim VT^{m_i} = (n_1 - m_i) + (n_2 - m_i) + \cdots + (n_{i-1} - m_i).$$

However, this contradicts the fact proved above that $\dim VT^{m_i} \geq (n_1 - m_i) + \cdots + (n_{i-1} - m_i) + (n_i - m_i)$, since $n_i - m_i > 0$.

Thus there is a *unique* set of integers $n_1 \geq n_2 \geq \cdots \geq n_r$ such that V is the direct sum of subspaces, cyclic with respect to T of dimensions $n_1, n_2, \ldots, n_r$. *Equivalently, we have shown that the invariants of T are unique.*

Matricially, the argument just carried out has proved that if $n_1 \geq n_2 \geq \cdots \geq n_r$ and $m_1 \geq m_2 \geq \cdots \geq m_s$, then the matrices

$$\begin{pmatrix} M_{n_1} & \cdots & 0 \\ 0 & & \\ \vdots & \ddots & \vdots \\ 0 & \cdots & M_{n_r} \end{pmatrix} \quad \text{and} \quad \begin{pmatrix} M_{m_1} & \cdots & 0 \\ 0 & & \\ \vdots & \ddots & \vdots \\ 0 & \cdots & M_{m_s} \end{pmatrix}$$

are similar only if $r = s$ and $n_1 = m_1, n_2 = m_2, \ldots, n_r = m_r$.

So far we have proved the more difficult half of

THEOREM 6.5.2 *Two nilpotent linear transformations are similar if and only if they have the same invariants.*

Proof. The discussion preceding the theorem has proved that if the two nilpotent linear transformations have different invariants, then they cannot be similar, for their respective matrices

$$\begin{pmatrix} M_{n_1} & \cdots & 0 \\ \vdots & \ddots & \vdots \\ 0 & \cdots & M_{n_r} \end{pmatrix} \quad \text{and} \quad \begin{pmatrix} M_{m_1} & \cdots & 0 \\ \vdots & \ddots & \vdots \\ 0 & \cdots & M_{m_s} \end{pmatrix}$$

cannot be similar.

In the other direction, if the two nilpotent linear transformations S and T have the same invariants $n_1 \geq \cdots \geq n_r$, by Theorem 6.5.1 there are bases $v_1, \ldots, v_n$ and $w_1, \ldots, w_n$ of V such that the matrix of S in $v_1, \ldots, v_n$ and that of T in $w_1, \ldots, w_n$, are each equal to

$$\begin{pmatrix} M_{n_1} & \cdots & 0 \\ \vdots & \ddots & \vdots \\ 0 & \cdots & M_{n_r} \end{pmatrix}.$$

But if A is the linear transformation defined on V by $v_iA = w_i$, then $S = ATA^{-1}$ (Prove! Compare with Problem 32 at the end of Section 6.3), whence S and T are similar.

Let us compute an example. Let

$$T = \begin{pmatrix} 0 & 1 & 1 \\ 0 & 0 & 0 \\ 0 & 0 & 0 \end{pmatrix} \in F_3$$

act on $F^{(3)}$ with basis $u_1 = (1, 0, 0)$, $u_2 = (0, 1, 0)$, $u_3 = (0, 0, 1)$. Let $v_1 = u_1$, $v_2 = u_1T = u_2 + u_3$, $v_3 = u_3$; in the basis v_1, v_2, v_3 the matrix of T is

$$\left(\begin{array}{cc|c} 0 & 1 & 0 \\ 0 & 0 & 0 \\ \hline 0 & 0 & 0 \end{array}\right),$$

so that the invariants of T are 2, 1. If A is the matrix of the change of basis, namely

$$\begin{pmatrix} 1 & 0 & 0 \\ 0 & 1 & 1 \\ 0 & 0 & 1 \end{pmatrix},$$

a simple computation shows that

$$ATA^{-1} = \begin{pmatrix} 0 & 1 & 0 \\ 0 & 0 & 0 \\ 0 & 0 & 0 \end{pmatrix}.$$

One final remark: the invariants of T determine a partition of n, the dimension of V. Conversely, any partition of n, $n_1 \geq \cdots \geq n_r$, $n_1 + n_2 + \cdots + n_r = n$, determines the invariants of the nilpotent linear transformation.

$$\begin{pmatrix} M_{n_1} & \cdots & 0 \\ \vdots & \ddots & \vdots \\ 0 & \cdots & M_{n_r} \end{pmatrix}.$$

Thus the number of distinct similarity classes of nilpotent $n \times n$ matrices is precisely $p(n)$, the number of partitions of n.

6.6 Canonical Forms: A Decomposition of V: Jordan Form

Let V be a finite-dimensional vector space over F and let T be an arbitrary element in $A_F(V)$. Suppose that V_1 is a subspace of V invariant under T. Therefore T induces a linear transformation T_1 on V_1 defined by $uT_1 = uT$ for every $u \in V_1$. Given any polynomial $q(x) \in F[x]$, we claim that the linear transformation induced by $q(T)$ on V_1 is precisely $q(T_1)$. (The proof of this is left as an exercise.) In particular, if $q(T) = 0$ then $q(T_1) = 0$. Thus T_1 satisfies any polynomial satisfied by T over F. What can be said in the opposite direction?

LEMMA 6.6.1 *Suppose that $V = V_1 \oplus V_2$, where V_1 and V_2 are subspaces of V invariant under T. Let T_1 and T_2 be the linear transformations induced by T on V_1 and V_2, respectively. If the minimal polynomial of T_1 over F is $p_1(x)$ while that of T_2 is $p_2(x)$, then the minimal polynomial for T over F is the least common multiple of $p_1(x)$ and $p_2(x)$.*

Proof. If $p(x)$ is the minimal polynomial for T over F, as we have seen above, both $p(T_1)$ and $p(T_2)$ are zero, whence $p_1(x) \mid p(x)$ and $p_2(x) \mid p(x)$. But then the least common multiple of $p_1(x)$ and $p_2(x)$ must also divide $p(x)$.

On the other hand, if $q(x)$ is the least common multiple of $p_1(x)$ and $p_2(x)$, consider $q(T)$. For $v_1 \in V_1$, since $p_1(x) \mid q(x)$, $v_1 q(T) = v_1 q(T_1) = 0$; similarly, for $v_2 \in V_2$, $v_2 q(T) = 0$. Given any $v \in V$, v can be written as $v = v_1 + v_2$, where $v_1 \in V_1$ and $v_2 \in V_2$, in consequence of which $vq(T) = (v_1 + v_2)q(T) = v_1 q(T) + v_2 q(T) = 0$. Thus $q(T) = 0$ and T satisfies $q(x)$. Combined with the result of the first paragraph, this yields the lemma.

COROLLARY *If $V = V_1 \oplus \cdots \oplus V_k$ where each V_i is invariant under T and if $p_i(x)$ is the minimal polynomial over F of T_i, the linear transformation induced by T on V_i, then the minimal polynomial of T over F is the least common multiple of $p_1(x), p_2(x), \ldots, p_k(x)$.*

We leave the proof of the corollary to the reader.

Let $T \in A_F(V)$ and suppose that $p(x)$ in $F[x]$ is the minimal polynomial of T over F. By Lemma 3.9.5, we can factor $p(x)$ in $F[x]$ in a unique way as $p(x) = q_1(x)^{l_1} q_2(x)^{l_2} \cdots q_k(x)^{l_k}$, where the $q_i(x)$ are distinct irreducible polynomials in $F[x]$ and where $l_1, l_2, \ldots, l_k$ are positive integers. Our objective is to decompose V as a direct sum of subspaces invariant under T such that on each of these the linear transformation induced by T has, as minimal polynomial, a power of an irreducible polynomial. If $k = 1$, V itself already does this for us. So, suppose that $k > 1$.

Let $V_1 = \{v \in V \mid vq_1(T)^{l_1} = 0\}$, $V_2 = \{v \in V \mid vq_2(T)^{l_2} = 0\}, \ldots$, $V_k = \{v \in V \mid vq_k(T)^{l_k} = 0\}$. It is a triviality that each V_i is a subspace of V. In addition, V_i is invariant under T, for if $u \in V_i$, since T and $q_i(T)$ commute, $(uT)q_i(T)^{l_i} = (uq_i(T)^{l_i})T = 0T = 0$. By the definition of V_i, this places uT in V_i. Let T_i be the linear transformation induced by T on V_i.

THEOREM 6.6.1 *For each $i = 1, 2, \ldots, k$, $V_i \neq (0)$ and $V = V_1 \oplus V_2 \oplus \cdots \oplus V_k$. The minimal polynomial of T_i is $q_i(x)^{l_i}$.*

Proof. If $k = 1$ then $V = V_1$ and there is nothing that needs proving. Suppose then that $k > 1$.

We first want to prove that each $V_i \neq (0)$. Towards this end, we introduce the k polynomials:

$$\begin{aligned}
h_1(x) &= q_2(x)^{l_2} q_3(x)^{l_3} \cdots q_k(x)^{l_k}, \\
h_2(x) &= q_1(x)^{l_1} q_3(x)^{l_3} \cdots q_k(x)^{l_k}, \ldots, \\
h_i(x) &= \prod_{j \neq i} q_j(x)^{l_j}, \ldots, \\
&\vdots \\
h_k(x) &= q_1(x)^{l_1} q_2(x)^{l_2} \cdots q_{k-1}(x)^{l_{k-1}}.
\end{aligned}$$

Since $k > 1$, $h_i(x) \neq p(x)$, whence $h_i(T) \neq 0$. Thus, given i, there is a $v \in V$ such that $w = vh_i(T) \neq 0$. But $wq_i(T)^{l_i} = v(h_i(T)q_i(T)^{l_i}) = vp(T)$

$= 0$. In consequence, $w \neq 0$ is in V_i and so $V_i \neq (0)$. In fact, we have shown a little more, namely, that $Vh_i(T) \neq (0)$ is in V_i. Another remark about the $h_i(x)$ is in order now: if $v_j \in V_j$ for $j \neq i$, since $q_j(x)^{l_j} \mid h_i(x)$, $v_j h_i(T) = 0$.

The polynomials $h_1(x), h_2(x), \ldots, h_k(x)$ are relatively prime. (Prove!) Hence by Lemma 3.9.4 we can find polynomials $a_1(x), \ldots, a_k(x)$ in $F[x]$ such that $a_1(x)h_1(x) + \cdots + a_k(x)h_k(x) = 1$. From this we get $a_1(T)h_1(T) + \cdots + a_k(T)h_k(T) = 1$, whence, given $v \in V$, $v = v1 = v(a_1(T)h_1(T) + \cdots + a_k(T)h_k(T)) = va_1(T)h_1(T) + \cdots + va_k(T)h_k(T)$. Now, each $va_i(T)h_i(T)$ is in $Vh_i(T)$, and since we have shown above that $Vh_i(T) \subset V_i$, we have now exhibited v as $v = v_1 + \cdots + v_k$, where each $v_i = va_i(T)h_i(T)$ is in V_i. Thus $V = V_1 + V_2 + \cdots + V_k$.

We must now verify that this sum is a direct sum. To show this, it is enough to prove that if $u_1 + u_2 + \cdots + u_k = 0$ with each $u_i \in V_i$, then each $u_i = 0$. So, suppose that $u_1 + u_2 + \cdots + u_k = 0$ and that some u_i, say u_1, is not 0. Multiply this relation by $h_1(T)$; we obtain $u_1h_1(T) + \cdots + u_kh_1(T) = 0h_1(T) = 0$. However, $u_jh_1(T) = 0$ for $j \neq 1$ since $u_j \in V_j$; the equation thus reduces to $u_1h_1(T) = 0$. But $u_1q_1(T)^{l_1} = 0$ and since $h_1(x)$ and $q_1(x)$ are relatively prime, we are led to $u_1 = 0$ (Prove!) which is, of course, inconsistent with the assumption that $u_1 \neq 0$. So far we have succeeded in proving that $V = V_1 \oplus V_2 \oplus \cdots \oplus V_k$.

To complete the proof of the theorem, we must still prove that the minimal polynomial of T_i on V_i is $q(x)^{l_i}$. By the definition of V_i, since $V_iq_i(T)^{l_i} = 0$, $q_i(T_i)^{l_i} = 0$, whence the minimal equation of T_i must be a divisor of $q_i(x)^{l_i}$, thus of the form $q_i(x)^{f_i}$ with $f_i \leq l_i$. By the corollary to Lemma 6.6.1 the minimal polynomial of T over F is the least common multiple of $q_1(x)^{f_1}, \ldots, q_k(x)^{f_k}$ and so must be $q_1(x)^{f_1} \cdots q_k(x)^{f_k}$. Since this minimal polynomial is in fact $q_1(x)^{l_1} \cdots q_k(x)^{l_k}$ we must have that $f_1 \geq l_1$, $f_2 \geq l_2, \ldots, f_k \geq l_k$. Combined with the opposite inequality above, this yields the desired result $l_i = f_i$ for $i = 1, 2, \ldots, k$ and so completes the proof of the theorem.

If all the characteristic roots of T should happen to lie in F, then the minimal polynomial of T takes on the especially nice form $q(x) = (x - \lambda_1)^{l_1} \cdots (x - \lambda_k)^{l_k}$ where $\lambda_1, \ldots, \lambda_k$ are the distinct characteristic roots of T. The irreducible factors $q_i(x)$ above are merely $q_i(x) = x - \lambda_i$. Note that on V_i, T_i only has λ_i as a characteristic root.

COROLLARY *If all the distinct characteristic roots $\lambda_i, \ldots, \lambda_k$ of T lie in F, then V can be written as $V = V_1 \oplus \cdots \oplus V_k$ where $V_i = \{v \in V \mid v(T - \lambda_i)^{l_i} = 0\}$ and where T_i has only one characteristic root, λ_i, on V_i.*

Let us go back to the theorem for a moment; we use the same notation

T_i, V_i as in the theorem. Since $V = V_1 \oplus \cdots \oplus V_k$, if $\dim V_i = n_i$, by Lemma 6.5.1 we can find a basis of V such that in this basis the matrix of T is of the form

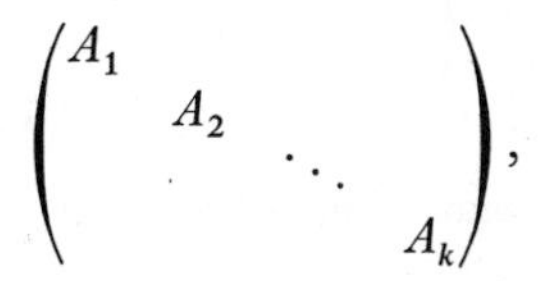

where each A_i is an $n_i \times n_i$ matrix and is in fact the matrix of T_i.

What exactly are we looking for? We want an element in the similarity class of T which we can distinguish in some way. In light of Theorem 6.3.2 this can be rephrased as follows: We seek a basis of V in which the matrix of T has an especially simple (and recognizable) form.

By the discussion above, this search can be limited to the linear transformations T_i; thus the general problem can be reduced from the discussion of general linear transformations to that of the special linear transformations whose minimal polynomials are powers of irreducible polynomials. For the special situation in which all the characteristic roots of T lie in F we do it below. The general case in which we put no restrictions on the characteristic roots of T will be done in the next section.

We are now in the happy position where all the pieces have been constructed and all we have to do is to put them together. This results in the highly important and useful theorem in which is exhibited what is usually called the *Jordan canonical form.* But first a definition.

DEFINITION The matrix

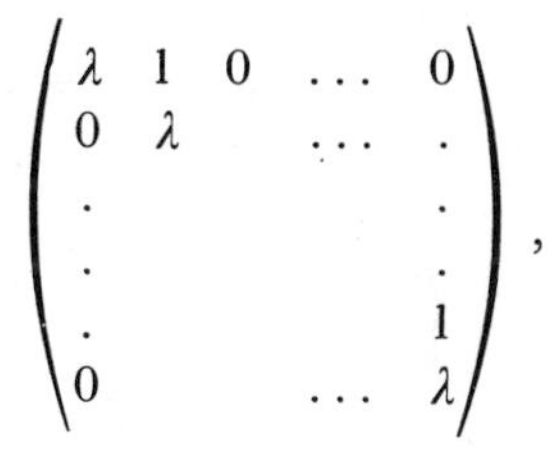

with λ's on the diagonal, 1's on the superdiagonal, and 0's elsewhere, is a basic *Jordan block belonging to* λ.

THEOREM 6.6.2 *Let $T \in A_F(V)$ have all its distinct characteristic roots, $\lambda_1, \ldots, \lambda_k$, in F. Then a basis of V can be found in which the matrix T is of the form*

$$\begin{pmatrix} J_1 & & & \\ & J_2 & & \\ & & \ddots & \\ & & & J_k \end{pmatrix}$$

where each

$$J_i = \begin{pmatrix} B_{i1} & & & \\ & B_{i2} & & \\ & & \ddots & \\ & & & B_{ir_i} \end{pmatrix}$$

and where $B_{i1}, \ldots, B_{ir_i}$ are basic Jordan blocks belonging to λ_i.

Proof. Before starting, note that an $m \times m$ basic Jordan block belonging to λ is merely $\lambda + M_m$, where M_m is as defined at the end of Lemma 6.5.2.

By the combinations of Lemma 6.5.1 and the corollary to Theorem 6.6.1, we can reduce to the case when T has only one characteristic root λ, that is, $T - \lambda$ is nilpotent. Thus $T = \lambda + (T - \lambda)$, and since $T - \lambda$ is nilpotent, by Theorem 6.5.1 there is a basis in which its matrix is of the form

$$\begin{pmatrix} M_{n_1} & & \\ & \ddots & \\ & & M_{n_r} \end{pmatrix}.$$

But then the matrix of T is of the form

$$\begin{pmatrix} \lambda & & & \\ & \lambda & & \\ & & \ddots & \\ & & & \lambda \end{pmatrix} + \begin{pmatrix} M_{n_1} & & \\ & \ddots & \\ & & M_{n_r} \end{pmatrix} = \begin{pmatrix} B_{n_1} & & \\ & \ddots & \\ & & B_{n_r} \end{pmatrix},$$

using the first remark made in this proof about the relation of a basic Jordan block and the M_m's. This completes the theorem.

Using Theorem 6.5.1 we could arrange things so that in each J_i the size of $B_{i1} \geq$ size of $B_{i2} \geq \cdots$. When this has been done, then the matrix

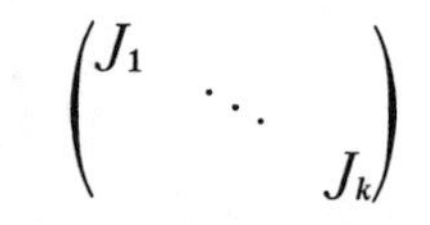

is called the *Jordan form* of T. Note that Theorem 6.6.2, for nilpotent matrices, reduces to Theorem 6.5.1.

We leave as an exercise the following: *Two linear transformations in $A_F(V)$ which have all their characteristic roots in F are similar if and only if they can be brought to the same Jordan form.*

Thus the Jordan form acts as a "determiner" for similarity classes of this type of linear transformation.

In matrix terms Theorem 6.6.2 can be stated as follows: *Let $A \in F_n$ and suppose that K is the splitting field of the minimal polynomial of A over F; then an invertible matrix $C \in K_n$ can be found so that CAC^{-1} is in Jordan form.*

We leave the few small points needed to make the transition from Theorem 6.6.2 to its matrix form, just given, to the reader.

One final remark: If $A \in F_n$ and if in K_n, where K is the splitting field of the minimal polynomial of A over F,

$$CAC^{-1} = \begin{pmatrix} J_1 & & & \\ & J_2 & & \\ & & \ddots & \\ & & & J_k \end{pmatrix}$$

where each J_i corresponds to a different characteristic root, λ_i, of A, then the *multiplicity of* λ_i as a characteristic root of A is defined to be n_i, where J_i is an $n_i \times n_i$ matrix. Note that the sum of the multiplicities is exactly n.

Clearly we can similarly define the multiplicity of a characteristic root of a linear transformation.

Problems

1. If S and T are nilpotent linear transformations which commute, prove that ST and $S + T$ are nilpotent linear transformations.
2. By a direct matrix computation, show that
$$\begin{pmatrix} 0 & 1 & 0 & 0 \\ 0 & 0 & 1 & 0 \\ 0 & 0 & 0 & 0 \\ 0 & 0 & 0 & 0 \end{pmatrix} \quad \text{and} \quad \begin{pmatrix} 0 & 1 & 0 & 0 \\ 0 & 0 & 1 & 0 \\ 0 & 0 & 0 & 1 \\ 0 & 0 & 0 & 0 \end{pmatrix}$$
are not similar.
3. If $n_1 \geq n_2$ and $m_1 \geq m_2$, by a direct matrix computation prove that
$$\begin{pmatrix} M_{n_1} & \\ & M_{n_2} \end{pmatrix} \quad \text{and} \quad \begin{pmatrix} M_{m_1} & \\ & M_{m_2} \end{pmatrix}$$
are similar if and only if $n_1 = m_1$, $n_2 = m_2$.

*4. If $n_1 \geq n_2 \geq n_3$ and $m_1 \geq m_2 \geq m_3$, by a direct matrix computation prove that
$$\begin{pmatrix} M_{n_1} & & \\ & M_{n_2} & \\ & & M_{n_3} \end{pmatrix} \quad \text{and} \quad \begin{pmatrix} M_{m_1} & & \\ & M_{m_2} & \\ & & M_{m_3} \end{pmatrix}$$
are similar if and only if $n_1 = m_1$, $n_2 = m_2$, $n_3 = m_3$.

5. (a) Prove that the matrix
$$\begin{pmatrix} 1 & 1 & 1 \\ -1 & -1 & -1 \\ 1 & 1 & 0 \end{pmatrix}$$
is nilpotent, and find its invariants and Jordan form.

(b) Prove that the matrix in part (a) is not similar to

$$\begin{pmatrix} 1 & 1 & 1 \\ -1 & -1 & -1 \\ 1 & 0 & 0 \end{pmatrix}.$$

6. Prove Lemma 6.6.1 and its corollary even if the sums involved are not direct sums.
7. Prove the statement made to the effect that two linear transformations in $A_F(V)$ all of whose characteristic roots lie in F are similar if and only if their Jordan forms are the same (except for a permutation in the ordering of the characteristic roots).
8. Complete the proof of the matrix version of Theorem 6.6.2, given in the text.
9. Prove that the $n \times n$ matrix

$$\begin{pmatrix} 0 & 0 & 0 & \cdots & 0 & 0 \\ 1 & 0 & 0 & \cdots & 0 & 0 \\ 0 & 1 & 0 & \cdots & 0 & 0 \\ 0 & 0 & 1 & \cdots & 0 & 0 \\ \vdots & & & \ddots & & \vdots \\ 0 & 0 & 0 & & 1 & 0 \end{pmatrix},$$

having entries 1's on the subdiagonal and 0's elsewhere, is similar to M_n.

10. If F has characteristic $p > 0$ prove that $A = \begin{pmatrix} 1 & \alpha \\ 0 & 1 \end{pmatrix}$ satisfies $A^p = 1$.

11. If F has characteristic 0 prove that $A = \begin{pmatrix} 1 & \alpha \\ 0 & 1 \end{pmatrix}$ satisfies $A^m = 1$, for $m > 0$, only if $\alpha = 0$.
12. Find all possible Jordan forms for
    (a) All 8×8 matrices having $x^2(x-1)^3$ as minimal polynomial.
    (b) All 10×10 matrices, over a field of characteristic different from 2, having $x^2(x-1)^2(x+1)^3$ as minimal polynomial.
13. Prove that the $n \times n$ matrix

$$A = \begin{pmatrix} 1 & 1 & 1 & \cdots & 1 \\ 1 & 1 & 1 & \cdots & 1 \\ \vdots & & & & \\ 1 & 1 & 1 & \cdots & 1 \end{pmatrix}$$

is similar to

$$\begin{pmatrix} n & 0 & 0 & \cdots & 0 \\ 0 & 0 & 0 & \cdots & 0 \\ \vdots & & & & \vdots \\ 0 & 0 & 0 & \cdots & 0 \end{pmatrix},$$

if the characteristic of F is 0 or if it is p and $p \nmid n$. What is the multiplicity of 0 as a characteristic root of A?

A matrix $A = (\alpha_{ij})$ is said to be a *diagonal* matrix if $\alpha_{ij} = 0$ for $i \neq j$, that is, if all the entries off the main diagonal are 0. A matrix (or linear transformation) is said to be *diagonalizable* if it is similar to a diagonal matrix (has a basis in which its matrix is diagonal).

14. If T is in $A(V)$ then T is diagonalizable (if all its characteristic roots are in F) if and only if whenever $v(T - \lambda)^m = 0$, for $v \in V$ and $\lambda \in F$, then $v(T - \lambda) = 0$.
15. Using the result of Problem 14, prove that if $E^2 = E$ then E is diagonalizable.
16. If $E^2 = E$ and $F^2 = F$ prove that they are similar if and only if they have the same rank.
17. If the multiplicity of each characteristic root of T is 1, and if all the characteristic roots of T are in F, prove that T is diagonalizable over F.
18. If the characteristic of F is 0 and if $T \in A_F(V)$ satisfies $T^m = 1$, prove that if the characteristic roots of T are in F then T is diagonalizable. (*Hint:* Use the Jordan form of T.)

*19. If $A, B \in F$ are diagonalizable and if they commute, prove that there is an element $C \in F_n$ such that both CAC^{-1} and CBC^{-1} are diagonal.

20. Prove that the result of Problem 19 is false if A and B do not commute.

6.7 Canonical Forms: Rational Canonical Form

The Jordan form is the one most generally used to prove theorems about linear transformations and matrices. Unfortunately, it has one distinct, serious drawback in that it puts requirements on the location of the characteristic roots. True, if $T \in A_F(V)$ (or $A \in F_n$) does not have its characteristic roots in F we need but go to a finite extension, K, of F in which all the characteristic roots of T lie and then to bring T to Jordan form over K. In fact, this is a standard operating procedure; however, it proves the result in K_n and not in F_n. Very often the result in F_n can be inferred from that in K_n, but there are many occasions when, after a result has been established for $A \in F_n$, considered as an element in K_n, we cannot go back from K_n to get the desired information in F_n.

Thus we need some canonical form for elements in $A_F(V)$ (or in F_n) which presumes nothing about the location of the characteristic roots of its elements, a canonical form and a set of invariants created in $A_F(V)$ itself using only its elements and operations. Such a canonical form is provided us by the *rational canonical form* which is described below in Theorem 6.7.1 and its corollary.

Let $T \in A_F(V)$; by means of T we propose to make V into a module over $F[x]$, the ring of polynomials in x over F. We do so by defining, for any polynomial $f(x)$ in $F[x]$, and any $v \in V$, $f(x)v = vf(T)$. We leave the verification to the reader that, under this definition of multiplication of elements of V by elements of $F[x]$, V becomes an $F[x]$-module.

Since V is finite-dimensional over F, it is finitely generated over F, hence, all the more so over $F[x]$ which contains F. Moreover, $F[x]$ is a Euclidean ring; thus as a finitely generated module over $F[x]$, by Theorem 4.5.1, V is the direct sum of a finite number of cyclic submodules. From the very way in which we have introduced the module structure on V, each of these cyclic submodules is invariant under T; moreover there is an element m_0, in such a submodule M, such that every element m, in M, is of the form $m = m_0 f(T)$ for some $f(x) \in F[x]$.

To determine the nature of T on V it will be, therefore, enough for us to know what T looks like on a cyclic submodule. This is precisely what we intend, shortly, to determine.

But first to carry out a preliminary decomposition of V, as we did in Theorem 6.6.1, according to the decomposition of the minimal polynomial of T as a product of irreducible polynomials.

Let the minimal polynomial of T over F be $p(x) = q_1(x)^{e_1} \cdots q_k(x)^{e_k}$, where the $q_i(x)$ are distinct irreducible polynomials in $F[x]$ and where each $e_i > 0$; then, as we saw earlier in Theorem 6.6.1, $V = V_1 \oplus V_2 \oplus \cdots \oplus V_k$ where each V_i is invariant under T and where the minimal polynomial of T on V_i is $q_i(x)^{e_i}$. To solve the nature of a cyclic submodule for an arbitrary T we see, from this discussion, that it suffices to settle it for a T whose minimal polynomial is a power of an irreducible one.

We prove the

LEMMA 6.7.1 *Suppose that T, in $A_F(V)$, has as minimal polynomial over F the polynomial $p(x) = \gamma_0 + \gamma_1 x + \cdots + \gamma_{r-1}x^{r-1} + x^r$. Suppose, further, that V, as a module (as described above), is a cyclic module (that is, is cyclic relative to T.) Then there is basis of V over F such that, in this basis, the matrix of T is*

$$\begin{pmatrix} 0 & 1 & 0 & \cdots & 0 \\ 0 & 0 & 1 & \cdots & 0 \\ \vdots & & & & \\ 0 & 0 & 0 & \cdots & 1 \\ -\gamma_0 & -\gamma_1 & \cdot & \cdots & -\gamma_{r-1} \end{pmatrix}.$$

Proof. Since V is cyclic relative to T, there exists a vector v in V such that every element w, in V, is of the form $w = vf(T)$ for some $f(x)$ in $F[x]$.

Now if for some polynomial $s(x)$ in $F[x]$, $vs(T) = 0$, then for any w in V, $ws(T) = (vf(T))s(T) = vs(T)f(T) = 0$; thus $s(T)$ annihilates all of V and so $s(T) = 0$. But then $p(x) \mid s(x)$ since $p(x)$ is the minimal polynomial of T. This remark implies that $v, vT, vT^2, \ldots, vT^{r-1}$ are linearly independent over F, for if not, then $\alpha_0 v + \alpha_1 vT + \cdots + \alpha_{r-1} vT^{r-1} = 0$ with $\alpha_0, \ldots, \alpha_{r-1}$ in F. But then $v(\alpha_0 + \alpha_1 T + \cdots + \alpha_{r-1} T^{r-1}) = 0$, hence by the above discussion $p(x) \mid (\alpha_0 + \alpha_1 x + \cdots + \alpha_{r-1} x^{r-1})$, which is impossible since $p(x)$ is of degree r unless

$$\alpha_0 = \alpha_1 = \cdots = \alpha_{r-1} = 0.$$

Since $T^r = -\gamma_0 - \gamma_1 T - \cdots - \gamma_{r-1} T^{r-1}$, we immediately have that T^{r+k}, for $k \geq 0$, is a linear combination of $1, T, \ldots, T^{r-1}$, and so $f(T)$, for any $f(x) \in F[x]$, is a linear combination of $1, T, \ldots, T^{r-1}$ over F. Since any w in V is of the form $w = vf(T)$ we get that w is a linear combination of $v, vT, \ldots, vT^{r-1}$.

We have proved, in the above two paragraphs, that the elements $v, vT, \ldots, vT^{r-1}$ form a basis of V over F. In this basis, as is immediately verified, the matrix of T is exactly as claimed

DEFINITION If $f(x) = \gamma_0 + \gamma_1 x + \cdots + \gamma_{r-1} x^{r-1} + x^r$ is in $F[x]$, then the $r \times r$ matrix

$$\begin{pmatrix} 0 & 1 & 0 & \cdots & 0 \\ 0 & 0 & 1 & \cdots & 0 \\ \vdots & & & & \\ 0 & 0 & 0 & \cdots & 1 \\ -\gamma_0 & -\gamma_1 & \cdot & \cdots & -\gamma_{r-1} \end{pmatrix}$$

is called the *companion matrix* of $f(x)$. We write it as $C(f(x))$.

Note that Lemma 6.7.1 says that *if V is cyclic relative to T and if the minimal polynomial of T in $F[x]$ is $p(x)$ then for some basis of V the matrix of T is $C(p(x))$.*

Note further that *the matrix $C(f(x))$, for any monic $f(x)$ in $F[x]$, satisfies $f(x)$ and has $f(x)$ as its minimal polynomial.* (See Problem 4 at the end of this section; also Problem 29 at the end of Section 6.1.)

We now prove the very important

THEOREM 6.7.1 *If T in $A_F(V)$ has as minimal polynomial $p(x) = q(x)^e$, where $q(x)$ is a monic, irreducible polynomial in $F[x]$, then a basis of V over F can be found in which the matrix of T is of the form*

$$\begin{pmatrix} C(q(x)^{e_1}) & & & \\ & C(q(x)^{e_2}) & & \\ & & \ddots & \\ & & & C(q(x)^{e_r}) \end{pmatrix}$$

where $e = e_1 \geq e_2 \geq \cdots \geq e_r$.

Proof. Since V, as a module over $F[x]$, is finitely generated, and since $F[x]$ is Euclidean, we can decompose V as $V = V_1 \oplus \cdots \oplus V_r$ where the

V_i are cyclic modules. The V_i are thus invariant under T; if T_i is the linear transformation induced by T on V_i, its minimal polynomial must be a divisor of $p(x) = q(x)^e$ so is of the form $q(x)^{e_i}$. We can renumber the spaces so that $e_1 \geq e_2 \geq \cdots \geq e_r$.

Now $q(T)^{e_1}$ annihilates each V_i, hence annihilates V, whence $q(T)^{e_1} = 0$. Thus $e_1 \geq e$; since e_1 is clearly at most e we get that $e_1 = e$.

By Lemma 6.7.1, since each V_i is cyclic relative to T, we can find a basis such that the matrix of the linear transformation of T_i on V_i is $C(q(x)^{e_i})$. Thus by Theorem 6.6.1 a basis of V can be found so that the matrix of T in this basis is

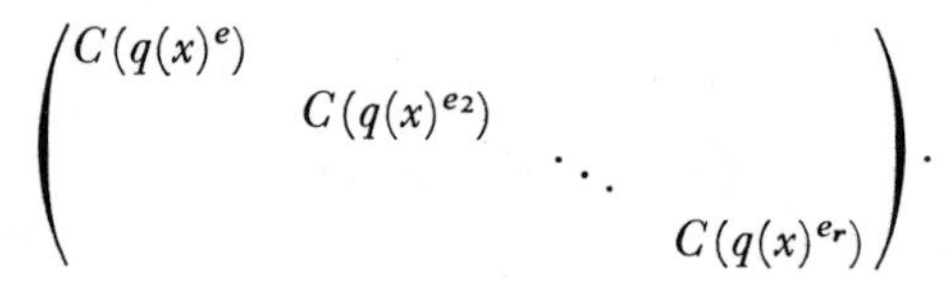

COROLLARY *If T in $A_F(V)$ has minimal polynomial $p(x) = q_1(x)^{l_1} \cdots q_k(x)^{l_k}$ over F, where $q_1(x), \ldots, q_k(x)$ are irreducible distinct polynomials in $F[x]$, then a basis of V can be found in which the matrix of T is of the form*

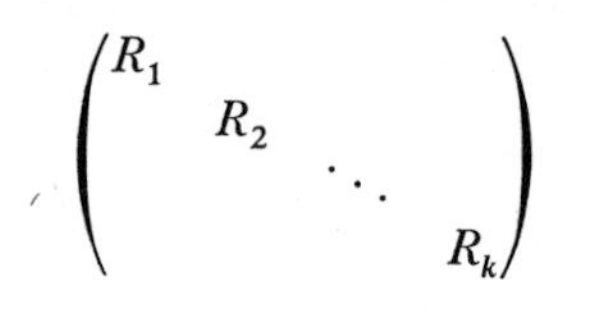

where each

$$R_i = \begin{pmatrix} C(q_i(x)^{e_{i1}}) & & \\ & \ddots & \\ & & C(q_i(x)^{e_{ir_i}}) \end{pmatrix}$$

where $e_i = e_{i1} \geq e_{i2} \geq \cdots \geq e_{ir_i}$.

Proof. By Theorem 6.5.1, V can be decomposed into the direct sum $V = V_1 \oplus \cdots \oplus V_k$, where each V_i is invariant under T and where the minimal polynomial of T_i, the linear transformation induced by T on V_i, has as minimal polynomial $q_i(x)^{e_i}$. Using Lemma 6.5.1 and the theorem just proved, we obtain the corollary. If the degree of $q_i(x)$ is d_i, note that the sum of all the $d_i e_{ij}$ is n, the dimension of V over F.

DEFINITION The matrix of T in the statement of the above corollary is called the *rational canonical form* of T.

DEFINITION The polynomials $q_1(x)^{e_{11}}, q_1(x)^{e_{12}}, \ldots, q_1(x)^{e_{1r_1}}, \ldots, q_k(x)^{e_{k1}}, \ldots, q_k(x)^{e_{kr_k}}$ in $F[x]$ are called the *elementary divisors* of T.

One more definition!

DEFINITION If $\dim_F(V) = n$, then the *characteristic polynomial* of T, $p_T(x)$, is the product of its elementary divisors.

We shall be able to identify the characteristic polynomial just defined with another polynomial which we shall explicitly construct in Section 6.9. The characteristic polynomial of T is a polynomial of degree n lying in $F[x]$. It has many important properties, one of which is contained in the

REMARK *Every linear transformation $T \in A_F(V)$ satisfies its characteristic polynomial. Every characteristic root of T is a root of $p_T(x)$.*

Note 1. The first sentence of this remark is the statement of a very famous theorem, the *Cayley-Hamilton theorem.* However, to call it that in the form we have given is a little unfair. The meat of the Cayley-Hamilton theorem is the fact that T satisfies $p_T(x)$ when $p_T(x)$ is given in a very specific, concrete form, easily constructible from T. However, even as it stands the remark does have some meat in it, for since the characteristic polynomial is a polynomial of degree n, we have shown that every element in $A_F(V)$ does satisfy a polynomial of degree n lying in $F[x]$. Until now, we had only proved this (in Theorem 6.4.2) for linear transformations having all their characteristic roots in F.

Note 2. As stated the second sentence really says nothing, for whenever T satisfies a polynomial then every characteristic root of T satisfies this same polynomial; thus $p_T(x)$ would be nothing special if what were stated in the theorem were all that held true for it. However, the actual story is the following: Every characteristic root of T is a root of $p_T(x)$, and conversely, *every root of $p_T(x)$ is a characteristic root of T; moreover, the multiplicity of any root of $p_T(x)$, as a root of the polynomial, equals its multiplicity as a characteristic root of T.* We could prove this now, but defer the proof until later when we shall be able to do it in a more natural fashion.

Proof of the Remark. We only have to show that T satisfies $p_T(x)$, but this beomes almost trivial. Since $p_T(x)$ is the product of $q_1(x)^{e_{11}}$, $q_1(x)^{e_{12}}$, $\ldots, q_k(x)^{e_{k1}}, \ldots,$ and since $e_{11} = e_1, e_{21} = e_2, \ldots, e_{k1} = e_k$, $p_T(x)$ is divisible by $p(x) = q_1(x)^{e_1} \cdots q_k(x)^{e_k}$, the minimal polynomial of T. Since $p(T) = 0$ it follows that $p_T(T) = 0$.

We have called the set of polynomials arising in the rational canonical form of T the elementary divisors of T. It would be highly desirable if these determined similarity in $A_F(V)$, for then the similarity classes in $A_F(V)$ would be in one-to-one correspondence with sets of polynomials in $F[x]$. We propose to do this, but first we establish a result which implies that two linear transformations have the same elementary divisors.

THEOREM 6.7.2 *Let V and W be two vector spaces over F and suppose that ψ*

is a vector space isomorphism of V onto W. Suppose that $S \in A_F(V)$ and $T \in A_F(W)$ are such that for any $v \in V$, $(vS)\psi = (v\psi)T$. Then S and T have the same elementary divisors.

Proof. We begin with a simple computation. If $v \in V$, then $(vS^2)\psi = ((vS)S)\psi = ((vS)\psi)T = ((v\psi)T)T = (v\psi)T^2$. Clearly, if we continue in this pattern we get $(vS^m)\psi = (v\psi)T^m$ for any integer $m \geq 0$ whence for any polynomial $f(x) \in F[x]$ and for any $v \in V$, $(vf(S))\psi = (v\psi)f(T)$.

If $f(S) = 0$ then $(v\psi)f(T) = 0$ for any $v \in V$, and since ψ maps V onto W, we would have that $Wf(T) = (0)$, in consequence of which $f(T) = 0$. Conversely, if $g(x) \in F[x]$ is such that $g(T) = 0$, then for any $v \in V$, $(vg(S))\psi = 0$, and since ψ is an isomorphism, this results in $vg(S) = 0$. This, of course, implies that $g(S) = 0$. Thus S and T satisfy the same set of polynomials in $F[x]$, hence *must have the same minimal polynomial.*

$$p(x) = q_1(x)^{e_1}q_2(x)^{e_2} \cdots q_k(x)^{e_k}$$

where $q_1(x), \ldots, q_k(x)$ are distinct irreducible polynomials in $F[x]$

If U is a subspace of V invariant under S, then $U\psi$ is a subspace of W invariant under T, for $(U\psi)T = (US)\psi \subset U\psi$. Since U and $U\psi$ are isomorphic, the minimal polynomial of S_1, the linear transformation induced by S on U is the same, by the remarks above, as the minimal polynomial of T_1, the linear transformation induced on $U\psi$ by T.

Now, since the minimal polynomial for S on V is $p(x) = q_1(x)^{e_1} \cdots q_k(x)^{e_k}$, as we have seen in Theorem 6.7.1 and its corollary, we can take as the first elementary divisor of S the polynomial $q_1(x)^{e_1}$ and we can find a subspace of V_1 of V which is invariant under S such that

1. $V = V_1 \oplus M$ where M is invariant under S.
2. The only elementary divisor of S_1, the linear transformation induced on V_1 by S, is $q_1(x)^{e_1}$.
3. The other elementary divisors of S are those of the linear transformation S_2 induced by S on M.

We now combine the remarks made above and assert

1. $W = W_1 \oplus N$ where $W_1 = V_1\psi$ and $N = M\psi$ are invariant under T.
2. The only elementary divisor of T_1, the linear transformation induced by T on W_1, is $q_1(x)^{e_1}$ (*which is an elementary divisor of T* since the minimal polynomial of T is $p(x) = q_1(x)^{e_1} \cdots q_k(x)^{e_k}$).
3. The other elementary divisors of T are those of the linear transformation T_2 induced by T on N.

Since $N = M\psi$, M and N are isomorphic vector spaces over F under the isomorphism ψ_2 induced by ψ. Moreover, if $u \in M$ then $(uS_2)\psi_2 =$

$(uS)\psi = (u\psi)T = (u\psi_2)T_2$, hence S_2 and T_2 are in the same relation vis-à-vis ψ_2 as S and T were vis-à-vis ψ. By induction on dimension (or repeating the argument) S_2 and T_2 have the same elementary divisors. But since the elementary divisors of S are merely $q_1(x)^{e_1}$ and those of S_2 while those of T are merely $q_1(x)^{e_1}$ and those of T_2, S, and T must have the same elementary divisors, thereby proving the theorem.

Theorem 6.7.1 and its corollary gave us the rational canonical form and gave rise to the elementary divisors. We should like to push this further and to be able to assert some uniqueness property. This we do in

THEOREM 6.7.3 *The elements S and T in $A_F(V)$ are similar in $A_F(V)$ if and only if they have the same elementary divisors.*

Proof. In one direction this is easy, for suppose that S and T have the same elementary divisors. Then there are two bases of V over F such that the matrix of S in the first basis equals the matrix of T in the second (and each equals the matrix of the rational canonical form). But as we have seen several times earlier, this implies that S and T are similar.

We now wish to go in the other direction. Here, too, the argument resembles closely that used in Section 6.5 in the proof of Theorem 6.5.2. Having been careful with details there, we can afford to be a little sketchier here.

We first remark that in view of Theorem 6.6.1 we may reduce from the general case to that of a linear transformation whose minimal polynomial is a power of an irreducible one. Thus without loss of generality we may suppose that the minimal polynomial of T is $q(x)^e$ where $q(x)$ is irreducible in $F[x]$ of degree d.

The rational canonical form tells us that we can decompose V as $V = V_1 \oplus \cdots \oplus V_r$, where the subspaces V_i are invariant under T and where the linear transformation induced by T on V_i has as matrix $C(q(x)^{e_i})$, the companion matrix of $q(x)^{e_i}$. We assume that what we are really trying to prove is the following: If $V = U_1 \oplus U_2 \oplus \cdots \oplus U_s$ where the U_j are invariant under T and where the linear transformation induced by T on U_j has as matrix $C(q(x)^{f_j})$, $f_1 \geq f_2 \geq \cdots \geq f_s$, then $r = s$ and $e_1 = f_1$, $e_2 = f_2, \ldots, e_r = f_r$. (Prove that the proof of this is equivalent to proving the theorem!)

Suppose then that we do have the two decompositions described above, $V = V_1 \oplus \cdots \oplus V_r$ and $V = U_1 \oplus \cdots \oplus U_s$, and that some $e_i \neq f_i$. Then there is a first integer m such that $e_m \neq f_m$, while $e_1 = f_1, \ldots, e_{m-1} = f_{m-1}$. We may suppose that $e_m > f_m$.

Now $g(T)^{f_m}$ annihilates $U_m, U_{m+1}, \ldots, U_s$, whence

$$Vq(T)^{f_m} = U_1 q(T)^{f_m} \oplus \cdots \oplus U_{m-1} q(T)^{f_m}.$$

However, it can be shown that the dimension of $U_i q(T)^{f_m}$ for $i \le m$ is $d(f_i - f_m)$ (Prove!) whence

$$\dim (Vq(T)^{f_m}) = d(f_1 - f_m) + \cdots + d(f_{m-1} - f_m).$$

On the other hand, $Vq(T)^{f_m} \supset V_1 q(T)^{f_m} \oplus \cdots \oplus \cdots \oplus V_m q(T)^{f_m}$ and since $V_i q(T)^{f_m}$ has dimension $d(e_i - f_m)$, for $i \le m$, we obtain that

$$\dim (Vq(T)^{f_m}) \ge d(e_i - f_m) + \cdots + d(e_m - f_m).$$

Since $e_1 = f_1, \ldots, e_{m-1} = f_{m-1}$ and $e_m > f_m$, this contradicts the equality proved above. We have thus proved the theorem.

COROLLARY 1 *Suppose the two matrices A, B in F_n are similar in K_n where K is an extension of F. Then A and B are already similar in F_n.*

Proof. Suppose that $A, B \in F_n$ are such that $B = C^{-1}AC$ with $C \in K_n$. We consider K_n as acting on $K^{(n)}$, the vector space of n-tuples over K. Thus $F^{(n)}$ is contained in $K^{(n)}$ and although *it is* a vector space over F *it is not* a vector space over K. The image of $F^{(n)}$, in $K^{(n)}$, under C *need not fall* back in $F^{(n)}$ but at any rate $F^{(n)}C$ is a subset of $K^{(n)}$ which *is a vector space over* F. (Prove!) Let V be the vector space $F^{(n)}$ over F, W the vector space $F^{(n)}C$ over F, and for $v \in V$ let $v\psi = vC$. Now $A \in A_F(V)$ and $B \in A_F(W)$ and for any $v \in V$, $(vA)\psi = vAC = vCB = (v\psi)B$ whence the conditions of Theorem 6.7.2 are satisfied. Thus A and B have the same elementary divisors; by Theorem 6.7.3, A and B must be similar in F_n.

A word of caution: The corollary *does not state* that if $A, B \in F_n$ are such that $B = C^{-1}AC$ with $C \in K_n$ then C must of necessity be in F_n; this is false. It merely states that if $A, B \in F_n$ are such that $B = C^{-1}AC$ with $C \in K_n$ then there exists a (possibly different) $D \in F_n$ such that $B = D^{-1}AD$.

Problems

1. Verify that V becomes an $F[x]$-module under the definition given.
2. In the proof of Theorem 6.7.3 provide complete proof at all points marked "(Prove)."

*3. (a) Prove that every root of the characteristic polynomial of T is a characteristic root of T.
   (b) Prove that the multiplicity of any root of $p_T(x)$ is equal to its multiplicity as a characteristic root of T.

4. Prove that for $f(x) \in F[x]$, $C(f(x))$ satisfies $f(x)$ and has $f(x)$ as its minimal polynomial. What is its characteristic polynomial?
5. If F is the field of rational numbers, find all possible rational canonical forms and elementary divisors for

(a) The 6×6 matrices in F_6 having $(x - 1)(x^2 + 1)^2$ as minimal polynomial.
(b) The 15×15 matrices in F_{15} having $(x^2 + x + 1)^2(x^3 + 2)^2$ as minimal polynomial.
(c) The 10×10 matrices in F_{10} having $(x^2 + 1)^2(x^3 + 1)$ as minimal polynomial.

6. (a) If K is an extension of F and if A is in K_n, prove that A can be written as $A = \lambda_1 A_1 + \cdots + \lambda_k A_k$ where $A_1, \ldots, A_k$ are in F_n and where $\lambda_1, \ldots, \lambda_k$ are in K and are linearly independent over F.
(b) With the notation as in part (a), prove that if $B \in F_n$ is such that $AB = 0$ then $A_1 B = A_2 B = \cdots = A_k B = 0$.
(c) If C in F_n commutes with A prove that C commutes with each of $A_1, A_2, \ldots, A_k$.

*7. If $A_1, \ldots, A_k$ are in F_n and are such that for some $\lambda_1, \ldots, \lambda_k$ in K, an extension of F, $\lambda_1 A_1 + \cdots + \lambda_k A_k$ is invertible in K_n, prove that *if F has an infinite number of elements* we can find $\alpha_1, \ldots, \alpha_k$ *in F* such that $\alpha_1 A_1 + \cdots + \alpha_k A_k$ is invertible in F_n.

*8. If F is a *finite field* prove the result of Problem 7 is false.

*9. Using the results of Problems 6(a) and 7 prove that if F has an infinite number of elements then whenever $A, B \in F_n$ are similar in K_n, where K is an extension of F, then they are familiar in F_n. (This provides us with a proof, independent of canonical forms of Corollary 1 to Theorem 6.7.3 in the special case when F is an infinite field.)

10. Using matrix computations (but following the lines laid out in Problem 9), prove that if F is the field of real numbers and K that of complex numbers, then two elements in F_2 which are similar with K_2 are already similar in F_2.

6.8 Trace and Transpose

After the rather heavy going of the previous few sections, the uncomplicated nature of the material to be treated now should come as a welcome respite.

Let F be a field and let A be a matrix in F_n.

DEFINITION The *trace* of A is the sum of the elements on the main diagonal of A.

We shall write the trace of A as tr A; if $A = (\alpha_{ij})$, then

$$\text{tr } A = \sum_{i=1}^{n} \alpha_{ii}.$$

The fundamental formal properties of the trace function are contained in

LEMMA 6.8.1 *For $A, B \in F_n$ and $\lambda \in F$,*

1. $\text{tr}\,(\lambda A) = \lambda\,\text{tr}\,A$.
2. $\text{tr}\,(A + B) = \text{tr}\,A + \text{tr}\,B$.
3. $\text{tr}\,(AB) = \text{tr}\,(BA)$.

Proof. To establish parts 1 and 2 (which assert that the trace is a linear functional on F_n) is straightforward and is left to the reader. We only present the proof of part 3 of the lemma.

If $A = (\alpha_{ij})$ and $B = (\beta_{ij})$ then $AB = (\gamma_{ij})$ where

$$\gamma_{ij} = \sum_{k=1}^{n} \alpha_{ik}\beta_{kj}$$

and $BA = (\mu_{ij})$ where

$$\mu_{ij} = \sum_{k=1}^{n} \beta_{ik}\alpha_{kj}.$$

Thus

$$\text{tr}\,(AB) = \sum_i \gamma_{ii} = \sum_i \left(\sum_k \alpha_{ik}\beta_{ki} \right);$$

if we interchange the order of summation in this last sum, we get

$$\text{tr}\,(AB) = \sum_{k=1}^{n}\sum_{i=1}^{n} \alpha_{ik}\beta_{ki} = \sum_{k=1}\left(\sum_{i=1} \beta_{ki}\alpha_{ik} \right) = \sum_{k=1}^{n} \mu_{kk} = \text{tr}\,(BA).$$

COROLLARY *If A is invertible then* $\text{tr}\,(ACA^{-1}) = \text{tr}\,C$.

Proof. Let $B = CA^{-1}$; then $\text{tr}\,(ACA^{-1}) = \text{tr}\,(AB) = \text{tr}\,(BA) = \text{tr}\,(CA^{-1}A) = \text{tr}\,C$.

This corollary has a twofold importance; first, it will allow us to define the trace of an arbitrary linear transformation; secondly, it will enable us to find an alternative expression for the trace of A.

DEFINITION If $T \in A(V)$ then tr T, the *trace* of T, is the trace of $m_1(T)$ where $m_1(T)$ is the matrix of T in some basis of V.

We claim that the definition is meaningful and depends only on T and not on any particular basis of V. For if $m_1(T)$ and $m_2(T)$ are the matrices of T in two different bases of V, by Theorem 6.3.2, $m_1(T)$ and $m_2(T)$ are similar matrices, so by the corollary to Lemma 6.8.1 they have the same trace.

LEMMA 6.8.2 *If $T \in A(V)$ then* tr T *is the sum of the characteristic roots of T (using each characteristic root as often as its multiplicity).*

Proof. We can assume that T is a matrix in F_n; if K is the splitting field for the minimal polynomial of T over F, then in K_n, by Theorem 6.6.2, T can be brought to its Jordan form, J. J is a matrix on whose diagonal appear the characteristic roots of T, each root appearing as often as its multiplicity. Thus $\text{tr } J$ = sum of the characteristic roots of T; however, since J is of the form ATA^{-1}, $\text{tr } J = \text{tr } T$, and this proves the lemma.

If T is nilpotent then all its characteristic roots are 0, whence by Lemma 6.8.2, $\text{tr } T = 0$. But if T is nilpotent, then so are $T^2, T^3, \ldots$; thus $\text{tr } T^i = 0$ for all $i \geq 1$.

What about other directions, namely, if $\text{tr } T^i = 0$ for $i = 1, 2, \ldots$ does it follow that T is nilpotent? In this generality the answer is no, for if F is a field of characteristic 2 then the unit matrix

$$\begin{pmatrix} 1 & 0 \\ 0 & 1 \end{pmatrix}$$

in F_2 has trace 0 (for $1 + 1 = 0$) as do all its powers, yet clearly the unit matrix is not nilpotent. However, if we restrict the characteristic of F to be 0, the result is indeed true.

LEMMA 6.8.3 *If F is a field of characteristic 0, and if $T \in A_F(V)$ is such that* $\text{tr } T^i = 0$ *for all $i \geq 1$ then T is nilpotent.*

Proof. Since $T \in A_F(V)$, T satisfies some minimal polynomial $p(x) = x^m + \alpha_1 x^{m-1} + \cdots + \alpha_m$; from $T^m + \alpha_1 T^{m-1} + \cdots + \alpha_{m-1} T + \alpha_m = 0$, taking traces of both sides yields

$$\text{tr } T^m + \alpha_1 \text{ tr } T^{m-1} + \cdots + \alpha_{m-1} \text{ tr } T + \text{tr } \alpha_m = 0.$$

However, by assumption, $\text{tr } T^i = 0$ for $i \geq 1$, thus we get $\text{tr } \alpha_m = 0$; if $\dim V = n$, $\text{tr } \alpha_m = n\alpha_m$ whence $n\alpha_m = 0$. But the characteristic of F is 0; therefore, $n \neq 0$, hence it follows that $\alpha_m = 0$. Since the constant term of the minimal polynomial of T is 0, by Theorem 6.1.2 T is singular and so 0 is a characteristic root of T.

We can consider T as a matrix in F_n and therefore also as a matrix in K_n, where K is an extension of F which in turn contains all the characteristic roots of T. In K_n, by Theorem 6.4.1, we can bring T to triangular form, and since 0 is a characteristic root of T, we can actually bring it to the form

$$\left(\begin{array}{c|cccc} 0 & 0 & & \cdots & 0 \\ \hline \beta_2 & \alpha_2 & 0 & . & 0 \\ \vdots & & & \ddots & \vdots \\ & & * & & \\ \beta_n & & & & \alpha_n \end{array}\right) = \left(\begin{array}{c|c} 0 & 0 \\ \hline * & T_2 \end{array}\right),$$

where

$$T_2 = \begin{pmatrix} \alpha_2 & 0 & 0 \\ & \ddots & \vdots \\ & * & \\ & & \alpha_n \end{pmatrix}$$

is an $(n - 1) \times (n - 1)$ matrix (the $*$'s indicate parts in which we are not interested in the explicit entries). Now

$$T^k = \left(\begin{array}{c|c} 0 & 0 \\ \hline * & T_2^k \end{array}\right)$$

hence $0 = \text{tr } T^k = \text{tr } T_2^k$. Thus T_2 is an $(n - 1) \times (n - 1)$ matrix with the property that $\text{tr } T_2^k = 0$ for all $k \geq 1$. Either using induction on n, or repeating the argument on T_2 used for T, we get, since $\alpha_2, \ldots, \alpha_n$ are the characteristic roots of T_2, that $\alpha_2 = \cdots = \alpha_n = 0$. Thus when T is brought to triangular form, all its entries on the main diagonal are 0, forcing T to be nilpotent. (Prove!)

This lemma, though it might seem to be special, will serve us in good stead often. We make immediate use of it to prove a result usually known as the *Jacobson lemma*.

LEMMA 6.8.4 *If F is of characteristic 0 and if S and T, in $A_F(V)$, are such that $ST - TS$ commutes with S, then $ST - TS$ is nilpotent.*

Proof. For any $k \geq 1$ we compute $(ST - TS)^k$. Now $(ST - TS)^k = (ST - TS)^{k-1}(ST - TS) = (ST - TS)^{k-1}ST - (ST - TS)^{k-1}TS$. Since $ST - TS$ commutes with S, the term $(ST - TS)^{k-1}ST$ can be written in the form $S((ST - TS)^{k-1}T)$. If we let $B = (ST - TS)^{k-1}T$, we see that $(ST - TS)^k = SB - BS$; hence $\text{tr } ((ST - TS)^k) = \text{tr } (SB - BS) = \text{tr } (SB) - \text{tr } (BS) = 0$ by Lemma 6.8.1. The previous lemma now tells us that $ST - TS$ must be nilpotent.

The trace provides us with an extremely useful linear functional on F_n (and so, on $A_F(V)$) into F. We now introduce an important mapping of F_n into itself.

DEFINITION If $A = (\alpha_{ij}) \in F_n$ then the *transpose* of A, written as A', is the matrix $A' = (\gamma_{ij})$ where $\gamma_{ji} = \alpha_{ji}$ for each i and j.

The transpose of A is the matrix obtained by interchanging the rows and columns of A. The basic formal properties of the transpose are contained in

LEMMA 6.8.5 *For all* $A, B \in F_n$,

1. $(A')' = A$.
2. $(A + B)' = A' + B'$.
3. $(AB)' = B'A'$.

Proof. The proofs of parts 1 and 2 are straightforward and are left to the reader; we content ourselves with proving part 3.

Suppose that $A = (\alpha_{ij})$ and $B = (\beta_{ij})$; then $AB = (\lambda_{ij})$ where

$$\lambda_{ij} = \sum_{k=1}^{n} \alpha_{ik}\beta_{kj}.$$

Therefore, by definition, $(AB)' = (\mu_{ij})$, where

$$\mu_{ij} = \lambda_{ji} = \sum_{k=1}^{n} \alpha_{jk}\beta_{ki}.$$

On the other hand, $A' = (\gamma_{ij})$ where $\gamma_{ij} = \alpha_{ji}$ and $B' = (\xi_{ij})$ where $\xi_{ij} = \beta_{ji}$, whence the (i, j) element of $B'A'$ is

$$\sum_{k=1}^{n} \xi_{ik}\gamma_{kj} = \sum_{k=1}^{n} \beta_{ki}\alpha_{jk} = \sum_{k=1}^{n} \alpha_{jk}\beta_{ki} = \mu_{ij}.$$

That is, $(AB)' = B'A'$ and we have verified part 3 of the lemma.

In part 3, if we specialize $A = B$ we obtain $(A^2)' = (A')^2$. Continuing, we obtain $(A^k)' = (A')^k$ for all positive integers k. When A is invertible, then $(A^{-1})' = (A')^{-1}$.

There is a further property enjoyed by the transpose, namely, if $\lambda \in F$ then $(\lambda A)' = \lambda A'$ for all $A \in F_n$. Now, if $A \in F_n$ satisfies a polynomial $\alpha_0 A^m + \alpha_1 A^{m-1} + \cdots + \alpha_m = 0$, we obtain $(\alpha_0 A^m + \cdots + \alpha_m)' = 0' = 0$. Computing out $(\alpha_0 A^m + \cdots + \alpha_m)'$ using the properties of the transpose, we obtain $\alpha_0 (A')^m + \alpha_1 (A')^{m-1} + \cdots + \alpha_m = 0$, that is to say, A' satisfies any polynomial over F which is satisfied by A. Since $A = (A')'$, by the same token, A satisfies any polynomial over F which is satisfied by A'. In particular, A and A' have the same minimal polynomial over F and so *they have the same characteristic roots*. One can show each root occurs with the same multiplicity in A and A'. This is evident once it is established that A and A' are actually similar (see Problem 14).

DEFINITION The matrix A is said to be a *symmetric matrix* if $A' = A$.

DEFINITION The matrix A is said to be a *skew-symmetric matrix* if $A' = -A$.

When the characteristic of F is 2, since $1 = -1$, we would not be able to distinguish between symmetric and skew-symmetric matrices. *We make* *the flat assumption for the remainder of this section that the characteristic of F is different from* 2.

Ready ways for producing symmetric and skew-symmetric matrices are available to us. For instance, if A is an arbitrary matrix, then $A + A'$ is symmetric and $A - A'$ is skew-symmetric. Noting that $A = \frac{1}{2}(A + A') + \frac{1}{2}(A - A')$, every matrix is a sum of a symmetric one and a skew-symmetric one. This decomposition is unique (see Problem 19). Another method of producing symmetric matrices is as follows: if A is an arbitrary matrix, then both AA' and $A'A$ are symmetric. (Note that these need not be equal.)

It is in the nature of a mathematician, once given an interesting concept arising from a particular situation, to try to strip this concept away from the particularity of its origins and to employ the key properties of the concept as a means of abstracting it. We proceed to do this with the transpose. We take, as the formal properties of greatest interest, those properties of the transpose contained in the statement of Lemma 6.8.5 which asserts that on F_n the transpose defines an anti-automorphism of period 2. This leads us to make the

DEFINITION A mapping $*$ from F_n into F_n is called an *adjoint* on F_n if

1. $(A^*)^* = A$;
2. $(A + B)^* = A^* + B^*$;
3. $(AB)^* = B^*A^*$;

for all $A, B \in F_n$.

Note that we do *not* insist that $(\lambda A)^* = \lambda A^*$ for $\lambda \in F$. In fact, in some of the most interesting adjoints used, this is not the case. We discuss one such now. Let F be the field of complex numbers; for $A = (\alpha_{ij}) \in F_n$, let $A^* = (\gamma_{ij})$ where $\gamma_{ij} = \bar{\alpha}_{ji}$ the complex conjugate of α_{ji}. In this case $*$ is usually called the *Hermitian adjoint* on F_n. A few sections from now, we shall make a fairly extensive study of matrices under the Hermitian adjoint.

Everything we said about transpose, e.g., symmetric, skew-symmetric, can be carried over to general adjoints, and we speak about elements symmetric under $*$ (i.e., $A^* = A$), skew-symmetric under $*$, etc. In the exercises at the end, there are many examples and problems referring to general adjoints.

However, now as a diversion let us play a little with the Hermitian adjoint. We do not call anything we obtain a theorem, not because it is not worthy of the title, but rather because we shall redo it later (and properly label it) from one central point of view.

So, let us suppose that F is the field of complex numbers and that the adjoint, $*$, on F_n is the Hermitian adjoint. The matrix A is called *Hermitian* if $A^* = A$.

First remark: If $A \neq 0 \in F_n$, then $\text{tr}(AA^*) > 0$. Second remark: As a consequence of the first remark, if $A_1, \ldots, A_k \in F_n$ and if $A_1A_1^* + A_2A_2^* + \cdots + A_kA_k^* = 0$, then $A_1 = A_2 = \cdots = A_k = 0$. Third remark: If λ is a scalar matrix then $\lambda^* = \bar{\lambda}$, the complex conjugate of λ.

Suppose that $A \in F_n$ is Hermitian and that the complex number $\alpha + \beta i$, where α and β are real and $i^2 = -1$, is a characteristic root of A. Thus $A - (\alpha + \beta i)$ is not invertible; but then $(A - (\alpha + \beta i))(A - (\alpha - \beta i)) = (A - \alpha)^2 + \beta^2$ is not invertible. However, if a matrix is singular, it must annihilate a nonzero matrix (Theorem 6.1.2, Corollary 2). There must therefore be a matrix $C \neq 0$ such that $C((A - \alpha)^2 + \beta^2) = 0$. We multiply this from the right by C^* and so obtain

$$C(A - \alpha)^2C^* + \beta^2CC^* = 0. \tag{1}$$

Let $D = C(A - \alpha)$ and $E = \beta C$. Since $A^* = A$ and α is real, $C(A - \alpha)^2C^* = DD^*$; since β is real, $\beta^2CC^* = EE^*$. Thus equation (1) becomes $DD^* + EE^* = 0$; by the remarks made above, this forces $D = 0$ and $E = 0$. We only exploit the relation $E = 0$. Since $0 = E = \beta C$ and since $C \neq 0$ we must have $\beta = 0$. What exactly have we proved? In fact, we have proved the pretty (and important) result *that if a complex number λ is a characteristic root of a Hermitian matrix, then λ must be real.* Exploiting properties of the field of complex numbers, one can actually restate this as follows: *The characteristic roots of a Hermitian matrix are all real.*

We continue a little farther in this vein. For $A \in F_n$, let $B = AA^*$; B is a Hermitian matrix. If the real number α is a characteristic root of B, can α be an arbitrary real number or must it be restricted in some way? Indeed, we claim that α must be nonnegative. For if α were negative then $\alpha = -\beta^2$, where β is a real number. But then $B - \alpha = B + \beta^2 = AA^* + \beta^2$ is not invertible, and there is a $C \neq 0$ such that $C(AA^* + \beta^2) = 0$. Multiplying by C^* from the right and arguing as before, we obtain $\beta = 0$, a contradiction. We have shown that any real characteristic root of AA^* must be nonnegative. In actuality, the "real" in this statement is superfluous and we could state: For any $A \in F_n$ all the characteristic roots of AA^* are nonnegative.

Problems

Unless otherwise specified, symmetric and skew-symmetric refer to transpose.

1. Prove that $\text{tr}(A + B) = \text{tr}\, A + \text{tr}\, B$ and that for $\lambda \in F$, $\text{tr}(\lambda A) = \lambda\, \text{tr}\, A$.
2. (a) Using a trace argument, prove that if the characteristic of F is 0 then it is impossible to find $A, B \in F_n$ such that $AB - BA = 1$.

(b) In part (a), prove, in fact, that $1 - (AB - BA)$ cannot be nilpotent.

3. (a) Let f be a function defined on F_n having its values in F such that
   1. $f(A + B) = f(A) + f(B)$;
   2. $f(\lambda A) = \lambda f(A)$;
   3. $f(AB) = f(BA)$;

   for all $A, B \in F_n$ and all $\lambda \in F$. Prove that there is an element $\alpha_0 \in F$ such that $f(A) = \alpha_0 \operatorname{tr} A$ for every A in F_n.

   (b) If the characteristic of F is 0 and if the f in part (a) satisfies the additional property that $f(1) = n$, prove that $f(A) = \operatorname{tr} A$ for all $A \in F_n$.

Note that Problem 3 characterizes the trace function.

*4. (a) If the field F has an infinite number of elements, prove that every element in F_n can be written as the sum of regular matrices.

   (b) If F has an infinite number of elements and if f, defined on F_n and having its values in F, satisfies
   1. $f(A + B) = f(A) + f(B)$;
   2. $f(\lambda A) = \lambda f(A)$;
   3. $f(BAB^{-1}) = f(A)$;

   for every $A \in F_n$, $\lambda \in F$ and invertible element B in F_n, prove that $f(A) = \alpha_0 \operatorname{tr} A$ for a particular $\alpha_0 \in F$ and all $A \in F_n$.

5. Prove the Jacobson lemma for elements $A, B \in F_n$ if n is less than the characteristic of F.

6. (a) If $C \in F_n$, define the mapping d_C on F_n, by $d_C(X) = XC - CX$ for $X \in F_n$. Prove that $d_C(XY) = (d_C(X))Y + X(d_C(Y))$. (Does this remind you of the derivative?)

   (b) Using (a), prove that if $AB - BA$ commutes with A, then for any polynomial $q(x) \in F[x]$, $q(A)B - Bq(A) = q'(A)(AB - BA)$, where $q'(x)$ is the derivative of $q(x)$.

*7. Use part (b) of Problem 6 to give a proof of the Jacobson lemma. (*Hint:* Let $p(x)$ be the minimal polynomial for A and consider $0 = p(A)B - Bp(A)$.)

8. (a) If A is a triangular matrix, prove that the entries on the diagonal of A are exactly all the characteristic roots of A.

   (b) If A is triangular and the elements on its main diagonal are 0, prove that A is nilpotent.

9. For any $A, B \in F_n$ and $\lambda \in F$ prove that $(A')' = A$, $(A + B)' = A' + B'$, and $(\lambda A)' = \lambda A'$.

10. If A is invertible, prove that $(A^{-1})' = (A')^{-1}$.

11. If A is skew-symmetric, prove that the elements on its main diagonal are all 0.

12. If A and B are symmetric matrices, prove that AB is symmetric if and only if $AB = BA$.
13. Give an example of an A such that $AA' \neq A'A$.
*14. Show that A and A' are similar.
15. The symmetric elements in F_n form a vector space; find its dimension and exhibit a basis for it.
*16. In F_n let S denote the set of symmetric elements; prove that the subring of F_n generated by S is all of F_n.
*17. If the characteristic of F is 0 and $A \in F_n$ has trace 0 ($\text{tr } A = 0$) prove that there is a $C \in F_n$ such that CAC^{-1} has only 0's on its main diagonal.
*18. If F is of characteristic 0 and $A \in F_n$ has trace 0, prove that there exist $B, C \in F_n$ such that $A = BC - CB$. (*Hint:* First step, assume, by result of Problem 17, that all the diagonal elements of A are 0.)
19. (a) If F is of characteristic not 2 and if $*$ is any adjoint on F_n, let $S = \{A \in F_n \mid A^* = A\}$ and let $K = \{A \in F_n \mid A^* = -A\}$. Prove that $S + K = F_n$.
    (b) If $A \in F_n$ and $A = B + C$ where $B \in S$ and $C \in K$, prove that B and C are unique and determine them.
20. (a) If $A, B \in S$ prove that $AB + BA \in S$.
    (b) If $A, B \in K$ prove that $AB - BA \in K$.
    (c) If $A \in S$ and $B \in K$ prove that $AB - BA \in S$ and that $AB + BA \in K$.
21. If ϕ is an automorphism of the field F we define the mapping Φ on F_n by: If $A = (\alpha_{ij})$ then $\Phi(A) = (\phi(\alpha_{ij}))$. Prove that $\Phi(A + B) = \Phi(A) + \Phi(B)$ and that $\Phi(AB) = \Phi(A)\Phi(B)$ for all $A, B \in F_n$.
22. If $*$ and $\circledast$ define two adjoints on F_n, prove that the mapping $\psi: A \to (A^*)^{\circledast}$ for every $A \in F_n$ satisfies $\psi(A + B) = \psi(A) + \psi(B)$ and $\psi(AB) = \psi(A)\psi(B)$ for every $A, B \in F_n$.
23. If $*$ is any adjoint on F_n and λ is a scalar matrix in F_n, prove that λ^* must also be a scalar matrix.
*24. Suppose we know the following theorem: If ψ is an automorphism of F_n (i.e., ψ maps F_n onto itself in such a way that $\psi(A + B) = \psi(A) + \psi(B)$ and $\psi(AB) = \psi(A)\psi(B)$) such that $\psi(\lambda) = \lambda$ for every scalar matrix λ, then there is an element $P \in F_n$ such that $\psi(A) = PAP^{-1}$ for every $A \in F_n$. On the basis of this theorem, prove: If $*$ is an adjoint of F_n such that $\lambda^* = \lambda$ for every scalar matrix λ then there exists a matrix $P \in F_n$ such that $A^* = PA'P^{-1}$ for every $A \in F_n$. Moreoever, $P^{-1}P'$ must be a scalar.
25. If $P \in F_n$ is such that $P^{-1}P' \neq 0$ is a scalar, prove that the mapping defined by $A^* = PA'P^{-1}$ is an adjoint on F_n.

*26. Assuming the theorem about automorphisms stated in Problem 24, prove the following: If $*$ is an adjoint on F_n there is an automorphism ϕ of F of period 2 and an element $P \in F_n$ such that $A^* = P(\Phi(A))'P^{-1}$ for all $A \in F_n$ (for notation, see Problem 21). Moreover, P must satisfy $P^{-1}\Phi(P)'$ is a scalar.

Problems 24 and 26 indicate that a general adjoint on F_n is not so far removed from the transpose as one would have guessed at first glance.

*27. If ψ is an automorphism of F_n such that $\psi(\lambda) = \lambda$ for all scalars, prove that there is a $P \in F_n$ such that $\psi(A) = PAP^{-1}$ for every $A \in F_n$.

In the remainder of the problems, F will be the field of complex numbers and $*$ the Hermitian adjoint on F_n.

28. If $A \in F_n$ prove that there are unique Hermitian matrices B and C such that $A = B + iC \quad (i^2 = -1)$.
29. Prove that tr $AA^* > 0$ if $A \neq 0$.
30. By directly computing the matrix entries, prove that if $A_1A_1^* + \cdots + A_kA_k^* = 0$, then $A_1 = A_2 = \cdots = A_k = 0$.
31. If A is in F_n and if $BAA^* = 0$, prove that $BA = 0$.
32. If A in F_n is Hermitian and $BA^k = 0$, prove that $BA = 0$.
33. If $A \in F_n$ is Hermitian and if λ, μ are two distinct (real) characteristic roots of A and if $C(A - \lambda) = 0$ and $D(A - \mu) = 0$, prove that $CD^* = DC^* = 0$.

*34. (a) Assuming that all the characteristic roots of the Hermitian matrix A are in the field of complex numbers, combining the results of Problems 32, 33, and the fact that the roots, then, must all be real and the result of the corollary to Theorem 6.6.1, prove that A can be brought to diagonal form; that is, there is a matrix P such that PAP^{-1} is diagonal.
(b) In part (a) prove that P could be chosen so that $PP^* = 1$.

35. Let $V_n = \{A \in F_n \mid AA^* = 1\}$. Prove that V_n is a group under matrix multiplication.
36. If A commutes with $AA^* - A^*A$ prove that $AA^* = A^*A$.

6.9 Determinants

The trace defines an important and useful function from the matrix ring F_n (and from $A_F(V)$) into F; its properties concern themselves, for the most part, with additive properties of matrices. We now shall introduce the even more important function, known as the determinant, which maps F_n into F.

Its properties are closely tied to the multiplicative properties of matrices.

Aside from its effectiveness as a tool in proving theorems, the determinant is valuable in "practical" ways. Given a matrix T, in terms of explicit determinants we can construct a concrete polynomial whose roots are the characteristic roots of T; even more, the multiplicity of a root of this polynomial corresponds to its multiplicity as a characteristic root of T. In fact, the characteristic polynomial of T, defined earlier, can be exhibited as this explicit, determinantal polynomial.

Determinants also play a key role in the solution of systems of linear equations. It is from this direction that we shall motivate their definition.

There are many ways to develop the theory of determinants, some very elegant and some deadly and ugly. We have chosen a way that is at neither of these extremes, but which for us has the advantage that we can reach the results needed for our discussion of linear transformations as quickly as possible.

In what follows F will be an arbitrary field, F_n the ring of $n \times n$ matrices over F, and $F^{(n)}$ the vector space of n-tuples over F. By a matrix we shall tacitly understand an element in F_n. As usual, Greek letters will indicate elements of F (unless otherwise defined).

Consider the system of equations

$$\alpha_{11}x_1 + \alpha_{12}x_2 = \beta_1,$$
$$\alpha_{21}x_1 + \alpha_{22}x_2 = \beta_2.$$

We ask: Under what conditions on the α_{ij} can we solve for x_1, x_2 given arbitrary β_1, β_2? Equivalently, given the matrix

$$A = \begin{pmatrix} \alpha_{11} & \alpha_{12} \\ \alpha_{21} & \alpha_{22} \end{pmatrix},$$

when does this map $F^{(2)}$ *onto* itself?

Proceeding as in high school, we eliminate x_1 between the two equations; the criterion for solvability then turns out to be $\alpha_{11}\alpha_{22} - \alpha_{12}\alpha_{21} \neq 0$.

We now try the system of three linear equations

$$\alpha_{11}x_1 + \alpha_{12}x_2 + \alpha_{13}x_3 = \beta_1,$$
$$\alpha_{21}x_1 + \alpha_{22}x_2 + \alpha_{23}x_3 = \beta_2,$$
$$\alpha_{31}x_1 + \alpha_{32}x_2 + \alpha_{33}x_3 = \beta_3,$$

and again ask for conditions for solvability given arbitrary $\beta_1, \beta_2, \beta_3$. Eliminating x_1 between these two-at-a-time, and then x_2 from the resulting two equations leads us to the criterion for solvability that

$$\alpha_{11}\alpha_{22}\alpha_{33} + \alpha_{12}\alpha_{23}\alpha_{31} + \alpha_{13}\alpha_{21}\alpha_{32} - \alpha_{12}\alpha_{21}\alpha_{33} - \alpha_{11}\alpha_{23}\alpha_{32} - \alpha_{13}\alpha_{22}\alpha_{31} \neq 0.$$

Using these two as models (and with the hindsight that all this will work) we shall make the broad jump to the general case and shall define the determinant of an arbitrary $n \times n$ matrix over F. But first a little notation!

Let S_n be the symmetric group of degree n; we consider elements in S_n to be acting on the set $\{1, 2, \ldots, n\}$. For $\sigma \in S_n$, $\sigma(i)$ will denote the image of i under σ. (We switch notation, writing the permutation as acting from the left rather than, as previously, from the right. We do so to facilitate writing subscripts.) The symbol $(-1)^\sigma$ for $\sigma \in S_n$ will mean $+1$ if σ is an *even* permutation and -1 if σ is an *odd* permutation.

DEFINITION If $A = (\alpha_{ij})$ then the *determinant of* A, written $\det A$, is the element $\sum_{\sigma \in S_n} (-1)^\sigma \alpha_{1\sigma(1)}\alpha_{2\sigma(2)} \cdots \alpha_{n\sigma(n)}$ in F.

We shall at times use the notation

$$\begin{vmatrix} \alpha_{11} & \cdots & \alpha_{1n} \\ \vdots & & \vdots \\ \alpha_{n1} & \cdots & \alpha_{nn} \end{vmatrix}$$

for the determinant of the matrix

$$\begin{pmatrix} \alpha_{11} & \cdots & \alpha_{1n} \\ \vdots & & \vdots \\ \alpha_{n1} & \cdots & \alpha_{nn} \end{pmatrix}.$$

Note that the determinant of a matrix A is the sum (neglecting, for the moment, signs) of all possible products of entries of A, one entry taken from each row and column of A. In general, it is a messy job to expand the determinant of a matrix—after all there are $n!$ terms in the expansion—but for at least one type of matrix we can do this expansion visually, namely,

LEMMA 6.9.1 *The determinant of a triangular matrix is the product of its entries on the main diagonal.*

Proof. Being triangular implies two possibilities, namely, either all the elements above the main diagonal are 0 or all the elements below the main diagonal are 0. We prove the result for A of the form

$$\begin{pmatrix} \alpha_{11} & 0 & \cdots & 0 \\ & \alpha_{22} & & \\ & * & \ddots & \vdots \\ & & & \alpha_{nn} \end{pmatrix}$$

and indicate the slight change in argument for the other kind of triangular matrices.

Since $\alpha_{1i} = 0$ unless $i = 1$, in the expansion of $\det A$ the only nonzero contribution comes in those terms where $\sigma(1) = 1$. Thus, since σ is a permutation, $\sigma(2) \neq 1$; however, if $\sigma(2) > 2$, $\alpha_{2\sigma(2)} = 0$, thus to get a nonzero contribution to det A, $\sigma(2) = 2$. Continuing in this way, we must have $\sigma(i) = i$ for all i, which is to say, in the expansion of det A the only nonzero term arises when σ is the identity element of S_n. Hence the sum of the $n!$ terms reduces to just one term, namely, $\alpha_{11}\alpha_{22} \cdots \alpha_{nn}$, which is the contention of the lemma.

If A is lower triangular we start at the opposite end, proving that for a nonzero contribution $\sigma(n) = n$, then $\sigma(n-1) = n - 1$, etc.

Some special cases are of interest:

1. If

$$A = \begin{pmatrix} \lambda_1 & & \\ & \ddots & \\ & & \lambda_n \end{pmatrix}$$

is diagonal, det $A = \lambda_1\lambda_2 \cdots \lambda_n$.

2. If

$$A = \begin{pmatrix} 1 & & & \\ & 1 & & \\ & & \ddots & \\ & & & 1 \end{pmatrix},$$

the identity matrix, then det $A = 1$.

3. If

$$A = \begin{pmatrix} \lambda & & & \\ & \lambda & & \\ & & \ddots & \\ & & & \lambda \end{pmatrix},$$

the scalar matrix, then det $A = \lambda^n$.

Note also that if a row (or column) of a matrix consists of 0*'s then the determinant is* 0, for each term of the expansion of the determinant would be a product in which one element, at least, is 0, hence each term is 0.

Given the matrix $A = (\alpha_{ij})$ in F_n we can consider its first row $v_1 = (\alpha_{11}, \alpha_{12}, \ldots, \alpha_{1n})$ as a vector in $F^{(n)}$; similarly, for its second row, v_2, and the others. We then can consider det A as a function of the n vectors $v_1, \ldots, v_n$. Many results are most succinctly stated in these terms so we shall often consider det $A = d(v_1, \ldots, v_n)$; in this *the notation is always meant to imply* that v_1 is the first row, v_2 the second, and so on, of A.

One further remark: Although we are working over a field, we could just as easily assume that we are working over a commutative ring, except in the obvious places where we divide by elements. This remark will only enter when we discuss determinants of matrices having polynomial entries, a little later in the section.

LEMMA 6.9.2 *If $A \in F_n$ and $\gamma \in F$ then $d(v_1, \ldots, v_{i-1}, \gamma v_i, v_{i+1}, \ldots, v_n) = \gamma d(v_1, \ldots, v_{i-1}, v_i, v_{i+1}, \ldots, v_n)$.*

Note that the lemma says that if all the elements in one row of A are multiplied by a fixed element γ in F then the determinant of A is itself multiplied by γ.

Proof. Since only the entries in the ith row are changed, the expansion of $d(v_1, \ldots, v_{i-1}, \gamma v_i, v_{i+1}, \ldots, v_n)$ is

$$\sum_{\sigma \in S_n} (-1)^\sigma \alpha_{1\sigma(1)} \cdots \alpha_{i-1,\sigma(i-1)}(\gamma \alpha_{i\sigma(i)})\alpha_{i+1,\sigma(i+1)} \cdots \alpha_{n\sigma(n)};$$

since this equals $\gamma \sum_{\sigma \in S_n} (-1)^\sigma \alpha_{1\sigma(1)} \cdots \alpha_{i\sigma(i)} \cdots \alpha_{n\sigma(n)}$, it does indeed equal $\gamma d(v_1, \ldots, v_n)$.

LEMMA 6.9.3

$$d(v_1, \ldots, v_{i-1}, v_i, v_{i+1}, \ldots, v_n) + d(v_1, \ldots, v_{i-1}, u_i, v_{i+1}, \ldots, v_n)$$
$$= d(v_1, \ldots, v_{i-1}, v_i + u_i, v_{i+1}, \ldots, v_n).$$

Before proving the result, let us see what it says and what it does not say. It does *not* say that $\det A + \det B = \det (A + B)$; this is false as is manifest in the example

$$A = \begin{pmatrix} 1 & 0 \\ 0 & 0 \end{pmatrix}, \qquad B = \begin{pmatrix} 0 & 0 \\ 0 & 1 \end{pmatrix},$$

where $\det A = \det B = 0$ while $\det (A + B) = 1$. It does say that if A and B are matrices equal everywhere but in the ith row then the new matrix obtained from A and B by using all the rows of A except the ith, and using as ith row the sum of the ith row of A and the ith row of B, has a determinant equal to $\det A + \det B$. If

$$A = \begin{pmatrix} 1 & 2 \\ 3 & 4 \end{pmatrix} \quad \text{and} \quad B = \begin{pmatrix} 1 & 1 \\ 3 & 4 \end{pmatrix},$$

then

$$\det A = -2, \quad \det B = 1, \quad \det \begin{pmatrix} 2 & 3 \\ 3 & 4 \end{pmatrix} = -1 = \det A + \det B.$$

Proof. If $v_1 = (\alpha_{11}, \ldots, \alpha_{1n}), \ldots, v_i = (\alpha_{i1}, \ldots, \alpha_{in}), \ldots, v_n = (\alpha_{n1}, \ldots, \alpha_{nn})$ and if $u_i = (\beta_{i1}, \ldots, \beta_{in})$, then

$$\begin{aligned}
d(v_1, \ldots, v_{i-1}, u_i + v_i, v_{i+1}, \ldots, v_n) \\
&= \sum_{\sigma \in S_n} (-1)^\sigma \alpha_{1\sigma(1)} \cdots \alpha_{i-1,\sigma(i-1)}(\alpha_{i\sigma(i)} + \beta_{i\sigma(i)})\alpha_{i+1,\sigma(i+1)} \cdots \alpha_{n\sigma(n)} \\
&= \sum_{\sigma \in S_n} (-1)^\sigma \alpha_{1\sigma(1)} \cdots \alpha_{i-1,\sigma(i-1)}\alpha_{i\sigma(i)} \cdots \alpha_{n\sigma(n)} \\
&\quad + \sum_{\sigma \in S_n} (-1)^\sigma \alpha_{1\sigma(1)} \cdots \alpha_{i-1,\sigma(i-1)}\beta_{i\sigma(i)} \cdots \alpha_{n\sigma(n)} \\
&= d(v_1, \ldots, v_i, \ldots, v_n) + d(v_1, \ldots, u_i, \ldots, v_n).
\end{aligned}$$

The properties embodied in Lemmas 6.9.1, 6.9.2, and 6.9.3, along with that in the next lemma, can be shown to characterize the determinant function (see Problem 13, end of this section). Thus, the formal property exhibited in the next lemma is basic in the theory of determinants.

LEMMA 6.9.4 *If two rows of A are equal (that is, $v_r = v_s$ for $r \neq s$), then* $\det A = 0$.

Proof. Let $A = (\alpha_{ij})$ and suppose that for some r, s where $r \neq s$, $\alpha_{rj} = \alpha_{sj}$ for all j. Consider the expansion

$$\det A = \sum_{\alpha \in S_n} (-1)^{\sigma} \alpha_{1\sigma(1)} \cdots \alpha_{r\sigma(r)} \cdots \alpha_{s\sigma(s)} \cdots \alpha_{n\sigma(n)}.$$

In the expansion we pair the terms as follows: For $\sigma \in S_n$ we pair the term $(-1)^{\sigma}\alpha_{1\sigma(1)} \cdots \alpha_{n\sigma(n)}$ with the term $(-1)^{\tau\sigma}\alpha_{1\tau\sigma(1)} \cdots \alpha_{n\tau\sigma(n)}$ where τ is the transposition $(\sigma(r), \sigma(s))$. Since τ is a transposition and $\tau^2 = 1$, this indeed gives us a pairing. However, since $\alpha_{r\sigma(r)} = \alpha_{s\sigma(r)}$, by assumption, and $\alpha_{r\sigma(r)} = \alpha_{s\tau\sigma(s)}$, we have that $\alpha_{r\sigma(r)} = \alpha_{s\tau\sigma(s)}$. Similarly, $\alpha_{s\sigma(s)} = \alpha_{r\tau\sigma(r)}$. On the other hand, for $i \neq r$ and $i \neq s$, since $\tau\sigma(i) = \sigma(i)$, $\alpha_{i\sigma(i)} = \alpha_{i\tau\sigma(i)}$. Thus the terms $\alpha_{1\sigma(1)} \cdots \alpha_{n\sigma(n)}$ and $\alpha_{1\tau\sigma(1)} \cdots \alpha_{n\tau\sigma(n)}$ are equal. The first occurs with the sign $(-1)^{\sigma}$ and the second with the sign $(-1)^{\tau\sigma}$ in the expansion of $\det A$. Since τ is a transposition and so an odd permutation, $(-1)^{\tau\sigma} = -(-1)^{\sigma}$. Therefore in the pairing, the paired terms cancel each other out in the sum, whence $\det A = 0$. (The proof does not depend on the characteristic of F and holds equally well even in the case of characteristic 2.)

From the results so far obtained we can determine the effect, on a determinant of a given matrix, of a given permutation of its rows.

LEMMA 6.9.5 *Interchanging two rows of A changes the sign of its determinant.*

Proof. Since two rows are equal, by Lemma 6.9.4, $d(v_1, \ldots, v_{i-1}, v_i + v_j, v_{i+1}, \ldots, v_{j-1}, v_i + v_j, v_{j+1}, \ldots, v_n) = 0$. Using Lemma 6.9.3 several times, we can expand this to obtain $d(v_1, \ldots, v_{i-1}, v_i, \ldots, v_{j-1}, v_j, \ldots, v_n) + d(v_1, \ldots, v_{i-1}, v_j, \ldots, v_{j-1}, v_i, \ldots, v_n) + d(v_1, \ldots, v_{i-1}, v_i, \ldots, v_{j-1}, v_i, \ldots, v_n) + d(v_1, \ldots, v_{i-1}, v_j, \ldots, v_{j-1}, v_j, \ldots, v_n) = 0$. However, each of the last two terms has in it two equal rows, whence, by Lemma 6.9.4, each is 0. The above relation then reduces to $d(v_1, \ldots, v_{i-1}, v_i, \ldots, v_{j-1}, v_j, \ldots, v_n) + d(v_1, \ldots, v_{i-1}, v_j, \ldots, v_{j-1}, v_i, \ldots, v_n) = 0$, which is precisely the assertion of the lemma.

COROLLARY *If the matrix B is obtained from A by a permutation of the rows of A then* $\det A = \pm\det B$, *the sign being $+1$ if the permutation is even, -1 if the permutation is odd.*

We are now in a position to collect pieces to prove the basic algebraic property of the determinant function, namely, that it preserves products. As a homomorphism of the multiplicative structure of F_n into F the determinant will acquire certain important characteristics.

THEOREM 6.9.1 *For $A, B \in F_n$, $\det(AB) = (\det A)(\det B)$.*

Proof. Let $A = (\alpha_{ij})$ and $B = (\beta_{ij})$; let the rows of B be the vectors $u_1, u_2, \ldots, u_n$. We introduce the n vectors $w_1, \ldots, w_n$ as follows:

$$\begin{aligned} w_1 &= \alpha_{11}u_1 + \alpha_{12}u_2 + \cdots + \alpha_{1n}u_n, \\ w_2 &= \alpha_{21}u_1 + \alpha_{22}u_2 + \cdots + \alpha_{2n}u_n, \\ &\vdots \\ w_n &= \alpha_{n1}u_1 + \alpha_{n2}u_2 + \cdots + \alpha_{nn}u_n. \end{aligned}$$

Consider $d(w_1, \ldots, w_n)$; expanding this out and making many uses of Lemmas 6.9.2 and 6.9.3, we obtain

$$d(w_1, \ldots, w_n) = \sum_{i_1, i_2, \ldots, i_n} \alpha_{1i_1}\alpha_{2i_2}\cdots\alpha_{ni_n} d(u_{i_1}, u_{i_2}, \ldots, u_{i_n}).$$

In this multiple sum $i_1, \ldots, i_n$ run independently from 1 to n. However, if any two $i_r = i_s$ then $u_{i_r} = u_{i_s}$ whence $d(u_{i_1}, \ldots, u_{i_r}, \ldots, u_{i_s}, \ldots, u_{i_n}) = 0$ by Lemma 6.9.4. In other words, the only terms in the sum that may give a nonzero contribution are those for which all of $i_1, i_2, \ldots, i_n$ are distinct, that is for which the mapping

$$\sigma = \begin{pmatrix} 1 & 2 & \cdots & n \\ i_1 & i_2 & \cdots & i_n \end{pmatrix}$$

is a permutation of $1, 2, \ldots, n$. Also any such permutation is possible. Finally note that by the corollary to Lemma 6.9.5, when

$$\sigma = \begin{pmatrix} 1 & 2 & \cdots & n \\ i_1 & i_2 & \cdots & i_n \end{pmatrix}$$

is a permutation, then $d(u_{i_1}, u_{i_2}, \ldots, u_{i_n}) = (-1)^\sigma d(u_1, \ldots, u_n) = (-1)^\sigma \det B$. Thus we get

$$\begin{aligned} d(w_1, \ldots, w_n) &= \sum_{\sigma \in S_n} \alpha_{1\sigma(1)} \cdots \alpha_{n\sigma(n)} (-1)^\sigma \det B \\ &= (\det B) \sum_{\sigma \in S_n} (-1)^\sigma \alpha_{1\sigma(1)} \cdots \alpha_{n\sigma(n)} \\ &= (\det B)(\det A). \end{aligned}$$

We now wish to identify $d(w_1, \ldots, w_n)$ as $\det(AB)$. However, since

$w_1 = \alpha_{11}u_1 + \cdots + \alpha_{1n}u_n$, $w_2 = \alpha_{21}u_1 + \cdots + \alpha_{2n}u_n, \ldots, w_n = \alpha_{n1}u_1 + \cdots + \alpha_{nn}u_n$

we get that $d(w_1, \ldots, w_n)$ is det C where the first row of C is w_1, the second is w_2, etc.

However, if we write out w_1, in terms of coordinates we obtain

$$\begin{aligned} w_1 &= \alpha_{11}u_1 + \cdots + \alpha_{1n}u_n = \alpha_{11}(\beta_{11}, \beta_{12}, \ldots, \beta_{1n}) \\ &\quad + \cdots + \alpha_{1n}(\beta_{n1}, \ldots, \beta_{nn}) \\ &= (\alpha_{11}\beta_{11} + \alpha_{12}\beta_{21} + \cdots + \alpha_{1n}\beta_{n1}, \alpha_{11}\beta_{12} + \cdots \\ &\quad + \alpha_{1n}\beta_{n2}, \ldots, \alpha_{11}\beta_{1n} + \cdots + \alpha_{1n}\beta_{nn}) \end{aligned}$$

which is the first row of AB. Similarly w_2 is the second row of AB, and so for the other rows. Thus we have $C = AB$. Since $\det(AB) = \det C = d(w_1, \ldots, w_n) = (\det A)(\det B)$, we have proved the theorem.

COROLLARY 1 *If A is invertible then* $\det A \neq 0$ *and* $\det(A^{-1}) = (\det A)^{-1}$.

Proof Since $AA^{-1} = 1$, $\det(AA^{-1}) = \det 1 = 1$. Thus by the theorem, $1 = \det(AA^{-1}) = (\det A)(\det A^{-1})$. This relation then states that $\det A \neq 0$ and $\det A^{-1} = 1/\det A$.

COROLLARY 2 *If A is invertible then for all B,* $\det(ABA^{-1}) = \det B$.

Proof. Using the theorem, as applied to $(AB)A^{-1}$, we get $\det((AB)A^{-1}) = \det(AB)\det(A^{-1}) = \det A \det B \det(A^{-1})$. Invoking Corollary 1, we reduce this further to $\det B$. Thus $\det(ABA^{-1}) = \det B$.

Corollary 2 allows us to define the determinant of a linear transformation. For, let $T \in A(V)$ and let $m_1(T)$ be the matrix of T in some basis of V. Given another basis, if $m_2(T)$ is the matrix of T in this second basis, then by Theorem 6.3.2, $m_2(T) = Cm_1(T)C^{-1}$, hence $\det(m_2(T)) = \det(m_1(T))$ by Corollary 2 above. That is, the matrix of T in any basis has the same determinant. *Thus the definition:* $\det T = \det m_1(T)$ *is in fact independent of the basis* and provides $A(V)$ with a determinant function.

In one of the earlier problems, it was the aim of the problem to prove that A', the transpose of A, is similar to A. Were this so (and it is), then A' and A, by Corollary 2, above would have the same determinant. Thus we should not be surprised that we can give a direct proof of this fact.

LEMMA 6.9.6 $\det A = \det(A')$.

Proof. Let $A = (\alpha_{ij})$ and $A' = (\beta_{ij})$; of course, $\beta_{ij} = \alpha_{ji}$. Now

$$\det A = \sum_{\sigma \in S_n} (-1)^\sigma \alpha_{1\sigma(1)} \cdots \alpha_{n\sigma(n)}$$

while

$$\det A' = \sum_{\sigma \in S_n} (-1)^\sigma \beta_{1\sigma(1)} \cdots \beta_{n\sigma(n)} = \sum_{\sigma \in S_n} (-1)^\sigma \alpha_{\sigma(1)1} \cdots \alpha_{\sigma(n)n}.$$

However, the term $(-1)^\sigma \alpha_{\sigma(1)1} \cdots \alpha_{\sigma(n)n}$ is equal to $(-1)^\sigma \alpha_{1\sigma^{-1}(1)} \cdots \alpha_{n\sigma^{-1}(n)}$. (Prove!) But σ and σ^{-1} are of the same parity, that is, if σ is odd, then so is σ^{-1}, whereas if σ is even then σ^{-1} is even. Thus

$$(-1)^\sigma \alpha_{1\sigma^{-1}(1)} \cdots \alpha_{n\sigma^{-1}(n)} = (-1)^{\sigma^{-1}} \alpha_{1\sigma^{-1}(1)} \cdots \alpha_{n\sigma^{-1}(n)}.$$

Finally as σ runs over S_n then σ^{-1} runs over S_n. Thus

$$\begin{aligned}\det A' &= \sum_{\sigma^{-1} \in S_n} (-1)^{\sigma^{-1}} \alpha_{1\sigma^{-1}(1)} \cdots \alpha_{n\sigma^{-1}(n)} \\ &= \sum_{\sigma \in S_n} (-1)^\sigma \alpha_{1\sigma(1)} \cdots \alpha_{n\sigma(n)} \\ &= \det A.\end{aligned}$$

In light of Lemma 6.9.6, interchanging the rows and columns of a matrix does not change its determinant. *But then Lemmas 6.9.2–6.9.5, which held for operations with rows of the matrix, hold equally for the columns of the same matrix.*

We make immediate use of the remark to derive *Cramer's rule* for solving a system of linear equations.

Given the system of linear equations

$$\begin{aligned}\alpha_{11}x_1 + \cdots + \alpha_{1n}x_n &= \beta_1 \\ &\vdots \\ \alpha_{n1}x_1 + \cdots + \alpha_{nn}x_n &= \beta_n,\end{aligned}$$

we call $A = (\alpha_{ij})$ the matrix of the system and $\Delta = \det A$ the *determinant of the system.*

Suppose that $\Delta \neq 0$; that is,

$$\Delta = \begin{vmatrix} \alpha_{11} & \cdots & \alpha_{1n} \\ \vdots & & \vdots \\ \alpha_{n1} & \cdots & \alpha_{nn} \end{vmatrix} \neq 0.$$

By Lemma 6.9.2 (as modified for columns instead of rows),

$$x_i\Delta = \begin{vmatrix} \alpha_{11} & \cdots & \alpha_{1i}x_i & \cdots & \alpha_{1n} \\ \vdots & & & & \vdots \\ \alpha_{n1} & \cdots & \alpha_{ni}x_i & \cdots & \alpha_{nn} \end{vmatrix}.$$

However, as a consequence of Lemmas 6.9.3, 6.9.4, we can add any multiple of a column to another without changing the determinant (see Problem 5). Add to the ith column of $x_i\Delta$, x_1 times the first column, x_2 times the second, $\ldots$, x_j times the jth column (for $j \neq i$). Thus

$$x_i\Delta = \begin{vmatrix} \alpha_{11} & \cdots & \alpha_{1,i-1} & (\alpha_{11}x_1 + \alpha_{12}x_2 + \cdots + \alpha_{1n}x_n) & \alpha_{1,i+1} & \cdots & \alpha_{1n} \\ \vdots & & \vdots & \vdots & & & \vdots \\ \alpha_{n1} & \cdots & \alpha_{n,i-1} & (\alpha_{n1}x_1 + \alpha_{n2}x_2 + \cdots + \alpha_{nn}x_n) & \alpha_{n,i+1} & \cdots & \alpha_{nn} \end{vmatrix}$$

and using $\alpha_{k1}x_1 + \cdots + \alpha_{kn}x_n = \beta_k$, we finally see that

$$x_i\Delta = \begin{vmatrix} \alpha_{11} & \cdots & \alpha_{1,i-1} & \beta_1 & \alpha_{1,i+1} & \cdots & \alpha_{1n} \\ \vdots & & \vdots & \vdots & \vdots & & \vdots \\ \alpha_{1n} & \cdots & \alpha_{n,i-1} & \beta_n & \alpha_{n,i+1} & \cdots & \alpha_{nn} \end{vmatrix} = \Delta_i, \quad \text{say.}$$

Hence, $x_i = \Delta_i/\Delta$. This is

THEOREM 6.9.2 (CRAMER'S RULE) *If the determinant, Δ, of the system of linear equations*

$$\begin{aligned} \alpha_{11}x_1 + \cdots + \alpha_{1n}x_n &= \beta_1 \\ &\vdots \\ \alpha_{n1}x_1 + \cdots + \alpha_{nn}x_n &= \beta_n \end{aligned}$$

is different from 0, *then the solution of the system is given by $x_i = \Delta_i/\Delta$, where Δ_i is the determinant obtained from Δ by replacing in Δ the ith column by $\beta_1, \beta_2, \ldots, \beta_n$.*

Example The system

$$\begin{aligned} x_1 + 2x_2 + 3x_3 &= -5, \\ 2x_1 + x_2 + x_3 &= -7, \\ x_1 + x_2 + x_3 &= 0, \end{aligned}$$

has determinant

$$\Delta = \begin{vmatrix} 1 & 2 & 3 \\ 2 & 1 & 1 \\ 1 & 1 & 1 \end{vmatrix} = 1 \neq 0,$$

hence

$$x_1 = \frac{\begin{vmatrix} -5 & 2 & 3 \\ -7 & 1 & 1 \\ 0 & 1 & 1 \end{vmatrix}}{\Delta}, \quad x_2 = \frac{\begin{vmatrix} 1 & -5 & 3 \\ 2 & -7 & 1 \\ 1 & 0 & 1 \end{vmatrix}}{\Delta}, \quad x_3 = \frac{\begin{vmatrix} 1 & 2 & -5 \\ 2 & 1 & -7 \\ 1 & 1 & 0 \end{vmatrix}}{\Delta}.$$

We can interrelate invertibility of a matrix (or linear transformation) with the value of its determinant. Thus the determinant provides us with a criterion for invertibility.

THEOREM 6.9.3 *A is invertible if and only if* $\det A \neq 0$.

Proof. If A is invertible, we have seen, in Corollary 1 to Theorem 6.9.1, that $\det A \neq 0$.

Suppose, on the other hand, that $\det A \neq 0$ where $A = (\alpha_{ij})$. By Cramer's rule we can solve the system

$$\begin{aligned} \alpha_{11}x_1 + \cdots + \alpha_{1n}x_n &= \beta_1 \\ &\vdots \\ \alpha_{n1}x_1 + \cdots + \alpha_{nn}x_n &= \beta_n \end{aligned}$$

for $x_1, \ldots, x_n$ given arbitrary $\beta_1, \ldots, \beta_n$. Thus, as a linear transformation on $F^{(n)}$, A' is onto; in fact the vector $(\beta_1, \ldots, \beta_n)$ is the image under A' of $\left(\frac{\Delta_1}{\Delta}, \ldots, \frac{\Delta_n}{\Delta}\right)$. Being onto, by Theorem 6.1.4, A' is invertible, hence A is invertible (Prove!).

We can see Theorem 6.9.3 from an alternative, and possibly more interesting, point of view. Given $A \in F_n$ we can embed it in K_n where K is an extension of F chosen so that in K_n, A can be brought to triangular form. Thus there is a $B \in K_n$ such that

$$BAB^{-1} = \begin{pmatrix} \lambda_1 & 0 & \cdots & 0 \\ & \lambda_2 & & \\ * & & \ddots & \vdots \\ & & & \lambda_n \end{pmatrix};$$

here $\lambda_1, \ldots, \lambda_n$ are all the characteristic roots of A, each occurring as often as its multiplicity as a characteristic root of A. Thus $\det A = \det(BAB^{-1}) = \lambda_1 \lambda_2 \cdots \lambda_n$ by Lemma 6.9.1. However, A is invertible if and only if none of its characteristic roots is 0; but $\det A \neq 0$ if and only if $\lambda_1 \lambda_2 \cdots \lambda_n \neq 0$, that is to say, if no characteristic root of A is 0. Thus A is invertible if and only if $\det A \neq 0$.

This alternative argument has some advantages, for in carrying it out we actually proved a subresult interesting in its own right, namely,

LEMMA 6.9.7 $\det A$ *is the product, counting multiplicities, of the characteristic roots of* A.

DEFINITION Given $A \in F_n$, the *secular equation* of A is the polynomial $\det(x - A)$ in $F[x]$.

Usually what we have called the secular equation of A is called the characteristic polynomial of A. However, we have already defined the characteristic polynomial of A to be the product of its elementary divisors. *It is a fact (see Problem 8) that the characteristic polynomial of A equals its secular equation*, but since we did not want to develop this explicitly in the text, we have introduced the term *secular equation.*

Let us compute and example. If

$$A = \begin{pmatrix} 1 & 2 \\ 3 & 0 \end{pmatrix},$$

then

$$x - A = \begin{pmatrix} x & 0 \\ 0 & x \end{pmatrix} - \begin{pmatrix} 1 & 2 \\ 3 & 0 \end{pmatrix} = \begin{pmatrix} x - 1 & -2 \\ -3 & x \end{pmatrix};$$

hence $\det(x - A) = (x - 1)x - (-2)(-3) = x^2 - x - 6$. Thus the secular equation of

$$\begin{pmatrix} 1 & 2 \\ 3 & 0 \end{pmatrix}$$

is $x^2 - x - 6$.

A few remarks about the secular equation: If λ is a root of $\det(x - A)$, then $\det(\lambda - A) = 0$; hence by Theorem 6.9.3, $\lambda - A$ is not invertible. Thus λ is a characteristic root of A. Conversely, if λ is a characteristic root of A, $\lambda - A$ is not invertible, whence $\det(\lambda - A) = 0$ and so λ is a root of $\det(x - A)$. Thus the explicit, computable polynomial, the secular equation of A, *provides us with a polynomial whose roots are exactly the characteristic roots of A.* We want to go one step further and to argue that a given root enters as a root of the secular equation precisely as often as it has multiplicity as a characteristic root of A. For if λ_i is the characteristic root of A with multiplicity m_i, we can bring A to triangular form so that we have the matrix shown in Figure 6.9.1, where each λ_i appears on the diagonal m_i

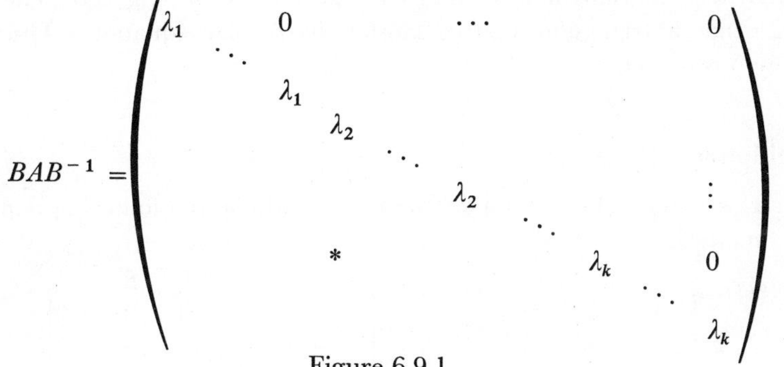

Figure 6.9.1

times. But as indicated by the matrix in Figure 6.9.2, $\det(x - A) = \det(B(x - A)B^{-1}) = (x - \lambda_1)^{m_1}(x - \lambda_2)^{m_2} \cdots (x - \lambda_k)^{m_k}$, and so each

$$B(x - A)B^{-1} = x - BAB^{-1} =$$

$$= \begin{pmatrix} x - \lambda_1 & 0 & & & & & \cdots & & 0 \\ & \ddots & & & & & & & \\ & & x - \lambda_1 & & & & & & \\ & & & x - \lambda_2 & & & & & \\ & & & & \ddots & & & & \vdots \\ & & & & & x - \lambda_2 & & & \\ & & * & & & & \ddots & & \\ & & & & & & & x - \lambda_k & \\ & & & & & & & & \ddots \\ & & & & & & & & & x - \lambda_k \end{pmatrix}$$

Figure 6.9.2

λ_i, whose multiplicity as a characteristic root of A is m_i is a root of the polynomial $\det(x - A)$ of multiplicity exactly m_i. We have proved

THEOREM 6.9.4 *The characteristic roots of A are the roots, with the correct multiplicity, of the secular equation,* $\det(x - A)$, *of A.*

We finish the section with the significant and historic *Cayley-Hamilton theorem.*

THEOREM 6.9.5 *Every $A \in F_n$ satisfies its secular equation.*

Proof. Given any invertible $B \in K_n$ for any extension K of F, $A \in F$ and BAB^{-1} satisfy the same polynomials. Also, since $\det(x - BAB^{-1}) = \det(B(x - A)B^{-1}) = \det(x - A)$, BAB^{-1} and A have the same secular equation. If we can show that some BAB^{-1} satisfies its secular equation, then it will follow that A does. But we can pick $K \supset F$ and $B \in K_n$ so that BAB^{-1} is triangular; in that case we have seen long ago (Theorem 6.4.2) that a triangular matrix satisfies its secular equation. Thus the theorem is proved.

Problems

1. If F is the field of complex numbers, evaluate the following determinants:

$$\text{(a)} \begin{vmatrix} 1 & i \\ 2 - i & 3 \end{vmatrix}. \quad \text{(b)} \begin{vmatrix} 1 & 2 & 3 \\ 4 & 5 & 6 \\ 7 & 8 & 9 \end{vmatrix}. \quad \text{(c)} \begin{vmatrix} 5 & 6 & 8 & -1 \\ 4 & 3 & 0 & 0 \\ 10 & 12 & 16 & -2 \\ 1 & 2 & 3 & 4 \end{vmatrix}.$$

2. For what characteristics of F are the following determinants 0:

$$\text{(a)} \begin{vmatrix} 1 & 2 & 3 & 0 \\ 3 & 2 & 1 & 0 \\ 1 & 1 & 1 & 1 \\ 2 & 4 & 5 & 6 \end{vmatrix}? \quad \text{(b)} \begin{vmatrix} 3 & 4 & 5 \\ 4 & 5 & 3 \\ 5 & 3 & 4 \end{vmatrix}?$$

3. If A is a matrix with integer entries such that A^{-1} is also a matrix with integer entries, what can the values of $\det A$ possibly be?

4. Prove that if you add the multiple of one row to another you do not change the value of the determinant.

*5. Given the matrix $A = (\alpha_{ij})$ let A_{ij} be the matrix obtained from A by removing the ith row and jth column. Let $M_{ij} = (-1)^{i+j} \det A_{ij}$. M_{ij} is called the *cofactor* of α_{ij}. Prove that $\det A = \alpha_{i1}M_{i1} + \cdots + \alpha_{in}M_{in}$.

6. (a) If A and B are square submatrices, prove that

$$\det \begin{pmatrix} A & C \\ 0 & B \end{pmatrix} = (\det A)(\det B).$$

(b) Generalize part (a) to

$$\det \begin{pmatrix} A_1 & & & * \\ & A_2 & & \\ & & \ddots & \\ 0 & & & A_n \end{pmatrix},$$

where each A_i is a square submatrix.

7. If $C(f)$ is the companion matrix of the polynomial $f(x)$, prove that the secular equation of $C(f)$ is $f(x)$.

8. Using Problems 6 and 7, prove that the secular equation of A is its characteristic polynomial. (See Section 6.7; this proves the remark made earlier that the roots of $p_T(x)$ occur with multiplicities equal to their multiplicities as characteristic roots of T.)

9. Using Problem 8, give an alternative proof of the Cayley-Hamilton theorem.

10. If F is the field of rational numbers, compute the secular equation, characteristic roots, and their multiplicities, of

(a) $\begin{pmatrix} 0 & 1 & 0 & 0 \\ 0 & 0 & 0 & 1 \\ 1 & 0 & 0 & 0 \\ 0 & 0 & 1 & 0 \end{pmatrix}$. (b) $\begin{pmatrix} 1 & 2 & 3 \\ 2 & 2 & 4 \\ 3 & 4 & 7 \end{pmatrix}$. (c) $\begin{pmatrix} 4 & 1 & 1 & 1 \\ 1 & 4 & 1 & 1 \\ 1 & 1 & 4 & 1 \\ 1 & 1 & 1 & 4 \end{pmatrix}$.

11. For each matrix in Problem 10 verify by direct matrix computation that it satisfies its secular equation.

*12. If the rank of A is r, prove that there is a square $r \times r$ submatrix of A of determinant different from 0, and if $r < n$, that there is no $(r+1) \times (r+1)$ submatrix of A with this property.

*13. Let f be a function on n variables from $F^{(n)}$ to F such that
(a) $f(v_1, \ldots, v_n) = 0$ for $v_i = v_j \in F^{(n)}$ for $i \neq j$.
(b) $f(v_1, \ldots, \alpha v_i, \ldots, v_n) = \alpha f(v_1, \ldots, v_n)$ for each i, and $\alpha \in F$.
(c) $f(v_1, \ldots, v_i + u_i, v_{i+1}, \ldots, v_n) = f(v_1, \ldots, v_{i-1}, v_i, v_{i+1}, \ldots, v_n) + f(v_1, \ldots, v_{i-1}, u_i, v_{i+1}, \ldots, v_n)$.
(d) $f(e_1, \ldots, e_n) = 1$, where $e_1 = (1, 0, \ldots, 0)$, $e_2 = (0, 1, 0, \ldots, 0)$, $\ldots, e_n = (0, 0, \ldots, 0, 1)$.
Prove that $f(v_1, \ldots, v_n) = \det A$ for any $A \in F_n$, where v_1 is the first row of A, v_2 the second, etc.

14. Use Problem 13 to prove that $\det A' = \det A$.

15. (a) Prove that AB and BA have the same secular (characteristic) equation.
    (b) Give an example where AB and BA do *not* have the same minimal polynomial.
16. If A is triangular prove by a direct computation that A satisfies its secular equation.
17. Use Cramer's rule to compute the solutions, in the real field, of the systems

    (a) $x + y + z = 1,$
    $2x + 3y + 4z = 1,$
    $x - y - z = 0.$

    (b) $x + y + z + w = 1,$
    $x + 2y + 3z + 4w = 0,$
    $x + y + 4z + 5w = 1,$
    $x + y + 5z + 6w = 0.$

18. (a) Let $GL(n, F)$ be the set of all elements in F_n whose determinant is different from 0. Prove $GL(n, F)$ is a group under matrix multiplication.
    (b) Let $D(n, F) = \{A \in GL(n, F) \mid \det A = 1\}$. Prove that $D(n, F)$ is a normal subgroup of $GL(n, F)$.
    (c) Prove that $GL(n, F)/D(n, F)$ is isomorphic to the group of non-zero elements of F under multiplication.
19. If K be an extension field of F, let $E(n, K, F) = \{A \in GL(n, K) \mid \det A \in F\}$.
    (a) Prove that $E(n, K, F)$ is a normal subgroup of $GL(n, K)$.
    *(b) Determine $GL(n, K)/E(n, K, F)$.

*20. If F is the field of rational numbers, prove that when N is a normal subgroup of $D(2, F)$ then either $N = D(2, F)$ or N consists only of scalar matrices.

6.10 Hermitian, Unitary, and Normal Transformations

In our previous considerations about linear transformations, the specific nature of the field F has played a relatively insignificant role. When it did make itself felt it was usually in regard to the presence or absence of characteristic roots. Now, for the first time, we shall restrict the field F—generally it will be the field of complex numbers but at times it may be the field of real numbers—and we shall make heavy use of the properties of real and complex numbers. *Unless explicitly stated otherwise, in all of this section F will denote the field of complex numbers.*

We shall also be making extensive and constant use of the notions and results of Section 4.4 about inner product spaces. The reader would be well advised to review and to digest thoroughly that material before proceeding.

One further remark about the complex numbers: Until now we have managed to avoid using results that were not proved in the book. Now, however, we are forced to deviate from this policy and to call on a basic fact about the field of complex numbers, often known as "*the fundamental theorem of algebra*," without establishing it ourselves. It displeases us to pull such a basic result out of the air, to state it as a fact, and then to make use of it. Unfortunately, it is essential for what follows and to digress to prove it here would take us too far afield. We hope that the majority of readers will have seen it proved in a course on complex variable theory.

FACT 1 *A polynomial with coefficients which are complex numbers has all its roots in the complex field.*

Equivalently, Fact 1 can be stated in the form that the only nonconstant irreducible polynomials over the field of complex numbers are those of degree 1.

FACT 2 *The only irreducible, nonconstant, polynomials over the field of real numbers are either of degree* 1 *or of degree* 2.

The formula for the roots of a quadratic equation allows us to prove easily the equivalence of Facts 1 and 2.

The immediate implication, for us, of Fact 1 will be that *every linear transformation which we shall consider will have all its characteristic roots in the field of complex numbers.*

In what follows, V will be a finite-dimensional inner-product space over F, the field of complex numbers; the inner product of two elements of V will be written, as it was before, as (v, w).

LEMMA 6.10.1 *If $T \in A(V)$ is such that $(vT, v) = 0$ for all $v \in V$, then $T = 0$.*

Proof. Since $(vT, v) = 0$ for $v \in V$, given $u, w \in V$, $((u + w)T, u + w) = 0$. Expanding this out and making use of $(uT, u) = (wT, w) = 0$, we obtain

$$(uT, w) + (wT, u) = 0 \text{ for all } u, w \in V. \tag{1}$$

Since equation (1) holds for arbitrary w in V, it still must hold if we replace in it w by iw where $i^2 = -1$; but $(uT, iw) = -i(uT, w)$ whereas $((iw)T, u) = i(wT, u)$. Substituting these values in (1) and canceling out i leads us to

$$-(uT, w) + (wT, u) = 0. \tag{2}$$

Adding (1) and (2) we get $(wT, u) = 0$ for all $u, w \in V$, whence, in particular, $(wT, wT) = 0$. By the defining properties of an inner-product space, this forces $wT = 0$ for all $w \in V$, hence $T = 0$. (*Note:* If V is an inner-product space over the real field, the lemma may be false. For example, let $V = \{(\alpha, \beta) \mid \alpha, \beta \text{ real}\}$, where the inner-product is the dot product. Let T be the linear transformation sending (α, β) into $(-\beta, \alpha)$. A simple check shows that $(vT, v) = 0$ for *all* $v \in V$, yet $T \neq 0$.)

DEFINITION The linear transformation $T \in A(V)$ is said to be *unitary* if $(uT, vT) = (u, v)$ for all $u, v \in V$.

A unitary transformation is one which preserves all the structure of V, its addition, its multiplication by scalars and *its inner product.* Note that a unitary transformation preserves length for

$$\|v\| = \sqrt{(v, v)} = \sqrt{(vT, vT)} = \|vT\|.$$

Is the converse true? The answer is provided us in

LEMMA 6.10.2 *If* $(vT, vT) = (v, v)$ *for all* $v \in V$ *then* T *is unitary.*

Proof. The proof is in the spirit of that of Lemma 6.10.1. Let $u, v \in V$: by assumption $((u + v)T, (u + v)T) = (u + v, u + v)$. Expanding this out and simplifying, we obtain

$$(uT, vT) + (vT, uT) = (u, v) + (v, u), \tag{1}$$

for $u, v \in V$. In (1) replace v by iv; computing the necessary parts, this yields

$$-(uT, vT) + (vT, uT) = -(u, v) + (v, u). \tag{2}$$

Adding (1) and (2) results in $(uT, vT) = (u, v)$ for all $u, v \in V$, hence T is unitary.

We characterize the property of being unitary in terms of action on a basis of V.

THEOREM 6.10.1 *The linear transformation* T *on* V *is unitary if and only if it takes an orthonormal basis of* V *into an orthonormal basis of* V.

Proof. Suppose that $\{v_1, \ldots, v_n\}$ is an orthonormal basis of V; thus $(v_i, v_j) = 0$ for $i \neq j$ while $(v_i, v_i) = 1$. We wish to show that if T is unitary, then $\{v_1T, \ldots, v_nT\}$ is also an orthonormal basis of V. But $(v_iT, v_jT) = (v_i, v_j) = 0$ for $i \neq j$ and $(v_iT, v_iT) = (v_i, v_i) = 1$, thus indeed $\{v_1T, \ldots, v_nT\}$ is an orthonormal basis of V.

On the other hand, if $T \in A(V)$ is such that both $\{v_1, \ldots, v_n\}$ and $\{v_1T, \ldots, v_nT\}$ are orthonormal bases of V, if $u, w \in V$ then

$$u = \sum_{i=1}^{n} \alpha_i v_i, \qquad w = \sum_{i=1}^{n} \beta_i v_i,$$

whence by the orthonormality of the v_i's,

$$(u, w) = \sum_{i=1}^{n} \alpha_i \bar{\beta}_i.$$

However,

$$uT = \sum_{i=1}^{n} \alpha_i v_i T \quad \text{and} \quad wT = \sum_{i=1}^{n} \beta_i v_i T$$

whence by the orthonormality of the $v_i T$'s,

$$(uT, wT) = \sum_{i=1}^{n} \alpha_i \bar{\beta}_i = (u, w),$$

proving that T is unitary.

Theorem 6.10.1 states that a change of basis from one orthonormal basis to another is accomplished by a unitary linear transformation.

LEMMA 6.10.3 *If $T \in A(V)$ then given any $v \in V$ there exists an element $w \in V$, depending on v and T, such that $(uT, v) = (u, w)$ for* all $u \in V$. *This element w is uniquely determined by v and T.*

Proof. To prove the lemma, it is sufficient to exhibit a $w \in V$ which works for all the elements of a basis of V. Let $\{u_1, \ldots, u_n\}$ be an orthonormal basis of V; we define

$$w = \sum_{i=1}^{n} \overline{(u_i T, v)} u_i.$$

An easy computation shows that $(u_i, w) = (u_i T, v)$ hence the element w has the desired property. That w is unique can be seen as follows: Suppose that $(uT, v) = (u, w_1) = (u, w_2)$; then $(u, w_1 - w_2) = 0$ for all $u \in V$ which forces, on putting $u = w_1 - w_2$, $w_1 = w_2$.

Lemma 6.10.3 allows us to make the

DEFINITION If $T \in A(V)$ then the *Hermitian adjoint* of T, written as T^*, is defined by $(uT, v) = (u, vT^*)$ for all $u, v \in V$.

Given $v \in V$ we have obtained above an explicit expression for vT^* (as w) and we could use this expression to prove the various desired properties of T^*. However, we prefer to do it in a "basis-free" way.

LEMMA 6.10.4 *If $T \in A(V)$ then $T^* \in A(V)$. Moreover,*

1. $(T^*)^* = T$;
2. $(S + T)^* = S^* + T^*$;
3. $(\lambda S)^* = \bar{\lambda} S^*$;
4. $(ST)^* = T^* S^*$;

for all $S, T \in A(V)$ and all $\lambda \in F$.

Proof. We must first prove that T^* is a linear transformation on V. If u, v, w are in V, then $(u, (v + w)T^*) = (uT, v + w) = (uT, v) + (uT, w) = (u, vT^*) + (u, wT^*) = (u, vT^* + wT^*)$, in consequence of which $(v + w)T^* = vT^* + wT^*$. Similarly, for $\lambda \in F$, $(u, (\lambda v)T^*) = (uT, \lambda v) = \bar{\lambda}(uT, v) = \bar{\lambda}(u, vT^*) = (u, \lambda(vT^*))$, whence $(\lambda v)T^* = \lambda(vT^*)$. We have thus proved that T^* is a linear transformation on V.

To see that $(T^*)^* = T$ notice that $(u, v(T^*)^*) = (uT^*, v) = \overline{(v, uT^*)} = \overline{(vT, u)} = (u, vT)$ for all $u, v \in V$ whence $v(T^*)^* = vT$ which implies that $(T^*)^* = T$. We leave the proofs of $(S + T)^* = S^* + T^*$ and of $(\lambda T)^* = \bar{\lambda}T^*$ to the reader. Finally, $(u, v(ST)^*) = (uST, v) = (uS, vT^*) = (u, vT^*S^*)$ for all $u, v \in V$; this forces $v(ST)^* = vT^*S^*$ for every $v \in V$ which results in $(ST)^* = T^*S^*$.

As a consequence of the lemma the Hermitian adjoint defines an adjoint, in the sense of Section 6.8, on $A(V)$.

The Hermitian adjoint allows us to give an alternative description for unitary transformations in terms of the relation of T and T^*.

LEMMA 6.10.5 *$T \in A(V)$ is unitary if and only if $TT^* = 1$.*

Proof. If T is unitary, then for all $u, v \in V$, $(u, vTT^*) = (uT, vT) = (u, v)$ hence $TT^* = 1$. On the other hand, if $TT^* = 1$, then $(u, v) = (u, vTT^*) = (uT, vT)$, which implies that T is unitary.

Note that a unitary transformation is nonsingular and its inverse is just its Hermitian adjoint. Note, too, that from $TT^* = 1$ we must have that $T^*T = 1$. We shall soon give an explicit matrix criterion that a linear transformation be unitary.

THEOREM 6.10.2 *If $\{v_1, \ldots, v_n\}$ is an orthonormal basis of V and if the matrix of $T \in A(V)$ in this basis is (α_{ij}) then the matrix of T^* in this basis is (β_{ij}), where $\beta_{ij} = \bar{\alpha}_{ji}$.*

Proof. Since the matrices of T and T^* in this basis are, respectively, (α_{ij}) and (β_{ij}), then

$$v_iT = \sum_{i=1}^{n} \alpha_{ij}v_j \quad \text{and} \quad v_iT^* = \sum_{i=1}^{n} \beta_{ij}v_j.$$

Now

$$\beta_{ij} = (v_iT^*, v_j) = (v_i, v_jT) = \left(v_i, \sum_{i=1}^{n} \alpha_{jk}v_k\right) = \bar{\alpha}_{ji}$$

by the orthonormality of the v_i's. This proves the theorem.

This theorem is very interesting to us in light of what we did earlier in Section 6.8. For the abstract Hermitian adjoint defined on the inner-product space V, when translated into matrices in an orthonormal basis of V, becomes nothing more than the explicit, concrete Hermitian adjoint we defined there for matrices.

Using the matrix representation in an orthonormal basis, we claim that $T \in A(V)$ is unitary if and only if, whenever (α_{ij}) is the matrix of T in this orthonormal basis, then

$$\sum_{i=1}^{n} \alpha_{ij}\bar{\alpha}_{ik} = 0 \qquad \text{for } j \neq k$$

while

$$\sum_{i=1}^{n} |\alpha_{ij}|^2 = 1.$$

In terms of dot products on complex vector spaces, it says that the rows of the matrix of T form an orthonormal set of vectors in $F^{(n)}$ under the dot product.

DEFINITION $T \in A(V)$ is called *self-adjoint* or *Hermitian* if $T^* = T$.

If $T^* = -T$ we call *skew-Hermitian*. Given any $S \in A(V)$,

$$S = \frac{S + S^*}{2} + i\left(\frac{S - S^*}{2i}\right),$$

and since $(S + S^*)/2$ and $(S - S^*)/2i$ are Hermitian, $S = A + iB$ where both A and B are Hermitian.

In Section 6.8, using matrix calculations, we proved that any complex characteristic root of a Hermitian matrix is real; in light of Fact 1, this can be changed to read: Every characteristic root of a Hermitian matrix is real. We now re-prove this from the more uniform point of view of an inner-product space.

THEOREM 6.10.3 *If $T \in A(V)$ is Hermitian, then all its characteristic roots are real.*

Proof. Let λ be a characteristic root of T; thus there is a $v \neq 0$ in V such that $vT = \lambda v$. We compute: $\lambda(v, v) = (\lambda v, v) = (vT, v) = (v, vT^*) = (v, vT) = (v, \lambda v) = \bar{\lambda}(v, v)$; since $(v, v) \neq 0$ we are left with $\lambda = \bar{\lambda}$ hence λ is real.

We want to describe canonical forms for unitary, Hermitian, and even more general types of linear transformations which will be even simpler than the Jordan form. This accounts for the next few lemmas which, although of independent interest, are for the most part somewhat technical in nature.

LEMMA 6.10.6 *If $S \in A(V)$ and if $vSS^* = 0$, then $vS = 0$.*

Proof. Consider (vSS^*, v); since $vSS^* = 0$, $0 = (vSS^*, v) = (vS, v(S^*)^*) = (vS, vS)$ by Lemma 6.10.4. In an inner-product space, this implies that $vS = 0$.

COROLLARY *If T is Hermitian and $vT^k = 0$ for $k \geq 1$ then $vT = 0$.*

Proof. We show that if $vT^{2^m} = 0$ then $vT = 0$; for if $S = T^{2^{m-1}}$, then $S^* = S$ and $SS^* = T^{2^m}$, whence $(vSS^*, v) = 0$ implies that $0 = vS = vT^{2^{m-1}}$. Continuing down in this way, we obtain $vT = 0$. If $vT^k = 0$, then $vT^{2^m} = 0$ for $2^m > k$, hence $vT = 0$.

We introduce a class of linear transformations which contains, as special cases, the unitary, Hermitian and skew-Hermitian transformations.

DEFINITION $T \in A(V)$ is said to be *normal* if $TT^* = T^*T$.

Instead of proving the theorems to follow for unitary and Hermitian transformations separately, we shall, instead, prove them for normal linear transformations and derive, as corollaries, the desired results for the unitary and Hermitian ones.

LEMMA 6.10.7 *If N is a normal linear transformation and if $vN = 0$ for $v \in V$, then $vN^* = 0$.*

Proof. Consider (vN^*, vN^*); by definition, $(vN^*, vN^*) = (vN^*N, v) = (vNN^*, v)$, since $NN^* = N^*N$. However, $vN = 0$, whence, certainly, $vNN^* = 0$. In this way we obtain that $(vN^*, vN^*) = 0$, forcing $vN^* = 0$.

COROLLARY 1 *If λ is a characteristic root of the normal transformation N and if $vN = \lambda v$ then $vN^* = \bar{\lambda} v$.*

Proof. Since N is normal, $NN^* = N^*N$, therefore, $(N - \lambda)(N - \lambda)^* = (N - \lambda)(N^* - \bar{\lambda}) = NN^* - \lambda N^* - \bar{\lambda} N + \lambda\bar{\lambda} = N^*N - \lambda N^* - \bar{\lambda} N + \lambda\bar{\lambda} = (N^* - \bar{\lambda})(N - \lambda) = (N - \lambda)^*(N - \lambda)$, that is to say, $N - \lambda$ is normal. Since $v(N - \lambda) = 0$ by the normality of $N - \lambda$, from the lemma, $v(N - \lambda)^* = 0$, hence $vN^* = \bar{\lambda} v$.

The corollary states the interesting fact that if λ is a characteristic root of the normal transformation N not only is $\bar{\lambda}$ a characteristic root of N^* but any characteristic vector of N belonging to λ is a characteristic vector of N^* belonging to $\bar{\lambda}$ and vice versa.

COROLLARY 2 *If T is unitary and if λ is a characteristic root of T, then $|\lambda| = 1$.*

Proof. Since T is unitary it is normal. Let λ be a characteristic root of T and suppose that $vT = \lambda v$ with $v \neq 0$ in V. By Corollary 1, $vT^* = \bar{\lambda}v$, thus $v = vTT^* = \lambda vT^* = \lambda\bar{\lambda}v$ since $TT^* = 1$. Thus we get $\lambda\bar{\lambda} = 1$, which, of course, says that $|\lambda| = 1$.

We pause to see where we are going. Our immediate goal is to prove that a normal transformation N can be brought to diagonal form by a unitary one. If $\lambda_1, \ldots, \lambda_k$ are the distinct characteristic roots of V, using Theorem 6.6.1 we can decompose V as $V = V_1 \oplus \cdots \oplus V_k$, where for $v_i \in V_i$, $v_i(N - \lambda_i)^{n_i} = 0$. Accordingly, we want to study two things, namely, the relation of vectors lying in different V_i's and the very nature of each V_i. When these have been determined, we will be able to assemble them to prove the desired theorem.

LEMMA 6.10.8 *If N is normal and if $vN^k = 0$, then $vN = 0$.*

Proof. Let $S = NN^*$; S is Hermitian, and by the normality of N, $vS^k = v(NN^*)^k = vN^k(N^*)^k = 0$. By the corollary to Lemma 6.10.6, we deduce that $vS = 0$, that is to say, $vNN^* = 0$. Invoking Lemma 6.10.6 itself yields $vN = 0$.

COROLLARY *If N is normal and if for $\lambda \in F$, $v(N - \lambda)^k = 0$, then $vN = \lambda v$.*

Proof. From the normality of N it follows that $N - \lambda$ is normal, whence by applying the lemma just proved to $N - \lambda$ we obtain the corollary.

In line with the discussion just preceding the last lemma, this corollary shows that *every vector in V_i is a characteristic vector of N belonging to the characteristic root λ_i.* We have determined the nature of V_i; now we proceed to investigate the interrelation between two distinct V_i's.

LEMMA 6.10.9 *Let N be a normal transformation and suppose that λ and μ are two distinct characteristic roots of N. If v, w are in V and are such that $vN = \lambda v$, $wN = \mu w$, then $(v, w) = 0$.*

Proof. We compute (vN, w) in two different ways. As a consequence of $vN = \lambda v$, $(vN, w) = (\lambda v, w) = \lambda(v, w)$. From $wN = \mu w$, using Lemma 6.10.7 we obtain that $wN^* = \bar{\mu}w$, whence $(vN, w) = (v, wN^*) = (v, \bar{\mu}w) = \mu(v, w)$. Comparing the two computations gives us $\lambda(v, w) = \mu(v, w)$ and since $\lambda \neq \mu$, this results in $(v, w) = 0$.

All the background work has been done to enable us to prove the basic and lovely

THEOREM 6.10.4 *If N is a normal linear transformation on V, then there exists an orthonormal basis, consisting of characteristic vectors of N, in which the matrix of N is diagonal. Equivalently, if N is a normal matrix there exists a unitary matrix U such that UNU^{-1} $(= UNU^*)$ is diagonal.*

Proof. We fill in the informal sketch we have made of the proof prior to proving Lemma 6.10.8.

Let N be normal and let $\lambda_1, \ldots, \lambda_k$ be the distinct characteristic roots of N. By the corollary to Theorem 6.6.1 we can decompose $V = V_1 \oplus \cdots \oplus V_k$ where every $v_i \in V_i$ is annihilated by $(N - \lambda_i)^{n_i}$. By the corollary to Lemma 6.10.8, V_i consists only of characteristic vectors of N belonging to the characteristic root λ_i. The inner product of V induces an inner product on V_i; by Theorem 4.4.2 we can find a basis of V_i orthonormal relative to this inner product.

By Lemma 6.10.9 elements lying in distinct V_i's are orthogonal. Thus putting together the orthonormal bases of the V_i's provides us with an orthonormal basis of V. This basis consists of characteristic vectors of N, hence in this basis the matrix of N is diagonal.

We do not prove the matrix equivalent, leaving it as a problem; we only point out that two facts are needed:

1. A change of basis from one orthonormal basis to another is accomplished by a unitary transformation (Theorem 6.10.1).
2. In a change of basis the matrix of a linear transformation is changed by conjugating by the matrix of the change of basis (Theorem 6.3.2).

Both corollaries to follow are very special cases of Theorem 6.10.4, but since each is so important in its own right we list them as corollaries in order to emphasize them.

COROLLARY 1 *If T is a unitary transformation, then there is an orthonormal basis in which the matrix of T is diagonal; equivalently, if T is a unitary matrix, then there is a unitary matrix U such that UTU^{-1} $(= UTU^*)$ is diagonal.*

COROLLARY 2 *If T is a Hermitian linear transformation, then there exists an orthonormal basis in which the matrix of T is diagonal; equivalently, if T is a Hermitian matrix, then there exists a unitary matrix U such that UTU^{-1} $(= UTU^*)$ is diagonal.*

The theorem proved is the basic result for normal transformations, for it sharply characterizes them as precisely those transformations which can be brought to diagonal form by unitary ones. It also shows that the distinction between normal, Hermitian, and unitary transformations is merely a distinction caused by the nature of their characteristic roots. This is made precise in

LEMMA 6.10.10 *The normal transformation N is*

1. *Hermitian if and only if its characteristic roots are real.*
2. *Unitary if and only if its characteristic roots are all of absolute value* 1.

Proof. We argue using matrices. If N is Hermitian, then it is normal and all its characteristic roots are real. If N is normal and has only real characteristic roots, then for some unitary matrix U, $UNU^{-1} = UNU^* = D$, where D is a diagonal matrix with real entries on the diagonal. Thus $D^* = D$; since $D^* = (UNU^*)^* = UN^*U^*$, the relation $D^* = D$ implies $UN^*U^* = UNU^*$, and since U is invertible we obtain $N^* = N$. Thus N is Hermitian.

We leave the proof of the part about unitary transformations to the reader.

If A is any linear transformation on V, then tr (AA^*) can be computed by using the matrix representation of A in any basis of V. We pick an orthonormal basis of V; in this basis, if the matrix of A is (α_{ij}) then that of A^* is (β_{ij}) where $\beta_{ij} = \bar{\alpha}_{ji}$. A simple computation then shows that tr $(AA^*) = \sum_{i,j} |\alpha_{ij}|^2$ and this is 0 if and only if each $\alpha_{ij} = 0$, that is, if and only if $A = 0$. In a word, tr $(AA^*) = 0$ *if and only if* $A = 0$. This is a useful criterion for showing that a given linear transformation is 0. This is illustrated in

LEMMA 6.10.11 *If N is normal and $AN = NA$, then $AN^* = N^*A$.*

Proof. We want to show that $X = AN^* - N^*A$ is 0; what we shall do is prove that tr $XX^* = 0$, and deduce from this that $X = 0$.

Since N commutes with A and with N^*, it must commute with $AN^* - N^*A$, thus $XX^* = (AN^* - N^*A)(NA^* - A^*N) = (AN^* - N^*A)NA^* - (AN^* - N^*A)A^*N = N\{(AN^* - N^*A)A^*\} - \{(AN^* - N^*A)A^*\}N$. Being of the form $NB - BN$, the trace of XX^* is 0. Thus $X = 0$, and $AN^* = N^*A$.

We have just seen that N^* commutes with all the linear transformations that commute with N, when N is normal; this is enough to force N^* to be a polynomial expression in N. However, this can be shown directly as a consequence of Theorem 6.10.4 (see Problem 14).

The linear transformation T is Hermitian if and only if (vT, v) is real for every $v \in V$. (See Problem 19.) Of special interest are those Hermitian linear transformations for which $(vT, v) \geq 0$ for all $v \in V$. We call these *nonnegative* linear transformations and denote the fact that a linear transformation is nonnegative by writing $T \geq 0$. If $T \geq 0$ and in addition $(vT, v) > 0$ for $v \neq 0$ then we call T *positive* (or *positive definite*) and write $T > 0$. We wish to distinguish these linear transformations by their characteristic roots.

LEMMA 6.10.12 *The Hermitian linear transformation T is nonnegative (positive) if and only if all of its characteristic roots are nonnegative (positive).*

Proof. Suppose that $T \geq 0$; if λ is a characteristic root of T, then $vT = \lambda v$ for some $v \neq 0$. Thus $0 \leq (vT, v) = (\lambda v, v) = \lambda(v, v)$; since $(v, v) > 0$ we deduce that $\lambda \geq 0$.

Conversely, if T is Hermitian with nonnegative characteristic roots, then we can find an orthonormal basis $\{v_1, \ldots, v_n\}$ consisting of characteristic vectors of T. For each v_i, $v_i T = \lambda_i v_i$, where $\lambda_i \geq 0$. Given $v \in V$, $v = \sum \alpha_i v_i$ hence $vT = \sum \alpha_i v_i T = \sum \lambda_i \alpha_i v_i$. But $(vT, v) = (\sum \lambda_i \alpha_i v_i, \sum \alpha_i v_i) = \sum \lambda_i \alpha_i \bar{\alpha}_i$ by the orthonormality of the v_i's. Since $\lambda_i \geq 0$ and $\alpha_i \bar{\alpha}_i \geq 0$, we get that $(vT, v) \geq 0$ hence $T \geq 0$.

The corresponding "positive" results are left as an exercise.

LEMMA 6.10.13 *$T \geq 0$ if and only if $T = AA^*$ for some A.*

Proof. We first show that $AA^* \geq 0$. Given $v \in V$, $(vAA^*, v) = (vA, vA) \geq 0$, hence $AA^* \geq 0$.

On the other hand, if $T \geq 0$ we can find a unitary matrix U such that

$$UTU^* = \begin{pmatrix} \lambda_1 & & \\ & \ddots & \\ & & \lambda_n \end{pmatrix}$$

where each λ_i is a characteristic root of T, hence each $\lambda_i \geq 0$. Let

$$S = \begin{pmatrix} \sqrt{\lambda_1} & & \\ & \ddots & \\ & & \sqrt{\lambda_n} \end{pmatrix};$$

since each $\lambda_i \geq 0$, each $\sqrt{\lambda_i}$ is real, whence S is Hermitian. Therefore, U^*SU is Hermitian; but

$$(U^*SU)^2 = U^*S^2U = U^* \begin{pmatrix} \lambda_1 & & \\ & \ddots & \\ & & \lambda_n \end{pmatrix} U = T.$$

We have represented T in the form AA^*, where $A = U^*SU$.

Notice that we have actually proved a little more; namely, if in constructing S above, we had chosen the nonnegative $\sqrt{\lambda_i}$ for each λ_i, then S, and U^*SU, would have been nonnegative. Thus $T \geq 0$ is the square of a nonnegative linear transformation; that is, every $T \geq 0$ has a nonnegative square root. This nonnegative square root can be shown to be unique (see Problem 24).

We close this section with a discussion of unitary and Hermitian matrices *over the real field*. In this case, the unitary matrices are called *orthogonal*, and satisfy $QQ' = 1$. The Hermitian ones are just symmetric, in this case.

We claim that a *real symmetric matrix can be brought to diagonal form by an orthogonal matrix.* Let A be a real symmetric matrix. We can consider A as acting on a real inner-product space V. Considered as a complex matrix, A is Hermitian and thus all its characteristic roots are real. If these are $\lambda_i, \ldots, \lambda_k$ then V can be decomposed as $V = V_1 \oplus \cdots \oplus V_k$ where $v_i(A - \lambda_i)^{n_i} = 0$ for $v_i \in V_i$. As in the proof of Lemma 6.10.8 this forces $v_iA = \lambda_i v_i$. Using exactly the same proof as was used in Lemma 6.10.9, we show that for $v_i \in V_i$, $v_j \in V_j$ with $i \neq j$, $(v_i, v_j) = 0$. Thus we can find an orthonormal basis of V consisting of characteristic vectors of A. The change of basis, from the orthonormal basis $\{(1, 0, \ldots, 0), (0, 1, 0, \ldots, 0), \ldots, (0, \ldots, 0, 1)\}$ to this new basis is accomplished by a real, unitary matrix, that is, by an orthogonal one. Thus A can be brought to diagonal form by an orthogonal matrix, proving our contention.

To determine canonical forms for the real orthogonal matrices over the real field is a little more complicated, both in its answer and its execution. We proceed to this now; but first we make a general remark about all unitary transformations.

If W is a subspace of V invariant under the unitary transformation T, is it true that W', the orthogonal complement of W, is also invariant under T? Let $w \in W$ and $x \in W'$; thus $(wT, xT) = (w, x) = 0$; since W is invariant under T and T is regular, $WT = W$, whence xT, for $x \in W'$, is orthogonal to all of W. Thus indeed $(W')T \subset W'$. Recall that $V = W \oplus W'$.

Let Q be a real orthogonal matrix; thus $T = Q + Q^{-1} = Q + Q'$ is symmetric, hence has real characteristic roots. If these are $\lambda_1, \ldots, \lambda_k$, then V can be decomposed as $V = V_1 \oplus \cdots \oplus V_k$, where $v_i \in V$ implies $v_iT = \lambda_i v_i$. The V_i's are mutually orthogonal. We claim each V_i is invariant under Q. (Prove!) Thus to discuss the action of Q on V, it is enough to describe it on each V_i.

On V_i, since $\lambda_i v_i = v_iT = v_i(Q + Q^{-1})$, multiplying by Q yields $v_i(Q^2 - \lambda_i Q + 1) = 0$. Two special cases present themselves, namely $\lambda_i = 2$ and $\lambda_i = -2$ (which may, of course, not occur), for then $v_i(Q \pm 1)^2 = 0$ leading to $v_i(Q \pm 1) = 0$. On these spaces Q acts as 1 or as -1.

If $\lambda_i \neq 2, -2$, then Q has no characteristic vectors on V_i, hence for $v \neq 0 \in V_i$, v, vQ are linearly independent. The subspace they generate, W, is invariant under Q, since $vQ^2 = \lambda_i vQ - v$. Now $V_i = W \oplus W'$ with W' invariant under Q. Thus we can get V_i as a direct sum of two-dimensional mutually orthogonal subspaces invariant under Q. To find canonical forms of Q on V_i (hence on V), we must merely settle the question for 2×2 real orthogonal matrices.

Let Q be a real 2×2 orthogonal matrix satisfying $Q^2 - \lambda Q + 1 = 0$; suppose that $Q = \begin{pmatrix} \alpha & \beta \\ \gamma & \delta \end{pmatrix}$. The orthogonality of Q implies

$$\alpha^2 + \beta^2 = 1; \tag{1}$$

$$\gamma^2 + \delta^2 = 1; \tag{2}$$

$$\alpha\gamma + \beta\delta = 0; \tag{3}$$

since $Q^2 - \lambda Q + 1 = 0$, the determinant of Q is 1, hence

$$\alpha\delta - \beta\gamma = 1. \tag{4}$$

We claim that equations (1)–(4) imply that $\alpha = \delta$, $\beta = -\gamma$. Since $\alpha^2 + \beta^2 = 1$, $|\alpha| \leq 1$, whence we can write $\alpha = \cos\theta$ for some real angle θ; in these terms $\beta = \sin\theta$. Therefore, the matrix Q looks like

$$\begin{pmatrix} \cos\theta & \sin\theta \\ -\sin\theta & \cos\theta \end{pmatrix}.$$

All the spaces used in all our decompositions were mutually orthogonal, thus by picking orthogonal bases of each of these we obtain an orthonormal basis of V. In this basis the matrix of Q is as shown in Figure 6.10.1.

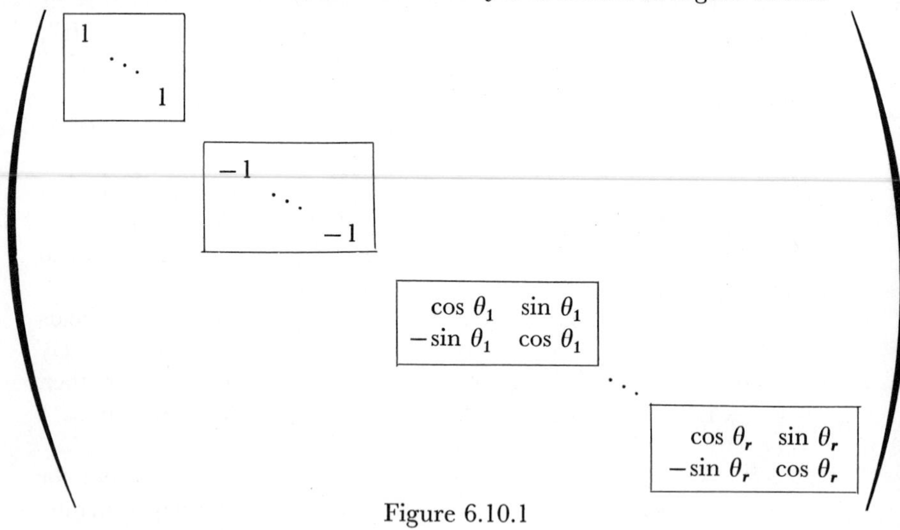

Figure 6.10.1

Since we have gone from one orthonormal basis to another, and since this is accomplished by an orthogonal matrix, given a real orthogonal matrix Q we can find an *orthogonal matrix* T *such that* TQT^{-1} $(= TQT^*)$ *is of the form just described.*

Problems

1. Determine which of the following matrices are unitary, Hermitian, normal.

(a) $\begin{pmatrix} 1 & 1 & 1 \\ 1 & 0 & 1 \\ 0 & 1 & 1 \end{pmatrix}$. (b) $\begin{pmatrix} 0 & i \\ i & 0 \end{pmatrix}$. (c) $\begin{pmatrix} 1 & 0 & 0 & 0 \\ 0 & 0 & 1 & 0 \\ 0 & 1 & 0 & 0 \\ 0 & 0 & 0 & 1 \end{pmatrix}$.

(d) $\begin{pmatrix} 1 & 2 - i \\ 2 - i & i \end{pmatrix}$. (e) $\begin{pmatrix} 3 & 0 & 0 \\ 0 & \dfrac{1}{\sqrt{2}} & -\dfrac{1}{\sqrt{2}} \\ 0 & \dfrac{1}{\sqrt{2}} & \dfrac{1}{\sqrt{2}} \end{pmatrix}$.

2. For those matrices in Problem 1 which are normal, find their characteristic roots and bring them to diagonal form by a unitary matrix.

3. If T is unitary, just using the definition $(vT, uT) = (v, u)$, prove that T is nonsingular.

4. If Q is a real orthogonal matrix, prove that $\det Q = \pm 1$.

5. If Q is a real symmetric matrix satisfying $Q^k = 1$ for $k \geq 1$, prove that $Q^2 = 1$.

6. Complete the proof of Lemma 6.10.4 by showing that $(S + T)^* = S^* + T^*$ and $(\lambda T)^* = \bar{\lambda} T^*$.

7. Prove the properties of $*$ in Lemma 6.10.4 by making use of the explicit form of $w = vT^*$ given in the proof of Lemma 6.10.3.

8. If T is skew-Hermitian, prove that all of its characteristic roots are pure imaginaries.

9. If T is a real, skew-symmetric $n \times n$ matrix, prove that if n is odd, then $\det T = 0$.

10. By a direct matrix calculation, prove that a real, 2×2 symmetric matrix can be brought to diagonal form by an orthogonal one.

11. Complete the proof outlined for the matrix-equivalent part of Theorem 6.10.4.

12. Prove that a normal transformation is unitary if and only if the characteristic roots are all of absolute value 1.

13. If $N_1, \ldots, N_k$ is a finite number of commuting normal transformations, prove that there exists a unitary transformation T such that all of TN_iT^{-1} are diagonal.

14. If N is normal, prove that $N^* = p(N)$ for some polynomial $p(x)$.
15. If N is normal and if $AN = 0$, prove that $AN^* = 0$.
16. Prove that A is normal if and only if A commutes with AA^*.
17. If N is normal prove that $N = \sum \lambda_i E_i$ where $E_i^2 = E_i$, $E_i^* = E_i$, and the λ_i's are the characteristic roots of N. (This is called the *spectral resolution* of N.)
18. If N is a normal transformation on V and if $f(x)$ and $g(x)$ are two relatively prime polynomials with real coefficients, prove that if $vf(N) = 0$ and $wg(N) = 0$, for v, w in V, then $(v, w) = 0$.
19. Prove that a linear transformation T on V is Hermitian if and only if (vT, v) is real for all $v \in V$.
20. Prove that $T > 0$ if and only if T is Hermitian and has all its characteristic roots positive.
21. If $A \geq 0$ and $(vA, v) = 0$, prove that $vA = 0$.
22. (a) If $A \geq 0$ and A^2 commutes with the Hermitian transformation B then A commutes with B.
    (b) Prove part (a) even if B is not Hermitian.
23. If $A \geq 0$ and $B \geq 0$ and $AB = BA$, prove that $AB \geq 0$.
24. Prove that if $A \geq 0$ then A has a *unique* nonnegative square root.
25. Let $A = (\alpha_{ij})$ be a real, symmetric $n \times n$ matrix. Let

$$A_s = \begin{pmatrix} \alpha_{11} & \cdots & \alpha_{1s} \\ \vdots & & \vdots \\ \alpha_{s1} & \cdots & \alpha_{ss} \end{pmatrix}.$$

    (a) If $A > 0$, prove that $A_s > 0$ for $s = 1, 2, \ldots, n$.
    (b) If $A > 0$ prove that $\det A_s > 0$ for $s = 1, 2, \ldots, n$.
    (c) If $\det A_s > 0$ for $s = 1, 2, \ldots, n$, prove that $A > 0$.
    (d) If $A \geq 0$ prove that $A_s \geq 0$ for $s = 1, 2, \ldots, n$.
    (e) If $A \geq 0$ prove that $\det A_s \geq 0$ for $s = 1, 2, \ldots, n$.
    (f) Give an example of an A such that $\det A_s \geq 0$ for all $s = 1, 2, \ldots, n$ yet A is *not* nonnegative.
26. Prove that any complex matrix can be brought to triangular form by a unitary matrix.

6.11 Real Quadratic Forms

We close the chapter with a brief discussion of quadratic forms over the field of real numbers.

Let V be a real, inner-product space and suppose that A is a (real) symmetric linear transformation on V. The real-valued function $Q(v)$ defined on V by $Q(v) = (vA, v)$ is called the *quadratic form associated with A*.

If we consider, as we may without loss of generality, that A is a real, $n \times n$ symmetric matrix (α_{ij}) acting on $F^{(n)}$ and that the inner product for $(\delta_1, \ldots, \delta_n)$ and $(\gamma_1, \ldots, \gamma_n)$ in $F^{(n)}$ is the real number $\delta_1\gamma_1 + \delta_2\gamma_2 + \cdots + \delta_n\gamma_n$, for an arbitrary vector $v = (x_1, \ldots, x_n)$ in $F^{(n)}$ a simple calculation shows that

$$Q(v) = (vA, v) = \alpha_{11}x_1^2 + \cdots + \alpha_{nn}x_n^2 + 2\sum_{i<j} \alpha_{ij}x_ix_j.$$

On the other hand, given any quadratic function in n-variables

$$\gamma_{11}x_1^2 + \cdots + \gamma_{nn}x_n^2 + 2\sum_{i<j} \gamma_{ij}x_ix_j,$$

with real coefficients γ_{ij}, we clearly can realize it as the quadratic form associated with the real symmetric matrix $C = (\gamma_{ij})$.

In real n-dimensional Euclidean space such quadratic functions serve to define the quadratic surfaces. For instance, in the real plane, the form $\alpha x^2 + \beta xy + \gamma y^2$ gives rise to a conic section (possibly with its major axis tilted). It is not too unnatural to expect that the geometric properties of this conic section should be intimately related with the symmetric matrix

$$\begin{pmatrix} \alpha & \beta/2 \\ \beta/2 & \gamma \end{pmatrix},$$

with which its quadratic form is associated.

Let us recall that in elementary analytic geometry one proves that by a suitable rotation of axes the equation $\alpha x^2 + \beta xy + \gamma y^2$ can, in the new coordinate system, assume the form $\alpha_1(x')^2 + \gamma_1(y')^2$. Recall that $\alpha_1 + \gamma_1 = \alpha + \gamma$ and $\alpha\gamma - \beta^2/4 = \alpha_1\gamma_1$. Thus α_1, γ_1 are the characteristic roots of the matrix

$$\begin{pmatrix} \alpha & \beta/2 \\ \beta/2 & \gamma \end{pmatrix};$$

the rotation of axes is just a change of basis by an orthogonal transformation, and what we did in the geometry was merely to bring the symmetric matrix to its diagonal form by an orthogonal matrix. The nature of $\alpha x^2 + \beta xy + \gamma y^2$ as a conic was basically determined by the size and sign of its characteristic roots α_1, γ_1.

A similar discussion can be carried out to classify quadric surfaces in 3-space, and, indeed quadric surfaces in n-space. What essentially determines the geometric nature of the quadric surface associated with

$$\alpha_{11}x_1^2 + \cdots + \alpha_{nn}x_n^2 + 2\sum_{i<j} \alpha_{ij}x_ix_j$$

is the size and sign of the characteristic roots of the matrix (α_{ij}). If we were not interested in the relative flatness of the quadric surface (e.g., if we consider an ellipse as a flattened circle), then we could ignore the size of the nonzero characteristic roots and the determining factor for the shape of the quadric surface would be the number of 0 characteristic roots and the number of positive (and negative) ones.

These things motivate, and at the same time will be clarified in, the discussion that follows, which culminates in *Sylvester's law of inertia*.

Let A be a real symmetric matrix and let us consider its associated quadratic form $Q(v) = (vA, v)$. If T is any nonsingular real linear transformation, given $v \in F^{(n)}$, $v = wT$ for some $w \in F^{(n)}$, whence $(vA, v) = (wTA, wT) = (wTAT', w)$. Thus A and TAT' effectively define the same quadratic form. This prompts the

DEFINITION Two real symmetric matrices A and B are *congruent* if there is a nonsingular real matrix T such that $B = TAT'$.

LEMMA 6.11.1 *Congruence is an equivalence relation.*

Proof. Let us write, when A is congruent to B, $A \cong B$.

1. $A \cong A$ for $A = 1A1'$.
2. If $A \cong B$ then $B = TAT'$ where T is nonsingular, hence $A = SBS'$ where $S = T^{-1}$. Thus $B \cong A$.
3. If $A \cong B$ and $B \cong C$ then $B = TAT'$ while $C = RBR'$, hence $C = RTAT'R' = (RT)A(RT)'$, and so $A \cong C$.

Since the relation satisfies the defining conditions for an equivalence relation, the lemma is proved.

The principal theorem concerning congruence is its characterization, contained in *Sylvester's law*.

THEOREM 6.11.1 *Given the real symmetric matrix A there is an invertible matrix T such that*

$$TAT' = \begin{pmatrix} I_r & & \\ & -I_s & \\ & & 0_t \end{pmatrix}$$

where I_r and I_s are respectively the $r \times r$ and $s \times s$ unit matrices and where 0_t is the $t \times t$ zero-matrix. The integers $r + s$, which is the rank of A, and $r - s$, which is the signature of A, characterize the congruence class of A. That is, two real symmetric matrices are congruent if and only if they have the same rank and signature.

Proof. Since A is real symmetric its characteristic roots are all real; let $\lambda_1, \ldots, \lambda_r$ be its positive characteristic roots, $-\lambda_{r+1}, \ldots, -\lambda_{r+s}$ its negative ones. By the discussion at the end of Section 6.10 we can find a real orthogonal matrix C such that

$$CAC^{-1} = CAC' = \begin{pmatrix} \lambda_1 & & & & & & \\ & \ddots & & & & & \\ & & \lambda_r & & & & \\ & & & -\lambda_{r+1} & & & \\ & & & & \ddots & & \\ & & & & & -\lambda_{r+s} & \\ & & & & & & 0_t \end{pmatrix}$$

where $t = n - r - s$. Let D be the real diagonal matrix shown in Figure 6.11.1.

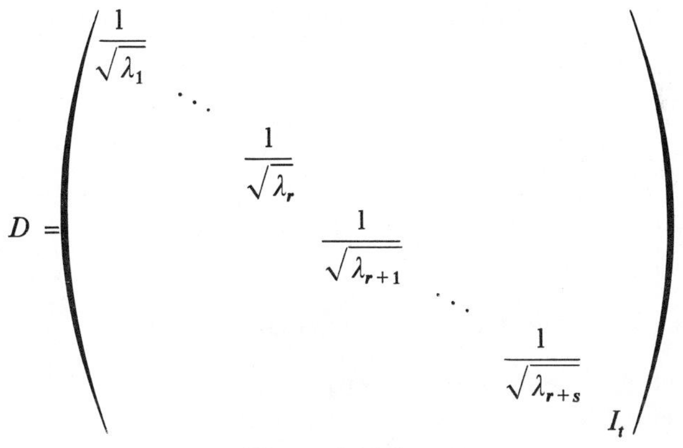

Figure 6.11.1

A simple computation shows that

$$DCAC'D' = \begin{pmatrix} I_r & & \\ & -I_s & \\ & & 0_t \end{pmatrix}.$$

Thus there is a matrix of the required form in the congruence class of A.

Our task is now to show that this is the only matrix in the congruence class of A of this form, or, equivalently, that

$$L = \begin{pmatrix} I_r & & \\ & -I_s & \\ & & 0_t \end{pmatrix} \quad \text{and} \quad M = \begin{pmatrix} I_{r'} & & \\ & -I_{s'} & \\ & & 0_{t'} \end{pmatrix}$$

are congruent only if $r = r'$, $s = s'$, and $t = t'$.

Suppose that $M = TLT'$ where T is invertible. By Lemma 6.1.3 the rank of M equals that of L; since the rank of M is $n - t'$ while that of L is $n - t$ we get $t = t'$.

Suppose that $r < r'$; since $n = r + s + t = r' + s' + t'$, and since $t = t'$, we must have $s > s'$. Let U be the subspace of $F^{(n)}$ of all vectors having the first r and last t coordinates 0; U is s-dimensional and for $u \neq 0$ in U, $(uL, u) < 0$.

Let W be the subspace of $F^{(n)}$ for which the $r' + 1, \ldots, r' + s'$ components are all 0; on W, $(wM, w) \geq 0$ for any $w \in W$. Since T is invertible, and since W is $(n - s')$-dimensional, WT is $(n - s')$-dimensional. For $w \in W$, $(wM, w) \geq 0$; hence $(wTLT', w) \geq 0$; that is, $(wTL, wT) \geq 0$. Therefore, on WT, $(wTL, wT) \geq 0$ for all elements. Now $\dim (WT) + \dim U = (n - s') + r = n + s - s' > n$; thus by the corollary to Lemma 4.2.6, $WT \cap U \neq 0$. This, however, is nonsense, for if $x \neq 0 \in WT \cap U$, on one hand, being in U, $(xL, x) < 0$, while on the other, being in WT, $(xL, x) \geq 0$. Thus $r = r'$ and so $s = s'$.

The rank, $r + s$, and signature, $r - s$, of course, determine r, s and so $t = (n - r - s)$, whence they determine the congruence class.

Problems

1. Determine the rank and signature of the following real quadratic forms:
   (a) $x_1^2 + 2x_1x_2 + x_2^2$.
   (b) $x_1^2 + x_1x_2 + 2x_1x_3 + 2x_2^2 + 4x_2x_3 + 2x_3^2$.
2. If A is a symmetric matrix with complex entries, prove we can find a complex invertible matrix B such that $BAB' = \begin{pmatrix} I_r & \\ & 0_t \end{pmatrix}$ and that r, the rank of A, determines the congruence class of A relative to complex congruence.
3. If F is a field of characteristic different from 2, given $A \in F_n$, prove that there exists a $B \in F_n$ such that BAB' is diagonal.
4. Prove the result of Problem 3 is false if the characteristic of F is 2.
5. How many congruence classes are there of $n \times n$ real symmetric matrices.

Supplementary Reading

HALMOS, PAUL R., *Finite-Dimensional Vector Spaces*, 2nd ed. Princeton, N.J.: D. Van Nostrand Company, 1958.

7

Selected Topics

In this final chapter we have set ourselves two objectives. Our first is to present some mathematical results which cut deeper than most of the material up to now, results which are more sophisticated, and are a little apart from the general development which we have followed. Our second goal is to pick results of this kind whose discussion, in addition, makes vital use of a large cross section of the ideas and theorems expounded earlier in the book. To this end we have decided on three items to serve as the focal points of this chapter.

The first of these is a celebrated theorem proved by Wedderburn in 1905 ("A Theorem on Finite Algebras," *Transactions of the American Mathematical Society*, Vol. 6 (1905), pages 349–352) which asserts that a division ring which has only a finite number of elements must be a commutative field. We shall give two proofs of this theorem, differing totally from each other. The first one will closely follow Wedderburn's original proof and will use a counting argument; it will lean heavily on results we developed in the chapter on group theory. The second one will use a mixture of group-theoretic and field-theoretic arguments, and will draw incisively on the material we developed in both these directions. The second proof has the distinct advantage that in the course of executing the proof certain side-results will fall out which will enable us to proceed to the proof, in the division ring case, of a beautiful theorem due to Jacobson ("Structure Theory for Algebraic Algebras of Bounded Degree," *Annals of Mathematics*, Vol. 46 (1945), pages 695–707) which is a far-reaching generalization of Wedderburn's theorem.

Our second high spot is a theorem due to Frobenius ("Über lineare Substitutionen und bilineare Formen," *Journal für die Reine und Angewandte Mathematik*, Vol. 84 (1877), especially pages 59–63) which states that the only division rings algebraic over the field of all real numbers are the field of real numbers, the field of complex numbers, and the division ring of real quaternions. The theorem points out a unique role for the quaternions, and makes it somewhat amazing that Hamilton should have discovered them in his somewhat ad hoc manner. Our proof of the Frobenius theorem, now quite elementary, is a variation of an approach laid out by Dickson and Albert; it will involve the theory of polynomials and fields.

Our third goal is the theorem that every positive integer can be represented as the sum of four squares. This famous result apparently was first conjectured by the early Greek mathematician Diophantos. Fermat grappled unsuccessfully with it and sadly announced his failure to solve it (in a paper where he did, however, solve the two-square theorem which we proved in Section 3.8). Euler made substantial inroads on the problem; basing his work on that of Euler, Lagrange in 1770 finally gave the first complete proof. Our approach will be entirely different from that of Lagrange. It is rooted in the work of Adolf Hurwitz and will involve a generalization of Euclidean rings. Using our ring-theoretic techniques on a certain ring of quaternions, the Lagrange theorem will drop out as a consequence.

En route to establishing these theorems many ideas and results, interesting in their own right, will crop up. This is characteristic of a good theorem—its proof invariably leads to side results of almost equal interest.

7.1 Finite Fields

Before we can enter into a discussion of Wedderburn's theorem and finite division rings, it is essential that we investigate the nature of fields having only a finite number of elements. Such fields are called *finite fields*. Finite fields do exist, for the ring J_p of integers modulo any prime p, provides us with an example of such. In this section we shall determine all possible finite fields and many of the important properties which they possess.

We begin with

LEMMA 7.1.1 *Let F be a finite field with q elements and suppose that $F \subset K$ where K is also a finite field. Then K has q^n elements where $n = [K:F]$.*

Proof. K is a vector space over F and since K is finite it is certainly finite-dimensional as a vector space over F. Suppose that $[K:F] = n$; then K has a basis of n elements over F. Let such a basis be $v_1, v_2, \ldots, v_n$. Then every element in K has a unique representation in the form $\alpha_1 v_1 + \alpha_2 v_2 + \cdots + \alpha_n v_n$ where $\alpha_1, \alpha_2, \ldots, \alpha_n$ are all in F. Thus the number of elements in K is the number of $\alpha_1 v_1 + \alpha_2 v_2 + \cdots + \alpha_n v_n$ as the α_1, $\alpha_2, \ldots, \alpha_n$ range over F. Since each coefficient can have q values K must clearly have q^n elements.

COROLLARY 1 *Let F be a finite field; then F has p^m elements where the prime number p is the characteristic of F.*

Proof. Since F has a finite number of elements, by Corollary 2 to Theorem 2.4.1, $f1 = 0$ where f is the number of elements in F. Thus F has characteristic p for some prime number p. Therefore F contains a field F_0 isomorphic to J_p. Since F_0 has p elements, F has p^m elements where $m = [F:F_0]$, by Lemma 7.1.1.

COROLLARY 2 *If the finite field F has p^m elements then every $a \in F$ satisfies $a^{p^m} = a$.*

Proof. If $a = 0$ the assertion of the corollary is trivially true.
On the other hand, the nonzero elements of F form a group under multiplication of order $p^m - 1$ thus by Corollary 2 to Theorem 2.4.1, $a^{p^m - 1} = 1$ for all $a \neq 0$ in F. Multiplying this relation by a we obtain that $a^{p^m} = a$.

From this last corollary we can easily pass to

LEMMA 7.1.2 *If the finite field F has p^m elements then the polynomial $x^{p^m} - x$ in $F[x]$ factors in $F[x]$ as $x^{p^m} - x = \prod_{\lambda \in F} (x - \lambda)$.*

Proof. By Lemma 5.3.2 the polynomial $x^{p^m} - x$ has at most p^{pm} roots in F. However, by Corollary 2 to Lemma 7.1.1 we know p^m such roots, namely all the elements of F. By the corollary to Lemma 5.3.1 we can conclude that $x^{p^m} - x = \prod_{\lambda \in F} (x - \lambda)$.

COROLLARY *If the field F has p^m elements then F is the splitting field of the polynomial $x^{p^m} - x$.*

Proof. By Lemma 7.1.2, $x^{p^m} - x$ certainly splits in F. However, it cannot split in any smaller field for that field would have to have all the roots of this polynomial and so would have to have at least p^m elements. Thus F is the splitting field of $x^{p^m} - x$.

As we have seen in Chapter 5 (Theorem 5.3.4) any two splitting fields over a given field of a given polynomial are isomorphic. In light of the corollary to Lemma 7.1.2 we can state

LEMMA 7.1.3 *Any two finite fields having the same number of elements are isomorphic.*

Proof. If these fields have p^m elements, by the above corollary they are both splitting fields of the polynomial $x^{p^m} - x$, over J_p whence they are isomorphic.

Thus for any integer m and any prime number p there is, up to isomorphism, at most one field having p^m elements. The purpose of the next lemma is to demonstrate that for any prime number p and any integer m there is a field having p^m elements. When this is done we shall know that there is exactly one field having p^m elements where p is an arbitrary prime and m an arbitrary integer.

LEMMA 7.1.4 *For every prime number p and every positive integer m there exists a field having p^m elements.*

Proof. Consider the polynomial $x^{p^m} - x$ in $J_p[x]$, the ring of polynomials in x over J_p, the field of integers mod p. Let K be the splitting field of this polynomial. In K let $F = \{a \in K \mid a^{p^m} = a\}$. The elements of F are thus the roots of $x^{p^m} - x$, which by Corollary 2 to Lemma 5.5.2 are distinct; whence F has p^m elements. We now claim that F is a field. If $a, b \in F$ then $a^{p^m} = a$, $b^{p^m} = b$ and so $(ab)^{p^m} = a^{p^m}b^{p^m} = ab$; thus $ab \in F$. Also since the characteristic is p, $(a \pm b)^{p^m} = a^{p^m} \pm b^{p^m} = a \pm b$, hence $a \pm b \in F$. Consequently F is a subfield of K and so is a field. Having exhibited the field F having p^m elements we have proved Lemma 7.1.4.

Combining Lemmas 7.1.3 and 7.1.4 we have

THEOREM 7.1.1 *For every prime number p and every positive integer m there is a unique field having p^m elements.*

We now return to group theory for a moment. The group-theoretic result we seek will determine the structure of any finite multiplicative subgroup of the group of nonzero elements of any field, and, in particular, it will determine the multiplicative structure of any finite field.

LEMMA 7.1.5 *Let G be a finite abelian group enjoying the property that the relation $x^n = e$ is satisfied by at most n elements of G, for every integer n. Then G is a cyclic group.*

Proof. If the order of G is a power of some prime number q then the result is very easy. For suppose that $a \in G$ is an element whose order is as large as possible; its order must be q^r for some integer r. The elements $e, a, a^2, \ldots, a^{q^r-1}$ give us q^r distinct solutions of the equation $x^{q^r} = e$, which, by our hypothesis, implies that these are all the solutions of this equation. Now if $b \in G$ its order is q^s where $s \leq r$, hence $b^{q^r} = (b^{q^s})^{q^{r-s}} = e$.

By the observation made above this forces $b = a^i$ for some i, and so G is cyclic.

The general finite abelian group G can be realized as $G = S_{q_1}S_{q_2}\cdots, S_{q_k}$ where the q_i are the distinct prime divisors of $o(G)$ and where the S_{q_i} are the Sylow subgroups of G. Moreover, every element $g \in G$ can be written in a *unique* way as $g = s_1s_2, \ldots, s_k$ where $s_i \in S_{q_i}$ (see Section 2.7). Any solution of $x^n = e$ in S_{q_i} is one of $x^n = e$ in G so that each S_{q_i} inherits the hypothesis we have imposed on G. By the remarks of the first paragraph of the proof, each S_{q_i} is a cyclic group; let a_i be a generator of S_{q_i}. We claim that $c = a_1a_2, \ldots, a_k$ is a cyclic generator of G. To verify this all we must do is prove that $o(G)$ divides m, the order of c. Since $c^m = e$, we have that $a_1{}^m a_2{}^m \cdots a_k{}^m = e$. By the uniqueness of representation of an element of G as a product of elements in the S_{q_i}, we conclude that each $a_i{}^m = e$. Thus $o(S_{q_i}) \mid m$ for every i. Thus $o(G) = o(S_{q_1})o(S_{q_2}) \cdots o(S_{q_k}) \mid m$. However, $m \mid o(G)$ and so $o(G) = m$. This proves that G is cyclic.

Lemma 7.1.5 has as an important consequence

LEMMA 7.1.6 *Let K be a field and let G be a finite subgroup of the multiplicative group of nonzero elements of K. Then G is a cyclic group.*

Proof. Since K is a field, any polynomial of degree n in $K[x]$ has at most n roots in K. Thus in particular, for any integer n, the polynomial $x^n - 1$ has at most n roots in K, and all the more so, at most n roots in G. The hypothesis of Lemma 7.1.5 is satisfied, so G is cyclic.

Even though the situation of a finite field is merely a special case of Lemma 7.1.6, it is of such widespread interest that we single it out as

THEOREM 7.1.2 *The multiplicative group of nonzero elements of a finite field is cyclic.*

Proof. Let F be a finite field. By merely applying Lemma 7.1.6 with $F = K$ and $G =$ the group of nonzero elements of F, the result drops out.

We conclude this section by using a counting argument to prove the existence of solutions of certain equations in a finite field. We shall need the result in one proof of the Wedderburn theorem.

LEMMA 7.1.7 *If F is a finite field and $\alpha \neq 0$, $\beta \neq 0$ are two elements of F then we can find elements a and b in F such that $1 + \alpha a^2 + \beta b^2 = 0$.*

Proof. If the characteristic of F is 2, F has 2^n elements and every element x in F satisfies $x^{2^n} = x$. Thus every element in F is a square. In particular $\alpha^{-1} = a^2$ for some $a \in F$. Using this a and $b = 0$, we have

$1 + \alpha a^2 + \beta b^2 = 1 + \alpha\alpha^{-1} + 0 = 1 + 1 = 0$, the last equality being a consequence of the fact that the characteristic of F is 2.

If the characteristic of F is an odd prime p, F has p^n elements. Let $W_\alpha = \{1 + \alpha x^2 \mid x \in F\}$. How many elements are there in W_a? We must check how often $1 + \alpha x^2 = 1 + \alpha y^2$. But this relation forces $\alpha x^2 = \alpha y^2$ and so, since $\alpha \neq 0$, $x^2 = y^2$. Finally this leads to $x = \pm y$. Thus for $x \neq 0$ we get from each pair x and $-x$ one element in W_α, and for $x = 0$ we get $1 \in W_\alpha$. Thus W_α has $1 + (p^n - 1)/2 = (p^n + 1)/2$ elements. Similarly $W_\beta = \{-\beta x^2 \mid x \in F\}$ has $(p^n + 1)/2$ elements. Since each of W_α and W_β has more than half the elements of F they must have a nonempty intersection. Let $c \in W_\alpha \cap W_\beta$. Since $c \in W_\alpha$, $c = 1 + \alpha a^2$ for some $a \in F$; since $c \in W_\beta$, $c = -\beta b^2$ for some $b \in F$. Therefore $1 + \alpha a^2 = -\beta b^2$, which, on transposing yields the desired result $1 + \alpha a^2 + \beta b^2 = 0$.

Problems

1. By Theorem 7.1.2 the nonzero elements of J_p form a cyclic group under multiplication. Any generator of this group is called a *primitive root* of p.
   (a) Find primitive roots of: 17, 23, 31.
   (b) How many primitive roots does a prime p have?
2. Using Theorem 7.1.2 prove that $x^2 \equiv -1 \bmod p$ is solvable if and only if the odd prime p is of the form $4n + 1$.
3. If a is an integer not divisible by the odd prime p, prove that $x^2 \equiv a \bmod p$ is solvable for some integer x if and only if $a^{(p-1)/2} \equiv 1 \bmod p$. (This is called the *Euler criterion* that a be a quadratic residue mod p.)
4. Using the result of Problem 3 determine if:
   (a) 3 is a square mod 17.
   (b) 10 is a square mod 13.
5. If the field F has p^n elements prove that the automorphisms of F form a cyclic group of order n.
6. If F is a finite field, by the quaternions over F we shall mean the set of all $\alpha_0 + \alpha_1 i + \alpha_2 j + \alpha_3 k$ where $\alpha_0, \alpha_1, \alpha_2, \alpha_3 \in F$ and where addition and multiplication are carried out as in the real quaternions (i.e., $i^2 = j^2 = k^2 = ijk = -1$, etc.). Prove that the quaternions over a finite field *do not* form a division ring.

7.2 Wedderburn's Theorem on Finite Division Rings

In 1905 Wedderburn proved the theorem, now considered a classic, that a finite division ring must be a commutative field. This result has caught the imagination of most mathematicians because it is so unexpected, interrelating two seemingly unrelated things, namely the number of elements in a certain algebraic system and the multiplication of that system. Aside from its intrinsic beauty the result has been very important and useful since it arises in so many contexts. To cite just one instance, the only known proof of the purely geometric fact that in a finite geometry the Desargues configuration implies that of Pappus (for the definition of these terms look in any good book on projective geometry) is to reduce the geometric problem to an algebraic one, and this algebraic question is then answered by invoking the Wedderburn theorem. For algebraists the Wedderburn theorem has served as a jumping-off point for a large area of research, in the 1940s and 1950s, concerned with the commutativity of rings.

THEOREM 7.2.1 (WEDDERBURN) *A finite division ring is necessarily a commutative field.*

First Proof. Let K be a finite division ring and let $Z = \{z \in K \mid zx = xz$ for all $x \in K\}$ be its center. If Z has q elements then, as in the proof of Lemma 7.1.1, it follows that K has q^n elements. Our aim is to prove that $Z = K$, or, equivalently, that $n = 1$.

If $a \in K$ let $N(a) = \{x \in K \mid xa = ax\}$. $N(a)$ clearly contains Z, and, as a simple check reveals, $N(a)$ is a subdivision ring of K. Thus $N(a)$ contains $q^{n(a)}$ elements for some integer $n(a)$. We claim that $n(a) \mid n$. For, the nonzero elements of $N(a)$ form a subgroup of order $q^{n(a)} - 1$ of the group of nonzero elements, under multiplication, of K which has $q^n - 1$ elements. By Lagrange's theorem (Theorem 2.4.1) $q^{n(a)} - 1$ is a divisor of $q^n - 1$; but this forces $n(a)$ to be a divisor of n (see Problem 1 at the end of this section).

In the group of nonzero elements of K we have the conjugacy relation used in Chapter 2, namely a is a conjugate of b if $a = x^{-1}bx$ for some $x \neq 0$ in K.

By Theorem 2.11.1 the number of elements in K conjugate to a is the index of the normalizer of a in the group of nonzero elements of K. Therefore the number of conjugates of a in K is $(q^n - 1)/(q^{n(a)} - 1)$. Now $a \in Z$ if and only if $n(a) = n$, thus by the class equation (see the corollary to Theorem 2.11.1)

$$q^n - 1 = q - 1 + \sum_{\substack{n(a) \mid n \\ n(a) \neq n}} \frac{q^n - 1}{q^{n(a)} - 1} \tag{1}$$

where the sum is carried out over one a in each conjugate class for a's *not* in the center.

The problem has been reduced to proving that no equation such as (1) can hold in the integers. Up to this point we have followed the proof in Wedderburn's original paper quite closely. He went on to rule out the possibility of equation (1) by making use of the following number-theoretic result due to Birkhoff and Vandiver: for $n > 1$ there exists a prime number which is a divisor of $q^n - 1$ but is not a divisor of *any* $q^m - 1$ where m is a proper divisor of n, with the exceptions of $2^6 - 1 = 63$ whose prime factors already occur as divisors of $2^2 - 1$ and $2^3 - 1$, and $n = 2$, and q a prime of the form $2^k - 1$. If we grant this result, how would we finish the proof? This prime number would be a divisor of the left-hand side of (1) and also a divisor of each term in the sum occurring on the right-hand side since it divides $q^n - 1$ but not $q^{n(a)} - 1$; thus this prime would then divide $q - 1$ giving us a contradiction. The case $2^6 - 1$ still would need ruling out but that is simple. In case $n = 2$, the other possibility not covered by the above argument, there can be no subfield between Z and K and this forces $Z = K$. (Prove!—See Problem 2.)

However, we do not want to invoke the result of Birkhoff and Vandiver without proving it, and its proof would be too large a digression here. So we look for another artifice. Our aim is to find an integer which divides $(q^n - 1)/(q^{n(a)} - 1)$, for all divisors $n(a)$ of n except $n(a) = n$, but does not divide $q - 1$. Once this is done, equation (1) will be impossible unless $n = 1$ and, therefore, Wedderburn's theorem will have been proved. The means to this end is the theory of cyclotomic polynomials. (These have been mentioned in the problems at the end of Section 5.6.)

Consider the polynomial $x^n - 1$ considered as an element of $C[x]$ where C is the field of complex numbers. In $C[x]$

$$x^n - 1 = \prod (x - \lambda), \tag{2}$$

where this product is taken over all λ satisfying $\lambda^n = 1$.

A complex number θ is said to be a *primitive nth root of unity* if $\theta^n = 1$ but $\theta^m \neq 1$ for any positive integer $m < n$. The complex numbers satisfying $x^n = 1$ form a finite subgroup, under multiplication, of the complex numbers, so by Theorem 7.1.2 this group is cyclic. Any cyclic generator of this group must then be a primitive nth root of unity, so we know that such primitive roots exist. (Alternatively, $\theta = e^{2\pi i/n}$ yields us a primitive nth root of unity.)

Let $\Phi_n(x) = \prod (x - \theta)$ where this product is taken over all the primitive nth roots of unity. This polynomial is called a *cyclotomic* polynomial. We list the first few cyclotomic polynomials: $\Phi_1(x) = x - 1$, $\Phi_2(x) = x + 1$, $\Phi_3(x) = x^2 + x + 1$, $\Phi_4(x) = x^2 + 1$, $\Phi_5(x) = x^4 + x^3 + x^2 + x + 1$, $\Phi_6(x) = x^2 - x + 1$. Notice that these are all monic polynomials with integer coefficients.

Our first aim is to prove that in general $\Phi_n(x)$ is a monic polynomial with integer coefficients. We regroup the factored form of $x^n - 1$ as given in (2), and obtain

$$x^n - 1 = \prod_{d \mid n} \Phi_d(x). \tag{3}$$

By induction we assume that $\Phi_d(x)$ is a monic polynomial with integer coefficients for $d \mid n$, $d \neq n$. Thus $x^n - 1 = \Phi_n(x)g(x)$ where $g(x)$ is a monic polynomial with integer coefficients. Therefore,

$$\Phi_n(x) = \frac{x^n - 1}{g(x)},$$

which, on actual division (or by comparing coefficients), tells us that $\Phi_n(x)$ is a monic polynomial with integer coefficients.

We now claim that for any divisor d of n, where $d \neq n$,

$$\Phi_n(x) \left| \frac{x^n - 1}{x^d - 1} \right.$$

in the sense that the quotient is a polynomial with integer coefficients. To see this, first note that

$$x^d - 1 = \prod_{k \mid d} \Phi_k(x),$$

and since every divisor of d is also a divisor of n, by regouping terms on the right-hand side of (3) we obtain $x^d - 1$ on the right-hand side; also since $d < n$, $x^d - 1$ does not involve $\Phi_n(x)$. Therefore, $x^n - 1 = \Phi_n(x)(x^d - 1)f(x)$ where

$$f(x) = \prod_{\substack{k \mid n \\ k \nmid d}} \Phi_k(x)$$

has integer coefficients, and so

$$\Phi_n(x) \left| \frac{x^n - 1}{x^d - 1} \right.$$

in the sense that the quotient is a polynomial with integer coefficients. This establishes our claim.

For any integer t, $\Phi_n(t)$ is an integer and from the above as an integer divides $(t^n - 1)/(t^d - 1)$. In particular, returning to equation (1),

$$\Phi_n(q) \left| \frac{q^n - 1}{q^{n(a)} - 1} \right.$$

and $\Phi_n(q) \mid (q^n - 1)$; thus by (1), $\Phi_n(q) \mid (q - 1)$. We claim, however, that if $n > 1$ then $|\Phi_n(q)| > q - 1$. For $\Phi_n(q) = \prod (q - \theta)$ where θ runs over all primitive nth roots of unity and $|q - \theta| > q - 1$ for all $\theta \neq 1$ a root of unity (Prove!) whence $|\Phi_n(q)| = \prod |q - \theta| > q - 1$. Clearly, then $\Phi_n(q)$ cannot divide $q - 1$, leading us to a contradiction. We must, therefore, assume that $n = 1$, forcing the truth of the Wedderburn theorem.

Second Proof. Before explicitly examining finite division rings again, we prove some preliminary lemmas.

LEMMA 7.2.1 *Let R be a ring and let $a \in R$. Let T_a be the mapping of R into itself defined by $xT_a = xa - ax$. Then*

$$xT_a^{\,m} = xa^m - maxa^{m-1} + \frac{m(m-1)}{2}a^2xa^{m-2} - \frac{m(m-1)(m-2)}{3!}a^3xa^{m-3} + \cdots.$$

Proof. What is $xT_a^{\,2}$? $xT_a^{\,2} = (xT_a)T_a = (xa - ax)T_a = (xa - ax)a - a(xa - ax) = xa^2 - 2axa + a^2x$. What about $xT_a^{\,3}$? $xT_a^{\,3} = (xT_a^{\,2})T_a = (xa^2 - 2axa + a^2x)a - a(xa^2 - 2axa + a^2x) = xa^3 - 3axa^2 + 3a^2xa - a^3x$. Continuing in this way, or by the use of induction, we get the result of Lemma 7.2.1.

COROLLARY *If R is a ring in which $px = 0$ for all $x \in R$, where p is a prime number, then $xT_a^{\,p^m} = xa^{p^m} - a^{p^m}x$.*

Proof. By the formula of Lemma 7.2.1, if $p = 2$, $xT_a^{\,2} = xa^2 - a^2x$, since $2axa = 0$. Thus, $xT_a^{\,4} = (xa^2 - a^2x)a^2 - a^2(xa^2 - a^2x) = xa^4 - a^4x$, and so on for $xT_a^{\,2^m}$.

If p is an odd prime, again by the formula of Lemma 7.2.1,

$$xT_a^{\,p} = xa^p - paxa^{p-1} + \frac{p(p-1)}{2}a^2xa^{p-2} + \cdots - a^px,$$

and since

$$p \,\Big|\, \frac{p(p-1)\cdots(p-i+1)}{i!}$$

for $i < p$, all the middle terms drop out and we are left with $xT_a^{\,p} = xa^p - a^px = xT_{a^p}$. Now $xT_a^{\,p^2} = x(T_{a^p})^p = xT_{a^{p^2}}$, and so on for the higher powers of p.

LEMMA 7.2.2 *Let D be a division ring of characteristic $p > 0$ with center Z, and let $P = \{0, 1, 2, \ldots, (p-1)\}$ be the subfield of Z isomorphic to J_p. Suppose that $a \in D$, $a \notin Z$ is such that $a^{p^n} = a$ for some $n \geq 1$. Then there exists an $x \in D$ such that*

1. $xax^{-1} \neq a$.
2. $xax^{-1} \in P(a)$ *the field obtained by adjoining a to P.*

Proof. Define the mapping T_a of D into itself by $yT_a = ya - ay$ for every $y \in D$.

$P(a)$ is a finite field, since a is algebraic over P and has, say, p^m elements. These all satisfy $u^{p^m} = u$. By the corollary to Lemma 7.2.1, $yT_a^{\,p^m} = ya^{p^m} - a^{p^m}y = ya - ay = yT_a$, and so $T_a^{\,p^m} = T_a$.

Now, if $\lambda \in P(a)$, $(\lambda x)T_a = (\lambda x)a - a(\lambda x) = \lambda xa - \lambda ax = \lambda(xa - ax) = \lambda(xT_a)$, since λ commutes with a. Thus the mapping λI of D into itself defined by $\lambda I: y \rightarrow \lambda y$ commutes with T_a for every $\lambda \in P(a)$. Now the polynomial

$$u^{p^m} - u = \prod_{\lambda \in P(a)} (u - \lambda)$$

by Lemma 7.2.1. *Since T_a commutes with λI for every $\lambda \in P(a)$*, and since $T_a^{p^m} = T_a$, we have that

$$0 = T_a^{p^m} - T_a = \prod_{\lambda \in P(a)} (T_a - \lambda I).$$

If for every $\lambda \neq 0$ in $P(a)$, $T_a - \lambda I$ annihilates no nonzero element in D (if $y(T_a - \lambda I) = 0$ implies $y = 0$), since $T_a(T_a - \lambda_1 I) \cdots (T_a - \lambda_k I) = 0$, where $\lambda_1, \ldots, \lambda_k$ are the nonzero elements of $P(a)$, we would get $T_a = 0$. That is, $0 = yT_a = ya - ay$ for every $y \in D$ forcing $a \in Z$ contrary to hypothesis. Thus there is a $\lambda \neq 0$ in $P(a)$ and an $x \neq 0$ in D such that $x(T_a - \lambda I) = 0$. Writing this out explicitly, $xa - ax - \lambda x = 0$; hence, $xax^{-1} = a + \lambda$ is in $P(a)$ and is not equal to a since $\lambda \neq 0$. This proves the lemma.

COROLLARY *In Lemma 7.2.2, $xax^{-1} = a^i \neq a$ for some integer i.*

Proof. Let a be of order s; then in the field $P(a)$ all the roots of the polynomial $u^s - 1$ are $1, a, a^2, \ldots, a^{s-1}$ since these are all distinct roots and they are s in number. Since $(xax^{-1})^s = xa^sx^{-1} = 1$, and since $xax^{-1} \in P(a)$, xax^{-1} is a root in $P(a)$ of $u^s - 1$, hence $xax^{-1} = a^i$.

We now have all the pieces that we need to carry out our second proof of Wedderburn's theorem.

Let D be a finite division ring and let Z be its center. By induction we may assume that any division ring having fewer elements than D is a commutative field.

We first remark that if $a, b \in D$ are such that $b^t a = ab^t$ but $ba \neq ab$, then $b^t \in Z$. For, consider $N(b^t) = \{x \in D \mid b^t x = xb^t\}$. $N(b^t)$ is a subdivision ring of D; if it were not D, by our induction hypothesis, it would be commutative. However, both a and b are in $N(b^t)$ and these do not commute; consequently, $N(b^t)$ is not commutative so must be all of D. Thus $b^t \in Z$.

Every nonzero element in D has finite order, so some positive power of it falls in Z. Given $w \in D$ let the *order of w relative to Z* be the smallest positive integer $m(w)$ such that $w^{m(w)} \in Z$. Pick an element a in D but not in Z having minimal possible order relative to Z, and let this order be r. *We claim that r is a prime number*, for if $r = r_1 r_2$ with $1 < r_1 < r$ then a^{r_1} is not in Z. Yet $(a^{r_1})^{r_2} = a^r \in Z$, implying that a^{r_1} has an order relative to Z smaller than that of a.

By the corollary to Lemma 7.2.2 there is an $x \in D$ such that $xax^{-1} = a^i \neq a$; thus $x^2ax^{-2} = x(xax^{-1})x^{-1} = xa^ix^{-1} = (xax^{-1})^i = (a^i)^i = a^{i^2}$. Similarly, we get $x^{r-1}ax^{-(r-1)} = a^{i^{r-1}}$. However, r is a prime number, thus by the little Fermat theorem (corollary to Theorem 2.4.1), $i^{r-1} = 1 + u_0r$, hence $a^{i^{r-1}} = a^{1+u_0r} = aa^{u_0r} = \lambda a$ where $\lambda = a^{u_0r} \in Z$. Thus $x^{r-1}a = \lambda ax^{r-1}$. Since $x \notin Z$, by the minimal nature of r, x^{r-1} cannot be in Z. By the remark of the earlier paragraph, since $xa \neq ax$, $x^{r-1}a \neq ax^{r-1}$ and so $\lambda \neq 1$. Let $b = x^{r-1}$; thus $bab^{-1} = \lambda a$; consequently, $\lambda^r a^r = (bab^{-1})^r = ba^rb^{-1} = a^r$ since $a^r \in Z$. This relation forces $\lambda^r = 1$.

We claim that if $y \in D$ then whenever $y^r = 1$, then $y = \lambda^i$ for some i, for in the *field* $Z(y)$ there are at most r roots of the polynomial $u^r - 1$; the elements $1, \lambda, \lambda^2, \ldots, \lambda^{r-1}$ in Z are all distinct since λ is of the prime order r and they already account for r roots of $u^r - 1$ in $Z(y)$, in consequence of which $y = \lambda^i$.

Since $\lambda^r = 1$, $b^r = \lambda^r b^r = (\lambda b)^r = (a^{-1}ba)^r = a^{-1}b^ra$ from which we get $ab^r = b^ra$. Since a commutes with b^r but does not commute with b, by the remark made earlier, b^r must be in Z. By Theorem 7.1.2 the multiplicative group of nonzero elements of Z is cyclic; let $\gamma \in Z$ by a generator. Thus $a^r = \gamma^j$, $b^r = \gamma^k$; if $j = sr$ then $a^r = \gamma^{sr}$, whence $(a/\gamma^s)^r = 1$; this would imply that $a/\gamma^s = \lambda^i$, leading to $a \in Z$, contrary to $a \notin Z$. Hence, $r \nmid j$; similarly $r \nmid k$. Let $a_1 = a^k$ and $b_1 = b^j$; a direct computation from $ba = \lambda ab$ leads to $a_1b_1 = \mu b_1a_1$ where $\mu = \lambda^{-jk} \in Z$. Since the prime number r which is the order of λ does not divide j or k, $\lambda^{jk} \neq 1$ hence $\mu \neq 1$. Note that $\mu^r = 1$.

Let us see where we are. We have produced two elements a_1, b_1 such that

1. $a_1{}^r = b_1{}^r = \alpha \in Z$.
2. $a_1b_1 = \mu b_1a_1$ with $\mu \neq 1$ in Z.
3. $\mu^r = 1$.

We compute $(a_1{}^{-1}b_1)^r$; $(a_1{}^{-1}b_1)^2 = a_1{}^{-1}b_1a_1{}^{-1}b_1 = a_1{}^{-1}(b_1a_1{}^{-1})b_1 = a_1{}^{-1}(\mu a_1{}^{-1}b_1)b_1 = \mu a_1{}^{-2}b_1{}^2$. If we compute $(a_1{}^{-1}b_1)^3$ we find it equal to $\mu^{1+2}a_1{}^{-3}b_1{}^3$. Continuing, we obtain $(a_1{}^{-1}b_1)^r = \mu^{1+2+\cdots+(r-1)}a_1{}^{-r}b_1{}^r = \mu^{1+2+\cdots+(r-1)} = \mu^{r(r-1)/2}$. If r is an odd prime, since $\mu^r = 1$, we get $\mu^{r(r-1)/2} = 1$, whence $(a_1{}^{-1}b_1)^r = 1$. Being a solution of $y^r = 1$, $a_1{}^{-1}b_1 = \lambda^i$ so that $b_1 = \lambda^i a_1$; but then $\mu b_1a_1 = a_1b_1 = b_1a_1$, contradicting $\mu \neq 1$. Thus if r is an odd prime number, the theorem is proved.

We must now rule out the case $r = 2$. In that special situation we have two elements $a_1, b_1 \in D$ such that $a_1{}^2 = b_1{}^2 = \alpha \in Z$, $a_1b_1 = \mu b_1a_1$ where $\mu^2 = 1$ and $\mu \neq 1$. Thus $\mu = -1$ and $a_1b_1 = -b_1a_1 \neq b_1a_1$; in consequence, the characteristic of D is *not* 2. By Lemma 7.1.7 we can find elements $\zeta, \eta \in Z$ such that $1 + \zeta^2 - \alpha\eta^2 = 0$. Consider $(a_1 + \zeta b_1 + \eta a_1b_1)^2$; on computing this out we find that $(a_1 + \zeta b_1 + \eta a_1b_1)^2 = \alpha(1 + \zeta^2 - \alpha\eta^2) = 0$. Being in a division ring this yields that $a_1 + \zeta b_1 + \eta a_1b_1 = 0$; thus $0 \neq$

$2a_1^{\,2} = a_1(a_1 + \zeta b_1 + \eta a_1 b_1) + (a_1 + \zeta b_1 + \eta a_1 b_1)a_1 = 0$. This contradiction finishes the proof and Wedderburn's theorem is established.

This second proof has some advantages in that we can use parts of it to proceed to a remarkable result due to Jacobson, namely,

THEOREM 7.2.2 (JACOBSON) *Let D be a division ring such that for every $a \in D$ there exists a positive integer $n(a) > 1$, depending on a, such that $a^{n(a)} = a$. Then D is a commutative field.*

Proof. If $a \neq 0$ is in D then $a^n = a$ and $(2a)^m = 2a$ for some integers $n, m > 1$. Let $s = (n-1)(m-1) + 1$; $s > 1$ and a simple calculation shows that $a^s = a$ and $(2a)^s = 2a$. But $(2a)^s = 2^s a^s = 2^s a$, whence $2^s a = 2a$ from which we get $(2^s - 2)a = 0$. Thus D has characteristic $p > 0$. If $P \subset Z$ is the field having p elements (isomorphic to J_p), since a is algebraic over P, $P(a)$ has a finite number of elements, in fact, p^h elements for some integer h. Thus, since $a \in P(a)$, $a^{p^h} = a$. Therefore, if $a \notin Z$ all the conditions of Lemma 7.2.2 are satisfied, hence there exists a $b \in D$ such that

$$bab^{-1} = a^{\mu} \neq a. \tag{1}$$

By the same argument, $b^{p^k} = b$ for some integer $k > 1$. Let

$$W = \left\{x \in D \mid x = \sum_{i=1}^{p^h} \sum_{j=1}^{p^k} p_{ij} a^i b^j \text{ where } p_{ij} \in P\right\}.$$

W is finite and is closed under addition. By virtue of (1) it is also closed under multiplication. (Verify!) Thus W is a finite ring, and being a subring of the division ring D, it itself must be a division ring (Problem 3). Thus W is a finite division ring; by Wedderburn's theorem it is commutative. But a and b are both in W; therefore, $ab = ba$ contrary to $a^{\mu}b = ba$. This proves the theorem.

Jacobson's theorem actually holds for *any* ring R satisfying $a^{n(a)} = a$ for every $a \in R$, not just for division rings. The transition from the division ring case to the general case, while not difficult, involves the axiom of choice, and to discuss it would take us too far afield.

Problems

1. If $t > 1$ is an integer and $(t^m - 1) | (t^n - 1)$, prove that $m \mid n$.
2. If D is a division ring, prove that its dimension (as a vector space) over its center cannot be 2.
3. Show that any finite subring of a division ring is a division ring.

4. (a) Let D be a division ring of characteristic $p \neq 0$ and let G be a finite subgroup of the group of nonzero elements of D under multiplication. Prove that G is abelian. (*Hint:* consider the subset $\{x \in D \mid x = \sum \lambda_i g_i,\ \lambda_i \in P,\ g_i \in G\}$.)
   (b) In part (a) prove that G is actually cyclic.

*5. (a) If R is a finite ring in which $x^n = x$, for all $x \in R$ where $n > 1$ prove that R is commutative.
   (b) If R is a finite ring in which $x^2 = 0$ implies that $x = 0$, prove that R is commutative.

*6. Let D be a division ring and suppose that $a \in D$ only has a finite number of conjugates (i.e., only a finite number of distinct $x^{-1}ax$). Prove that a has only one conjugate and must be in the center of D.

7. Use the result of Problem 6 to prove that if a polynomial of degree n having coefficients in the center of a division ring has $n + 1$ roots in the division ring then it has an infinite number of roots in that division ring.

*8. Let D be a division ring and K a subdivision ring of D such that $xKx^{-1} \subset K$ for every $x \neq 0$ in D. Prove that either $K \subset Z$, the center of D or $K = D$. (This result is known as the *Brauer-Cartan-Hua theorem.*)

*9. Let D be a division ring and K a subdivision ring of D. Suppose that the group of nonzero elements of K is a subgroup of finite index in the group (under multiplication) of nonzero elements of D. Prove that either D is finite or $K = D$.

10. If $\theta \neq 1$ is a root of unity and if q is a positive integer, prove that $|q - \theta| > q - 1$.

7.3 A Theorem of Frobenius

In 1877 Frobenius classified all division rings having the field of real numbers in their center and satisfying, in addition, one other condition to be described below. The aim of this section is to present this result of Frobenius.

In Chapter 6 we brought attention to two important facts about the field of complex numbers. We recall them here:

FACT 1 Every polynomial of degree n over the field of complex numbers has all its n roots in the field of complex numbers.

FACT 2 The only irreducible polynomials over the field of real numbers are of degree 1 or 2.

DEFINITION A division algebra D is said to be *algebraic over a field F* if

1. F is contained in the center of D;
2. every $a \in D$ satisfies a nontrivial polynomial with coefficients in F.

If D, as a vector space, is finite-dimensional over the field F which is contained in its center, it can easily be shown that D is algebraic over F (see Problem 1, end of this section). However, it can happen that D is algebraic over F yet is not finite-dimensional over F.

We start our investigation of division rings algebraic over the real field by first finding those algebraic over the complex field.

LEMMA 7.3.1 *Let C be the field of complex numbers and suppose that the division ring D is algebraic over C. Then $D = C$.*

Proof. Suppose that $a \in D$. Since D is algebraic over C, $a^n + \alpha_1 a^{n-1} + \cdots + \alpha_{n-1}a + \alpha_n = 0$ for some $\alpha_1, \alpha_2, \ldots, \alpha_n$ in C.

Now the polynomial $p(x) = x^n + \alpha_1 x^{n-1} + \cdots + \alpha_{n-1}x + \alpha_n$ in $C[x]$, by Fact 1, can be factored, in $C[x]$, into a product of linear factors; that is, $p(x) = (x - \lambda_1)(x - \lambda_2) \cdots (x - \lambda_n)$, where $\lambda_1, \lambda_2, \ldots, \lambda_n$ are all in C. Since C is in the center of D, every element of C commutes with a, hence $p(a) = (a - \lambda_1)(a - \lambda_2) \cdots (a - \lambda_n)$. But, by assumption, $p(a) = 0$, thus $(a - \lambda_1)(a - \lambda_2) \cdots (a - \lambda_n) = 0$. Since a product in a division ring is zero only if one of the terms of the product is zero, we conclude that $a - \lambda_k = 0$ for some k, hence $a = \lambda_k$, from which we get that $a \in C$. Therefore, every element of D is in C; since $C \subset D$, we obtain $D = C$.

We are now in a position to prove the classic result of Frobenius, namely,

THEOREM 7.3.1 (FROBENIUS) *Let D be a division ring algebraic over F, the field of real numbers. Then D is isomorphic to one of: the field of real numbers, the field of complex numbers, or the division ring of real quaternions.*

Proof. The proof consists of three parts. In the first, and easiest, we dispose of the commutative case; in the second, assuming that D is not commutative, we construct a replica of the real quaternions in D; in the third part we show that this replica of the quaternions fills out all of D.

Suppose that $D \neq F$ and that a is in D but not in F. By our assumptions, a satisfies some polynomial over F, hence some irreducible polynomial over F. In consequence of Fact 2, a satisfies either a linear or quadratic equation over F. If this equation is linear, a must be in F contrary to assumption. So we may suppose that $a^2 - 2\alpha a + \beta = 0$ where $\alpha, \beta \in F$. Thus $(a - \alpha)^2 = \alpha^2 - \beta$; we claim that $\alpha^2 - \beta < 0$ for, otherwise, it would have a real square root δ and we would have $a - \alpha = \pm\delta$ and so a would be in F. Since $\alpha^2 - \beta < 0$ it can be written as $-\gamma^2$ where $\gamma \in F$. Consequently $(a - \alpha)^2 = -\gamma^2$, whence $[(a - \alpha)/\gamma]^2 = -1$. *Thus if $a \in D$, $a \notin F$ we can find real α, γ such that $[(a - \alpha)/\gamma]^2 = -1$.*

If D is commutative, pick $a \in D$, $a \notin F$ and let $i = (a - \alpha)/\gamma$ where α, γ in F are chosen so as to make $i^2 = -1$. Therefore D contains $F(i)$, a field isomorphic to the field of complex numbers. Since D is commutative and algebraic over F it is, all the more so, algebraic over $F(i)$. By Lemma 7.3.1 we conclude that $D = F(i)$. Thus if D is commutative it is either F or $F(i)$.

Assume, then, that D is *not* commutative. We claim that the center of D must be exactly F. If not, there is an a in the center, a not in F. But then for some $\alpha, \gamma \in F$, $[(a - \alpha)/\gamma]^2 = -1$ so that the center contains a field isomorphic to the complex numbers. However, by Lemma 7.3.1 if the complex numbers (or an isomorph of them) were in the center of D then $D = C$ forcing D to be commutative. Hence F is the center of D.

Let $a \in D$, $a \notin F$; for some $\alpha, \gamma \in F$, $i = (a - \alpha)/\gamma$ satisfies $i^2 = -1$. Since $i \notin F$, i is not in the center of F. Therefore there is an element $b \in D$ such that $c = bi - ib \neq 0$. We compute $ic + ci$; $ic + ci = i(bi - ib) + (bi - ib)i = ibi - i^2b + bi^2 - ibi = 0$ since $i^2 = -1$. Thus $ic = -ci$; from this we get $ic^2 = -c(ic) = -c(-ci) = c^2i$, and so c^2 commutes with i. Now c satisfies some quadratic equation over F, $c^2 + \lambda c + \mu = 0$. Since c^2 and μ commute with i, λc must commute with i; that is, $\lambda ci = i\lambda c = \lambda ic = -\lambda ci$, hence $2\lambda ci = 0$, and since $2ci \neq 0$ we have that $\lambda = 0$. Thus $c^2 = -\mu$; since $c \notin F$ (for $ci = -ic \neq ic$) we can say, as we have before, that μ is positive and so $\mu = \nu^2$ where $\nu \in F$. Therefore $c^2 = -\nu^2$; let $j = c/\nu$. Then j satisfies

1. $j^2 = \dfrac{c^2}{\nu^2} = -1.$

2. $ji + ij = \dfrac{c}{\nu}i + i\dfrac{c}{\nu} = \dfrac{ci + ic}{\nu} = 0.$

Let $k = ij$. The i, j, k we have constructed behave like those for the quaternions, whence $T = \{\alpha_0 + \alpha_1 i + \alpha_2 j + \alpha_3 k \mid \alpha_0, \alpha_1, \alpha_2, \alpha_3 \in F\}$ forms a subdivision ring of D isomorphic to the real quaternions. We have produced a replica, T, of the division ring of real quaternions in D!

Our last objective is to demonstrate that $T = D$.

If $r \in D$ satisfies $r^2 = -1$ let $N(r) = \{x \in D \mid xr = rx\}$. $N(r)$ is a subdivision ring of D; moreover r, and so all $\alpha_0 + \alpha_1 r$, $\alpha_0, \alpha_1 \in F$, are in the center of $N(r)$. By Lemma 7.3.1 it follows that $N(r) = \{\alpha_0 + \alpha_1 r \mid \alpha_0, \alpha_1 \in F\}$. Thus if $xr = rx$ then $x = \alpha_0 + \alpha_1 r$ for some α_0, α_1 in F.

Suppose that $u \in D$, $u \notin F$. For some $\alpha, \beta \in F$, $w = (u - \alpha)/\beta$ satisfies $w^2 = -1$. We claim that $wi + iw$ commutes with both i and w; for $i(wi + iw) = iwi + i^2w = iwi + wi^2 = (iw + wi)i$ since $i^2 = -1$. Similarly $w(wi + iw) = (wi + iw)w$. By the remark of the preceding paragraph, $wi + iw = \alpha_0' + \alpha_1' i = \alpha_0 + \alpha_1 w$. If $w \notin T$ this last relation forces $\alpha_1 = 0$ (for otherwise we could solve for w in terms of i). Thus $wi + iw = \alpha_0 \in F$. Similarly $wj + jw = \beta_0 \in F$ and $wk + kw = \gamma_0 \in F$. Let

$$z = w + \frac{\alpha_0}{2}i + \frac{\beta_0}{2}j + \frac{\gamma_0}{2}k.$$

Then

$$zi + iz = wi + iw + \frac{\alpha_0}{2}(i^2 + i^2) + \frac{\beta_0}{2}(ji + ij) + \frac{\gamma_0}{2}(ki + ik)$$
$$= \alpha_0 - \alpha_0 = 0;$$

similarly $zj + jz = 0$ and $zk + kz = 0$. We claim these relations force z to be 0. For $0 = zk + kz = zij + ijz = (zi + iz)j + i(jz - zj) = i(jz - zj)$ since $zi + iz = 0$. However $i \neq 0$, and since we are in a division ring, it follows that $jz - zj = 0$. But $jz + zj = 0$. Thus $2jz = 0$, and since $2j \neq 0$ we have that $z = 0$. Going back to the expression for z we get

$$w + \frac{\alpha_0}{2}i + \frac{\beta_0}{2}j + \frac{\gamma_0}{2}k = 0,$$

hence $w \in T$, contradicting $w \notin T$. Thus, indeed, $w \in T$. Since $w = (u - \alpha)/\beta$, $u = \beta w + \alpha$ and so $u \in T$. We have proved that any element in D is in T. Since $T \subset D$ we conclude that $D = T$; because T is isomorphic to the real quaternions we now get that D is isomorphic to the division ring of real quaternions. This, however, is just the statement of the theorem.

Problems

1. If the division ring D is finite-dimensional, as a vector space, over the field F contained in the center of D, prove that D is algebraic over F.
2. Give an example of a field K algebraic over another field F but not finite-dimensional over F.
3. If A is a ring algebraic over a field F and A has no zero divisors prove that A is a division ring.

7.4 Integral Quaternions and the Four-Square Theorem

In Chapter 3 we considered a certain special class of integral domains called Euclidean rings. When the results about this class of rings were applied to the ring of Gaussian integers, we obtained, as a consequence, the famous result of Fermat that every prime number of the form $4n + 1$ is the sum of two squares.

We shall now consider a particular subring of the quaternions which, in all ways except for its lack of commutativity, will look like a Euclidean ring. Because of this it will be possible to explicitly characterize all its left-ideals. This characterization of the left-ideals will lead us quickly to a proof of the classic theorem of Lagrange that every positive integer is a sum of four squares.

Let Q be the division ring of real quaternions. In Q we now proceed to introduce an adjoint operation, $*$, by making the

DEFINITION For $x = \alpha_0 + \alpha_1 i + \alpha_2 j + \alpha_3 k$ in Q the *adjoint* of x, denoted by x^*, is defined by $x^* = \alpha_0 - \alpha_1 i - \alpha_2 j - \alpha_3 k$.

LEMMA 7.4.1 *The adjoint in Q satisfies*

1. $x^{**} = x$;
2. $(\delta x + \gamma y)^* = \delta x^* + \gamma y^*$;
3. $(xy)^* = y^* x^*$;

for all x, y in Q and all real δ and γ.

Proof. If $x = \alpha_0 + \alpha_1 i + \alpha_2 j + \alpha_3 k$ then $x^* = \alpha_0 - \alpha_1 i - \alpha_2 j - \alpha_3 k$, whence $x^{**} = (x^*)^* = \alpha_0 + \alpha_1 i + \alpha_2 j + \alpha_3 k$, proving part 1.

Let $x = \alpha_0 + \alpha_1 i + \alpha_2 j + \alpha_3 k$ and $y = \beta_0 + \beta_1 i + \beta_2 j + \beta_3 k$ be in Q and let δ and γ be arbitrary real numbers. Thus $\delta x + \gamma y = (\delta\alpha_0 + \gamma\beta_0) + (\delta\alpha_1 + \gamma\beta_1)i + (\delta\alpha_2 + \gamma\beta_2)j + (\delta\alpha_3 + \gamma\beta_3)k$; therefore by the definition of the $*$, $(\delta x + \gamma y)^* = (\delta\alpha_0 + \gamma\beta_0) - (\delta\alpha_1 + \gamma\beta_1)i - (\delta\alpha_2 + \gamma\beta_2)j - (\delta\alpha_3 + \gamma\beta_3)k = \delta(\alpha_0 - \alpha_1 i - \alpha_2 j - \alpha_3 k) + \gamma(\beta_0 - \beta_1 i - \beta_2 j - \beta_3 k) = \delta x^* + \gamma y^*$. This, of course, proves part 2.

In light of part 2, to prove 3 it is enough to do so for a basis of Q over the reals. We prove it for the particular basis $1, i, j, k$. Now $ij = k$, hence $(ij)^* = k^* = -k = ji = (-j)(-i) = j^* i^*$. Similarly $(ik)^* = k^* i^*$, $(jk)^* = k^* j^*$. Also $(i^2)^* = (-1)^* = -1 = (i^*)^2$, and similarly for j and k. Since part 3 is true for the basis elements and part 2 holds, 3 is true for all linear combinations of the basis elements with real coefficients, hence 3 holds for arbitrary x and y in Q.

DEFINITION If $x \in Q$ then the *norm* of x, denoted by $N(x)$, is defined by $N(x) = xx^*$.

Note that if $x = \alpha_0 + \alpha_1 i + \alpha_2 j + \alpha_3 k$ then $N(x) = xx^* = (\alpha_0 + \alpha_1 i + \alpha_2 j + \alpha_3 k)(\alpha_0 - \alpha_1 i - \alpha_2 j - \alpha_3 k) = \alpha_0^2 + \alpha_1^2 + \alpha_2^2 + \alpha_3^2$; therefore $N(0) = 0$ and $N(x)$ is a *positive* real number for $x \neq 0$ in Q. In particular, for any real number α, $N(\alpha) = \alpha^2$. If $x \neq 0$ note that $x^{-1} = [1/N(x)]x^*$.

LEMMA 7.4.2 *For all $x, y \in Q$, $N(xy) = N(x)N(y)$.*

Proof. By the very definition of norm, $N(xy) = (xy)(xy)^*$; by part 3 of Lemma 7.4.1, $(xy)^* = y^* x^*$ and so $N(xy) = xyy^* x^*$. However, $yy^* = N(y)$ is a real number, and thereby it is in the center of Q; in particular it must commute with x^*. Consequently $N(xy) = x(yy^*)x^* = (xx^*)(yy^*) = N(x)N(y)$.

As an immediate consequence of Lemma 7.4.2 we obtain

LEMMA 7.4.3 (LAGRANGE IDENTITY) *If* $\alpha_0, \alpha_1, \alpha_2, \alpha_3$ *and* $\beta_0, \beta_1, \beta_2, \beta_3$ *are real numbers then* $(\alpha_0^2 + \alpha_1^2 + \alpha_2^2 + \alpha_3^2)(\beta_0^2 + \beta_1^2 + \beta_2^2 + \beta_3^2) = (\alpha_0\beta_0 - \alpha_1\beta_1 - \alpha_2\beta_2 - \alpha_3\beta_3)^2 + (\alpha_0\beta_1 + \alpha_1\beta_0 + \alpha_2\beta_3 - \alpha_3\beta_2)^2 + (\alpha_0\beta_2 - \alpha_1\beta_3 + \alpha_2\beta_0 + \alpha_3\beta_1)^2 + (\alpha_0\beta_3 + \alpha_1\beta_2 - \alpha_2\beta_1 + \alpha_3\beta_0)^2$.

Proof. Of course there is one obvious proof of this result, namely, multiply everything out and compare terms.

However, an easier way both to reconstruct the result at will and, at the same time, to prove it, is to notice that the left-hand side is $N(x)N(y)$ while the right-hand side is $N(xy)$ where $x = \alpha_0 + \alpha_1 i + \alpha_2 j + \alpha_3 k$ and $y = \beta_0 + \beta_1 i + \beta_2 j + \beta_3 k$. By Lemma 7.4.2, $N(x)N(y) = N(xy)$, ergo the Lagrange identity.

The Lagrange identity says that the sum of four squares times the sum of four squares is again, in a very specific way, the sum of four squares. A very striking result of Adolf Hurwitz says that if the sum of n squares times the sum of n squares is again a sum of n squares, where this last sum has terms computed bilinearly from the other two sums, then $n = 1, 2, 4$, or 8. There is, in fact, an identity for the product of sums of eight squares but it is too long and cumbersome to write down here.

Now is the appropriate time to introduce the Hurwitz ring of integral quaternions. Let $\zeta = \frac{1}{2}(1 + i + j + k)$ and let

$$H = \{m_0\zeta + m_1 i + m_2 j + m_3 k \mid m_0, m_1, m_2, m_3 \text{ integers}\}.$$

LEMMA 7.4.4 *H is a subring of Q. If $x \in H$ then $x^* \in H$ and $N(x)$ is a positive* integer *for every nonzero x in H.*

We leave the proof of Lemma 7.4.4 to the reader. It should offer no difficulties.

In some ways H might appear to be a rather contrived ring. Why use the quaternions ζ? Why not merely consider the more natural ring $Q_0 = \{m_0 + m_1 i + m_2 j + m_3 k \mid m_0, m_1, m_2, m_3 \text{ are integers}\}$? The answer is that Q_0 is not large enough, whereas H is, for the key lemma which follows to hold in it. But we want this next lemma to be true in the ring at our disposal for it allows us to characterize its left-ideals. This, perhaps, indicates why we (or rather Hurwitz) chose to work in H rather than in Q_0.

LEMMA 7.4.5 (LEFT-DIVISION ALGORITHM) *Let a and b be in H with $b \neq 0$. Then there exist two elements c and d in H such that $a = cb + d$ and $N(d) < N(b)$.*

Proof. Before proving the lemma, let's see what it tells us. If we look back in the section in Chapter 3 which deals with Euclidean rings, we can see that Lemma 7.4.5 assures us that except for its lack of commutativity H has all the properties of a Euclidean ring. The fact that elements in H may fail to commute will not bother us. True, we must be a little careful not to jump to erroneous conclusions; for instance $a = cb + d$ but we have no right to assume that a is also equal to $bc + d$, for b and c might not commute. But this will not influence any argument that we shall use.

In order to prove the lemma we first do so for a very special case, namely, that one in which a is an arbitrary element of H but b is a positive integer n. Suppose that $a = t_0\zeta + t_1 i + t_2 j + t_3 k$ where t_0, t_1, t_2, t_3 are integers and that $b = n$ where n is a positive integer. Let $c = x_0\zeta + x_1 i + x_2 j + x_3 k$ where x_0, x_1, x_2, x_3 are integers yet to be determined. We want to choose them in such a manner as to force $N(a - cn) < N(n) = n^2$. But

$$\begin{aligned} a - cn &= \left(t_0\left(\frac{1 + i + j + k}{2}\right) + t_1 i + t_2 j + t_3 k\right) \\ &\quad - nx_0\left(\frac{1 + i + j + k}{2}\right) - nx_1 i - nx_2 j - nx_3 k \\ &= \tfrac{1}{2}(t_0 - nx_0) + \tfrac{1}{2}(t_0 + 2t_1 - n(t_0 + 2x_1))i \\ &\quad + \tfrac{1}{2}(t_0 + 2t_1 - n(t_0 + 2x_2))j + \tfrac{1}{2}(t_0 + 2t_3 - n(t_0 + 2x_3))k. \end{aligned}$$

If we could choose the integers x_0, x_1, x_2, x_3 in such a way as to make $|t_0 - nx_0| \le \frac{1}{2}n$, $|t_0 + 2t_1 - n(t_0 + 2x_1)| \le n$, $|t_0 + 2t_2 - n(t_0 + 2x_2)| \le n$ and $|t_0 + 2t_3 - n(t_0 + 2x_3)| \le n$ then we would have

$$\begin{aligned} N(a - cn) &= \frac{(t_0 - nx_0)^2}{4} + \frac{(t_0 + 2t_1 - n(t_0 + 2x_1))^2}{4} + \cdots \\ &\le \tfrac{1}{16}n^2 + \tfrac{1}{4}n^2 + \tfrac{1}{4}n^2 + \tfrac{1}{4}n^2 < n^2 = N(n), \end{aligned}$$

which is the desired result. But now we claim this can always be done:

1. There is an integer x_0 such that $t_0 = x_0 n + r$ where $-\frac{1}{2}n \le r \le \frac{1}{2}n$; for this x_0, $|t_0 - x_0 n| = |r| \le \frac{1}{2}n$.
2. There is an integer k such that $t_0 + 2t_1 = kn + r$ and $0 \le r \le n$. If $k - t_0$ is even, put $2x_1 = k - t_0$; then $t_0 + 2t_1 = (2x_1 + t_0)n + r$ and $|t_0 + 2t_1 - (2x_1 + t_0)n| = r < n$. If, on the other hand, $k - t_0$ is odd, put $2x_1 = k - t_0 + 1$; thus $t_0 + 2t_1 = (2x_1 + t_0 - 1)n + r = (2x_1 + t_0)n + r - n$, whence $|t_0 + 2t_1 - (2x_1 + t_0)n| = |r - n| \le n$ since $0 \le r < n$. Therefore we can find an integer x_1 satisfying $|t_0 + 2t_1 - (2x_1 + t_0)n| \le n$.
3. As in part 2, we can find integers x_2 and x_3 which satisfy $|t_0 + 2t_2 - (2x_2 + t_0)n| \le n$ and $|t_0 + 2t_3 - (2x_3 + t_0)n| \le n$, respectively.

In the special case in which a is an arbitrary element of H and b is a positive integer we have now shown the lemma to be true.

We go to the general case wherein a and b are arbitrary elements of H and $b \neq 0$. By Lemma 7.4.4, $n = bb^*$ is a positive integer; thus there exists a $c \in H$ such that $ab^* = cn + d_1$ where $N(d_1) < N(n)$. Thus $N(ab^* - cn) < N(n)$; but $n = bb^*$ whence we get $N(ab^* - cbb^*) < N(n)$, and so $N((a - cb)b^*) < N(n) = N(bb^*)$. By Lemma 7.4.2 this reduces to $N(a - cb)N(b^*) < N(b)N(b^*)$; since $N(b^*) > 0$ we get $N(a - cb) < N(b)$. Putting $d = a - cb$ we have $a = cb + d$ where $N(d) < N(b)$. This completely proves the lemma.

As in the commutative case we are able to deduce from Lemma 7.4.5

LEMMA 7.4.6 *Let L be a left-ideal of H. Then there exists an element $u \in L$ such that every element in L is a left-multiple of u; in other words, there exists $u \in L$ such that every $x \in L$ is of the form $x = ru$ where $r \in H$.*

Proof. If $L = (0)$ there is nothing to prove, merely put $u = 0$.

Therefore we may assume that L has nonzero elements. The norms of the nonzero elements are positive integers (Lemma 7.4.4) whence there is an element $u \neq 0$ in L whose norm is minimal over the nonzero elements of L. If $x \in L$, by Lemma 7.4.5, $x = cu + d$ where $N(d) < N(u)$. However d is in L because both x and u, and so cu, are in L which is a left-ideal. Thus $N(d) = 0$ and so $d = 0$. From this $x = cu$ is a consequence.

Before we can prove the four-square theorem, which is the goal of this section, we need one more lemma, namely

LEMMA 7.4.7 *If $a \in H$ then $a^{-1} \in H$ if and only if $N(a) = 1$.*

Proof. If both a and a^{-1} are in H, then by Lemma 7.4.4 both $N(a)$ and $N(a^{-1})$ are positive integers. However, $aa^{-1} = 1$, hence, by Lemma 7.4.2, $N(a)N(a^{-1}) = N(aa^{-1}) = N(1) = 1$. This forces $N(a) = 1$.

On the other hand, if $a \in H$ and $N(a) = 1$, then $aa^* = N(a) = 1$ and so $a^{-1} = a^*$. But, by Lemma 7.4.4, since $a \in H$ we have that $a^* \in H$, and so $a^{-1} = a^*$ is also in H.

We now have determined enough of the structure of H to use it effectively to study properties of the integers. We prove the famous classical theorem of Lagrange,

THEOREM 7.4.1 *Every positive integer can be expressed as the sum of squares of four integers.*

Proof. Given a positive integer n we claim in the theorem that $n = x_0^2 + x_1^2 + x_2^2 + x_3^2$ for four integers x_0, x_1, x_2, x_3. Since every integer factors into a product of prime numbers, if every prime number were realizable as a sum of four squares, in view of Lagrange's identity (Lemma 7.4.3) every integer would be expressible as a sum of four squares. We have reduced the problem to consider only prime numbers n. Certainly the prime number 2 can be written as $1^2 + 1^2 + 0^2 + 0^2$ as a sum of four squares.

Thus, without loss of generality, we may assume that n is an *odd prime number*. As is customary we denote it by p.

Consider the quaternions W_p over J_p, the integers mod p; $W_p = \{\alpha_0 + \alpha_1 i + \alpha_2 j + \alpha_3 k \mid \alpha_0, \alpha_1, \alpha_2, \alpha_3 \in J_p\}$. W_p is a finite ring; moreover, since $p \neq 2$ it is not commutative for $ij = -ji \neq ji$. Thus, by Wedderburn's theorem it cannot be a division ring, hence by Problem 1 at the end of Section 3.5, it must have a left-ideal which is neither (0) nor W_p.

But then the two-sided ideal V in H defined by $V = \{x_0\zeta + x_1 i + x_2 j + x_3 k \mid p$ divides all of $x_0, x_1, x_2, x_3\}$ cannot be a maximal left-ideal of H, since H/V is isomorphic to W_p. (Prove!) (If V were a maximal left-ideal in H, H/V, and so W_p, would have no left-ideals other than (0) and H/V).

Thus there is a left-ideal L of H satisfying: $L \neq H$, $L \neq V$, and $L \supset V$. By Lemma 7.4.6, there is an element $u \in L$ such that every element in L is a left-multiple of u. Since $p \in V$, $p \in L$, whence $p = cu$ for some $c \in H$. Since $u \notin V$, c cannot have an inverse in H, otherwise $u = c^{-1}p$ would be in V. Thus $N(c) > 1$ by Lemma 7.4.7. Since $L \neq H$, u cannot have an inverse in H, whence $N(u) > 1$. Since $p = cu$, $p^2 = N(p) = N(cu) = N(c)N(u)$. But $N(c)$ and $N(u)$ are integers, since both c and u are in H, both are larger than 1 and both divide p^2. The only way this is possible is that $N(c) = N(u) = p$.

Since $u \in H$, $u = m_0\zeta + m_1 i + m_2 j + m_3 k$ where m_0, m_1, m_2, m_3 are integers; thus $2u = 2m_0\zeta + 2m_1 i + 2m_2 j + 2m_3 k = (m_0 + m_0 i + m_0 j + m_0 k) + 2m_1 i + 2m_2 j + 2m_3 k = m_0 + (2m_1 + m_0)i + (2m_2 + m_0)j + (2m_3 + m_0)k$. Therefore $N(2u) = m_0^2 + (2m_1 + m_0)^2 + (2m_2 + m_0)^2 + (2m_3 + m_0)^2$. But $N(2u) = N(2)N(u) = 4p$ since $N(2) = 4$ and $N(u) = p$. We have shown that $4p = m_0^2 + (2m_1 + m_0)^2 + (2m_2 + m_0)^2 + (2m_3 + m_0)^2$. We are almost done.

To finish the proof we introduce an old trick of Euler's: If $2a = x_0^2 + x_1^2 + x_2^2 + x_3^2$ where a, x_0, x_1, x_2 and x_3 are integers, then $a = y_0^2 + y_1^2 + y_2^2 + y_3^2$ for some integers y_0, y_1, y_2, y_3. To see this note that, since $2a$ is even, the x's are all even, all odd or two are even and two are odd. At any rate in all three cases we can renumber the x's and pair them in such a way that

$$y_0 = \frac{x_0 + x_1}{2}, \quad y_1 = \frac{x_0 - x_1}{2}, \quad y_2 = \frac{x_2 + x_3}{2}, \quad \text{and} \quad y_3 = \frac{x_2 - x_3}{2}$$

are all integers. But

$$y_0^2 + y_1^2 + y_2^2 + y_3^2$$

$$= \left(\frac{x_0 + x_1}{2}\right)^2 + \left(\frac{x_0 - x_1}{2}\right)^2 + \left(\frac{x_2 + x_3}{2}\right)^2 + \left(\frac{x_2 - x_3}{2}\right)^2$$

$$= \tfrac{1}{2}(x_0^2 + x_1^2 + x_2^2 + x_3^2)$$

$$= \tfrac{1}{2}(2a)$$

$$= a.$$

Since $4p$ is a sum of four squares, by the remark just made $2p$ also is; since $2p$ is a sum of four squares, p also must be such a sum. Thus $p = a_0^2 + a_1^2 + a_2^2 + a_3^2$ for some integers a_0, a_1, a_2, a_3 and Lagrange's theorem is established.

This theorem itself is the starting point of a large research area in number theory, the so-called *Waring problem.* This asks if every integer can be written as a sum of a fixed number of kth powers. For instance it can be shown that every integer is a sum of nine cubes, nineteen fourth powers, etc. The Waring problem was shown to have an affirmative answer, in this century, by the great mathematician Hilbert.

Problems

1. Prove Lemma 7.4.4.
2. Find all the elements a in Q_0 such that a^{-1} is also in Q_0.
3. Prove that there are exactly 24 elements a in H such that a^{-1} is also in H. Determine all of them.
4. Give an example of an a and b, $b \neq 0$, in Q_0 such that it is impossible to find c and d in Q_0 satisfying $a = cb + d$ where $N(d) < N(b)$.
5. Prove that if $a \in H$ then there exist integers α, β such that $a^2 + \alpha a + \beta = 0$.
6. Prove that there is a positive integer which cannot be written as the sum of three squares.

*7. Exhibit an infinite number of positive integers which cannot be written as the sum of three squares.

Supplementary Reading

For a deeper discussion of finite fields: ALBERT, A. A., *Fundamental Concepts of Higher Algebra.* Chicago: University of Chicago Press, 1956.

For many proofs of the four-square theorem and a discussion of the Waring problem: HARDY, G. H., and WRIGHT, E. M., *An Introduction to the Theory of Numbers*, 4th ed. New York: Oxford University Press, 1960.

For another proof of the Wedderburn theorem: ARTIN, E., "Über einen Satz von Herrn J. H. M. Wedderburn," *Abhandlungen, Hamburg Mathematisches Seminar*, Vol. 5 (1928), pages 245–50.

Index